Student Solutions Manual

for

MULTIVARIABLE CALCULUS

FIFTH EDITION

DAN CLEGG
Palomar College

BARBARA FRANK
St. Andrews Presbyterian College

THOMSON

BROOKS/COLE

Australia □ Canada □ Mexico □ Singapore □ Spain □ United Kingdom □ United States

Printed in Canada

1 2 3 4 5 6 7   07 06 05 04 03

Printer: Webcom

ISBN 0-534-39360-8

For more information about our products,
contact us at:
Thomson Learning Academic Resource Center
1-800-423-0563

For permission to use material from this text,
contact us by:
Phone: 1-800-730-2214
Fax: 1-800-730-2215
Web: http://www.thomsonrights.com

Brooks/Cole–Thomson Learning
10 Davis Drive
Belmont, CA 94002
USA

Asia
Thomson Learning
5 Shenton Way #01-01
UIC Building
Singapore 068808

Australia/New Zealand
Nelson Thomson Learning
102 Dodds Street
Southbank, Victoria 3006
Australia

Canada
Nelson
1120 Birchmount Road
Toronto, Ontario M1K 5G4
Canada

Europe/Middle East/Africa
Thomson Learning
High Holborn House
50/51 Bedford Row
London WC1R 4LR
United Kingdom

Latin America
Thomson Learning
Seneca, 53
Colonia Polanco
11560 Mexico D.F.
Mexico

Spain
Paraninfo
Calle/Magallanes, 25
28015 Madrid, Spain

☐ PREFACE

This *Student Solutions Manual* contains detailed solutions to selected exercises in the texts *Multivariable Calculus, Fifth Edition* and *Multivariable Calculus: Early Transcendentals, Fifth Edition* (Chapters 11–18 of *Calculus, Fifth Edition* and Chapters 10–17 of *Calculus: Early Transcendentals, Fifth Edition*) by James Stewart. Specifically, it includes solutions to the odd-numbered exercises in each chapter section, review section, True-False Quiz, and Problems Plus section. Also included are all solutions to the Concept Check questions.

The *Early Transcendentals* version of the text uses different chapter and page numbers; consequently, all section numbers and references are given in a dual format. Readers of the *Early Transcendentals* text should use the references denoted by "ET."

Each solution is presented in the context of the corresponding section of the text. In general, solutions to the initial exercises involving a new concept illustrate that concept in more detail; this knowledge is then utilized in subsequent solutions. Thus, while the intermediate steps of a solution are given, you may need to refer back to earlier exercises in the section or prior sections for additional explanation of the concepts involved. Note that, in many cases, different routes to an answer may exist which are equally valid; also, answers can be expressed in different but equivalent forms. Thus, the goal of this manual is not to give the definitive solution to each exercise, but rather to assist you as a student in understanding the concepts of the text and learning how to apply them to the challenge of solving a problem.

We would like to thank James Stewart for entrusting us with the writing of this manual and offering suggestions, Kathi Townes and Stephanie Kuhns of TECH-arts for typesetting and producing this manual, and Brian Betsill of TECH-arts for creating the illustrations. We also thank Bob Pirtle and Stacy Green of Brooks/Cole for their trust, assistance, and patience.

DAN CLEGG
Palomar College

BARBARA FRANK
St. Andrews Presbyterian College

☐ ABBREVIATIONS AND SYMBOLS

CD	concave downward
CU	concave upward
D	the domain of f
FDT	First Derivative Test
HA	horizontal asymptote(s)
I	interval of convergence
IP	inflection points(s)
R	radius of convergence
VA	vertical asymptote(s)
$\overset{CAS}{=}$	indicates the use of a computer algebra system
$\overset{H}{=}$	indicates the use of l'Hospital's Rule
$\overset{j}{=}$	indicates the use of Formula j in the Table of Integrals given on Reference Pages 6–10
$\overset{s}{=}$	indicates the use of the substitution $\{u = \sin x,\ du = \cos x\,dx\}$
$\overset{c}{=}$	indicates the use of the substitution $\{u = \cos x,\ du = -\sin x\,dx\}$

CONTENTS

□ 17 VECTOR CALCULUS 277   ET 16

□ 18 SECOND-ORDER DIFFERENTIAL EQUATIONS 315   ET 17

□ APPENDIX

11 □ PARAMETRIC EQUATIONS AND POLAR COORDINATES

11.1 Curves Defined by Parametric Equations

1. $x = 1 + \sqrt{t}, \quad y = t^2 - 4t, \quad 0 \le t \le 5$

t	0	1	2	3	4	5
x	1	2	$1 + \sqrt{2}$	$1 + \sqrt{3}$	3	$1 + \sqrt{5}$
			2.41	2.73		3.24
y	0	-3	-4	-3	0	5

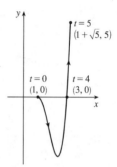

3. $x = 5\sin t, \quad y = t^2, \quad -\pi \le t \le \pi$

t	$-\pi$	$-\pi/2$	0	$\pi/2$	π
x	0	-5	0	5	0
y	π^2	$\pi^2/4$	0	$\pi^2/4$	π^2
	9.87	2.47		2.47	9.87

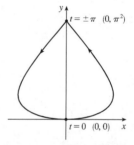

5. $x = 3t - 5, \quad y = 2t + 1$

(a)

t	-2	-1	0	1	2	3	4
x	-11	-8	-5	-2	1	4	7
y	-3	-1	1	3	5	7	9

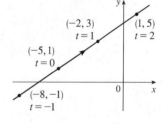

(b) $x = 3t - 5 \implies 3t = x + 5 \implies t = \frac{1}{3}(x + 5) \implies$

$y = 2 \cdot \frac{1}{3}(x + 5) + 1$, so $y = \frac{2}{3}x + \frac{13}{3}$.

7. $x = t^2 - 2, \quad y = 5 - 2t, \quad -3 \le t \le 4$

(a)

t	-3	-2	-1	0	1	2	3	4
x	7	2	-1	-2	-1	2	7	14
y	11	9	7	5	3	1	-1	-3

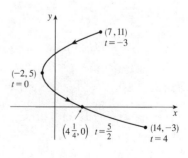

(b) $y = 5 - 2t \implies 2t = 5 - y \implies t = \frac{1}{2}(5 - y) \implies$

$x = \left[\frac{1}{2}(5 - y)\right]^2 - 2$, so $x = \frac{1}{4}(5 - y)^2 - 2$,

$-3 \le y \le 11$.

9. (a) $x = \sqrt{t}$, $y = 1 - t$

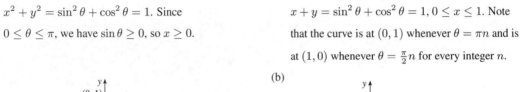

(0, 1) $t = 0$

(1, 0) $t = 1$

(2, −3) $t = 4$

t	0	1	2	3	4
x	0	1	1.414	1.732	2
y	1	0	−1	−2	−3

(b) $x = \sqrt{t} \;\Rightarrow\; t = x^2 \;\Rightarrow\; y = 1 - t = 1 - x^2$.

Since $t \geq 0$, $x \geq 0$.

11. (a) $x = \sin\theta$, $y = \cos\theta$, $0 \leq \theta \leq \pi$.

$x^2 + y^2 = \sin^2\theta + \cos^2\theta = 1$. Since

$0 \leq \theta \leq \pi$, we have $\sin\theta \geq 0$, so $x \geq 0$.

13. (a) $x = \sin^2\theta$, $y = \cos^2\theta$.

$x + y = \sin^2\theta + \cos^2\theta = 1$, $0 \leq x \leq 1$. Note

that the curve is at $(0, 1)$ whenever $\theta = \pi n$ and is

at $(1, 0)$ whenever $\theta = \frac{\pi}{2}n$ for every integer n.

(b)

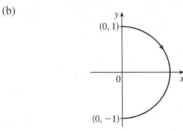

(0, 1)

(0, −1)

(b)

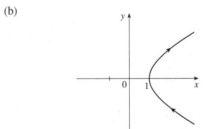

(0, 1)

(1, 0)

15. (a) $x = e^t$, $y = e^{-t}$.

$y = 1/e^t = 1/x$, $x > 0$

(b)

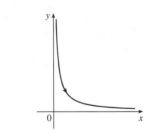

17. (a) $x = \cosh t$, $y = \sinh t$,

$x^2 - y^2 = \cosh^2 t - \sinh^2 t = 1$, $x \geq 1$

(b)

19. $x^2 + y^2 = \cos^2\pi t + \sin^2\pi t = 1$, $1 \leq t \leq 2$, so the particle moves counterclockwise along the circle $x^2 + y^2 = 1$

from $(-1, 0)$ to $(1, 0)$, along the lower half of the circle.

21. $\left(\frac{1}{2}x\right)^2 + \left(\frac{1}{3}y\right)^2 = \sin^2 t + \cos^2 t = 1$, so the particle moves once clockwise along the ellipse $\frac{1}{4}x^2 + \frac{1}{9}y^2 = 1$,

starting and ending at $(0, 3)$.

23. We must have $1 \leq x \leq 4$ and $2 \leq y \leq 3$. So the graph of the curve must be contained in the rectangle $[1, 4]$

by $[2, 3]$.

25. When $t = -1$, $(x, y) = (0, -1)$. As t increases to 0, x decreases to

-1 and y increases to 0. As t increases from 0 to 1, x increases to 0

and y increases to 1. As t increases beyond 1, both x and y increase.

For $t < -1$, x is positive and decreasing and y is negative and

increasing. We could achieve greater accuracy by estimating x- and

y-values for selected values of t from the given graphs and plotting

the corresponding points.

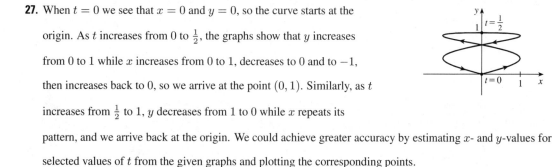

27. When $t = 0$ we see that $x = 0$ and $y = 0$, so the curve starts at the

origin. As t increases from 0 to $\frac{1}{2}$, the graphs show that y increases

from 0 to 1 while x increases from 0 to 1, decreases to 0 and to -1,

then increases back to 0, so we arrive at the point $(0, 1)$. Similarly, as t

increases from $\frac{1}{2}$ to 1, y decreases from 1 to 0 while x repeats its

pattern, and we arrive back at the origin. We could achieve greater accuracy by estimating x- and y-values for

selected values of t from the given graphs and plotting the corresponding points.

29. As in Example 5, we let $y = t$ and $x = t - 3t^3 + t^5$ and use a t-interval of $[-2\pi, 2\pi]$.

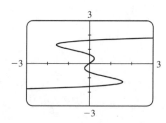

31. (a) $x = x_1 + (x_2 - x_1)t$, $y = y_1 + (y_2 - y_1)t$, $0 \le t \le 1$. Clearly the curve passes through $P_1(x_1, y_1)$ when

$t = 0$ and through $P_2(x_2, y_2)$ when $t = 1$. For $0 < t < 1$, x is strictly between x_1 and x_2 and y is strictly

between y_1 and y_2. For every value of t, x and y satisfy the relation $y - y_1 = \dfrac{y_2 - y_1}{x_2 - x_1}(x - x_1)$, which is the

equation of the line through $P_1(x_1, y_1)$ and $P_2(x_2, y_2)$.

Finally, any point (x, y) on that line satisfies $\dfrac{y - y_1}{y_2 - y_1} = \dfrac{x - x_1}{x_2 - x_1}$; if we call that common value t, then the

given parametric equations yield the point (x, y); and any (x, y) on the line between $P_1(x_1, y_1)$ and $P_2(x_2, y_2)$

yields a value of t in $[0, 1]$. So the given parametric equations exactly specify the line segment from $P_1(x_1, y_1)$

to $P_2(x_2, y_2)$.

(b) $x = -2 + [3 - (-2)]t = -2 + 5t$ and $y = 7 + (-1 - 7)t = 7 - 8t$ for $0 \le t \le 1$.

33. The circle $x^2 + y^2 = 4$ can be represented parametrically by $x = 2\cos t$, $y = 2\sin t$; $0 \leq t \leq 2\pi$. The circle

$x^2 + (y-1)^2 = 4$ can be represented by $x = 2\cos t$, $y = 1 + 2\sin t$; $0 \leq t \leq 2\pi$. This representation gives us the

circle with a counterclockwise orientation starting at $(2, 1)$.

(a) To get a clockwise orientation, we could change the equations to $x = 2\cos t$, $y = 1 - 2\sin t$, $0 \leq t \leq 2\pi$.

(b) To get three times around in the counterclockwise direction, we use the original equations $x = 2\cos t$,
$y = 1 + 2\sin t$ with the domain expanded to $0 \leq t \leq 6\pi$.

(c) To start at $(0, 3)$ using the original equations, we must have $x_1 = 0$; that is, $2\cos t = 0$. Hence, $t = \frac{\pi}{2}$. So we
use $x = 2\cos t$, $y = 1 + 2\sin t$; $\frac{\pi}{2} \leq t \leq \frac{3\pi}{2}$.
  Alternatively, if we want t to start at 0, we could change the equations of the curve. For example, we could use
$x = -2\sin t$, $y = 1 + 2\cos t$, $0 \leq t \leq \pi$.

35. (a) Let $x^2/a^2 = \sin^2 t$ and $y^2/b^2 = \cos^2 t$ to obtain   (b) The equations are $x = 3\sin t$ and
  $x = a\sin t$ and $y = b\cos t$ with $0 \leq t \leq 2\pi$ as possible $y = b\cos t$ for $b \in \{1, 2, 4, 8\}$.
  parametric equations for the ellipse
  $x^2/a^2 + y^2/b^2 = 1$.

(c) As b increases, the ellipse stretches vertically.

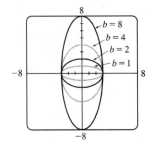

37. The case $\frac{\pi}{2} < \theta < \pi$ is illustrated. C has coordinates $(r\theta, r)$ as in Example 6,
and Q has coordinates $(r\theta, r + r\cos(\pi - \theta)) = (r\theta, r(1 - \cos\theta))$ [since
$\cos(\pi - \alpha) = \cos\pi\cos\alpha + \sin\pi\sin\alpha = -\cos\alpha$], so P has coordinates
$(r\theta - r\sin(\pi - \theta), r(1 - \cos\theta)) = (r(\theta - \sin\theta), r(1 - \cos\theta))$ [since
$\sin(\pi - \alpha) = \sin\pi\cos\alpha - \cos\pi\sin\alpha = \sin\alpha$]. Again we have the
parametric equations $x = r(\theta - \sin\theta)$, $y = r(1 - \cos\theta)$.

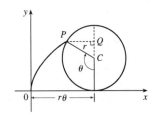

39. It is apparent that $x = |OQ|$ and $y = |QP| = |ST|$. From the
diagram, $x = |OQ| = a\cos\theta$ and $y = |ST| = b\sin\theta$. Thus, the
parametric equations are $x = a\cos\theta$ and $y = b\sin\theta$. To eliminate θ
we rearrange: $\sin\theta = y/b \implies \sin^2\theta = (y/b)^2$ and
$\cos\theta = x/a \implies \cos^2\theta = (x/a)^2$. Adding the two equations:
$\sin^2\theta + \cos^2\theta = 1 = x^2/a^2 + y^2/b^2$. Thus, we have an ellipse.

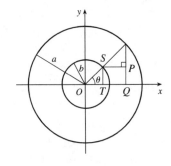

41. $C = (2a \cot \theta, 2a)$, so the x-coordinate of P is $x = 2a \cot \theta$. Let $B = (0, 2a)$. Then $\angle OAB$ is a right angle and
$\angle OBA = \theta$, so $|OA| = 2a \sin \theta$ and $A = ((2a \sin \theta) \cos \theta, (2a \sin \theta) \sin \theta)$. Thus, the y-coordinate of P is
$y = 2a \sin^2 \theta$.

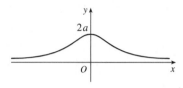

43. (a)

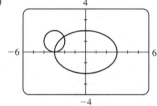

There are 2 points of intersection:
$(-3, 0)$ and approximately $(-2.1, 1.4)$.

(b) A collision point occurs when $x_1 = x_2$ and $y_1 = y_2$ for the same t. So solve the equations:

$$3 \sin t = -3 + \cos t \quad \textbf{(1)}$$
$$2 \cos t = 1 + \sin t \quad \textbf{(2)}$$

From **(2)**, $\sin t = 2 \cos t - 1$. Substituting into **(1)**, we get $3(2 \cos t - 1) = -3 + \cos t \quad \Rightarrow$

$5 \cos t = 0 \; (*) \quad \Rightarrow \quad \cos t = 0 \quad \Rightarrow \quad t = \frac{\pi}{2} \text{ or } \frac{3\pi}{2}$. We check that $t = \frac{3\pi}{2}$ satisfies **(1)** and **(2)** but $t = \frac{\pi}{2}$ does

not. So the only collision point occurs when $t = \frac{3\pi}{2}$, and this gives the point $(-3, 0)$. [We could check our work

by graphing x_1 and x_2 together as functions of t and, on another plot, y_1 and y_2 as functions of t. If we do so,

we see that the only value of t for which *both* pairs of graphs intersect is $t = \frac{3\pi}{2}$.]

(c) The circle is centered at $(3, 1)$ instead of $(-3, 1)$. There are still 2 intersection points: $(3, 0)$ and $(2.1, 1.4)$, but

there are no collision points, since $(*)$ in part (b) becomes $5 \cos t = 6 \quad \Rightarrow \quad \cos t = \frac{6}{5} > 1$.

45. $x = t^2, y = t^3 - ct$. We use a graphing device to produce the graphs for various values of c with $-\pi \le t \le \pi$.
Note that all the members of the family are symmetric about the x-axis. For $c < 0$, the graph does not cross itself,
but for $c = 0$ it has a cusp at $(0, 0)$ and for $c > 0$ the graph crosses itself at $x = c$, so the loop grows larger as c
increases.

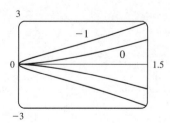 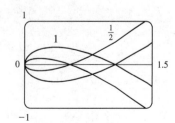

47. Note that all the Lissajous figures are symmetric about the x-axis. The parameters a and b simply stretch the graph in the x- and y-directions respectively. For $a = b = n = 1$ the graph is simply a circle with radius 1. For $n = 2$ the graph crosses itself at the origin and there are loops above and below the x-axis. In general, the figures have $n - 1$ points of intersection, all of which are on the y-axis, and a total of n closed loops.

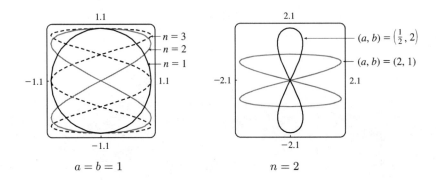

$a = b = 1$ $n = 2$

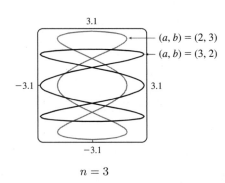

$n = 3$

11.2 Calculus with Parametric Curves   ET 10.2

1. $x = t - t^3$, $y = 2 - 5t$ $\Rightarrow$ $\dfrac{dy}{dt} = -5$, $\dfrac{dx}{dt} = 1 - 3t^2$, and $\dfrac{dy}{dx} = \dfrac{dy/dt}{dx/dt} = \dfrac{-5}{1 - 3t^2}$ or $\dfrac{5}{3t^2 - 1}$.

3. $x = t^4 + 1$, $y = t^3 + t$; $t = -1$. $\dfrac{dy}{dt} = 3t^2 + 1$, $\dfrac{dx}{dt} = 4t^3$, and $\dfrac{dy}{dx} = \dfrac{dy/dt}{dx/dt} = \dfrac{3t^2 + 1}{4t^3}$.

When $t = -1$, $(x, y) = (2, -2)$ and $dy/dx = \frac{4}{-4} = -1$, so an equation of the tangent to the curve at the point corresponding to $t = -1$ is $y - (-2) = (-1)(x - 2)$, or $y = -x$.

5. $x = e^{\sqrt{t}}$, $y = t - \ln t^2$; $t = 1$. $\dfrac{dy}{dt} = 1 - \dfrac{2t}{t^2} = 1 - \dfrac{2}{t}$, $\dfrac{dx}{dt} = \dfrac{e^{\sqrt{t}}}{2\sqrt{t}}$, and

$\dfrac{dy}{dx} = \dfrac{dy/dt}{dx/dt} = \dfrac{1 - 2/t}{e^{\sqrt{t}}/(2\sqrt{t})} \cdot \dfrac{2t}{2t} = \dfrac{2t - 4}{\sqrt{t}e^{\sqrt{t}}}$. When $t = 1$, $(x, y) = (e, 1)$ and $\dfrac{dy}{dx} = -\dfrac{2}{e}$, so an equation of the tangent line is $y - 1 = -\frac{2}{e}(x - e)$, or $y = -\frac{2}{e}x + 3$.

7. (a) $x = e^t$, $y = (t-1)^2$; $(1,1)$. $\dfrac{dy}{dt} = 2(t-1)$, $\dfrac{dx}{dt} = e^t$, and $\dfrac{dy}{dx} = \dfrac{dy/dt}{dx/dt} = \dfrac{2(t-1)}{e^t}$.

At $(1,1)$, $t = 0$ and $\dfrac{dy}{dx} = -2$, so an equation of the tangent is $y - 1 = -2(x-1)$, or $y = -2x + 3$.

(b) $x = e^t \;\Rightarrow\; t = \ln x$, so $y = (t-1)^2 = (\ln x - 1)^2$ and $\dfrac{dy}{dx} = 2(\ln x - 1)\left(\dfrac{1}{x}\right)$. When $x = 1$,

$\dfrac{dy}{dx} = 2(-1)(1) = -2$, so an equation of the tangent is $y = -2x + 3$, as in part (a).

9. $x = 2\sin 2t$, $y = 2\sin t$; $(\sqrt{3}, 1)$.

$\dfrac{dy}{dx} = \dfrac{dy/dt}{dx/dt} = \dfrac{2\cos t}{2 \cdot 2\cos 2t} = \dfrac{\cos t}{2\cos 2t}$. The point $(\sqrt{3}, 1)$ corresponds

to $t = \frac{\pi}{6}$, so the slope of the tangent at that point is

$\dfrac{\cos\frac{\pi}{6}}{2\cos\frac{\pi}{3}} = \dfrac{\frac{\sqrt{3}}{2}}{2 \cdot \frac{1}{2}} = \dfrac{\sqrt{3}}{2}$. An equation of the tangent is therefore

$(y - 1) = \frac{\sqrt{3}}{2}\left(x - \sqrt{3}\right)$, or $y = \frac{\sqrt{3}}{2}x - \frac{1}{2}$.

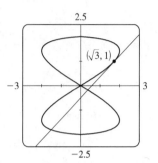

11. $x = 4 + t^2$, $y = t^2 + t^3$ $\;\Rightarrow\;$ $\dfrac{dy}{dx} = \dfrac{dy/dt}{dx/dt} = \dfrac{2t + 3t^2}{2t} = 1 + \dfrac{3}{2}t$ $\;\Rightarrow\;$

$\dfrac{d^2y}{dx^2} = \dfrac{d}{dx}\left(\dfrac{dy}{dx}\right) = \dfrac{d(dy/dx)/dt}{dx/dt} = \dfrac{(d/dt)\left(1 + \frac{3}{2}t\right)}{2t} = \dfrac{3/2}{2t} = \dfrac{3}{4t}$. The curve is CU when $\dfrac{d^2y}{dx^2} > 0$, that is,

when $t > 0$.

13. $x = t - e^t$, $y = t + e^{-t}$ $\;\Rightarrow\;$

$\dfrac{dy}{dx} = \dfrac{dy/dt}{dx/dt} = \dfrac{1 - e^{-t}}{1 - e^t} = \dfrac{1 - \frac{1}{e^t}}{1 - e^t} = \dfrac{\frac{e^t - 1}{e^t}}{1 - e^t} = -e^{-t}$ $\;\Rightarrow\;$ $\dfrac{d^2y}{dx^2} = \dfrac{\frac{d}{dt}\left(\frac{dy}{dx}\right)}{dx/dt} = \dfrac{\frac{d}{dt}(-e^{-t})}{dx/dt} = \dfrac{e^{-t}}{1 - e^t}$.

The curve is CU when $e^t < 1$ [since $e^{-t} > 0$] $\;\Rightarrow\;$ $t < 0$.

15. $x = 2\sin t$, $y = 3\cos t$, $0 < t < 2\pi$.

$\dfrac{dy}{dx} = \dfrac{dy/dt}{dx/dt} = \dfrac{-3\sin t}{2\cos t} = -\dfrac{3}{2}\tan t$, so $\dfrac{d^2y}{dx^2} = \dfrac{\frac{d}{dt}\left(\frac{dy}{dx}\right)}{dx/dt} = \dfrac{-\frac{3}{2}\sec^2 t}{2\cos t} = -\dfrac{3}{4}\sec^3 t$.

The curve is CU when $\sec^3 t < 0$ $\;\Rightarrow\;$ $\sec t < 0$ $\;\Rightarrow\;$ $\cos t < 0$ $\;\Rightarrow\;$ $\frac{\pi}{2} < t < \frac{3\pi}{2}$.

17. $x = 10 - t^2$, $y = t^3 - 12t$.

$dy/dt = 3t^2 - 12 = 3(t+2)(t-2)$, so

$dy/dt = 0 \;\Leftrightarrow\; t = \pm 2 \;\Leftrightarrow\;$

$(x, y) = (6, \mp 16)$. $dx/dt = -2t$, so $dx/dt = 0$

$\Leftrightarrow\; t = 0 \;\Leftrightarrow\; (x, y) = (10, 0)$. The curve has

horizontal tangents at $(6, \pm 16)$ and a vertical

tangent at $(10, 0)$.

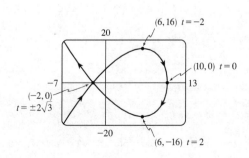

19. $x = 2\cos\theta$, $y = \sin 2\theta$.

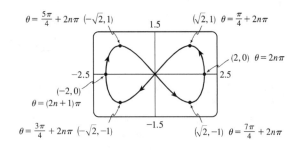

$dy/d\theta = 2\cos 2\theta$, so $dy/d\theta = 0$ $\Leftrightarrow$

$2\theta = \frac{\pi}{2} + n\pi$ (n an integer) $\Leftrightarrow$ $\theta = \frac{\pi}{4} + \frac{\pi}{2}n$

$\Leftrightarrow$ $(x, y) = (\pm\sqrt{2}, \pm 1)$. Also,

$dx/d\theta = -2\sin\theta$, so $dx/d\theta = 0$ $\Leftrightarrow$ $\theta = n\pi$

$\Leftrightarrow$ $(x, y) = (\pm 2, 0)$. The curve has horizontal

tangents at $(\pm\sqrt{2}, \pm 1)$ (four points), and vertical

tangents at $(\pm 2, 0)$.

21. From the graph, it appears that the leftmost point on the curve $x = t^4 - t^2$,

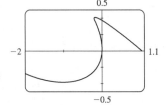

$y = t + \ln t$ is about $(-0.25, 0.36)$. To find the exact coordinates, we find

the value of t for which the graph has a vertical tangent, that is,

$0 = dx/dt = 4t^3 - 2t$ $\Leftrightarrow$ $2t(2t^2 - 1) = 0$ $\Leftrightarrow$

$2t(\sqrt{2}\,t + 1)(\sqrt{2}\,t - 1) = 0$ $\Leftrightarrow$ $t = 0$ or $\pm\frac{1}{\sqrt{2}}$. The negative and

0 roots are inadmissible since $y(t)$ is only defined for $t > 0$, so the leftmost point must be

$$\left(x\left(\tfrac{1}{\sqrt{2}}\right), y\left(\tfrac{1}{\sqrt{2}}\right)\right) = \left(\left(\tfrac{1}{\sqrt{2}}\right)^4 - \left(\tfrac{1}{\sqrt{2}}\right)^2, \tfrac{1}{\sqrt{2}} + \ln\tfrac{1}{\sqrt{2}}\right) = \left(-\tfrac{1}{4}, \tfrac{1}{\sqrt{2}} - \tfrac{1}{2}\ln 2\right)$$

23. We graph the curve $x = t^4 - 2t^3 - 2t^2$,

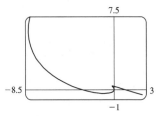

$y = t^3 - t$ in the viewing rectangle

$[-2, 1.1]$ by $[-0.5, 0.5]$. This rectangle

corresponds approximately to

$t \in [-1, 0.8]$. We estimate that the curve

has horizontal tangents at about

$(-1, -0.4)$ and $(-0.17, 0.39)$ and vertical tangents at about $(0, 0)$ and $(-0.19, 0.37)$. We calculate

$\dfrac{dy}{dx} = \dfrac{dy/dt}{dx/dt} = \dfrac{3t^2 - 1}{4t^3 - 6t^2 - 4t}$. The horizontal tangents occur when $dy/dt = 3t^2 - 1 = 0$ $\Leftrightarrow$ $t = \pm\frac{1}{\sqrt{3}}$, so

both horizontal tangents are shown in our graph. The vertical tangents occur when $dx/dt = 2t(2t^2 - 3t - 2) = 0$

$\Leftrightarrow$ $2t(2t + 1)(t - 2) = 0$ $\Leftrightarrow$ $t = 0, -\frac{1}{2}$ or 2. It seems that we have missed one vertical tangent, and indeed if

we plot the curve on the t-interval $[-1.2, 2.2]$ we see that there is another vertical tangent at $(-8, 6)$.

25. $x = \cos t$, $y = \sin t \cos t$. $\frac{dx}{dt} = -\sin t$,

$\frac{dy}{dt} = -\sin^2 t + \cos^2 t = \cos 2t$. $(x, y) = (0, 0)$ $\Leftrightarrow$ $\cos t = 0$ $\Leftrightarrow$

t is an odd multiple of $\frac{\pi}{2}$. When $t = \frac{\pi}{2}$, $\frac{dx}{dt} = -1$ and $\frac{dy}{dt} = -1$, so

$\frac{dy}{dx} = 1$. When $t = \frac{3\pi}{2}$, $\frac{dx}{dt} = 1$ and $\frac{dy}{dt} = -1$. So $\frac{dy}{dx} = -1$. Thus,

$y = x$ and $y = -x$ are both tangent to the curve at $(0, 0)$.

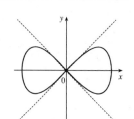

27. (a) $x = r\theta - d\sin\theta$, $y = r - d\cos\theta$; $\dfrac{dx}{d\theta} = r - d\cos\theta$, $\dfrac{dy}{d\theta} = d\sin\theta$. So $\dfrac{dy}{dx} = \dfrac{d\sin\theta}{r - d\cos\theta}$.

(b) If $0 < d < r$, then $|d\cos\theta| \le d < r$, so $r - d\cos\theta \ge r - d > 0$. This shows that $dx/d\theta$ never vanishes, so the trochoid can have no vertical tangent if $d < r$.

29. The line with parametric equations $x = -7t$, $y = 12t - 5$ is $y = 12\left(-\frac{1}{7}x\right) - 5$, which has slope $-\frac{12}{7}$. The curve

$x = t^3 + 4t$, $y = 6t^2$ has slope $\dfrac{dy}{dx} = \dfrac{dy/dt}{dx/dt} = \dfrac{12t}{3t^2 + 4}$. This equals $-\frac{12}{7}$ $\Leftrightarrow$ $3t^2 + 4 = -7t$ $\Leftrightarrow$

$(3t + 4)(t + 1) = 0$ $\Leftrightarrow$ $t = -1$ or $t = -\frac{4}{3}$ $\Leftrightarrow$ $(x, y) = (-5, 6)$ or $\left(-\frac{208}{27}, \frac{32}{3}\right)$.

31. By symmetry of the ellipse about the x- and y-axes,

$$A = 4\int_0^a y\,dx = 4\int_{\pi/2}^0 b\sin\theta\,(-a\sin\theta)\,d\theta = 4ab\int_0^{\pi/2}\sin^2\theta\,d\theta = 4ab\int_0^{\pi/2}\tfrac{1}{2}(1 - \cos 2\theta)\,d\theta$$

$$= 2ab\left[\theta - \tfrac{1}{2}\sin 2\theta\right]_0^{\pi/2} = 2ab\left(\tfrac{\pi}{2}\right) = \pi ab$$

33. $A = \int_0^1 (y - 1)\,dx = \int_{\pi/2}^0 (e^t - 1)(-\sin t)\,dt = \int_0^{\pi/2}(e^t\sin t - \sin t)\,dt \overset{98}{=} \left[\tfrac{1}{2}e^t(\sin t - \cos t) + \cos t\right]_0^{\pi/2}$

$= \tfrac{1}{2}(e^{\pi/2} - 1)$

35. $A = \int_0^{2\pi r} y\,dx = \int_0^{2\pi}(r - d\cos\theta)(r - d\cos\theta)\,d\theta = \int_0^{2\pi}(r^2 - 2dr\cos\theta + d^2\cos^2\theta)\,d\theta$

$= \left[r^2\theta - 2dr\sin\theta + \tfrac{1}{2}d^2\left(\theta + \tfrac{1}{2}\sin 2\theta\right)\right]_0^{2\pi} = 2\pi r^2 + \pi d^2$

37. $x = t - t^2$, $y = \frac{4}{3}t^{3/2}$, $1 \le t \le 2$. $dx/dt = 1 - 2t$ and $dy/dt = 2t^{1/2}$, so

$(dx/dt)^2 + (dy/dt)^2 = (1 - 2t)^2 + (2t^{1/2})^2 = 1 - 4t + 4t^2 + 4t = 1 + 4t^2$. Thus,

$L = \int_a^b \sqrt{(dx/dt)^2 + (dy/dt)^2}\,dt = \int_1^2 \sqrt{1 + 4t^2}\,dt$.

39. $x = t + \cos t$, $y = t - \sin t$, $0 \le t \le 2\pi$. $dx/dt = 1 - \sin t$ and $dy/dt = 1 - \cos t$, so

$(dx/dt)^2 + (dy/dt)^2 = (1 - \sin t)^2 + (1 - \cos t)^2 = (1 - 2\sin t + \sin^2 t) + (1 - 2\cos t + \cos^2 t)$

$$= 3 - 2\sin t - 2\cos t$$

Thus, $L = \int_a^b \sqrt{(dx/dt)^2 + (dy/dt)^2}\,dt = \int_0^{2\pi} \sqrt{3 - 2\sin t - 2\cos t}\,dt$.

41. $x = 1 + 3t^2$, $y = 4 + 2t^3$, $0 \le t \le 1$. $dx/dt = 6t$ and $dy/dt = 6t^2$, so $(dx/dt)^2 + (dy/dt)^2 = 36t^2 + 36t^4$.

Thus, $L = \int_0^1 \sqrt{36t^2 + 36t^4}\,dt = \int_0^1 6t\sqrt{1 + t^2}\,dt = 6\int_1^2 \sqrt{u}\left(\frac{1}{2}du\right)$ $[u = 1 + t^2,\ du = 2t\,dt]$

$\qquad = 3\left[\frac{2}{3}u^{3/2}\right]_1^2 = 2(2^{3/2} - 1) = 2\left(2\sqrt{2} - 1\right)$

43. $x = \dfrac{t}{1+t}$, $y = \ln(1+t)$, $0 \le t \le 2$. $\dfrac{dx}{dt} = \dfrac{(1+t)\cdot 1 - t\cdot 1}{(1+t)^2} = \dfrac{1}{(1+t)^2}$ and $\dfrac{dy}{dt} = \dfrac{1}{1+t}$, so

$\left(\dfrac{dx}{dt}\right)^2 + \left(\dfrac{dy}{dt}\right)^2 = \dfrac{1}{(1+t)^4} + \dfrac{1}{(1+t)^2} = \dfrac{1}{(1+t)^4}\left[1 + (1+t)^2\right] = \dfrac{t^2 + 2t + 2}{(1+t)^4}$. Thus,

$L = \int_0^2 \dfrac{\sqrt{t^2 + 2t + 2}}{(1+t)^2}\,dt = \int_1^3 \dfrac{\sqrt{u^2 + 1}}{u^2}\,du$ $[u = t + 1,\ du = dt]$ $\overset{24}{=} \left[-\dfrac{\sqrt{u^2 + 1}}{u} + \ln\left(u + \sqrt{u^2 + 1}\right)\right]_1^3$

$\qquad = -\dfrac{\sqrt{10}}{3} + \ln\left(3 + \sqrt{10}\right) + \sqrt{2} - \ln\left(1 + \sqrt{2}\right)$

45. $x = e^t \cos t$, $y = e^t \sin t$, $0 \le t \le \pi$.

$\left(\dfrac{dx}{dt}\right)^2 + \left(\dfrac{dy}{dt}\right)^2 = \left[e^t(\cos t - \sin t)\right]^2 + \left[e^t(\sin t + \cos t)\right]^2$

$\qquad = \left(e^t\right)^2\left(\cos^2 t - 2\cos t \sin t + \sin^2 t\right)$

$\qquad\qquad + \left(e^t\right)^2\left(\sin^2 t + 2\sin t \cos t + \cos^2 t\right)$

$\qquad = e^{2t}\left(2\cos^2 t + 2\sin^2 t\right) = 2e^{2t}$

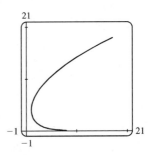

Thus, $L = \int_0^\pi \sqrt{2e^{2t}}\,dt = \int_0^\pi \sqrt{2}\,e^t\,dt = \sqrt{2}\left[e^t\right]_0^\pi = \sqrt{2}\left(e^\pi - 1\right)$.

47. $x = e^t - t$, $y = 4e^{t/2}$, $-8 \le t \le 3$.

$(dx/dt)^2 + (dy/dt)^2 = (e^t - 1)^2 + \left(2e^{t/2}\right)^2 = e^{2t} - 2e^t + 1 + 4e^t$

$\qquad = e^{2t} + 2e^t + 1 = (e^t + 1)^2$

Thus,

$\qquad L = \int_{-8}^3 \sqrt{(e^t + 1)^2}\,dt = \int_{-8}^3 (e^t + 1)\,dt = \left[e^t + t\right]_{-8}^3$

$\qquad\quad = (e^3 + 3) - (e^{-8} - 8) = e^3 - e^{-8} + 11$.

49. $x = t - e^t$, $y = t + e^t$, $-6 \le t \le 6$.

$\left(\frac{dx}{dt}\right)^2 + \left(\frac{dy}{dt}\right)^2 = (1 - e^t)^2 + (1 + e^t)^2 = (1 - 2e^t + e^{2t}) + (1 + 2e^t + e^{2t}) = 2 + 2e^{2t}$, so

$L = \int_{-6}^6 \sqrt{2 + 2e^{2t}}\,dt$. Set $f(t) = \sqrt{2 + 2e^{2t}}$. Then by Simpson's Rule with $n = 6$ and $\Delta t = \frac{6 - (-6)}{6} = 2$,

we get $L \approx \frac{2}{3}[f(-6) + 4f(-4) + 2f(-2) + 4f(0) + 2f(2) + 4f(4) + f(6)] \approx 612.3053$.

51. $x = \sin^2 t$, $y = \cos^2 t$, $0 \le t \le 3\pi$.

$$(dx/dt)^2 + (dy/dt)^2 = (2\sin t \cos t)^2 + (-2\cos t \sin t)^2 = 8\sin^2 t \cos^2 t = 2\sin^2 2t \quad \Rightarrow$$

$$\text{Distance} = \int_0^{3\pi} \sqrt{2}\,|\sin 2t|\,dt = 6\sqrt{2}\int_0^{\pi/2} \sin 2t\,dt \ \text{[by symmetry]} \ = -3\sqrt{2}\left[\cos 2t\right]_0^{\pi/2}$$

$$= -3\sqrt{2}\,(-1-1) = 6\sqrt{2}$$

The full curve is traversed as t goes from 0 to $\frac{\pi}{2}$, because the curve is the segment of $x + y = 1$ that lies in the first quadrant (since x, $y \ge 0$), and this segment is completely traversed as t goes from 0 to $\frac{\pi}{2}$.

Thus, $L = \int_0^{\pi/2} \sin 2t\,dt = \sqrt{2}$, as above.

53. $x = a\sin\theta$, $y = b\cos\theta$, $0 \le \theta \le 2\pi$.

$$\left(\frac{dx}{d\theta}\right)^2 + \left(\frac{dy}{d\theta}\right)^2 = (a\cos\theta)^2 + (-b\sin\theta)^2 = a^2\cos^2\theta + b^2\sin^2\theta = a^2(1 - \sin^2\theta) + b^2\sin^2\theta$$

$$= a^2 - (a^2 - b^2)\sin^2\theta = a^2 - c^2\sin^2\theta = a^2\left(1 - \frac{c^2}{a^2}\sin^2\theta\right) = a^2(1 - e^2\sin^2\theta)$$

So $L = 4\int_0^{\pi/2}\sqrt{a^2(1 - e^2\sin^2\theta)}\,d\theta$ [by symmetry] $= 4a\int_0^{\pi/2}\sqrt{1 - e^2\sin^2\theta}\,d\theta$.

55. (a) Notice that $0 \le t \le 2\pi$ does not give the complete curve because $x(0) \ne x(2\pi)$. In fact, we must take $t \in [0, 4\pi]$ in order to obtain the complete curve, since the first term in each of the parametric equations has period 2π and the second has period $\frac{2\pi}{11/2} = \frac{4\pi}{11}$, and the least common integer multiple of these two numbers is 4π.

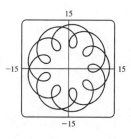

(b) We use the CAS to find the derivatives dx/dt and dy/dt, and then use Formula 1 to find the arc length. Recent versions of Maple express the integral $\int_0^{4\pi}\sqrt{(dx/dt)^2 + (dy/dt)^2}\,dt$ as $88E(2\sqrt{2}\,i)$, where $E(x)$ is the elliptic integral $\int_0^1 \frac{\sqrt{1 - x^2 t^2}}{\sqrt{1 - t^2}}\,dt$ and i is the imaginary number $\sqrt{-1}$. Some earlier versions of Maple (as well as Mathematica) cannot do the integral exactly, so we use the command

`evalf(Int(sqrt(diff(x,t)^2+diff(y,t)^2),t=0..4*Pi));` to estimate the length, and find that the arc length is approximately 294.03. Derive's `Para_arc_length` function in the utility file `Int_apps` simplifies the integral to $11\int_0^{4\pi}\sqrt{-4\cos t \cos\left(\frac{11t}{2}\right) - 4\sin t \sin\left(\frac{11t}{2}\right) + 5}\,dt$.

57. $x = t - t^2$, $\ y = \frac{4}{3}t^{3/2}$, $\ 1 \le t \le 2$. $\left(\dfrac{dx}{dt}\right)^2 + \left(\dfrac{dy}{dt}\right)^2 = (1-2t)^2 + (2t^{1/2})^2 = 1 - 4t + 4t^2 + 4t = 1 + 4t^2$,

so $S = \int_1^2 2\pi y\, ds = \int_1^2 2\pi \cdot \frac{4}{3}t^{3/2}\sqrt{1+4t^2}\, dt = \int_1^2 \frac{8\pi}{3}t^{3/2}\sqrt{1+4t^2}\, dt$.

59. $x = t^3$, $y = t^2$, $0 \le t \le 1$. $\ \left(\dfrac{dx}{dt}\right)^2 + \left(\dfrac{dy}{dt}\right)^2 = (3t^2)^2 + (2t)^2 = 9t^4 + 4t^2$.

$$S = \int_0^1 2\pi y \sqrt{\left(\dfrac{dx}{dt}\right)^2 + \left(\dfrac{dy}{dt}\right)^2}\, dt = \int_0^1 2\pi t^2 \sqrt{9t^4 + 4t^2}\, dt = 2\pi \int_0^1 t^2 \sqrt{t^2(9t^2+4)}\, dt$$

$$= 2\pi \int_4^{13} \left(\dfrac{u-4}{9}\right)\sqrt{u}\left(\tfrac{1}{18}du\right)\quad \begin{bmatrix} u = 9t^2 + 4,\ t^2 = (u-4)/9 \\ du = 18t\, dt,\ \text{so}\ t\, dt = \frac{1}{18}\, du \end{bmatrix}\quad = \dfrac{2\pi}{9\cdot 18}\int_4^{13}\left(u^{3/2} - 4u^{1/2}\right)du$$

$$= \tfrac{\pi}{81}\left[\tfrac{2}{5}u^{5/2} - \tfrac{8}{3}u^{3/2}\right]_4^{13} = \tfrac{\pi}{81}\cdot\tfrac{2}{15}[3u^{5/2} - 20u^{3/2}]_4^{13}$$

$$= \tfrac{2\pi}{1215}\left[\left(3\cdot 13^2\sqrt{13} - 20\cdot 13\sqrt{13}\right) - (3\cdot 32 - 20\cdot 8)\right]$$

$$= \tfrac{2\pi}{1215}\left(247\sqrt{13} + 64\right)$$

61. $x = a\cos^3\theta$, $y = a\sin^3\theta$, $0 \le \theta \le \tfrac{\pi}{2}$.

$$\left(\dfrac{dx}{d\theta}\right)^2 + \left(\dfrac{dy}{d\theta}\right)^2 = \left(-3a\cos^2\theta\sin\theta\right)^2 + \left(3a\sin^2\theta\cos\theta\right)^2 = 9a^2\sin^2\theta\cos^2\theta.$$

$$S = \int_0^{\pi/2} 2\pi\cdot a\sin^3\theta\cdot 3a\sin\theta\cos\theta\, d\theta = 6\pi a^2\int_0^{\pi/2}\sin^4\theta\cos\theta\, d\theta = \tfrac{6}{5}\pi a^2\left[\sin^5\theta\right]_0^{\pi/2} = \tfrac{6}{5}\pi a^2$$

63. $x = t + t^3$, $y = t - \dfrac{1}{t^2}$, $1 \le t \le 2$. $\dfrac{dx}{dt} = 1 + 3t^2$ and $\dfrac{dy}{dt} = 1 + \dfrac{2}{t^3}$, so

$$\left(\dfrac{dx}{dt}\right)^2 + \left(\dfrac{dy}{dt}\right)^2 = (1+3t^2)^2 + \left(1 + \dfrac{2}{t^3}\right)^2 \text{ and}$$

$$S = \int 2\pi y\, ds = \int_1^2 2\pi\left(t - \dfrac{1}{t^2}\right)\sqrt{(1+3t^2)^2 + \left(1 + \dfrac{2}{t^3}\right)^2}\, dt \approx 59.101.$$

65. $x = 3t^2$, $y = 2t^3$, $0 \le t \le 5$ $\ \Rightarrow\ $ $\left(\dfrac{dx}{dt}\right)^2 + \left(\dfrac{dy}{dt}\right)^2 = (6t)^2 + (6t^2)^2 = 36t^2(1+t^2)$ $\ \Rightarrow$

$$S = \int_0^5 2\pi x\sqrt{(dx/dt)^2 + (dy/dt)^2}\, dt = \int_0^5 2\pi(3t^2)6t\sqrt{1+t^2}\, dt = 18\pi\int_0^5 t^2\sqrt{1+t^2}\, 2t\, dt$$

$$= 18\pi\int_1^{26}(u-1)\sqrt{u}\, du \ \text{[where } u = 1+t^2,\ du = 2t\, dt]\ = 18\pi\int_1^{26}\left(u^{3/2} - u^{1/2}\right)du$$

$$= 18\pi\left[\tfrac{2}{5}u^{5/2} - \tfrac{2}{3}u^{3/2}\right]_1^{26} = 18\pi\left[\left(\tfrac{2}{5}\cdot 676\sqrt{26} - \tfrac{2}{3}\cdot 26\sqrt{26}\right) - \left(\tfrac{2}{5} - \tfrac{2}{3}\right)\right]$$

$$= \tfrac{24}{5}\pi\left(949\sqrt{26} + 1\right)$$

67. If f' is continuous and $f'(t) \neq 0$ for $a \leq t \leq b$, then either $f'(t) > 0$ for all t in $[a, b]$ or $f'(t) < 0$ for all t in $[a, b]$.

Thus, f is monotonic (in fact, strictly increasing or strictly decreasing) on $[a, b]$. It follows that f has an inverse.

Set $F = g \circ f^{-1}$, that is, define F by $F(x) = g(f^{-1}(x))$. Then $x = f(t) \Rightarrow f^{-1}(x) = t$, so

$y = g(t) = g(f^{-1}(x)) = F(x)$.

69. (a) $\phi = \tan^{-1}\left(\dfrac{dy}{dx}\right) \Rightarrow \dfrac{d\phi}{dt} = \dfrac{d}{dt}\tan^{-1}\left(\dfrac{dy}{dx}\right) = \dfrac{1}{1 + (dy/dx)^2}\left[\dfrac{d}{dt}\left(\dfrac{dy}{dx}\right)\right]$. But $\dfrac{dy}{dx} = \dfrac{dy/dt}{dx/dt} = \dfrac{\dot{y}}{\dot{x}} \Rightarrow$

$\dfrac{d}{dt}\left(\dfrac{dy}{dx}\right) = \dfrac{d}{dt}\left(\dfrac{\dot{y}}{\dot{x}}\right) = \dfrac{\ddot{y}\dot{x} - \ddot{x}\dot{y}}{\dot{x}^2} \Rightarrow \dfrac{d\phi}{dt} = \dfrac{1}{1 + (\dot{y}/\dot{x})^2}\left(\dfrac{\ddot{y}\dot{x} - \ddot{x}\dot{y}}{\dot{x}^2}\right) = \dfrac{\dot{x}\ddot{y} - \ddot{x}\dot{y}}{\dot{x}^2 + \dot{y}^2}$.

Using the Chain Rule, and the fact that $s = \displaystyle\int_0^t \sqrt{\left(\dfrac{dx}{dt}\right)^2 + \left(\dfrac{dy}{dt}\right)^2}\, dt \Rightarrow$

$\dfrac{ds}{dt} = \sqrt{\left(\dfrac{dx}{dt}\right)^2 + \left(\dfrac{dy}{dt}\right)^2} = (\dot{x}^2 + \dot{y}^2)^{1/2}$, we have that

$\dfrac{d\phi}{ds} = \dfrac{d\phi/dt}{ds/dt} = \left(\dfrac{\dot{x}\ddot{y} - \ddot{x}\dot{y}}{\dot{x}^2 + \dot{y}^2}\right)\dfrac{1}{(\dot{x}^2 + \dot{y}^2)^{1/2}} = \dfrac{\dot{x}\ddot{y} - \ddot{x}\dot{y}}{(\dot{x}^2 + \dot{y}^2)^{3/2}}$. So

$\kappa = \left|\dfrac{d\phi}{ds}\right| = \left|\dfrac{\dot{x}\ddot{y} - \ddot{x}\dot{y}}{(\dot{x}^2 + \dot{y}^2)^{3/2}}\right| = \dfrac{|\dot{x}\ddot{y} - \ddot{x}\dot{y}|}{(\dot{x}^2 + \dot{y}^2)^{3/2}}$.

(b) $x = x$ and $y = f(x) \Rightarrow \dot{x} = 1, \ddot{x} = 0$ and $\dot{y} = \dfrac{dy}{dx}, \ddot{y} = \dfrac{d^2y}{dx^2}$.

So $\kappa = \dfrac{|1 \cdot (d^2y/dx^2) - 0 \cdot (dy/dx)|}{[1 + (dy/dx)^2]^{3/2}} = \dfrac{|d^2y/dx^2|}{[1 + (dy/dx)^2]^{3/2}}$.

71. $x = \theta - \sin\theta \Rightarrow \dot{x} = 1 - \cos\theta \Rightarrow \ddot{x} = \sin\theta$, and $y = 1 - \cos\theta \Rightarrow \dot{y} = \sin\theta \Rightarrow \ddot{y} = \cos\theta$.

Therefore, $\kappa = \dfrac{|\cos\theta - \cos^2\theta - \sin^2\theta|}{[(1 - \cos\theta)^2 + \sin^2\theta]^{3/2}} = \dfrac{|\cos\theta - (\cos^2\theta + \sin^2\theta)|}{(1 - 2\cos\theta + \cos^2\theta + \sin^2\theta)^{3/2}} = \dfrac{|\cos\theta - 1|}{(2 - 2\cos\theta)^{3/2}}$. The top

of the arch is characterized by a horizontal tangent, and from Example 2(b) in Section 11.2 [ET 10.2], the tangent is

horizontal when $\theta = (2n - 1)\pi$, so take $n = 1$ and substitute $\theta = \pi$ into the expression for κ:

$\kappa = \dfrac{|\cos\pi - 1|}{(2 - 2\cos\pi)^{3/2}} = \dfrac{|-1 - 1|}{[2 - 2(-1)]^{3/2}} = \dfrac{1}{4}$.

73. The coordinates of T are $(r\cos\theta, r\sin\theta)$. Since TP was unwound from

arc TA, TP has length $r\theta$. Also $\angle PTQ = \angle PTR - \angle QTR = \frac{1}{2}\pi - \theta$,

so P has coordinates $x = r\cos\theta + r\theta\cos(\frac{1}{2}\pi - \theta) = r(\cos\theta + \theta\sin\theta)$,

$y = r\sin\theta - r\theta\sin(\frac{1}{2}\pi - \theta) = r(\sin\theta - \theta\cos\theta)$.

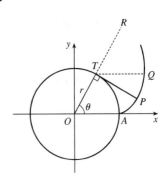

11.3 Polar Coordinates ET 10.3

1. (a) By adding 2π to $\frac{\pi}{2}$, we obtain the

point $\left(1, \frac{5\pi}{2}\right)$. The direction

opposite $\frac{\pi}{2}$ is $\frac{3\pi}{2}$, so $\left(-1, \frac{3\pi}{2}\right)$ is a

point that satisfies the $r < 0$

requirement.

(b) $\left(-2, \frac{\pi}{4}\right)$

(c) $(3, 2)$

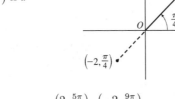

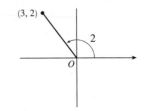

$\left(1, \frac{\pi}{2}\right)$

$\left(2, \frac{5\pi}{4}\right), \left(-2, \frac{9\pi}{4}\right)$

$(3, 2 + 2\pi), (-3, 2 + \pi)$

3. (a)

(b)

(c)

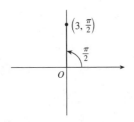

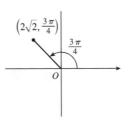

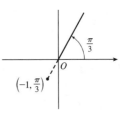

$x = 3\cos\frac{\pi}{2} = 3(0) = 0$ and

$y = 3\sin\frac{\pi}{2} = 3(1) = 3$ give us

the Cartesian coordinates $(0, 3)$.

$x = 2\sqrt{2}\cos\frac{3\pi}{4}$

$= 2\sqrt{2}\left(-\frac{1}{\sqrt{2}}\right) = -2$ and

$y = 2\sqrt{2}\sin\frac{3\pi}{4} = 2\sqrt{2}\left(\frac{1}{\sqrt{2}}\right) = 2$

give us $(-2, 2)$.

$x = -1\cos\frac{\pi}{3} = -\frac{1}{2}$ and

$y = -1\sin\frac{\pi}{3} = -\frac{\sqrt{3}}{2}$ give

us $\left(-\frac{1}{2}, -\frac{\sqrt{3}}{2}\right)$.

5. (a) $x = 1$ and $y = 1$ $\Rightarrow$ $r = \sqrt{1^2 + 1^2} = \sqrt{2}$ and $\theta = \tan^{-1}\left(\frac{1}{1}\right) = \frac{\pi}{4}$. Since $(1, 1)$ is in the first quadrant, the polar coordinates are (i) $\left(\sqrt{2}, \frac{\pi}{4}\right)$ and (ii) $\left(-\sqrt{2}, \frac{5\pi}{4}\right)$.

(b) $x = 2\sqrt{3}$ and $y = -2$ $\Rightarrow$ $r = \sqrt{\left(2\sqrt{3}\right)^2 + (-2)^2} = \sqrt{12 + 4} = \sqrt{16} = 4$ and $\theta = \tan^{-1}\left(-\frac{2}{2\sqrt{3}}\right) = \tan^{-1}\left(-\frac{1}{\sqrt{3}}\right) = -\frac{\pi}{6}$. Since $\left(2\sqrt{3}, -2\right)$ is in the fourth quadrant and $0 \le \theta \le 2\pi$, the polar coordinates are (i) $\left(4, \frac{11\pi}{6}\right)$ and (ii) $\left(-4, \frac{5\pi}{6}\right)$.

7. The curves $r = 1$ and $r = 2$ represent circles with center O and radii 1 and 2. The region in the plane satisfying $1 \le r \le 2$ consists of both circles and the shaded region between them in the figure.

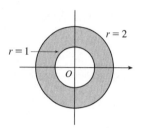

9. The region satisfying $0 \le r < 4$ and $-\pi/2 \le \theta < \pi/6$ does not include the circle $r = 4$ nor the line $\theta = \frac{\pi}{6}$.

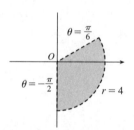

11. $2 < r < 3$, $\frac{5\pi}{3} \le \theta \le \frac{7\pi}{3}$

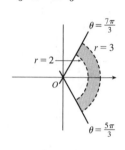

13. $(r, \theta) = \left(1, \frac{\pi}{6}\right)$ $\Rightarrow$ $x = 1 \cos \frac{\pi}{6} = \frac{\sqrt{3}}{2}$ and $y = 1 \sin \frac{\pi}{6} = \frac{1}{2}$.

$(r, \theta) = \left(3, \frac{3\pi}{4}\right)$ $\Rightarrow$ $x = 3 \cos \frac{3\pi}{4} = -\frac{3\sqrt{2}}{2}$ and $y = 3 \sin \frac{3\pi}{4} = \frac{3\sqrt{2}}{2}$. The distance between them is

$$\sqrt{\left[\frac{\sqrt{3}}{2} - \left(-\frac{3\sqrt{2}}{2}\right)\right]^2 + \left(\frac{1}{2} - \frac{3\sqrt{2}}{2}\right)^2} = \sqrt{\frac{1}{4}\left(\sqrt{3} + 3\sqrt{2}\right)^2 + \frac{1}{4}\left(1 - 3\sqrt{2}\right)^2}$$

$$= \sqrt{\frac{1}{4}\sqrt{\left(3 + 6\sqrt{6} + 18\right) + \left(1 - 6\sqrt{2} + 18\right)}} = \frac{1}{2}\sqrt{40 + 6\sqrt{6} - 6\sqrt{2}}$$

15. $r = 2$ $\Leftrightarrow$ $\sqrt{x^2 + y^2} = 2$ $\Leftrightarrow$ $x^2 + y^2 = 4$, a circle of radius 2 centered at the origin.

17. $r = 3 \sin \theta$ $\Rightarrow$ $r^2 = 3r \sin \theta$ $\Leftrightarrow$ $x^2 + y^2 = 3y$ $\Leftrightarrow$ $x^2 + \left(y - \frac{3}{2}\right)^2 = \left(\frac{3}{2}\right)^2$, a circle of radius $\frac{3}{2}$ centered at $\left(0, \frac{3}{2}\right)$. The first two equations are actually equivalent since $r^2 = 3r \sin \theta$ $\Rightarrow$ $r(r - 3 \sin \theta) = 0$ $\Rightarrow$ $r = 0$ or $r = 3 \sin \theta$. But $r = 3 \sin \theta$ gives the point $r = 0$ (the pole) when $\theta = 0$. Thus, the single equation $r = 3 \sin \theta$ is equivalent to the compound condition ($r = 0$ or $r = 3 \sin \theta$).

19. $r = \csc\theta$ $\Leftrightarrow$ $r = \dfrac{1}{\sin\theta}$ $\Leftrightarrow$ $r\sin\theta = 1$ $\Leftrightarrow$ $y = 1$, a horizontal line 1 unit above the x-axis.

21. $x = 3$ $\Leftrightarrow$ $r\cos\theta = 3$ $\Leftrightarrow$ $r = 3/\cos\theta$ $\Leftrightarrow$ $r = 3\sec\theta$.

23. $x = -y^2$ $\Leftrightarrow$ $r\cos\theta = -r^2\sin^2\theta$ $\Leftrightarrow$ $\cos\theta = -r\sin^2\theta$ $\Leftrightarrow$ $r = -\dfrac{\cos\theta}{\sin^2\theta} = -\cot\theta\csc\theta$.

25. $x^2 + y^2 = 2cx$ $\Leftrightarrow$ $r^2 = 2cr\cos\theta$ $\Leftrightarrow$ $r^2 - 2cr\cos\theta = 0$ $\Leftrightarrow$ $r(r - 2c\cos\theta) = 0$ $\Leftrightarrow$ $r = 0$ or

$r = 2c\cos\theta$. $r = 0$ is included in $r = 2c\cos\theta$ when $\theta = \frac{\pi}{2} + n\pi$, so the curve is represented by the single

equation $r = 2c\cos\theta$.

27. (a) The description leads immediately to the polar equation $\theta = \frac{\pi}{6}$, and the Cartesian equation $\tan\theta = y/x$ $\Rightarrow$

$y = \left(\tan\frac{\pi}{6}\right)x = \frac{1}{\sqrt{3}}x$ is slightly more difficult to derive.

(b) The easier description here is the Cartesian equation $x = 3$.

29. $\theta = -\pi/6$

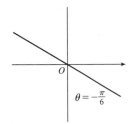

31. $r = \sin\theta$ $\Leftrightarrow$ $r^2 = r\sin\theta$ $\Leftrightarrow$ $x^2 + y^2 = y$

$\Leftrightarrow$ $x^2 + \left(y - \frac{1}{2}\right)^2 = \left(\frac{1}{2}\right)^2$. The reasoning here is

the same as in Exercise 17. This is a circle of radius $\frac{1}{2}$

centered at $\left(0, \frac{1}{2}\right)$.

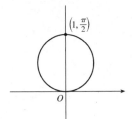

33. $r = 2(1 - \sin\theta)$. This curve is a cardioid.

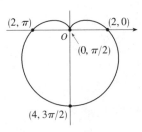

35. $r = \theta$, $\theta \geq 0$

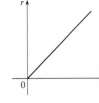
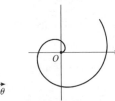

37. $r = \sin 2\theta$

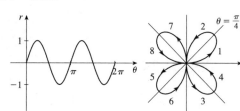

39. $r = 2\cos 4\theta$

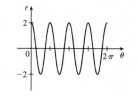

41. $r^2 = 4\cos 2\theta$

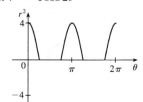 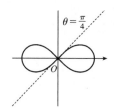

43. $r = 2\cos\left(\frac{3}{2}\theta\right)$

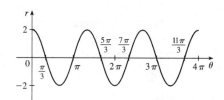 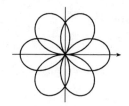

45. $r = 1 + 2\cos 2\theta$

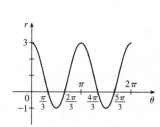 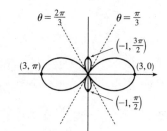

47. For $\theta = 0$, π, and 2π, r has its minimum value of about 0.5. For $\theta = \frac{\pi}{2}$ and $\frac{3\pi}{2}$, r attains its maximum value of 2.

We see that the graph has a similar shape for $0 \le \theta \le \pi$ and $\pi \le \theta \le 2\pi$.

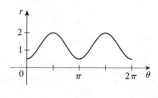 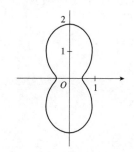

49. $x = r \cos \theta = (4 + 2 \sec \theta) \cos \theta = 4 \cos \theta + 2$. Now, $r \to \infty$ $\Rightarrow$

$(4 + 2 \sec \theta) \to \infty$ $\Rightarrow$ $\theta \to \left(\frac{\pi}{2}\right)^-$ or $\theta \to \left(\frac{3\pi}{2}\right)^+$ (since we need only

consider $0 \le \theta < 2\pi$), so $\lim\limits_{r \to \infty} x = \lim\limits_{\theta \to \pi/2^-} (4 \cos \theta + 2) = 2$. Also, $r \to -\infty$

$\Rightarrow$ $(4 + 2 \sec \theta) \to -\infty$ $\Rightarrow$ $\theta \to \left(\frac{\pi}{2}\right)^+$ or $\theta \to \left(\frac{3\pi}{2}\right)^-$, so

$\lim\limits_{r \to -\infty} x = \lim\limits_{\theta \to \pi/2^+} (4 \cos \theta + 2) = 2$. Therefore, $\lim\limits_{r \to \pm\infty} x = 2$ $\Rightarrow$ $x = 2$ is a

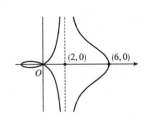

vertical asymptote.

51. To show that $x = 1$ is an asymptote we must prove $\lim\limits_{r \to \pm\infty} x = 1$.

$x = r \cos \theta = (\sin \theta \tan \theta) \cos \theta = \sin^2 \theta$. Now, $r \to \infty$ $\Rightarrow$ $\sin \theta \tan \theta \to \infty$

$\Rightarrow$ $\theta \to \left(\frac{\pi}{2}\right)^-$, so $\lim\limits_{r \to \infty} x = \lim\limits_{\theta \to \pi/2^-} \sin^2 \theta = 1$. Also, $r \to -\infty$ $\Rightarrow$

$\sin \theta \tan \theta \to -\infty$ $\Rightarrow$ $\theta \to \left(\frac{\pi}{2}\right)^+$, so $\lim\limits_{r \to -\infty} x = \lim\limits_{\theta \to \pi/2^+} \sin^2 \theta = 1$.

Therefore, $\lim\limits_{r \to \pm\infty} x = 1$ $\Rightarrow$ $x = 1$ is a vertical asymptote. Also notice that $x = \sin^2 \theta \ge 0$ for all θ, and

$x = \sin^2 \theta \le 1$ for all θ. And $x \ne 1$, since the curve is not defined at odd multiples of $\frac{\pi}{2}$. Therefore, the curve lies

entirely within the vertical strip $0 \le x < 1$.

53. (a) We see that the curve crosses itself at the origin, where $r = 0$ (in fact the inner loop corresponds to negative

r-values), so we solve the equation of the limaçon for $r = 0$ $\Leftrightarrow$ $c \sin \theta = -1$ $\Leftrightarrow$ $\sin \theta = -1/c$. Now if

$|c| < 1$, then this equation has no solution and hence there is no inner loop. But if $c < -1$, then on the interval

$(0, 2\pi)$ the equation has the two solutions $\theta = \sin^{-1}(-1/c)$ and $\theta = \pi - \sin^{-1}(-1/c)$, and if $c > 1$, the

solutions are $\theta = \pi + \sin^{-1}(1/c)$ and $\theta = 2\pi - \sin^{-1}(1/c)$. In each case, $r < 0$ for θ between the two

solutions, indicating a loop.

(b) For $0 < c < 1$, the dimple (if it exists) is characterized by the fact that y has a local maximum at $\theta = \frac{3\pi}{2}$. So we

determine for what c-values $\dfrac{d^2 y}{d\theta^2}$ is negative at $\theta = \frac{3\pi}{2}$, since by the Second Derivative Test this indicates a

maximum: $y = r \sin \theta = \sin \theta + c \sin^2 \theta$ $\Rightarrow$ $\dfrac{dy}{d\theta} = \cos \theta + 2c \sin \theta \cos \theta = \cos \theta + c \sin 2\theta$ $\Rightarrow$

$\dfrac{d^2 y}{d\theta^2} = -\sin \theta + 2c \cos 2\theta$. At $\theta = \frac{3\pi}{2}$, this is equal to $-(-1) + 2c(-1) = 1 - 2c$, which is negative only for

$c > \frac{1}{2}$. A similar argument shows that for $-1 < c < 0$, y only has a local minimum at $\theta = \frac{\pi}{2}$ (indicating a

dimple) for $c < -\frac{1}{2}$.

55. $r = 2 \sin \theta$ $\Rightarrow$ $x = r \cos \theta = 2 \sin \theta \cos \theta = \sin 2\theta$, $y = r \sin \theta = 2 \sin^2 \theta$ $\Rightarrow$

$$\frac{dy}{dx} = \frac{dy/d\theta}{dx/d\theta} = \frac{2 \cdot 2 \sin \theta \cos \theta}{\cos 2\theta \cdot 2} = \frac{\sin 2\theta}{\cos 2\theta} = \tan 2\theta$$

When $\theta = \frac{\pi}{6}$, $\dfrac{dy}{dx} = \tan\left(2 \cdot \frac{\pi}{6}\right) = \tan \frac{\pi}{3} = \sqrt{3}$.

57. $r = 1/\theta \quad \Rightarrow \quad x = r\cos\theta = (\cos\theta)/\theta,\ y = r\sin\theta = (\sin\theta)/\theta \quad \Rightarrow$

$$\frac{dy}{dx} = \frac{dy/d\theta}{dx/d\theta} = \frac{\sin\theta(-1/\theta^2) + (1/\theta)\cos\theta}{\cos\theta(-1/\theta^2) - (1/\theta)\sin\theta} \cdot \frac{\theta^2}{\theta^2} = \frac{-\sin\theta + \theta\cos\theta}{-\cos\theta - \theta\sin\theta}$$

When $\theta = \pi$, $\dfrac{dy}{dx} = \dfrac{-0 + \pi(-1)}{-(-1) - \pi(0)} = \dfrac{-\pi}{1} = -\pi.$

59. $r = 1 + \cos\theta \quad \Rightarrow \quad x = r\cos\theta = \cos\theta + \cos^2\theta,\ y = r\sin\theta = \sin\theta + \sin\theta\cos\theta \quad \Rightarrow$

$$\frac{dy}{dx} = \frac{dy/d\theta}{dx/d\theta} = \frac{\cos\theta + \cos^2\theta - \sin^2\theta}{-\sin\theta - 2\cos\theta\sin\theta} = \frac{\cos\theta + \cos 2\theta}{-\sin\theta - \sin 2\theta}$$

When $\theta = \dfrac{\pi}{6}$, $\dfrac{dy}{dx} = \dfrac{\frac{\sqrt{3}}{2} + \frac{1}{2}}{-\frac{1}{2} - \frac{\sqrt{3}}{2}} = \dfrac{\frac{\sqrt{3}}{2} + \frac{1}{2}}{-\left(\frac{1}{2} + \frac{\sqrt{3}}{2}\right)} = -1.$

61. $r = 3\cos\theta \quad \Rightarrow \quad x = r\cos\theta = 3\cos\theta\cos\theta,\ y = r\sin\theta = 3\cos\theta\sin\theta \quad \Rightarrow$

$dy/d\theta = -3\sin^2\theta + 3\cos^2\theta = 3\cos 2\theta = 0 \quad \Rightarrow \quad 2\theta = \frac{\pi}{2}$ or $\frac{3\pi}{2} \quad \Leftrightarrow \quad \theta = \frac{\pi}{4}$ or $\frac{3\pi}{4}$. So the tangent is

horizontal at $\left(\frac{3}{\sqrt{2}}, \frac{\pi}{4}\right)$ and $\left(-\frac{3}{\sqrt{2}}, \frac{3\pi}{4}\right)$ $\left[\text{same as } \left(\frac{3}{\sqrt{2}}, -\frac{\pi}{4}\right)\right]$. $dx/d\theta = -6\sin\theta\cos\theta = -3\sin 2\theta = 0 \quad \Rightarrow$

$2\theta = 0$ or $\pi \quad \Leftrightarrow \quad \theta = 0$ or $\frac{\pi}{2}$. So the tangent is vertical at $(3, 0)$ and $\left(0, \frac{\pi}{2}\right)$.

63. $r = 1 + \cos\theta \quad \Rightarrow \quad x = r\cos\theta = \cos\theta(1 + \cos\theta),\ y = r\sin\theta = \sin\theta(1 + \cos\theta) \quad \Rightarrow$

$dy/d\theta = (1 + \cos\theta)\cos\theta - \sin^2\theta = 2\cos^2\theta + \cos\theta - 1 = (2\cos\theta - 1)(\cos\theta + 1) = 0 \quad \Rightarrow \quad \cos\theta = \frac{1}{2}$ or

$-1 \quad \Rightarrow \quad \theta = \frac{\pi}{3}, \pi$, or $\frac{5\pi}{3} \quad \Rightarrow \quad$ horizontal tangent at $\left(\frac{3}{2}, \frac{\pi}{3}\right), (0, \pi)$ [the pole], and $\left(\frac{3}{2}, \frac{5\pi}{3}\right)$.

$dx/d\theta = -(1 + \cos\theta)\sin\theta - \cos\theta\sin\theta = -\sin\theta(1 + 2\cos\theta) = 0 \quad \Rightarrow \quad \sin\theta = 0$ or $\cos\theta = -\frac{1}{2} \quad \Rightarrow$

$\theta = 0, \pi, \frac{2\pi}{3}$, or $\frac{4\pi}{3} \quad \Rightarrow \quad$ vertical tangent at $(2, 0), \left(\frac{1}{2}, \frac{2\pi}{3}\right)$, and $\left(\frac{1}{2}, \frac{4\pi}{3}\right)$. Note that the tangent is horizontal, not

vertical when $\theta = \pi$, since $\displaystyle\lim_{\theta \to \pi} \frac{dy/d\theta}{dx/d\theta} = 0.$

65. $r = \cos 2\theta \quad \Rightarrow \quad x = r\cos\theta = \cos 2\theta\cos\theta,\ y = r\sin\theta = \cos 2\theta\sin\theta \quad \Rightarrow$

$$dy/d\theta = -2\sin 2\theta\sin\theta + \cos 2\theta\cos\theta = -4\sin^2\theta\cos\theta + \left(\cos^3\theta - \sin^2\theta\cos\theta\right)$$

$$= \cos\theta\left(\cos^2\theta - 5\sin^2\theta\right) = \cos\theta\left(1 - 6\sin^2\theta\right) = 0 \quad \Rightarrow$$

$\cos\theta = 0$ or $\sin\theta = \pm\frac{1}{\sqrt{6}} \quad \Rightarrow \quad \theta = \frac{\pi}{2}, \frac{3\pi}{2}, \alpha, \pi - \alpha, \pi + \alpha$, or $2\pi - \alpha$ (where $\alpha = \sin^{-1}\frac{1}{\sqrt{6}}$).

So the tangent is horizontal at $\left(-1, \frac{\pi}{2}\right), \left(-1, \frac{3\pi}{2}\right), \left(\frac{2}{3}, \alpha\right), \left(\frac{2}{3}, \pi - \alpha\right), \left(\frac{2}{3}, \pi + \alpha\right)$, and $\left(\frac{2}{3}, 2\pi - \alpha\right)$.

$$dx/d\theta = -2\sin 2\theta\cos\theta - \cos 2\theta\sin\theta = -4\sin\theta\cos^2\theta - \left(2\cos^2\theta - 1\right)\sin\theta$$

$$= \sin\theta\left(1 - 6\cos^2\theta\right) = 0 \quad \Rightarrow$$

$\sin\theta = 0$ or $\cos\theta = \pm\frac{1}{\sqrt{6}} \quad \Rightarrow \quad \theta = 0, \pi, \beta, \pi - \beta, \pi + \beta$, or $2\pi - \beta$ (where $\beta = \cos^{-1}\frac{1}{\sqrt{6}}$).

So the tangent is vertical at $(1, 0), (1, \pi), \left(-\frac{2}{3}, \beta\right), \left(-\frac{2}{3}, \pi - \beta\right), \left(-\frac{2}{3}, \pi + \beta\right)$, and $\left(-\frac{2}{3}, 2\pi - \beta\right)$.

67. $r = a\sin\theta + b\cos\theta \Rightarrow r^2 = ar\sin\theta + br\cos\theta \Rightarrow x^2 + y^2 = ay + bx \Rightarrow$

$x^2 - bx + \left(\frac{1}{2}b\right)^2 + y^2 - ay + \left(\frac{1}{2}a\right)^2 = \left(\frac{1}{2}b\right)^2 + \left(\frac{1}{2}a\right)^2 \Rightarrow \left(x - \frac{1}{2}b\right)^2 + \left(y - \frac{1}{2}a\right)^2 = \frac{1}{4}\left(a^2 + b^2\right)$, and this

is a circle with center $\left(\frac{1}{2}b, \frac{1}{2}a\right)$ and radius $\frac{1}{2}\sqrt{a^2 + b^2}$.

Note for Exercises 69–74: Maple is able to plot polar curves using the `polarplot` command, or using the

`coords=polar` option in a regular `plot` command. In Mathematica, use `PolarPlot`. In Derive, change to `Polar` under

`Options State`. If your graphing device cannot plot polar equations, you must convert to parametric equations. For example,

in Exercise 69, $x = r\cos\theta = [1 + 2\sin(\theta/2)]\cos\theta$, $y = r\sin\theta = [1 + 2\sin(\theta/2)]\sin\theta$.

69. $r = 1 + 2\sin(\theta/2)$. The parameter interval
is $[0, 4\pi]$.

71. $r = e^{\sin\theta} - 2\cos(4\theta)$. The parameter interval
is $[0, 2\pi]$.

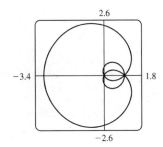

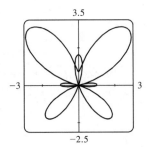

73. $r = 2 - 5\sin(\theta/6)$. The parameter interval is $[-6\pi, 6\pi]$.

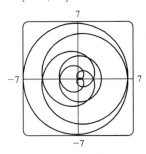

75.

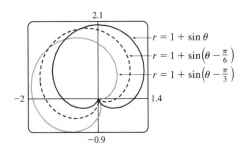

$r = 1 + \sin\theta$

$r = 1 + \sin\left(\theta - \frac{\pi}{6}\right)$

$r = 1 + \sin\left(\theta - \frac{\pi}{3}\right)$

It appears that the graph of $r = 1 + \sin\left(\theta - \frac{\pi}{6}\right)$ is the same shape as the graph of $r = 1 + \sin\theta$, but rotated

counterclockwise about the origin by $\frac{\pi}{6}$. Similarly, the graph of $r = 1 + \sin\left(\theta - \frac{\pi}{3}\right)$ is rotated by $\frac{\pi}{3}$. In general, the graph of $r = f(\theta - \alpha)$ is the same shape as that of $r = f(\theta)$, but rotated counterclockwise through α about the origin. That is, for any point (r_0, θ_0) on the curve $r = f(\theta)$, the point $(r_0, \theta_0 + \alpha)$ is on the curve $r = f(\theta - \alpha)$, since $r_0 = f(\theta_0) = f((\theta_0 + \alpha) - \alpha)$.

77. (a) $r = \sin n\theta$. From the graphs, it seems that when n is even, the number of loops in the curve (called a rose) is $2n$, and when n is odd, the number of loops is simply n.

This is because in the case of n odd, every point on the graph is traversed twice, due to the fact that

$$r(\theta + \pi) = \sin\left[n(\theta + \pi)\right] = \sin n\theta \cos n\pi + \cos n\theta \sin n\pi = \begin{cases} \sin n\theta & \text{if } n \text{ is even} \\ -\sin n\theta & \text{if } n \text{ is odd} \end{cases}$$

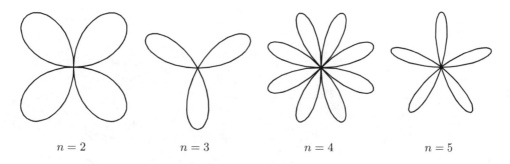

$n = 2$ $n = 3$ $n = 4$ $n = 5$

(b) The graph of $r = |\sin n\theta|$ has $2n$ loops whether n is odd or even, since $r(\theta + \pi) = r(\theta)$.

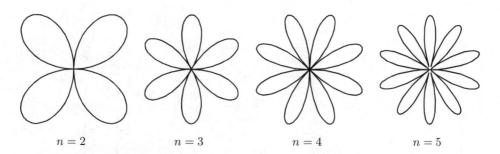

$n = 2$ $n = 3$ $n = 4$ $n = 5$

79. $r = \dfrac{1 - a\cos\theta}{1 + a\cos\theta}$. We start with $a = 0$, since in this case the curve is simply the circle $r = 1$.

As a increases, the graph moves to the left, and its right side becomes flattened. As a increases through about 0.4, the right side seems to grow a dimple, which upon closer investigation (with narrower θ-ranges) seems to appear at $a \approx 0.42$ (the actual value is $\sqrt{2} - 1$). As $a \to 1$, this dimple becomes more pronounced, and the curve begins to stretch out horizontally, until at $a = 1$ the denominator vanishes at $\theta = \pi$, and the dimple becomes an actual cusp. For $a > 1$ we must choose our parameter interval carefully, since $r \to \infty$ as $1 + a\cos\theta \to 0$ ⇔

$\theta \to \pm \cos^{-1}(-1/a)$. As a increases from 1, the curve splits into two parts. The left part has a loop, which grows larger as a increases, and the right part grows broader vertically, and its left tip develops a dimple when $a \approx 2.42$ (actually, $\sqrt{2} + 1$). As a increases, the dimple grows more and more pronounced. If $a < 0$, we get the same graph as we do for the corresponding positive a-value, but with a rotation through π about the pole, as happened when c was replaced with $-c$ in Exercise 78.

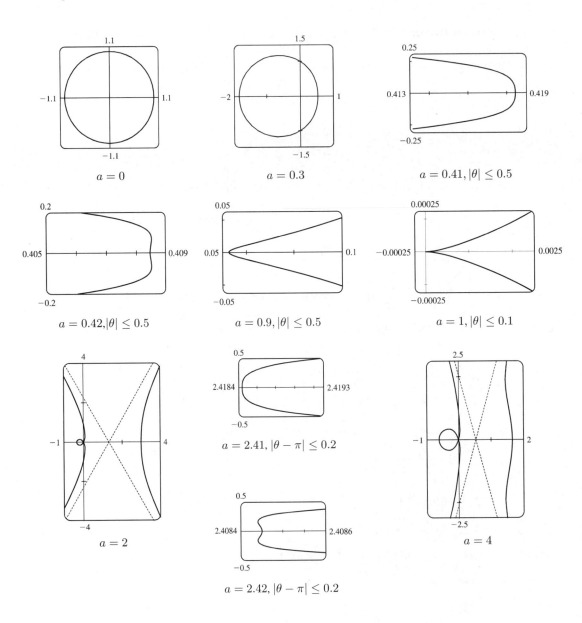

81.
$$\tan \psi = \tan(\phi - \theta) = \frac{\tan \phi - \tan \theta}{1 + \tan \phi \tan \theta} = \frac{\dfrac{dy}{dx} - \tan \theta}{1 + \dfrac{dy}{dx} \tan \theta} = \frac{\dfrac{dy/d\theta}{dx/d\theta} - \tan \theta}{1 + \dfrac{dy/d\theta}{dx/d\theta} \tan \theta}$$

$$= \frac{\dfrac{dy}{d\theta} - \dfrac{dx}{d\theta} \tan \theta}{\dfrac{dx}{d\theta} + \dfrac{dy}{d\theta} \tan \theta} = \frac{\left(\dfrac{dr}{d\theta} \sin \theta + r \cos \theta\right) - \tan \theta \left(\dfrac{dr}{d\theta} \cos \theta - r \sin \theta\right)}{\left(\dfrac{dr}{d\theta} \cos \theta - r \sin \theta\right) + \tan \theta \left(\dfrac{dr}{d\theta} \sin \theta + r \cos \theta\right)}$$

$$= \frac{r \cos \theta + r \cdot \dfrac{\sin^2 \theta}{\cos \theta}}{\dfrac{dr}{d\theta} \cos \theta + \dfrac{dr}{d\theta} \cdot \dfrac{\sin^2 \theta}{\cos \theta}} = \frac{r \cos^2 \theta + r \sin^2 \theta}{\dfrac{dr}{d\theta} \cos^2 \theta + \dfrac{dr}{d\theta} \sin^2 \theta} = \frac{r}{dr/d\theta}$$

11.4 Areas and Lengths in Polar Coordinates ET 10.4

1. $r = \sqrt{\theta}, \, 0 \le \theta \le \frac{\pi}{4}$. $A = \int_0^{\pi/4} \frac{1}{2} r^2 \, d\theta = \int_0^{\pi/4} \frac{1}{2} \left(\sqrt{\theta}\right)^2 d\theta = \int_0^{\pi/4} \frac{1}{2} \theta \, d\theta = \left[\frac{1}{4} \theta^2\right]_0^{\pi/4} = \frac{1}{64} \pi^2$

3. $r = \sin \theta, \, \frac{\pi}{3} \le \theta \le \frac{2\pi}{3}$.

$$A = \int_{\pi/3}^{2\pi/3} \frac{1}{2} \sin^2 \theta \, d\theta = \frac{1}{4} \int_{\pi/3}^{2\pi/3} (1 - \cos 2\theta) \, d\theta = \frac{1}{4} \left[\theta - \frac{1}{2} \sin 2\theta\right]_{\pi/3}^{2\pi/3}$$

$$= \frac{1}{4} \left[\frac{2\pi}{3} - \frac{1}{2} \sin \frac{4\pi}{3} - \frac{\pi}{3} + \frac{1}{2} \sin \frac{2\pi}{3}\right] = \frac{1}{4} \left[\frac{2\pi}{3} - \frac{1}{2}\left(-\frac{\sqrt{3}}{2}\right) - \frac{\pi}{3} + \frac{1}{2}\left(\frac{\sqrt{3}}{2}\right)\right] = \frac{1}{4}\left(\frac{\pi}{3} + \frac{\sqrt{3}}{2}\right) = \frac{\pi}{12} + \frac{\sqrt{3}}{8}$$

5. $r = \theta, \, 0 \le \theta \le \pi$. $A = \int_0^{\pi} \frac{1}{2} \theta^2 \, d\theta = \left[\frac{1}{6} \theta^3\right]_0^{\pi} = \frac{1}{6} \pi^3$

7. $r = 4 + 3 \sin \theta, \, -\frac{\pi}{2} \le \theta \le \frac{\pi}{2}$.

$$A = \int_{-\pi/2}^{\pi/2} \frac{1}{2}(4 + 3 \sin \theta)^2 d\theta = \frac{1}{2} \int_{-\pi/2}^{\pi/2} (16 + 24 \sin \theta + 9 \sin^2 \theta) \, d\theta$$

$$= \frac{1}{2} \int_{-\pi/2}^{\pi/2} (16 + 9 \sin^2 \theta) \, d\theta \quad \text{[by Theorem 5.5.6(b) [ET 5.5.7(b)]]}$$

$$= \frac{1}{2} \cdot 2 \int_0^{\pi/2} \left[16 + 9 \cdot \frac{1}{2}(1 - \cos 2\theta)\right] d\theta \quad \text{[by Theorem 5.5.6(a) [ET 5.5.7(a)]]}$$

$$= \int_0^{\pi/2} \left(\frac{41}{2} - \frac{9}{2} \cos 2\theta\right) d\theta = \left[\frac{41}{2} \theta - \frac{9}{4} \sin 2\theta\right]_0^{\pi/2} = \left(\frac{41\pi}{4} - 0\right) - (0 - 0) = \frac{41\pi}{4}$$

9. The area above the polar axis is bounded by $r = 3 \cos \theta$ for
$\theta = 0$ to $\theta = \pi/2$ (*not* π). By symmetry,

$$A = 2 \int_0^{\pi/2} \frac{1}{2} r^2 \, d\theta = \int_0^{\pi/2} (3 \cos \theta)^2 d\theta$$

$$= 3^2 \int_0^{\pi/2} \cos^2 \theta \, d\theta = 9 \int_0^{\pi/2} \frac{1}{2}(1 + \cos 2\theta) \, d\theta$$

$$= \frac{9}{2} \left[\theta + \frac{1}{2} \sin 2\theta\right]_0^{\pi/2} = \frac{9}{2}\left[\left(\frac{\pi}{2} + 0\right) - (0 + 0)\right] = \frac{9\pi}{4}.$$

$r = 3 \cos \theta$
$(3, 0)$
O

Also, note that this is a circle with radius $\frac{3}{2}$, so its area is $\pi\left(\frac{3}{2}\right)^2 = \frac{9\pi}{4}$.

11. The curve $r^2 = 4\cos 2\theta$ goes through the pole when

$\theta = \pi/4$, so we'll find the area for $0 \le \theta \le \pi/4$ and

multiply it by 4.

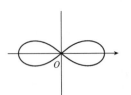

$A = 4\int_0^{\pi/4} \frac{1}{2}r^2\,d\theta = 2\int_0^{\pi/4}(4\cos 2\theta)\,d\theta$

$\quad = 8\int_0^{\pi/4}\cos 2\theta\,d\theta = 4\left[\sin 2\theta\right]_0^{\pi/4} = 4(1-0) = 4$

13. One-sixth of the area lies above the polar axis and is bounded

by the curve $r = 2\cos 3\theta$ for $\theta = 0$ to $\theta = \pi/6$.

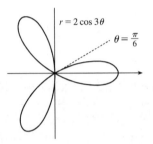

$A = 6\int_0^{\pi/6}\frac{1}{2}(2\cos 3\theta)^2\,d\theta = 12\int_0^{\pi/6}\cos^2 3\theta\,d\theta$

$\quad = \frac{12}{2}\int_0^{\pi/6}(1+\cos 6\theta)\,d\theta$

$\quad = 6\left[\theta + \frac{1}{6}\sin 6\theta\right]_0^{\pi/6} = 6\left(\frac{\pi}{6}\right) = \pi$

15. $A = \int_0^{2\pi}\frac{1}{2}(1+2\sin 6\theta)^2\,d\theta = \frac{1}{2}\int_0^{2\pi}(1+4\sin 6\theta + 4\sin^2 6\theta)\,d\theta$

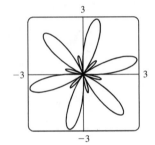

$\quad = \frac{1}{2}\int_0^{2\pi}\left[1 + 4\sin 6\theta + 4\cdot\frac{1}{2}(1-\cos 12\theta)\right]d\theta$

$\quad = \frac{1}{2}\int_0^{2\pi}(3 + 4\sin 6\theta - 2\cos 12\theta)\,d\theta$

$\quad = \frac{1}{2}\left[3\theta - \frac{2}{3}\cos 6\theta - \frac{1}{6}\sin 12\theta\right]_0^{2\pi}$

$\quad = \frac{1}{2}\left[\left(6\pi - \frac{2}{3} - 0\right) - \left(0 - \frac{2}{3} - 0\right)\right] = 3\pi.$

17. The shaded loop is traced out from $\theta = 0$

to $\theta = \pi/2$.

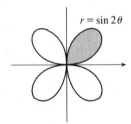

$\qquad A = \int_0^{\pi/2}\frac{1}{2}r^2\,d\theta = \frac{1}{2}\int_0^{\pi/2}\sin^2 2\theta\,d\theta$

$\qquad\quad = \frac{1}{2}\int_0^{\pi/2}\frac{1}{2}(1-\cos 4\theta)\,d\theta$

$\qquad\quad = \frac{1}{4}\left[\theta - \frac{1}{4}\sin 4\theta\right]_0^{\pi/2} = \frac{1}{4}\left(\frac{\pi}{2}\right) = \frac{\pi}{8}$

19. $r = 0 \;\Rightarrow\; 3\cos 5\theta = 0 \;\Rightarrow\; 5\theta = \frac{\pi}{2} \;\Rightarrow\; \theta = \frac{\pi}{10}.$

$A = \int_{-\pi/10}^{\pi/10}\frac{1}{2}(3\cos 5\theta)^2\,d\theta = \int_0^{\pi/10}9\cos^2 5\theta\,d\theta = \frac{9}{2}\int_0^{\pi/10}(1+\cos 10\theta)\,d\theta = \frac{9}{2}\left[\theta + \frac{1}{10}\sin 10\theta\right]_0^{\pi/10} = \frac{9\pi}{20}$

21.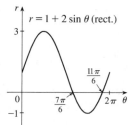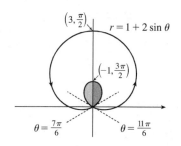

This is a limaçon, with inner loop traced out between $\theta = \frac{7\pi}{6}$ and $\frac{11\pi}{6}$ [found by solving $r = 0$].

$$A = 2 \int_{7\pi/6}^{3\pi/2} \tfrac{1}{2}(1 + 2\sin\theta)^2 \, d\theta = \int_{7\pi/6}^{3\pi/2} (1 + 4\sin\theta + 4\sin^2\theta) \, d\theta$$

$$= \int_{7\pi/6}^{3\pi/2} \left[1 + 4\sin\theta + 4 \cdot \tfrac{1}{2}(1 - \cos 2\theta)\right] d\theta = \left[\theta - 4\cos\theta + 2\theta - \sin 2\theta\right]_{7\pi/6}^{3\pi/2}$$

$$= \left(\tfrac{9\pi}{2}\right) - \left(\tfrac{7\pi}{2} + 2\sqrt{3} - \tfrac{\sqrt{3}}{2}\right) = \pi - \tfrac{3\sqrt{3}}{2}$$

23. $4\sin\theta = 2 \iff \sin\theta = \tfrac{1}{2} \iff \theta = \tfrac{\pi}{6}$ or $\tfrac{5\pi}{6}$ (for $0 \le \theta \le 2\pi$). We'll

subtract the unshaded area from the shaded area for $\pi/6 \le \theta \le \pi/2$ and double

that value.

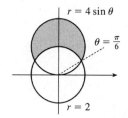

$$A = 2 \int_{\pi/6}^{\pi/2} \tfrac{1}{2}(4\sin\theta)^2 \, d\theta - 2 \int_{\pi/6}^{\pi/2} \tfrac{1}{2}(2)^2 \, d\theta = 2 \int_{\pi/6}^{\pi/2} \tfrac{1}{2}\left[(4\sin\theta)^2 - 2^2\right] d\theta$$

$$= \int_{\pi/6}^{\pi/2} \left(16\sin^2\theta - 4\right) d\theta = \int_{\pi/6}^{\pi/2} \left[8(1 - \cos 2\theta) - 4\right] d\theta$$

$$= \int_{\pi/6}^{\pi/2} (4 - 8\cos 2\theta) \, d\theta = \left[4\theta - 4\sin 2\theta\right]_{\pi/6}^{\pi/2}$$

$$= (2\pi - 0) - \left(\tfrac{2\pi}{3} - 4 \cdot \tfrac{\sqrt{3}}{2}\right) = \tfrac{4}{3}\pi + 2\sqrt{3}$$

25. To find the area inside the leminiscate $r^2 = 8\cos 2\theta$ and outside the

circle $r = 2$, we first note that the two curves intersect when

$r^2 = 8\cos 2\theta$ and $r = 2$; i.e., when $\cos 2\theta = \tfrac{1}{2}$. For $-\pi < \theta \le \pi$,

$\cos 2\theta = \tfrac{1}{2} \iff 2\theta = \pm\pi/3$ or $\pm 5\pi/3 \iff \theta = \pm\pi/6$ or

$\pm 5\pi/6$. The figure shows that the desired area is 4 times the area

between the curves from 0 to $\pi/6$. Thus,

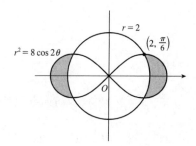

$$A = 4 \int_0^{\pi/6} \left[\tfrac{1}{2}(8\cos 2\theta) - \tfrac{1}{2}(2)^2\right] d\theta = 8 \int_0^{\pi/6} (2\cos 2\theta - 1) \, d\theta$$

$$= 8 \left[\sin 2\theta - \theta\right]_0^{\pi/6} = 8\left(\sqrt{3}/2 - \pi/6\right) = 4\sqrt{3} - 4\pi/3$$

27. $3\cos\theta = 1 + \cos\theta \iff \cos\theta = \frac{1}{2} \implies \theta = \frac{\pi}{3}$ or $-\frac{\pi}{3}$.

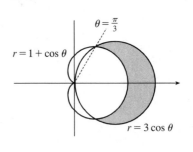

$$A = 2\int_0^{\pi/3} \frac{1}{2}\left[(3\cos\theta)^2 - (1+\cos\theta)^2\right]d\theta$$

$$= \int_0^{\pi/3}\left(8\cos^2\theta - 2\cos\theta - 1\right)d\theta$$

$$= \int_0^{\pi/3}\left[4(1+\cos 2\theta) - 2\cos\theta - 1\right]d\theta$$

$$= \int_0^{\pi/3}\left(3 + 4\cos 2\theta - 2\cos\theta\right)d\theta$$

$$= \left[3\theta + 2\sin 2\theta - 2\sin\theta\right]_0^{\pi/3}$$

$$= \pi + \sqrt{3} - \sqrt{3} = \pi$$

29. $A = 2\int_0^{\pi/4} \frac{1}{2}\sin^2\theta\, d\theta = \int_0^{\pi/4} \frac{1}{2}\left(1 - \cos 2\theta\right)d\theta$

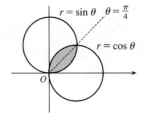

$$= \frac{1}{2}\left[\theta - \frac{1}{2}\sin 2\theta\right]_0^{\pi/4} = \frac{1}{2}\left[\left(\frac{\pi}{4} - \frac{1}{2}\cdot 1\right) - (0-0)\right]$$

$$= \frac{1}{8}\pi - \frac{1}{4}$$

31. $\sin 2\theta = \cos 2\theta \implies \dfrac{\sin 2\theta}{\cos 2\theta} = 1 \implies \tan 2\theta = 1 \implies$

$2\theta = \frac{\pi}{4} \implies \theta = \frac{\pi}{8} \implies$

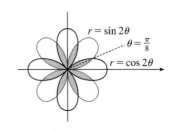

$$A = 8\cdot 2\int_0^{\pi/8}\frac{1}{2}\sin^2 2\theta\, d\theta = 8\int_0^{\pi/8}\frac{1}{2}(1-\cos 4\theta)\, d\theta$$

$$= 4\left[\theta - \frac{1}{4}\sin 4\theta\right]_0^{\pi/8} = 4\left(\frac{\pi}{8} - \frac{1}{4}\cdot 1\right) = \frac{1}{2}\pi - 1$$

33. $A = 2\left[\int_{-\pi/2}^{-\pi/6}\frac{1}{2}(3+2\sin\theta)^2\, d\theta + \int_{-\pi/6}^{\pi/2}\frac{1}{2}2^2\, d\theta\right]$

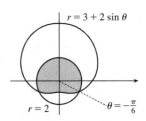

$$= \int_{-\pi/2}^{-\pi/6}\left(9 + 12\sin\theta + 4\sin^2\theta\right)d\theta + \left[4\theta\right]_{-\pi/6}^{\pi/2}$$

$$= \left[9\theta - 12\cos\theta + 2\theta - \sin 2\theta\right]_{-\pi/2}^{-\pi/6} + \frac{8\pi}{3} = \frac{19\pi}{3} - \frac{11\sqrt{3}}{2}$$

35. The darker shaded region (from $\theta = 0$ to $\theta = 2\pi/3$) represents $\frac{1}{2}$ of the desired area plus $\frac{1}{2}$ of the area of the inner loop. From this area, we'll subtract $\frac{1}{2}$ of the area of the inner loop (the lighter shaded region from $\theta = 2\pi/3$

to $\theta = \pi$), and then double that difference to obtain the desired area.

$$A = 2\left[\int_0^{2\pi/3} \tfrac{1}{2}\left(\tfrac{1}{2} + \cos\theta\right)^2 d\theta - \int_{2\pi/3}^{\pi} \tfrac{1}{2}\left(\tfrac{1}{2} + \cos\theta\right)^2 d\theta\right]$$

$$= \int_0^{2\pi/3} \left(\tfrac{1}{4} + \cos\theta + \cos^2\theta\right) d\theta - \int_{2\pi/3}^{\pi} \left(\tfrac{1}{4} + \cos\theta + \cos^2\theta\right) d\theta$$

$$= \int_0^{2\pi/3} \left[\tfrac{1}{4} + \cos\theta + \tfrac{1}{2}(1 + \cos 2\theta)\right] d\theta$$

$$\qquad\qquad - \int_{2\pi/3}^{\pi} \left[\tfrac{1}{4} + \cos\theta + \tfrac{1}{2}(1 + \cos 2\theta)\right] d\theta$$

$$= \left[\frac{\theta}{4} + \sin\theta + \frac{\theta}{2} + \frac{\sin 2\theta}{4}\right]_0^{2\pi/3} - \left[\frac{\theta}{4} + \sin\theta + \frac{\theta}{2} + \frac{\sin 2\theta}{4}\right]_{2\pi/3}^{\pi}$$

$$= \left(\tfrac{\pi}{6} + \tfrac{\sqrt{3}}{2} + \tfrac{\pi}{3} - \tfrac{\sqrt{3}}{8}\right) - \left(\tfrac{\pi}{4} + \tfrac{\pi}{2}\right) + \left(\tfrac{\pi}{6} + \tfrac{\sqrt{3}}{2} + \tfrac{\pi}{3} - \tfrac{\sqrt{3}}{8}\right)$$

$$= \tfrac{\pi}{4} + \tfrac{3}{4}\sqrt{3} = \tfrac{1}{4}\left(\pi + 3\sqrt{3}\right)$$

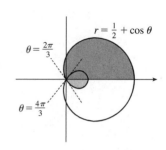

37. The two circles intersect at the pole since $(0,0)$ satisfies the first
equation and $\left(0, \tfrac{\pi}{2}\right)$ the second. The other intersection point
$\left(\tfrac{1}{\sqrt{2}}, \tfrac{\pi}{4}\right)$ occurs where $\sin\theta = \cos\theta$.

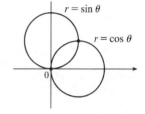

39. The curves intersect at the pole since $\left(0, \tfrac{\pi}{2}\right)$ satisfies $r = \cos\theta$ and $(0,0)$ satisfies $r = 1 - \cos\theta$. Now
$\cos\theta = 1 - \cos\theta \;\Rightarrow\; 2\cos\theta = 1 \;\Rightarrow\; \cos\theta = \tfrac{1}{2} \;\Rightarrow\; \theta = \tfrac{\pi}{3}$ or $\tfrac{5\pi}{3} \;\Rightarrow$
the other intersection points are $\left(\tfrac{1}{2}, \tfrac{\pi}{3}\right)$ and $\left(\tfrac{1}{2}, \tfrac{5\pi}{3}\right)$.

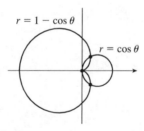

41. The pole is a point of intersection. $\sin\theta = \sin 2\theta = 2\sin\theta\cos\theta \;\Leftrightarrow\; \sin\theta(1 - 2\cos\theta) = 0 \;\Leftrightarrow\; \sin\theta = 0$ or
$\cos\theta = \tfrac{1}{2} \;\Rightarrow\; \theta = 0, \pi, \tfrac{\pi}{3}, -\tfrac{\pi}{3} \;\Rightarrow\; \left(\tfrac{\sqrt{3}}{2}, \tfrac{\pi}{3}\right)$ and $\left(\tfrac{\sqrt{3}}{2}, \tfrac{2\pi}{3}\right)$ (by symmetry) are the other intersection points.

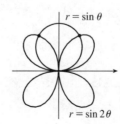

43.

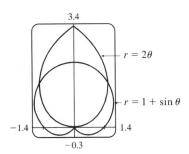
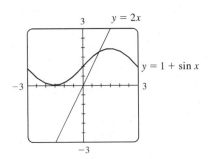

From the first graph, we see that the pole is one point of intersection. By zooming in or using the cursor, we find the θ-values of the intersection points to be $\alpha \approx 0.88786 \approx 0.89$ and $\pi - \alpha \approx 2.25$. (The first of these values may be more easily estimated by plotting $y = 1 + \sin x$ and $y = 2x$ in rectangular coordinates; see the second graph.) By symmetry, the total area contained is twice the area contained in the first quadrant, that is,

$$A = 2 \int_0^\alpha \tfrac{1}{2}(2\theta)^2 \, d\theta + 2 \int_\alpha^{\pi/2} \tfrac{1}{2}(1 + \sin\theta)^2 \, d\theta = \int_0^\alpha 4\theta^2 \, d\theta + \int_\alpha^{\pi/2} \left[1 + 2\sin\theta + \tfrac{1}{2}(1 - \cos 2\theta)\right] d\theta$$

$$= \left[\tfrac{4}{3}\theta^3\right]_0^\alpha + \left[\theta - 2\cos\theta + \left(\tfrac{1}{2}\theta - \tfrac{1}{4}\sin 2\theta\right)\right]_\alpha^{\pi/2}$$

$$= \tfrac{4}{3}\alpha^3 + \left[\left(\tfrac{\pi}{2} + \tfrac{\pi}{4}\right) - \left(\alpha - 2\cos\alpha + \tfrac{1}{2}\alpha - \tfrac{1}{4}\sin 2\alpha\right)\right] \approx 3.4645$$

45. $L = \int_a^b \sqrt{r^2 + (dr/d\theta)^2} \, d\theta = \int_0^{\pi/3} \sqrt{(3\sin\theta)^2 + (3\cos\theta)^2} \, d\theta = \int_0^{\pi/3} \sqrt{9(\sin^2\theta + \cos^2\theta)} \, d\theta$

$$= 3 \int_0^{\pi/3} d\theta = 3[\theta]_0^{\pi/3} = 3\left(\tfrac{\pi}{3}\right) = \pi.$$

As a check, note that the circumference of a circle with radius $\tfrac{3}{2}$ is $2\pi\left(\tfrac{3}{2}\right) = 3\pi$, and since $\theta = 0$ to $\pi = \tfrac{\pi}{3}$ traces out $\tfrac{1}{3}$ of the circle (from $\theta = 0$ to $\theta = \pi$), $\tfrac{1}{3}(3\pi) = \pi$.

47. $L = \int_a^b \sqrt{r^2 + (dr/d\theta)^2} \, d\theta = \int_0^{2\pi} \sqrt{(\theta^2)^2 + (2\theta)^2} \, d\theta = \int_0^{2\pi} \sqrt{\theta^4 + 4\theta^2} \, d\theta$

$$= \int_0^{2\pi} \sqrt{\theta^2(\theta^2 + 4)} \, d\theta = \int_0^{2\pi} \theta\sqrt{\theta^2 + 4} \, d\theta$$

Now let $u = \theta^2 + 4$, so that $du = 2\theta \, d\theta \ \left[\theta \, d\theta = \tfrac{1}{2} \, du\right]$ and

$$\int_0^{2\pi} \theta\sqrt{\theta^2 + 4} \, d\theta = \int_4^{4\pi^2 + 4} \tfrac{1}{2}\sqrt{u} \, du = \tfrac{1}{2} \cdot \tfrac{2}{3}\left[u^{3/2}\right]_4^{4(\pi^2 + 1)} = \tfrac{1}{3}\left[4^{3/2}(\pi^2 + 1)^{3/2} - 4^{3/2}\right]$$

$$= \tfrac{8}{3}\left[(\pi^2 + 1)^{3/2} - 1\right]$$

49. The curve $r = 3\sin 2\theta$ is completely traced with $0 \le \theta \le 2\pi$. $r^2 + \left(\tfrac{dr}{d\theta}\right)^2 = (3\sin 2\theta)^2 + (6\cos 2\theta)^2 \ \Rightarrow$

$$L = \int_0^{2\pi} \sqrt{9\sin^2 2\theta + 36\cos^2 2\theta} \, d\theta \approx 29.0653$$

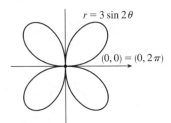

51. The curve $r = \sin\left(\frac{\theta}{2}\right)$ is completely traced with $0 \le \theta \le 4\pi$. $r^2 + \left(\frac{dr}{d\theta}\right)^2 = \sin^2\left(\frac{\theta}{2}\right) + \left[\frac{1}{2}\cos\left(\frac{\theta}{2}\right)\right]^2$ $\Rightarrow$

$$L = \int_0^{4\pi} \sqrt{\sin^2\left(\frac{\theta}{2}\right) + \frac{1}{4}\cos^2\left(\frac{\theta}{2}\right)}\, d\theta$$

$$\approx 9.6884$$

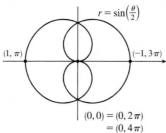

53. The curve $r = \cos^4(\theta/4)$ is completely traced with $0 \le \theta \le 4\pi$.

$$r^2 + (dr/d\theta)^2 = [\cos^4(\theta/4)]^2 + \left[4\cos^3(\theta/4) \cdot (-\sin(\theta/4)) \cdot \frac{1}{4}\right]^2$$

$$= \cos^8(\theta/4) + \cos^6(\theta/4)\sin^2(\theta/4)$$

$$= \cos^6(\theta/4)\left[\cos^2(\theta/4) + \sin^2(\theta/4)\right]$$

$$= \cos^6(\theta/4)$$

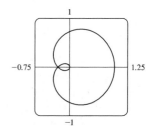

$$L = \int_0^{4\pi} \sqrt{\cos^6(\theta/4)}\, d\theta = \int_0^{4\pi} \left|\cos^3(\theta/4)\right|\, d\theta$$

$$= 2\int_0^{2\pi} \cos^3(\theta/4)\, d\theta \quad [\text{since } \cos^3(\theta/4) \ge 0 \text{ for } 0 \le \theta \le 2\pi] \quad = 8\int_0^{\pi/2} \cos^3 u\, du \quad \left[u = \frac{1}{4}\theta\right]$$

$$\overset{68}{=} 8\left[\frac{1}{3}(2 + \cos^2 u)\sin u\right]_0^{\pi/2} = \frac{8}{3}[(2 \cdot 1) - (3 \cdot 0)] = \frac{16}{3}$$

55. (a) From (11.2.7 [ET 10.2.7]),

$$S = \int_a^b 2\pi y \sqrt{(dx/d\theta)^2 + (dy/d\theta)^2}\, d\theta$$

$$= \int_a^b 2\pi y \sqrt{r^2 + (dr/d\theta)^2}\, d\theta \qquad [\text{from the derivation of Equation 11.4.5 [ET 10.4.5]}]$$

$$= \int_a^b 2\pi r\sin\theta \sqrt{r^2 + (dr/d\theta)^2}\, d\theta$$

(b) The curve $r^2 = \cos 2\theta$ goes through the pole when

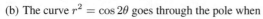

$\cos 2\theta = 0 \quad \Rightarrow \quad 2\theta = \dfrac{\pi}{2} \quad \Rightarrow \quad \theta = \dfrac{\pi}{4}$. We'll rotate the curve

from $\theta = 0$ to $\theta = \dfrac{\pi}{4}$ and double this value to obtain the total surface

area generated. $r^2 = \cos 2\theta \quad \Rightarrow \quad 2r\dfrac{dr}{d\theta} = -2\sin 2\theta \quad \Rightarrow$

$$\left(\frac{dr}{d\theta}\right)^2 = \frac{\sin^2 2\theta}{r^2} = \frac{\sin^2 2\theta}{\cos 2\theta}.$$

$$S = 2\int_0^{\pi/4} 2\pi \sqrt{\cos 2\theta}\, \sin\theta \sqrt{\cos 2\theta + (\sin^2 2\theta)/\cos 2\theta}\, d\theta$$

$$= 4\pi \int_0^{\pi/4} \sqrt{\cos 2\theta}\, \sin\theta \sqrt{\frac{\cos^2 2\theta + \sin^2 2\theta}{\cos 2\theta}}\, d\theta = 4\pi \int_0^{\pi/4} \sqrt{\cos 2\theta}\, \sin\theta \frac{1}{\sqrt{\cos 2\theta}}\, d\theta$$

$$= 4\pi \int_0^{\pi/4} \sin\theta\, d\theta = 4\pi \left[-\cos\theta\right]_0^{\pi/4} = -4\pi\left(\frac{\sqrt{2}}{2} - 1\right) = 2\pi\left(2 - \sqrt{2}\right)$$

11.5 Conic Sections ET 10.5

1. $x = 2y^2 \Rightarrow y^2 = \frac{1}{2}x$. $4p = \frac{1}{2}$, so $p = \frac{1}{8}$. The

vertex is $(0, 0)$, the focus is $\left(\frac{1}{8}, 0\right)$, and the

directrix is $x = -\frac{1}{8}$.

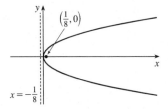

3. $4x^2 = -y \Rightarrow x^2 = -\frac{1}{4}y$. $4p = -\frac{1}{4}$, so

$p = -\frac{1}{16}$. The vertex is $(0, 0)$, the focus is

$\left(0, -\frac{1}{16}\right)$, and the directrix is $y = \frac{1}{16}$.

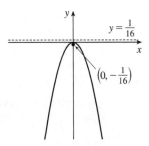

5. $(x + 2)^2 = 8(y - 3)$. $4p = 8$, so $p = 2$. The vertex is $(-2, 3)$, the focus is $(-2, 5)$, and the directrix is $y = 1$.

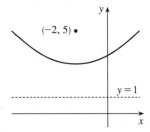

7. $y^2 + 2y + 12x + 25 = 0 \Rightarrow$

$y^2 + 2y + 1 = -12x - 24 \Rightarrow (y + 1)^2 = -12(x + 2)$. $4p = -12$, so $p = -3$. The vertex is $(-2, -1)$, the

focus is $(-5, -1)$, and the directrix is $x = 1$.

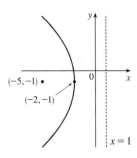

9. The equation has the form $y^2 = 4px$, where $p < 0$. Since the parabola passes through $(-1, 1)$, we have

$1^2 = 4p(-1)$, so $4p = -1$ and an equation is $y^2 = -x$ or $x = -y^2$. $4p = -1$, so $p = -\frac{1}{4}$ and the focus is

$\left(-\frac{1}{4}, 0\right)$ while the directrix is $x = \frac{1}{4}$.

11. $\dfrac{x^2}{9} + \dfrac{y^2}{5} = 1 \;\Rightarrow\; a = \sqrt{9} = 3, b = \sqrt{5}, c = \sqrt{a^2 - b^2} = \sqrt{9 - 5} = 2.$ The ellipse is centered at $(0,0)$, with

vertices at $(\pm 3, 0)$. The foci are $(\pm 2, 0)$.

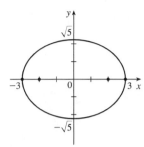

13. $4x^2 + y^2 = 16 \;\Rightarrow\; \dfrac{x^2}{4} + \dfrac{y^2}{16} = 1 \;\Rightarrow\; a = \sqrt{16} = 4, b = \sqrt{4} = 2, c = \sqrt{a^2 - b^2} = \sqrt{16 - 4} = 2\sqrt{3}.$ The

ellipse is centered at $(0,0)$, with vertices at $(0, \pm 4)$. The foci are $\left(0, \pm 2\sqrt{3}\right)$.

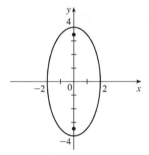

15. $9x^2 - 18x + 4y^2 = 27 \;\Leftrightarrow\; 9(x^2 - 2x + 1) + 4y^2 = 27 + 9 \;\Leftrightarrow$

$9(x - 1)^2 + 4y^2 = 36 \;\Leftrightarrow\; \dfrac{(x-1)^2}{4} + \dfrac{y^2}{9} = 1 \;\Rightarrow\; a = 3, b = 2,$

$c = \sqrt{5} \;\Rightarrow\;$ center $(1, 0)$, vertices $(1, \pm 3)$, foci $\left(1, \pm\sqrt{5}\right)$

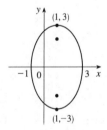

17. The center is $(0,0)$, $a = 3$, and $b = 2$, so an equation is $\dfrac{x^2}{4} + \dfrac{y^2}{9} = 1$. $c = \sqrt{a^2 - b^2} = \sqrt{5}$, so the foci are

$\left(0, \pm\sqrt{5}\right)$.

19. $\dfrac{x^2}{144} - \dfrac{y^2}{25} = 1 \;\Rightarrow\; a = 12, b = 5, c = \sqrt{144 + 25} = 13 \;\Rightarrow$

center $(0,0)$, vertices $(\pm 12, 0)$, foci $(\pm 13, 0)$,

asymptotes $y = \pm\frac{5}{12}x$.

Note: It is helpful to draw a $2a$-by-$2b$ rectangle whose center is the

center of the hyperbola. The asymptotes are the extended diagonals

of the rectangle.

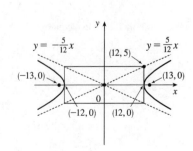

21. $y^2 - x^2 = 4 \iff \dfrac{y^2}{4} - \dfrac{x^2}{4} = 1 \implies a = \sqrt{4} = 2 = b,$

$c = \sqrt{4+4} = 2\sqrt{2} \implies$ center $(0,0)$, vertices $(0, \pm 2)$,

foci $(0, \pm 2\sqrt{2})$, asymptotes $y = \pm x$

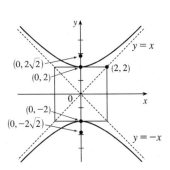

23. $2y^2 - 4y - 3x^2 + 12x = -8 \iff$

$2(y^2 - 2y + 1) - 3(x^2 - 4x + 4) = -8 + 2 - 12 \iff$

$2(y-1)^2 - 3(x-2)^2 = -18 \iff \dfrac{(x-2)^2}{6} - \dfrac{(y-1)^2}{9} = 1$

$\implies a = \sqrt{6}, b = 3, c = \sqrt{15} \implies$ center $(2,1)$, vertices

$\left(2 \pm \sqrt{6}, 1\right)$, foci $\left(2 \pm \sqrt{15}, 1\right)$, asymptotes $y - 1 = \pm \dfrac{3}{\sqrt{6}} (x-2)$

or $y - 1 = \pm \dfrac{\sqrt{6}}{2}(x-2)$

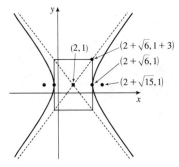

25. $x^2 = y + 1 \iff x^2 = 1(y+1)$. This is an equation of a *parabola* with $4p = 1$, so $p = \frac{1}{4}$. The vertex is $(0, -1)$

and the focus is $\left(0, -\frac{3}{4}\right)$.

27. $x^2 = 4y - 2y^2 \iff x^2 + 2y^2 - 4y = 0 \iff x^2 + 2(y^2 - 2y + 1) = 2 \iff x^2 + 2(y-1)^2 = 2 \iff$

$\dfrac{x^2}{2} + \dfrac{(y-1)^2}{1} = 1$. This is an equation of an *ellipse* with vertices at $\left(\pm\sqrt{2}, 1\right)$. The foci are at

$\left(\pm\sqrt{2-1}, 1\right) = (\pm 1, 1)$.

29. $y^2 + 2y = 4x^2 + 3 \iff y^2 + 2y + 1 = 4x^2 + 4 \iff (y+1)^2 - 4x^2 = 4 \iff \dfrac{(y+1)^2}{4} - x^2 = 1$. This is

an equation of a *hyperbola* with vertices $(0, -1 \pm 2) = (0, 1)$ and $(0, -3)$. The foci are at

$\left(0, -1 \pm \sqrt{4+1}\right) = \left(0, -1 \pm \sqrt{5}\right)$.

31. The parabola with vertex $(0,0)$ and focus $(0, -2)$ opens downward and has $p = -2$, so its equation is

$x^2 = 4py = -8y$.

33. The distance from the focus $(-4, 0)$ to the directrix $x = 2$ is $2 - (-4) = 6$, so the distance from the focus to the

vertex is $\frac{1}{2}(6) = 3$ and the vertex is $(-1, 0)$. Since the focus is to the left of the vertex, $p = -3$. An equation is

$y^2 = 4p(x+1) \implies y^2 = -12(x+1)$.

35. The parabola must have equation $y^2 = 4px$, so $(-4)^2 = 4p(1) \implies p = 4 \implies y^2 = 16x$.

37. The ellipse with foci $(\pm 2, 0)$ and vertices $(\pm 5, 0)$ has center $(0, 0)$ and a horizontal major axis, with $a = 5$ and

$c = 2$, so $b = \sqrt{a^2 - c^2} = \sqrt{21}$. An equation is $\dfrac{x^2}{25} + \dfrac{y^2}{21} = 1$.

39. Since the vertices are $(0, 0)$ and $(0, 8)$, the ellipse has center $(0, 4)$ with a vertical axis and $a = 4$. The foci at $(0, 2)$

and $(0, 6)$ are 2 units from the center, so $c = 2$ and $b = \sqrt{a^2 - c^2} = \sqrt{4^2 - 2^2} = \sqrt{12}$. An equation is

$\dfrac{(x-0)^2}{b^2} + \dfrac{(y-4)^2}{a^2} = 1 \implies \dfrac{x^2}{12} + \dfrac{(y-4)^2}{16} = 1$.

41. Center $(2, 2)$, $c = 2$, $a = 3$ $\Rightarrow$ $b = \sqrt{5}$ $\Rightarrow$ $\frac{1}{9}(x - 2)^2 + \frac{1}{5}(y - 2)^2 = 1$

43. Center $(0, 0)$, vertical axis, $c = 3$, $a = 1$ $\Rightarrow$ $b = \sqrt{8} = 2\sqrt{2}$ $\Rightarrow$ $y^2 - \frac{1}{8}x^2 = 1$

45. Center $(4, 3)$, horizontal axis, $c = 3$, $a = 2$ $\Rightarrow$ $b = \sqrt{5}$ $\Rightarrow$ $\frac{1}{4}(x - 4)^2 - \frac{1}{5}(y - 3)^2 = 1$

47. Center $(0, 0)$, horizontal axis, $a = 3$, $\frac{b}{a} = 2$ $\Rightarrow$ $b = 6$ $\Rightarrow$ $\frac{1}{9}x^2 - \frac{1}{36}y^2 = 1$

49. In Figure 8, we see that the point on the ellipse closest to a focus is the closer vertex (which is a distance $a - c$ from it) while the farthest point is the other vertex (at a distance of $a + c$). So for this lunar orbit,

$(a - c) + (a + c) = 2a = (1728 + 110) + (1728 + 314)$, or $a = 1940$; and

$(a + c) - (a - c) = 2c = 314 - 110$, or $c = 102$. Thus, $b^2 = a^2 - c^2 = 3,753,196$, and the equation is

$$\frac{x^2}{3,763,600} + \frac{y^2}{3,753,196} = 1.$$

51. (a) Set up the coordinate system so that A is $(-200, 0)$ and B is $(200, 0)$.

$|PA| - |PB| = (1200)(980) = 1,176,000 \text{ ft} = \frac{2450}{11} \text{ mi} = 2a$ $\Rightarrow$ $a = \frac{1225}{11}$, and $c = 200$ so

$b^2 = c^2 - a^2 = \frac{3,339,375}{121}$ $\Rightarrow$ $\frac{121x^2}{1,500,625} - \frac{121y^2}{3,339,375} = 1.$

(b) Due north of B $\Rightarrow$ $x = 200$ $\Rightarrow$ $\frac{(121)(200)^2}{1,500,625} - \frac{121y^2}{3,339,375} = 1$ $\Rightarrow$ $y = \frac{133,575}{539} \approx 248 \text{ mi}$

53. The function whose graph is the upper branch of this hyperbola is concave upward. The

function is $y = f(x) = a\sqrt{1 + \dfrac{x^2}{b^2}} = \dfrac{a}{b}\sqrt{b^2 + x^2}$, so $y' = \dfrac{a}{b}x(b^2 + x^2)^{-1/2}$ and

$y'' = \dfrac{a}{b}\left[(b^2 + x^2)^{-1/2} - x^2(b^2 + x^2)^{-3/2}\right] = ab(b^2 + x^2)^{-3/2} > 0$ for all x, and so f is concave upward.

55. (a) If $k > 16$, then $k - 16 > 0$, and $\dfrac{x^2}{k} + \dfrac{y^2}{k - 16} = 1$ is an *ellipse* since it is the sum of two squares on the left side.

(b) If $0 < k < 16$, then $k - 16 < 0$, and $\dfrac{x^2}{k} + \dfrac{y^2}{k - 16} = 1$ is a *hyperbola* since it is the difference of two squares on the left side.

(c) If $k < 0$, then $k - 16 < 0$, and there is *no curve* since the left side is the sum of two negative terms, which cannot equal 1.

(d) In case (a), $a^2 = k$, $b^2 = k - 16$, and $c^2 = a^2 - b^2 = 16$, so the foci are at $(\pm 4, 0)$. In case (b), $k - 16 < 0$, so $a^2 = k$, $b^2 = 16 - k$, and $c^2 = a^2 + b^2 = 16$, and so again the foci are at $(\pm 4, 0)$.

57. Use the parametrization $x = 2\cos t$, $y = \sin t$, $0 \le t \le 2\pi$ to get

$$L = 4\int_0^{\pi/2}\sqrt{(dx/dt)^2 + (dy/dt)^2}\, dt = 4\int_0^{\pi/2}\sqrt{4\sin^2 t + \cos^2 t}\, dt = 4\int_0^{\pi/2}\sqrt{3\sin^2 t + 1}\, dt$$

Using Simpson's Rule with $n = 10$, $\Delta t = \frac{\pi/2 - 0}{10} = \frac{\pi}{20}$, and $f(t) = \sqrt{3\sin^2 t + 1}$, we get

$$L \approx \frac{4}{3}\left(\frac{\pi}{20}\right)\left[f(0) + 4f\left(\frac{\pi}{20}\right) + 2f\left(\frac{2\pi}{20}\right) + \cdots + 2f\left(\frac{8\pi}{20}\right) + 4f\left(\frac{9\pi}{20}\right) + f\left(\frac{\pi}{2}\right)\right] \approx 9.69$$

59. $\dfrac{x^2}{a^2} + \dfrac{y^2}{b^2} = 1 \;\Rightarrow\; \dfrac{2x}{a^2} + \dfrac{2yy'}{b^2} = 0 \;\Rightarrow\; y' = -\dfrac{b^2x}{a^2y}$ $(y \neq 0)$. Thus, the slope of the tangent line at P is

$-\dfrac{b^2x_1}{a^2y_1}$. The slope of F_1P is $\dfrac{y_1}{x_1+c}$ and of F_2P is $\dfrac{y_1}{x_1-c}$. By the formula from Problems Plus, we have

$$\tan\alpha = \dfrac{\dfrac{y_1}{x_1+c} + \dfrac{b^2x_1}{a^2y_1}}{1 - \dfrac{b^2x_1y_1}{a^2y_1(x_1+c)}} = \dfrac{a^2y_1^2 + b^2x_1(x_1+c)}{a^2y_1(x_1+c) - b^2x_1y_1} = \dfrac{a^2b^2 + b^2cx_1}{c^2x_1y_1 + a^2cy_1} \quad \begin{bmatrix} \text{using } b^2x_1^2 + a^2y_1^2 = a^2b^2 \\ \text{and } a^2 - b^2 = c^2 \end{bmatrix}$$

$$= \dfrac{b^2(cx_1 + a^2)}{cy_1(cx_1 + a^2)} = \dfrac{b^2}{cy_1}$$

and

$$\tan\beta = \dfrac{-\dfrac{y_1}{x_1-c} - \dfrac{b^2x_1}{a^2y_1}}{1 - \dfrac{b^2x_1y_1}{a^2y_1(x_1-c)}} = \dfrac{-a^2y_1^2 - b^2x_1(x_1-c)}{a^2y_1(x_1-c) - b^2x_1y_1} = \dfrac{-a^2b^2 + b^2cx_1}{c^2x_1y_1 - a^2cy_1} = \dfrac{b^2(cx_1 - a^2)}{cy_1(cx_1 - a^2)} = \dfrac{b^2}{cy_1}$$

So $\alpha = \beta$.

11.6 Conic Sections in Polar Coordinates ET 10.6

1. The directrix $y = 6$ is above the focus at the origin, so we use the form with "$+\,e\sin\theta$" in the denominator. (See

Theorem 6 and Figure 2.) $r = \dfrac{ed}{1 + e\sin\theta} = \dfrac{\frac{7}{4}\cdot 6}{1 + \frac{7}{4}\sin\theta} = \dfrac{42}{4 + 7\sin\theta}$

3. The directrix $x = -5$ is to the left of the focus at the origin, so we use the form with "$-\,e\cos\theta$" in the

denominator. $r = \dfrac{ed}{1 - e\cos\theta} = \dfrac{\frac{3}{4}\cdot 5}{1 - \frac{3}{4}\cos\theta} = \dfrac{15}{4 - 3\cos\theta}$

5. The vertex $(4, 3\pi/2)$ is 4 units below the focus at the origin, so the directrix is 8 units below the focus $(d = 8)$, and

we use the form with "$-e\sin\theta$" in the denominator. $e = 1$ for a parabola, so an equation is

$r = \dfrac{ed}{1 - e\sin\theta} = \dfrac{1(8)}{1 - 1\sin\theta} = \dfrac{8}{1 - \sin\theta}$.

7. The directrix $r = 4\sec\theta$ (equivalent to $r\cos\theta = 4$ or $x = 4$) is to the right of the focus at the origin, so we will use

the form with "$+e\cos\theta$" in the denominator. The distance from the focus to the directrix is $d = 4$, so an equation is

$r = \dfrac{ed}{1 + e\cos\theta} = \dfrac{0.5(4)}{1 + 0.5\cos\theta} \cdot \dfrac{2}{2} = \dfrac{4}{2 + \cos\theta}$.

9. $r = \dfrac{1}{1 + \sin\theta} = \dfrac{ed}{1 + e\sin\theta}$, where $d = e = 1$.

   (a) Eccentricity $= e = 1$

   (b) Since $e = 1$, the conic is a parabola.

   (c) Since "$+e\sin\theta$" appears in the denominator, the directrix is above

      the focus at the origin. $d = |Fl| = 1$, so an equation of the directrix

      is $y = 1$.

   (d) The vertex is at $\left(\frac{1}{2}, \frac{\pi}{2}\right)$, midway between the focus and the directrix.

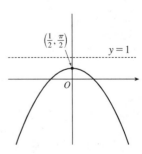

11. $r = \dfrac{12}{4 - \sin\theta} \cdot \dfrac{1/4}{1/4} = \dfrac{3}{1 - \frac{1}{4}\sin\theta}$, where $e = \frac{1}{4}$ and $ed = 3 \ \Rightarrow \ d = 12$.

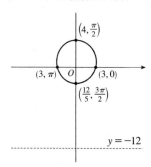

(a) Eccentricity $= e = \frac{1}{4}$

(b) Since $e = \frac{1}{4} < 1$, the conic is an ellipse.

(c) Since "$-e\sin\theta$" appears in the denominator, the directrix is below

the focus at the origin. $d = |Fl| = 12$, so an equation of the directrix

is $y = -12$.

(d) The vertices are $\left(4, \frac{\pi}{2}\right)$ and $\left(\frac{12}{5}, \frac{3\pi}{2}\right)$, so the center is midway

between them, that is, $\left(\frac{4}{5}, \frac{\pi}{2}\right)$.

13. $r = \dfrac{9}{6 + 2\cos\theta} \cdot \dfrac{1/6}{1/6} = \dfrac{3/2}{1 + \frac{1}{3}\cos\theta}$, where $e = \frac{1}{3}$ and $ed = \frac{3}{2} \ \Rightarrow \ d = \frac{9}{2}$.

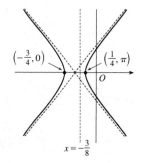

(a) Eccentricity $= e = \frac{1}{3}$

(b) Since $e = \frac{1}{3} < 1$, the conic is an ellipse.

(c) Since "$+e\cos\theta$" appears in the denominator, the directrix is to the

right of the focus at the origin. $d = |Fl| = \frac{9}{2}$, so an equation of the

directrix is $x = \frac{9}{2}$.

(d) The vertices are $\left(\frac{9}{8}, 0\right)$ and $\left(\frac{9}{4}, \pi\right)$, so the center is midway between

them, that is, $\left(\frac{9}{16}, \pi\right)$.

15. $r = \dfrac{3}{4 - 8\cos\theta} \cdot \dfrac{1/4}{1/4} = \dfrac{3/4}{1 - 2\cos\theta}$, where $e = 2$ and $ed = \frac{3}{4} \ \Rightarrow \ d = \frac{3}{8}$.

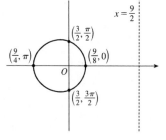

(a) Eccentricity $= e = 2$

(b) Since $e = 2 > 1$, the conic is a hyperbola.

(c) Since "$-e\cos\theta$" appears in the denominator, the directrix is to the

left of the focus at the origin. $d = |Fl| = \frac{3}{8}$, so an equation of the

directrix is $x = -\frac{3}{8}$.

(d) The vertices are $\left(-\frac{3}{4}, 0\right)$ and $\left(\frac{1}{4}, \pi\right)$, so the center is midway

between them, that is, $\left(\frac{1}{2}, \pi\right)$.

17. (a) The equation is $r = \dfrac{1}{4 - 3\cos\theta} = \dfrac{1/4}{1 - \frac{3}{4}\cos\theta}$, so $e = \frac{3}{4}$ and

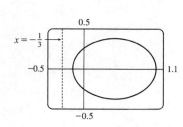

$ed = \frac{1}{4} \ \Rightarrow \ d = \frac{1}{3}$. The conic is an ellipse, and the equation of its

directrix is $x = r\cos\theta = -\frac{1}{3} \ \Rightarrow \ r = -\dfrac{1}{3\cos\theta}$. We must be

careful in our choice of parameter values in this equation

$(-1 \le \theta \le 1$ works well).

(b) The equation is obtained by replacing θ with $\theta - \frac{\pi}{3}$ in the equation of

the original conic (see Example 4), so $r = \dfrac{1}{4 - 3\cos\left(\theta - \frac{\pi}{3}\right)}$.

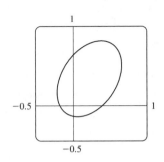

19. For $e < 1$ the curve is an ellipse. It is nearly circular when e is close to 0.
As e increases, the graph is stretched out to the right, and grows larger
(that is, its right-hand focus moves to the right while its left-hand focus
remains at the origin.) At $e = 1$, the curve becomes a parabola with focus
at the origin.

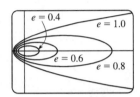

21. $|PF| = e\,|Pl| \;\Rightarrow\; r = e[d - r\cos(\pi - \theta)] = e(d + r\cos\theta) \;\Rightarrow$

$r(1 - e\cos\theta) = ed \;\Rightarrow\; r = \dfrac{ed}{1 - e\cos\theta}$

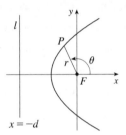

23. $|PF| = e\,|Pl| \;\Rightarrow\; r = e[d - r\sin(\theta - \pi)] = e(d + r\sin\theta) \;\Rightarrow$

$r(1 - e\sin\theta) = ed \;\Rightarrow\; r = \dfrac{ed}{1 - e\sin\theta}$

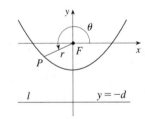

25. (a) If the directrix is $x = -d$, then $r = \dfrac{ed}{1 - e\cos\theta}$ [see Figure 2(b)], and, from (4), $a^2 = \dfrac{e^2 d^2}{(1 - e^2)^2} \;\Rightarrow$

$ed = a(1 - e^2)$. Therefore, $r = \dfrac{a(1 - e^2)}{1 - e\cos\theta}$.

(b) $e = 0.017$ and the major axis $= 2a = 2.99 \times 10^8 \;\Rightarrow\; a = 1.495 \times 10^8$.

Therefore $r = \dfrac{1.495 \times 10^8 \left[1 - (0.017)^2\right]}{1 - 0.017\cos\theta} \approx \dfrac{1.49 \times 10^8}{1 - 0.017\cos\theta}$.

27. Here $2a = $ length of major axis $= 36.18$ AU $\;\Rightarrow\; a = 18.09$ AU and $e = 0.97$. By Exercise 25(a), the equation

of the orbit is $r = \dfrac{18.09\left[1 - (0.97)^2\right]}{1 - 0.97\cos\theta} \approx \dfrac{1.07}{1 - 0.97\cos\theta}$. By Exercise 26(a), the maximum distance from the

comet to the sun is $18.09(1 + 0.97) \approx 35.64$ AU or about 3.314 billion miles.

29. The minimum distance is at perihelion, where

$4.6 \times 10^7 = r = a(1 - e) = a(1 - 0.206) = a(0.794)$ ⇒ $a = 4.6 \times 10^7 / 0.794$. So the maximum distance,

which is at aphelion, is $r = a(1 + e) = (4.6 \times 10^7 / 0.794)(1.206) \approx 7.0 \times 10^7$ km.

31. From Exercise 29, we have $e = 0.206$ and $a(1 - e) = 4.6 \times 10^7$ km. Thus, $a = 4.6 \times 10^7 / 0.794$. From

Exercise 25, we can write the equation of Mercury's orbit as $r = a\dfrac{1 - e^2}{1 - e\cos\theta}$. So since

$\dfrac{dr}{d\theta} = \dfrac{-a(1 - e^2)e\sin\theta}{(1 - e\cos\theta)^2}$ ⇒

$$r^2 + \left(\frac{dr}{d\theta}\right)^2 = \frac{a^2(1 - e^2)^2}{(1 - e\cos\theta)^2} + \frac{a^2(1 - e^2)^2 e^2\sin^2\theta}{(1 - e\cos\theta)^4} = \frac{a^2(1 - e^2)^2}{(1 - e\cos\theta)^4}(1 - 2e\cos\theta + e^2)$$

the length of the orbit is

$$L = \int_0^{2\pi} \sqrt{r^2 + (dr/d\theta)^2}\, d\theta = a(1 - e^2) \int_0^{2\pi} \frac{\sqrt{1 + e^2 - 2e\cos\theta}}{(1 - e\cos\theta)^2}\, d\theta \approx 3.6 \times 10^8 \text{ km}$$

This seems reasonable, since Mercury's orbit is nearly circular, and the circumference of a circle of radius a is

$2\pi a \approx 3.6 \times 10^8$ km.

11 Review

ET 10

─────────────────────────── CONCEPT CHECK ───────────────────────────

1. (a) A parametric curve is a set of points of the form $(x, y) = (f(t), g(t))$, where f and g are continuous functions

of a variable t.

(b) Sketching a parametric curve, like sketching the graph of a function, is difficult to do in general. We can plot

points on the curve by finding $f(t)$ and $g(t)$ for various values of t, either by hand or with a calculator or

computer. Sometimes, when f and g are given by formulas, we can eliminate t from the equations $x = f(t)$ and

$y = g(t)$ to get a Cartesian equation relating x and y. It may be easier to graph that equation than to work with

the original formulas for x and y in terms of t.

2. (a) You can find $\dfrac{dy}{dx}$ as a function of t by calculating $\dfrac{dy}{dx} = \dfrac{dy/dt}{dx/dt}$ (if $dx/dt \neq 0$).

(b) Calculate the area as $\int_a^b y\, dx = \int_\alpha^\beta g(t)f'(t)dt$ [or $\int_\beta^\alpha g(t)f'(t)dt$ if the leftmost point is $(f(\beta), g(\beta))$ rather

than $(f(\alpha), g(\alpha))$].

3. (a) $L = \int_\alpha^\beta \sqrt{(dx/dt)^2 + (dy/dt)^2}\, dt = \int_\alpha^\beta \sqrt{[f'(t)]^2 + [g'(t)]^2}\, dt$

(b) $S = \int_\alpha^\beta 2\pi y \sqrt{(dx/dt)^2 + (dy/dt)^2}\, dt = \int_\alpha^\beta 2\pi g(t)\sqrt{[f'(t)]^2 + [g'(t)]^2}\, dt$

4. (a) See Figure 5 in Section 11.3 [ET 10.3].

(b) $x = r\cos\theta,\ y = r\sin\theta$

(c) To find a polar representation (r, θ) with $r \geq 0$ and $0 \leq \theta < 2\pi$, first calculate $r = \sqrt{x^2 + y^2}$. Then θ is

specified by $\cos\theta = x/r$ and $\sin\theta = y/r$.

5. (a) Calculate $\dfrac{dy}{dx} = \dfrac{\dfrac{dy}{d\theta}}{\dfrac{dx}{d\theta}} = \dfrac{\dfrac{d}{d\theta}(y)}{\dfrac{d}{d\theta}(x)} = \dfrac{\dfrac{d}{d\theta}(r\sin\theta)}{\dfrac{d}{d\theta}(r\cos\theta)} = \dfrac{\left(\dfrac{dr}{d\theta}\right)\sin\theta + r\cos\theta}{\left(\dfrac{dr}{d\theta}\right)\cos\theta - r\sin\theta}$, where $r = f(\theta)$.

(b) Calculate $A = \int_a^b \frac{1}{2} r^2 \, d\theta = \int_a^b \frac{1}{2}\, [f(\theta)]^2 \, d\theta$

(c) $L = \int_a^b \sqrt{(dx/d\theta)^2 + (dy/d\theta)^2}\, d\theta = \int_a^b \sqrt{r^2 + (dr/d\theta)^2}\, d\theta = \int_a^b \sqrt{[f(\theta)]^2 + [f'(\theta)]^2}\, d\theta$

6. (a) A parabola is a set of points in a plane whose distances from a fixed point F (the focus) and a fixed line l (the directrix) are equal.

(b) $x^2 = 4py; \ y^2 = 4px$

7. (a) An ellipse is a set of points in a plane the sum of whose distances from two fixed points (the foci) is a constant.

(b) $\dfrac{x^2}{a^2} + \dfrac{y^2}{a^2 - c^2} = 1$.

8. (a) A hyperbola is a set of points in a plane the difference of whose distances from two fixed points (the foci) is a constant. This difference should be interpreted as the larger distance minus the smaller distance.

(b) $\dfrac{x^2}{a^2} - \dfrac{y^2}{c^2 - a^2} = 1$

(c) $y = \pm \dfrac{\sqrt{c^2 - a^2}}{a}\, x$

9. (a) If a conic section has focus F and corresponding directrix l, then the eccentricity e is the fixed ratio $|PF|\,/\,|Pl|$ for points P of the conic section.

(b) $e < 1$ for an ellipse; $e > 1$ for a hyperbola; $e = 1$ for a parabola.

(c) $x = d$: $r = \dfrac{ed}{1 + e\cos\theta}$. $x = -d$: $r = \dfrac{ed}{1 - e\cos\theta}$. $y = d$: $r = \dfrac{ed}{1 + e\sin\theta}$. $y = -d$: $r = \dfrac{ed}{1 - e\sin\theta}$.

———————————————— TRUE-FALSE QUIZ ————————————————

1. False. Consider the curve defined by $x = f(t) = (t-1)^3$ and $y = g(t) = (t-1)^2$. Then $g'(t) = 2(t-1)$, so $g'(1) = 0$, but its graph has a *vertical* tangent when $t = 1$. *Note:* The statement is true if $f'(1) \neq 0$ when $g'(1) = 0$.

3. False. For example, if $f(t) = \cos t$ and $g(t) = \sin t$ for $0 \le t \le 4\pi$, then the curve is a circle of radius 1, hence its length is 2π, but

$$\int_0^{4\pi} \sqrt{[f'(t)]^2 + [g'(t)]^2}\, dt = \int_0^{4\pi} \sqrt{(-\sin t)^2 + (\cos t)^2}\, dt$$

$$= \int_0^{4\pi} 1\, dt = 4\pi,$$

since as t increases from 0 to 4π, the circle is traversed twice.

5. True. The curve $r = 1 - \sin 2\theta$ is unchanged if we rotate it through $180°$ about O because $1 - \sin 2(\theta + \pi) = 1 - \sin(2\theta + 2\pi) = 1 - \sin 2\theta$. So it's unchanged if we replace r by $-r$. (See the discussion after Example 8 in Section 11.3 [ET 10.3].) In other words, it's the same curve as $r = -(1 - \sin 2\theta) = \sin 2\theta - 1$.

7. False.   The first pair of equations yields the portion of the parabola $y = x^2$ with $x \geq 0$, whereas the second pair of equations traces out the whole parabola $y = x^2$.

9. True.   By rotating and translating the parabola, we can assume it has an equation of the form $y = cx^2$, where $c > 0$. The tangent at the point (a, ca^2) is the line $y - ca^2 = 2ca(x - a)$; i.e., $y = 2cax - ca^2$. This tangent meets the parabola at the points (x, cx^2) where $cx^2 = 2cax - ca^2$. This equation is equivalent to $x^2 = 2ax - a^2$ (since $c > 0$). But $x^2 = 2ax - a^2 \iff x^2 - 2ax + a^2 = 0 \iff (x - a)^2 = 0 \iff x = a \iff (x, cx^2) = (a, ca^2)$. This shows that each tangent meets the parabola at exactly one point.

———————————————————— EXERCISES ————————————————————

1. $x = t^2 + 4t$, $y = 2 - t$, $-4 \leq t \leq 1$. $t = 2 - y$, so $x = (2 - y)^2 + 4(2 - y) = 4 - 4y + y^2 + 8 - 4y = y^2 - 8y + 12 \iff x + 4 = y^2 - 8y + 16 = (y - 4)^2$. This is part of a parabola with vertex $(-4, 4)$, opening to the right.

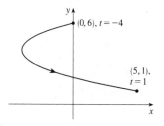

3. $x = \tan\theta$, $y = \cot\theta$. $y = 1/\tan\theta = 1/x$. The whole curve is traced out as θ ranges over the open interval $\left(-\frac{\pi}{2}, \frac{\pi}{2}\right)$ [or any open interval of the form $\left(-\frac{\pi}{2} + n\pi, \frac{\pi}{2} + n\pi\right)$, where n is an integer].

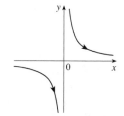

5. Three different sets of parametric equations for the curve $y = \sqrt{x}$ are

(i) $x = t$, $y = \sqrt{t}$, $t \geq 0$

(ii) $x = t^4$, $y = t^2$

(iii) $x = \tan^2 t$, $y = \tan t$, $0 \leq t < \pi/2$

   There are many other sets of equations that also give this curve.

7. $r = 1 - \cos\theta$. This cardiod is symmetric about the polar axis.

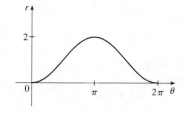

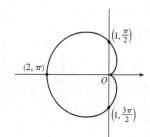

9. $r = 1 + \cos 2\theta$. The curve is symmetric about the pole and both the horizontal and vertical axes.

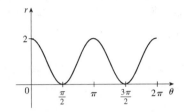

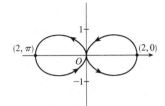

11. $r^2 = \sec 2\theta \quad \Rightarrow$
$r^2 \cos 2\theta = 1 \quad \Rightarrow$
$r^2 \left(\cos^2 \theta - \sin^2 \theta\right) = 1 \quad \Rightarrow$
$r^2 \cos^2 \theta - r^2 \sin^2 \theta = 1 \quad \Rightarrow$
$x^2 - y^2 = 1$, a hyperbola

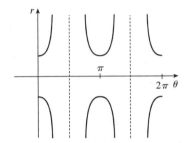

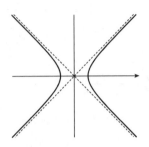

13. $r = \dfrac{1}{1 + \cos \theta} \quad \Rightarrow \quad e = 1 \quad \Rightarrow \quad$ parabola; $d = 1 \quad \Rightarrow \quad$ directrix $x = 1$

and vertex $\left(\frac{1}{2}, 0\right)$; y-intercepts are $\left(1, \frac{\pi}{2}\right)$ and $\left(1, \frac{3\pi}{2}\right)$.

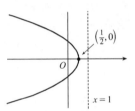

15. $x + y = 2 \quad \Leftrightarrow \quad r \cos \theta + r \sin \theta = 2 \quad \Leftrightarrow \quad r(\cos \theta + \sin \theta) = 2 \quad \Leftrightarrow \quad r = \dfrac{2}{\cos \theta + \sin \theta}$

17. $r = (\sin \theta)/\theta$. As $\theta \to \pm\infty$, $r \to 0$. As $\theta \to 0, r \to 1$. In the first figure, there are an infinite number of x-intercepts at $x = \pi n, n$ a nonzero integer. These correspond to pole points in the second figure.

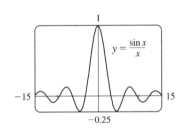

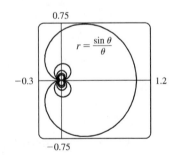

19. $x = \ln t, y = 1 + t^2; t = 1$. $\dfrac{dy}{dt} = 2t$ and $\dfrac{dx}{dt} = \dfrac{1}{t}$, so $\dfrac{dy}{dx} = \dfrac{dy/dt}{dx/dt} = \dfrac{2t}{1/t} = 2t^2$. When $t = 1, (x, y) = (0, 2)$

and $dy/dx = 2$.

21. $r = e^{-\theta} \quad \Rightarrow \quad y = r \sin \theta = e^{-\theta} \sin \theta$ and $x = r \cos \theta = e^{-\theta} \cos \theta \quad \Rightarrow$

$\dfrac{dy}{dx} = \dfrac{dy/d\theta}{dx/d\theta} = \dfrac{\frac{dr}{d\theta} \sin \theta + r \cos \theta}{\frac{dr}{d\theta} \cos \theta - r \sin \theta} = \dfrac{-e^{-\theta} \sin \theta + e^{-\theta} \cos \theta}{-e^{-\theta} \cos \theta - e^{-\theta} \sin \theta} \cdot \dfrac{-e^{\theta}}{-e^{\theta}} = \dfrac{\sin \theta - \cos \theta}{\cos \theta + \sin \theta}$. When $\theta = \pi$,

$\dfrac{dy}{dx} = \dfrac{0 - (-1)}{-1 + 0} = \dfrac{1}{-1} = -1$.

23. $x = t \cos t$, $y = t \sin t$. $\dfrac{dy}{dx} = \dfrac{dy/dt}{dx/dt} = \dfrac{t \cos t + \sin t}{-t \sin t + \cos t}$. $\dfrac{d^2y}{dx^2} = \dfrac{\frac{d}{dt}\left(\frac{dy}{dx}\right)}{dx/dt}$, where

$$\frac{d}{dt}\left(\frac{dy}{dx}\right) = \frac{(-t \sin t + \cos t)(-t \sin t + 2 \cos t) - (t \cos t + \sin t)(-t \cos t - 2 \sin t)}{(-t \sin t + \cos t)^2}$$

$$= \frac{t^2 + 2}{(-t \sin t + \cos t)^2} \quad \Rightarrow \quad \frac{d^2y}{dx^2} = \frac{t^2 + 2}{(-t \sin t + \cos t)^3}.$$

25. We graph the curve $x = t^3 - 3t$, $y = t^2 + t + 1$ for $-2.2 \le t \le 1.2$. By

zooming in or using a cursor, we find that the lowest point is about

$(1.4, 0.75)$. To find the exact values, we find the t-value at which

$dy/dt = 2t + 1 = 0 \quad \Leftrightarrow \quad t = -\frac{1}{2} \quad \Leftrightarrow \quad (x, y) = \left(\frac{11}{8}, \frac{3}{4}\right)$.

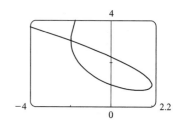

27. $x = 2a \cos t - a \cos 2t \quad \Rightarrow \quad \dfrac{dx}{dt} = -2a \sin t + 2a \sin 2t = 2a \sin t(2 \cos t - 1) = 0 \quad \Leftrightarrow$

$\sin t = 0$ or $\cos t = \frac{1}{2} \quad \Rightarrow \quad t = 0, \frac{\pi}{3}, \pi$, or $\frac{5\pi}{3}$.

$y = 2a \sin t - a \sin 2t \quad \Rightarrow$

$\dfrac{dy}{dt} = 2a \cos t - 2a \cos 2t = 2a(1 + \cos t - 2 \cos^2 t) = 2a(1 - \cos t)(1 + 2 \cos t) = 0 \quad \Rightarrow \quad t = 0, \frac{2\pi}{3}$, or $\frac{4\pi}{3}$.

Thus the graph has vertical tangents where

$t = \frac{\pi}{3}, \pi$ and $\frac{5\pi}{3}$, and horizontal tangents where

$t = \frac{2\pi}{3}$ and $\frac{4\pi}{3}$. To determine what the slope is

where $t = 0$, we use l'Hospital's Rule to evaluate

$\displaystyle\lim_{t \to 0} \frac{dy/dt}{dx/dt} = 0$, so there is a horizontal tangent

there.

t	x	y
0	a	0
$\frac{\pi}{3}$	$\frac{3}{2}a$	$\frac{\sqrt{3}}{2}a$
$\frac{2\pi}{3}$	$-\frac{1}{2}a$	$\frac{3\sqrt{3}}{2}a$
π	$-3a$	0
$\frac{4\pi}{3}$	$-\frac{1}{2}a$	$-\frac{3\sqrt{3}}{2}a$
$\frac{5\pi}{3}$	$\frac{3}{2}a$	$-\frac{\sqrt{3}}{2}a$

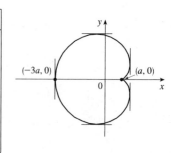

29. The curve $r^2 = 9 \cos 5\theta$ has 10 "petals." For instance, for $-\frac{\pi}{10} \le \theta \le \frac{\pi}{10}$, there are two petals, one with $r > 0$ and one with $r < 0$.

$$A = 10 \int_{-\pi/10}^{\pi/10} \tfrac{1}{2} r^2 \, d\theta = 5 \int_{-\pi/10}^{\pi/10} 9 \cos 5\theta \, d\theta = 5 \cdot 9 \cdot 2 \int_0^{\pi/10} \cos 5\theta \, d\theta = 18 \left[\sin 5\theta\right]_0^{\pi/10} = 18$$

31. The curves intersect when $4 \cos \theta = 2 \quad \Rightarrow \quad \cos \theta = \frac{1}{2} \quad \Rightarrow \quad \theta = \pm\frac{\pi}{3}$

for $-\pi \le \theta \le \pi$. The points of intersection are $\left(2, \frac{\pi}{3}\right)$ and $\left(2, -\frac{\pi}{3}\right)$.

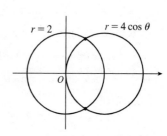

33. The curves intersect where $2\sin\theta = \sin\theta + \cos\theta$ $\Rightarrow$

$\sin\theta = \cos\theta$ $\Rightarrow$ $\theta = \frac{\pi}{4}$, and also at the origin (at which $\theta = \frac{3\pi}{4}$ on the

second curve).

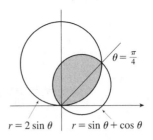

$r = 2\sin\theta$ | $r = \sin\theta + \cos\theta$

$$A = \int_0^{\pi/4} \tfrac{1}{2}(2\sin\theta)^2\,d\theta + \int_{\pi/4}^{3\pi/4} \tfrac{1}{2}(\sin\theta + \cos\theta)^2\,d\theta$$

$$= \int_0^{\pi/4}(1 - \cos 2\theta)\,d\theta + \tfrac{1}{2}\int_{\pi/4}^{3\pi/4}(1 + \sin 2\theta)\,d\theta$$

$$= \left[\theta - \tfrac{1}{2}\sin 2\theta\right]_0^{\pi/4} + \left[\tfrac{1}{2}\theta - \tfrac{1}{4}\cos 2\theta\right]_{\pi/4}^{3\pi/4} = \tfrac{1}{2}(\pi - 1)$$

35. $x = 3t^2$, $y = 2t^3$.

$$L = \int_0^2 \sqrt{(dx/dt)^2 + (dy/dt)^2}\,dt = \int_0^2 \sqrt{(6t)^2 + (6t^2)^2}\,dt = \int_0^2 \sqrt{36t^2 + 36t^4}\,dt$$

$$= \int_0^2 \sqrt{36t^2}\sqrt{1 + t^2}\,dt = \int_0^2 6\,|t|\,\sqrt{1 + t^2}\,dt = 6\int_0^2 t\sqrt{1 + t^2}\,dt$$

$$= 6\int_1^5 u^{1/2}\left(\tfrac{1}{2}du\right) \quad \left[u = 1 + t^2,\, du = 2t\,dt\right]$$

$$= 6 \cdot \tfrac{1}{2} \cdot \tfrac{2}{3}\left[u^{3/2}\right]_1^5 = 2\left(5^{3/2} - 1\right) = 2\left(5\sqrt{5} - 1\right)$$

37. $L = \int_\pi^{2\pi} \sqrt{r^2 + (dr/d\theta)^2}\,d\theta = \int_\pi^{2\pi} \sqrt{(1/\theta)^2 + (-1/\theta^2)^2}\,d\theta = \displaystyle\int_\pi^{2\pi} \frac{\sqrt{\theta^2 + 1}}{\theta^2}\,d\theta$

$$\overset{24}{=} \left[-\frac{\sqrt{\theta^2 + 1}}{\theta} + \ln\left(\theta + \sqrt{\theta^2 + 1}\right)\right]_\pi^{2\pi} = \frac{\sqrt{\pi^2 + 1}}{\pi} - \frac{\sqrt{4\pi^2 + 1}}{2\pi} + \ln\left(\frac{2\pi + \sqrt{4\pi^2 + 1}}{\pi + \sqrt{\pi^2 + 1}}\right)$$

$$= \frac{2\sqrt{\pi^2 + 1} - \sqrt{4\pi^2 + 1}}{2\pi} + \ln\left(\frac{2\pi + \sqrt{4\pi^2 + 1}}{\pi + \sqrt{\pi^2 + 1}}\right)$$

39. $x = 4\sqrt{t}$, $y = \dfrac{t^3}{3} + \dfrac{1}{2t^2}$, $1 \le t \le 4$ $\Rightarrow$

$$S = \int_1^4 2\pi y\sqrt{(dx/dt)^2 + (dy/dt)^2}\,dt = \int_1^4 2\pi\left(\tfrac{1}{3}t^3 + \tfrac{1}{2}t^{-2}\right)\sqrt{\left(2/\sqrt{t}\right)^2 + (t^2 - t^{-3})^2}\,dt$$

$$= 2\pi\int_1^4\left(\tfrac{1}{3}t^3 + \tfrac{1}{2}t^{-2}\right)\sqrt{(t^2 + t^{-3})^2}\,dt = 2\pi\int_1^4\left(\tfrac{1}{3}t^5 + \tfrac{5}{6} + \tfrac{1}{2}t^{-5}\right)dt = 2\pi\left[\tfrac{1}{18}t^6 + \tfrac{5}{6}t - \tfrac{1}{8}t^{-4}\right]_1^4 = \frac{471{,}295}{1024}\pi$$

41. For all c except -1, the curve is asymptotic to the line $x = 1$. For

$c < -1$, the curve bulges to the right near $y = 0$. As c increases, the

bulge becomes smaller, until at $c = -1$ the curve is the straight line $x = 1$.

As c continues to increase, the curve bulges to the left, until at $c = 0$ there

is a cusp at the origin. For $c > 0$, there is a loop to the left of the origin,

whose size and roundness increase as c increases. Note that the x-intercept

of the curve is always $-c$.

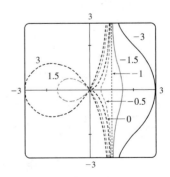

43. $\dfrac{x^2}{9} + \dfrac{y^2}{8} = 1$ is an ellipse with center $(0,0)$.

$a = 3, b = 2\sqrt{2}, c = 1 \;\Rightarrow$

foci $(\pm 1, 0)$, vertices $(\pm 3, 0)$.

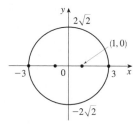

45. $6y^2 + x - 36y + 55 = 0 \;\Leftrightarrow$

$6(y^2 - 6y + 9) = -(x+1) \;\Leftrightarrow$

$(y-3)^2 = -\frac{1}{6}(x+1)$, a parabola with vertex

$(-1, 3)$, opening to the left, $p = -\frac{1}{24} \;\Rightarrow\;$ focus

$\left(-\frac{25}{24}, 3\right)$ and directrix $x = -\frac{23}{24}$.

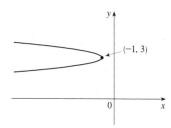

47. The parabola opens upward with vertex $(0, 4)$ [midway between the focus $(0, 6)$ and the directrix $y = 2$] and $p = 2$, so its equation is $(x - 0)^2 = 4 \cdot 2(y - 4) \;\Leftrightarrow\; x^2 = 8(y - 4)$.

49. The hyperbola has center $(0, 0)$ and foci on the x-axis. $c = 3$ and $b/a = \frac{1}{2}$ (from the asymptotes)

$\Rightarrow\quad 9 = c^2 = a^2 + b^2 = (2b)^2 + b^2 = 5b^2 \;\Rightarrow\; b = \frac{3}{\sqrt{5}} \;\Rightarrow\; a = \frac{6}{\sqrt{5}} \;\Rightarrow\;$ an equation of the hyperbola is

$\dfrac{x^2}{36/5} - \dfrac{y^2}{9/5} = 1 \;\Leftrightarrow\; 5x^2 - 20y^2 = 36.$

51. $x^2 = -(y - 100)$ has its vertex at $(0, 100)$, so one of the vertices of the ellipse is $(0, 100)$. Another form of the

equation of a parabola is $x^2 = 4p(y - 100)$ so $4p(y - 100) = -(y - 100) \;\Rightarrow\; 4p = -1 \;\Rightarrow\; p = -\frac{1}{4}.$

Therefore the shared focus is found at $\left(0, \frac{399}{4}\right)$ so $2c = \frac{399}{4} - 0 \;\Rightarrow\; c = \frac{399}{8}$ and the center of the ellipse is

$\left(0, \frac{399}{8}\right)$. So $a = 100 - \frac{399}{8} = \frac{401}{8}$ and $b^2 = a^2 - c^2 = \frac{401^2 - 399^2}{8^2} = 25$. So the equation of the ellipse is

$\dfrac{x^2}{b^2} + \dfrac{\left(y - \frac{399}{8}\right)^2}{a^2} = 1 \;\Rightarrow\; \dfrac{x^2}{25} + \dfrac{\left(y - \frac{399}{8}\right)^2}{\left(\frac{401}{8}\right)^2} = 1$ or $\dfrac{x^2}{25} + \dfrac{(8y - 399)^2}{160{,}801} = 1.$

53. Directrix $x = 4 \;\Rightarrow\; d = 4$, so $e = \frac{1}{3} \;\Rightarrow\; r = \dfrac{ed}{1 + e\cos\theta} = \dfrac{4}{3 + \cos\theta}.$

55. In polar coordinates, an equation for the circle is $r = 2a\sin\theta$. Thus, the coordinates of Q are

$x = r\cos\theta = 2a\sin\theta\cos\theta$ and $y = r\sin\theta = 2a\sin^2\theta$. The coordinates of R are $x = 2a\cot\theta$ and $y = 2a$.

Since P is the midpoint of QR, we use the midpoint formula to get $x = a(\sin\theta\cos\theta + \cot\theta)$ and

$y = a(1 + \sin^2\theta).$

◻ PROBLEMS PLUS

1. $x = \int_1^t \dfrac{\cos u}{u}\,du$, $y = \int_1^t \dfrac{\sin u}{u}\,du$, so by FTC1, we have $\dfrac{dx}{dt} = \dfrac{\cos t}{t}$ and $\dfrac{dy}{dt} = \dfrac{\sin t}{t}$. Vertical tangent lines

occur when $\dfrac{dx}{dt} = 0 \iff \cos t = 0$. The parameter value corresponding to $(x, y) = (0, 0)$ is $t = 1$, so the

nearest vertical tangent occurs when $t = \frac{\pi}{2}$. Therefore, the arc length between these points is

$$L = \int_1^{\pi/2} \sqrt{\left(\frac{dx}{dt}\right)^2 + \left(\frac{dy}{dt}\right)^2}\,dt = \int_1^{\pi/2} \sqrt{\frac{\cos^2 t}{t^2} + \frac{\sin^2 t}{t^2}}\,dt = \int_1^{\pi/2} \frac{dt}{t} = [\ln t]_1^{\pi/2} = \ln \frac{\pi}{2}$$

3. In terms of x and y, we have $x = r\cos\theta = (1 + c\sin\theta)\cos\theta = \cos\theta + c\sin\theta\cos\theta = \cos\theta + \frac{1}{2}c\sin 2\theta$ and

$y = r\sin\theta = (1 + c\sin\theta)\sin\theta = \sin\theta + c\sin^2\theta$. Now $-1 \le \sin\theta \le 1 \Rightarrow$

$-1 \le \sin\theta + c\sin^2\theta \le 1 + c \le 2$, so $-1 \le y \le 2$. Furthermore, $y = 2$ when $c = 1$ and $\theta = \frac{\pi}{2}$, while $y = -1$

for $c = 0$ and $\theta = \frac{3\pi}{2}$. Therefore, we need a viewing rectangle with $-1 \le y \le 2$.

To find the x-values, look at the equation $x = \cos\theta + \frac{1}{2}c\sin 2\theta$ and use the fact that $\sin 2\theta \ge 0$ for $0 \le \theta \le \frac{\pi}{2}$

and $\sin 2\theta \le 0$ for $-\frac{\pi}{2} \le \theta \le 0$. [Because $r = 1 + c\sin\theta$ is symmetric about the y-axis, we only need to consider

$-\frac{\pi}{2} \le \theta \le \frac{\pi}{2}$.] So for $-\frac{\pi}{2} \le \theta \le 0$, x has a maximum value when $c = 0$ and then $x = \cos\theta$ has a maximum value

of 1 at $\theta = 0$. Thus, the maximum value of x must occur on $\left[0, \frac{\pi}{2}\right]$ with $c = 1$. Then $x = \cos\theta + \frac{1}{2}\sin 2\theta \Rightarrow$

$\frac{dx}{d\theta} = -\sin\theta + \cos 2\theta = -\sin\theta + 1 - 2\sin^2\theta \Rightarrow \frac{dx}{d\theta} = -(2\sin\theta - 1)(\sin\theta + 1) = 0$ when $\sin\theta = -1$ or $\frac{1}{2}$

(but $\sin\theta \ne -1$ for $0 \le \theta \le \frac{\pi}{2}$). If $\sin\theta = \frac{1}{2}$, then $\theta = \frac{\pi}{6}$ and

$x = \cos\frac{\pi}{6} + \frac{1}{2}\sin\frac{\pi}{3} = \frac{3}{4}\sqrt{3}$. Thus, the maximum value of x is

$\frac{3}{4}\sqrt{3}$, and, by symmetry, the minimum value is $-\frac{3}{4}\sqrt{3}$. Therefore,

the smallest viewing rectangle that contains every member of the

family of polar curves $r = 1 + c\sin\theta$, where $0 \le c \le 1$, is

$\left[-\frac{3}{4}\sqrt{3}, \frac{3}{4}\sqrt{3}\right] \times [-1, 2]$.

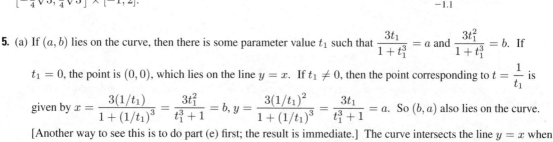

5. (a) If (a, b) lies on the curve, then there is some parameter value t_1 such that $\dfrac{3t_1}{1 + t_1^3} = a$ and $\dfrac{3t_1^2}{1 + t_1^3} = b$. If

$t_1 = 0$, the point is $(0, 0)$, which lies on the line $y = x$. If $t_1 \ne 0$, then the point corresponding to $t = \dfrac{1}{t_1}$ is

given by $x = \dfrac{3(1/t_1)}{1 + (1/t_1)^3} = \dfrac{3t_1^2}{t_1^3 + 1} = b$, $y = \dfrac{3(1/t_1)^2}{1 + (1/t_1)^3} = \dfrac{3t_1}{t_1^3 + 1} = a$. So (b, a) also lies on the curve.

[Another way to see this is to do part (e) first; the result is immediate.] The curve intersects the line $y = x$ when

$\dfrac{3t}{1 + t^3} = \dfrac{3t^2}{1 + t^3} \Rightarrow t = t^2 \Rightarrow t = 0$ or 1, so the points are $(0, 0)$ and $\left(\frac{3}{2}, \frac{3}{2}\right)$.

(b) $\dfrac{dy}{dt} = \dfrac{(1+t^3)(6t) - 3t^2(3t^2)}{(1+t^3)^2} = \dfrac{6t - 3t^4}{(1+t^3)^2} = 0$ when $6t - 3t^4 = 3t(2 - t^3) = 0 \;\Rightarrow\; t = 0$ or $t = \sqrt[3]{2}$,

so there are horizontal tangents at $(0,0)$ and $\left(\sqrt[3]{2}, \sqrt[3]{4}\right)$. Using the symmetry from part (a), we see that there are vertical tangents at $(0,0)$ and $\left(\sqrt[3]{4}, \sqrt[3]{2}\right)$.

(c) Notice that as $t \to -1^+$, we have $x \to -\infty$ and $y \to \infty$. As $t \to -1^-$, we have $x \to \infty$ and $y \to -\infty$. Also

$$y - (-x - 1) = y + x + 1 = \dfrac{3t + 3t^2 + (1+t^3)}{1+t^3} = \dfrac{(t+1)^3}{1+t^3} = \dfrac{(t+1)^2}{t^2 - t + 1} \to 0 \text{ as } t \to -1. \text{ So}$$

$y = -x - 1$ is a slant asymptote.

(d) $\dfrac{dx}{dt} = \dfrac{(1+t^3)(3) - 3t(3t^2)}{(1+t^3)^2} = \dfrac{3 - 6t^3}{(1+t^3)^2}$ and from part (b) we have $\dfrac{dy}{dt} = \dfrac{6t - 3t^4}{(1+t^3)^2}$.

So $\dfrac{dy}{dx} = \dfrac{dy/dt}{dx/dt} = \dfrac{t(2 - t^3)}{1 - 2t^3}$. Also

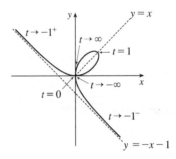

$$\dfrac{d^2y}{dx^2} = \dfrac{\dfrac{d}{dt}\left(\dfrac{dy}{dx}\right)}{dx/dt} = \dfrac{2(1+t^3)^4}{3(1 - 2t^3)^3} > 0 \;\Leftrightarrow\; t < \dfrac{1}{\sqrt[3]{2}}. \text{ So}$$

the curve is concave upward there and has a minimum point at $(0,0)$ and a maximum point at $\left(\sqrt[3]{2}, \sqrt[3]{4}\right)$. Using this together with the information from parts (a), (b), and (c), we sketch the curve.

(e) $x^3 + y^3 = \left(\dfrac{3t}{1+t^3}\right)^3 + \left(\dfrac{3t^2}{1+t^3}\right)^3 = \dfrac{27t^3 + 27t^6}{(1+t^3)^3} = \dfrac{27t^3(1+t^3)}{(1+t^3)^3} = \dfrac{27t^3}{(1+t^3)^2}$ and

$3xy = 3\left(\dfrac{3t}{1+t^3}\right)\left(\dfrac{3t^2}{1+t^3}\right) = \dfrac{27t^3}{(1+t^3)^2}$, so $x^3 + y^3 = 3xy$.

(f) We start with the equation from part (e) and substitute $x = r\cos\theta$, $y = r\sin\theta$. Then $x^3 + y^3 = 3xy \;\Rightarrow\;$

$r^3\cos^3\theta + r^3\sin^3\theta = 3r^2\cos\theta\sin\theta$. For $r \neq 0$, this gives $r = \dfrac{3\cos\theta\sin\theta}{\cos^3\theta + \sin^3\theta}$. Dividing numerator and

denominator by $\cos^3\theta$, we obtain $r = \dfrac{3\left(\dfrac{1}{\cos\theta}\right)\dfrac{\sin\theta}{\cos\theta}}{1 + \dfrac{\sin^3\theta}{\cos^3\theta}} = \dfrac{3\sec\theta\tan\theta}{1 + \tan^3\theta}$.

(g) The loop corresponds to $\theta \in \left(0, \dfrac{\pi}{2}\right)$, so its area is

$$A = \int_0^{\pi/2} \dfrac{r^2}{2}\,d\theta = \dfrac{1}{2}\int_0^{\pi/2}\left(\dfrac{3\sec\theta\tan\theta}{1 + \tan^3\theta}\right)^2 d\theta = \dfrac{9}{2}\int_0^{\pi/2}\dfrac{\sec^2\theta\tan^2\theta}{(1 + \tan^3\theta)^2}\,d\theta$$

$$= \dfrac{9}{2}\int_0^{\infty}\dfrac{u^2\,du}{(1 + u^3)^2}\ \text{[let } u = \tan\theta] = \lim_{b\to\infty}\dfrac{9}{2}\left[-\dfrac{1}{3}(1 + u^3)^{-1}\right]_0^b = \dfrac{3}{2}$$

(h) By symmetry, the area between the folium and the line $y = -x - 1$ is equal to the enclosed area in the third quadrant, plus twice the enclosed area in the fourth quadrant. The area in the third quadrant is $\dfrac{1}{2}$, and since

$y = -x - 1 \;\Rightarrow\; r\sin\theta = -r\cos\theta - 1 \;\Rightarrow\; r = -\dfrac{1}{\sin\theta + \cos\theta}$, the area in the fourth quadrant is

$$\dfrac{1}{2}\int_{-\pi/2}^{-\pi/4}\left[\left(-\dfrac{1}{\sin\theta + \cos\theta}\right)^2 - \left(\dfrac{3\sec\theta\tan\theta}{1 + \tan^3\theta}\right)^2\right]d\theta \overset{\text{CAS}}{=} \dfrac{1}{2}. \text{ Therefore, the total area is } \dfrac{1}{2} + 2\left(\dfrac{1}{2}\right) = \dfrac{3}{2}.$$

12 ☐ INFINITE SEQUENCES AND SERIES

12.1 Sequences

1. (a) A sequence is an ordered list of numbers. It can also be defined as a function whose domain is the set of positive integers.

(b) The terms a_n approach 8 as n becomes large. In fact, we can make a_n as close to 8 as we like by taking n sufficiently large.

(c) The terms a_n become large as n becomes large. In fact, we can make a_n as large as we like by taking n sufficiently large.

3. $a_n = 1 - (0.2)^n$, so the sequence is $\{0.8, 0.96, 0.992, 0.9984, 0.99968, \dots\}$.

5. $a_n = \dfrac{3(-1)^n}{n!}$, so the sequence is $\left\{\dfrac{-3}{1}, \dfrac{3}{2}, \dfrac{-3}{6}, \dfrac{3}{24}, \dfrac{-3}{120}, \dots\right\} = \left\{-3, \dfrac{3}{2}, -\dfrac{1}{2}, \dfrac{1}{8}, -\dfrac{1}{40}, \dots\right\}$.

7. $a_1 = 3$, $a_{n+1} = 2a_n - 1$. Each term is defined in terms of the preceding term.
$a_2 = 2a_1 - 1 = 2(3) - 1 = 5$. $a_3 = 2a_2 - 1 = 2(5) - 1 = 9$. $a_4 = 2a_3 - 1 = 2(9) - 1 = 17$.
$a_5 = 2a_4 - 1 = 2(17) - 1 = 33$. The sequence is $\{3, 5, 9, 17, 33, \dots\}$.

9. The numerators are all 1 and the denominators are powers of 2, so $a_n = \dfrac{1}{2^n}$.

11. $\{2, 7, 12, 17, \dots\}$. Each term is larger than the preceding one by 5, so
$a_n = a_1 + d(n-1) = 2 + 5(n-1) = 5n - 3$.

13. $\left\{1, -\dfrac{2}{3}, \dfrac{4}{9}, -\dfrac{8}{27}, \dots\right\}$. Each term is $-\dfrac{2}{3}$ times the preceding one, so $a_n = \left(-\dfrac{2}{3}\right)^{n-1}$.

15. $a_n = n(n-1)$. $a_n \to \infty$ as $n \to \infty$, so the sequence diverges.

17. $a_n = \dfrac{3 + 5n^2}{n + n^2} = \dfrac{(3 + 5n^2)/n^2}{(n + n^2)/n^2} = \dfrac{5 + 3/n^2}{1 + 1/n}$, so $a_n \to \dfrac{5 + 0}{1 + 0} = 5$ as $n \to \infty$. Converges

19. $a_n = \dfrac{2^n}{3^{n+1}} = \dfrac{1}{3}\left(\dfrac{2}{3}\right)^n$, so $\lim\limits_{n\to\infty} a_n = \dfrac{1}{3}\lim\limits_{n\to\infty}\left(\dfrac{2}{3}\right)^n = \dfrac{1}{3}\cdot 0 = 0$ by (8) with $r = \dfrac{2}{3}$. Converges

21. $a_n = \dfrac{(-1)^{n-1}n}{n^2 + 1} = \dfrac{(-1)^{n-1}}{n + 1/n}$, so $0 \le |a_n| = \dfrac{1}{n + 1/n} \le \dfrac{1}{n} \to 0$ as $n \to \infty$, so $a_n \to 0$ by the Squeeze Theorem and Theorem 6. Converges

23. $a_n = \cos(n/2)$. This sequence diverges since the terms don't approach any particular real number as $n \to \infty$. The terms take on values between -1 and 1.

25. $a_n = \dfrac{(2n-1)!}{(2n+1)!} = \dfrac{(2n-1)!}{(2n+1)(2n)(2n-1)!} = \dfrac{1}{(2n+1)(2n)} \to 0$ as $n \to \infty$. Converges

27. $a_n = \dfrac{e^n + e^{-n}}{e^{2n} - 1} \cdot \dfrac{e^{-n}}{e^{-n}} = \dfrac{1 + e^{-2n}}{e^n - e^{-n}} \to \dfrac{1 + 0}{e^n - 0} \to 0$ as $n \to \infty$. Converges

29. $a_n = n^2 e^{-n} = \dfrac{n^2}{e^n}$. Since $\lim\limits_{x\to\infty}\dfrac{x^2}{e^x} \overset{\text{H}}{=} \lim\limits_{x\to\infty}\dfrac{2x}{e^x} \overset{\text{H}}{=} \lim\limits_{x\to\infty}\dfrac{2}{e^x} = 0$, it follows from Theorem 3 that $\lim\limits_{n\to\infty} a_n = 0$. Converges

31. $0 \le \dfrac{\cos^2 n}{2^n} \le \dfrac{1}{2^n}$ [since $0 \le \cos^2 n \le 1$], so since $\lim\limits_{n\to\infty}\dfrac{1}{2^n} = 0$, $\left\{\dfrac{\cos^2 n}{2^n}\right\}$ converges to 0 by the Squeeze Theorem.

33. $a_n = n \sin(1/n) = \dfrac{\sin(1/n)}{1/n}$. Since $\displaystyle\lim_{x \to \infty} \dfrac{\sin(1/x)}{1/x} = \lim_{t \to 0^+} \dfrac{\sin t}{t}$ [where $t = 1/x$] $= 1$, it follows from

Theorem 3 that $\{a_n\}$ converges to 1.

35. $a_n = \left(1 + \dfrac{2}{n}\right)^{1/n} \implies \ln a_n = \dfrac{1}{n} \ln\left(1 + \dfrac{2}{n}\right)$. As $n \to \infty$, $\dfrac{1}{n} \to 0$ and $\ln\left(1 + \dfrac{2}{n}\right) \to 0$, so $\ln a_n \to 0$.

Thus, $a_n \to e^0 = 1$ as $n \to \infty$. Converges

37. $\{0, 1, 0, 0, 1, 0, 0, 0, 1, \dots\}$ diverges since the sequence takes on only two values, 0 and 1, and never stays arbitrarily close to either one (or any other value) for n sufficiently large.

39. $a_n = \dfrac{n!}{2^n} = \dfrac{1}{2} \cdot \dfrac{2}{2} \cdot \dfrac{3}{2} \cdot \ldots \cdot \dfrac{(n-1)}{2} \cdot \dfrac{n}{2} \geq \dfrac{1}{2} \cdot \dfrac{n}{2}$ [for $n > 1$] $= \dfrac{n}{4} \to \infty$ as $n \to \infty$, so $\{a_n\}$ diverges.

41.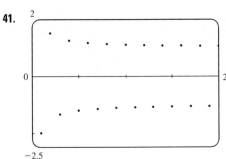

From the graph, we see that the sequence $\left\{(-1)^n \dfrac{n+1}{n}\right\}$ is divergent, since it oscillates between 1 and -1 (approximately).

43.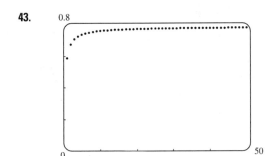

From the graph, it appears that the sequence converges to about 0.78.

$$\lim_{n \to \infty} \frac{2n}{2n+1} = \lim_{n \to \infty} \frac{2}{2+1/n} = 1, \text{ so}$$

$$\lim_{n \to \infty} \arctan\left(\frac{2n}{2n+1}\right) = \arctan 1 = \frac{\pi}{4}.$$

45.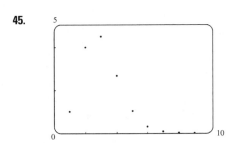

From the graph, it appears that the sequence converges to 0.

$$0 < a_n = \frac{n^3}{n!} = \frac{n}{n} \cdot \frac{n}{(n-1)} \cdot \frac{n}{(n-2)} \cdot \frac{1}{(n-3)} \cdot \ldots \cdot \frac{1}{3} \cdot \frac{1}{2} \cdot \frac{1}{1}$$

$$\leq \frac{n^2}{(n-1)(n-2)(n-3)} \quad [\text{for } n \geq 4]$$

$$= \frac{1/n}{(1 - 1/n)(1 - 2/n)(1 - 3/n)} \to 0 \text{ as } n \to \infty$$

So by the Squeeze Theorem, $\{n^3/n!\}$ converges to 0.

47.

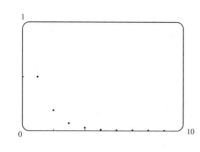

From the graph, it appears that the sequence approaches 0.

$$0 < a_n = \frac{1 \cdot 3 \cdot 5 \cdot \cdots \cdot (2n-1)}{(2n)^n} = \frac{1}{2n} \cdot \frac{3}{2n} \cdot \frac{5}{2n} \cdot \cdots \cdot \frac{2n-1}{2n}$$

$$\leq \frac{1}{2n} \cdot (1) \cdot (1) \cdot \cdots \cdot (1) = \frac{1}{2n} \to 0 \text{ as } n \to \infty$$

So by the Squeeze Theorem, $\left\{ \dfrac{1 \cdot 3 \cdot 5 \cdot \cdots \cdot (2n-1)}{(2n)^n} \right\}$ converges to 0.

49. (a) $a_n = 1000(1.06)^n \Rightarrow a_1 = 1060, a_2 = 1123.60, a_3 = 1191.02, a_4 = 1262.48,$ and $a_5 = 1338.23$.

(b) $\lim\limits_{n \to \infty} a_n = 1000 \lim\limits_{n \to \infty} (1.06)^n$, so the sequence diverges by (8) with $r = 1.06 > 1$.

51. If $|r| \geq 1$, then $\{r^n\}$ diverges by (8), so $\{nr^n\}$ diverges also, since $|nr^n| = n \, |r^n| \geq |r^n|$. If $|r| < 1$ then

$$\lim_{x \to \infty} xr^x = \lim_{x \to \infty} \frac{x}{r^{-x}} \overset{\text{H}}{=} \lim_{x \to \infty} \frac{1}{(-\ln r) \, r^{-x}} = \lim_{x \to \infty} \frac{r^x}{-\ln r} = 0, \text{ so } \lim_{n \to \infty} nr^n = 0, \text{ and hence } \{nr^n\} \text{ converges}$$

whenever $|r| < 1$.

53. Since $\{a_n\}$ is a decreasing sequence, $a_n > a_{n+1}$ for all $n \geq 1$. Because all of its terms lie between 5 and 8, $\{a_n\}$ is a bounded sequence. By the Monotonic Sequence Theorem, $\{a_n\}$ is convergent; that is, $\{a_n\}$ has a limit L. L must be less than 8 since $\{a_n\}$ is decreasing, so $5 \leq L < 8$.

55. $a_n = \dfrac{1}{2n+3}$ is decreasing since $a_{n+1} = \dfrac{1}{2(n+1)+3} = \dfrac{1}{2n+5} < \dfrac{1}{2n+3} = a_n$ for each $n \geq 1$. The sequence is bounded since $0 < a_n \leq \frac{1}{5}$ for all $n \geq 1$. Note that $a_1 = \frac{1}{5}$.

57. $a_n = \cos(n\pi/2)$ is not monotonic. The first few terms are $0, -1, 0, 1, 0, -1, 0, 1, \ldots$. In fact, the sequence consists of the terms $0, -1, 0, 1$ repeated over and over again in that order. The sequence is bounded since $|a_n| \leq 1$ for all $n \geq 1$.

59. $a_n = \dfrac{n}{n^2 + 1}$ defines a decreasing sequence since for $f(x) = \dfrac{x}{x^2 + 1}$,

$$f'(x) = \frac{(x^2+1)(1) - x(2x)}{(x^2+1)^2} = \frac{1 - x^2}{(x^2+1)^2} \leq 0 \text{ for } x \geq 1. \text{ The sequence is bounded since } 0 < a_n \leq \frac{1}{2} \text{ for all}$$

$n \geq 1$.

61. $a_1 = 2^{1/2}, a_2 = 2^{3/4}, a_3 = 2^{7/8}, \ldots$, so $a_n = 2^{(2^n - 1)/2^n} = 2^{1 - (1/2^n)}$. $\lim\limits_{n \to \infty} a_n = \lim\limits_{n \to \infty} 2^{1 - (1/2^n)} = 2^1 = 2$.

Alternate solution: Let $L = \lim\limits_{n \to \infty} a_n$. (We could show the limit exists by showing that $\{a_n\}$ is bounded and increasing.) Then L must satisfy $L = \sqrt{2 \cdot L} \Rightarrow L^2 = 2L \Rightarrow L(L-2) = 0$. $L \neq 0$ since the sequence increases, so $L = 2$.

63. We show by induction that $\{a_n\}$ is increasing and bounded above by 3.

Let P_n be the proposition that $a_{n+1} > a_n$ and $0 < a_n < 3$. Clearly P_1 is true. Assume that P_n is true.

Then $a_{n+1} > a_n \Rightarrow \dfrac{1}{a_{n+1}} < \dfrac{1}{a_n} \Rightarrow -\dfrac{1}{a_{n+1}} > -\dfrac{1}{a_n}$.

Now $a_{n+2} = 3 - \dfrac{1}{a_{n+1}} > 3 - \dfrac{1}{a_n} = a_{n+1} \Leftrightarrow P_{n+1}$. This proves that $\{a_n\}$ is increasing and bounded above by 3, so $1 = a_1 < a_n < 3$, that is, $\{a_n\}$ is bounded, and hence convergent by the Monotonic Sequence Theorem.

If $L = \lim\limits_{n \to \infty} a_n$, then $\lim\limits_{n \to \infty} a_{n+1} = L$ also, so L must satisfy $L = 3 - 1/L \Rightarrow L^2 - 3L + 1 = 0 \Rightarrow$

$L = \frac{3 \pm \sqrt{5}}{2}$. But $L > 1$, so $L = \frac{3 + \sqrt{5}}{2}$.

65. (a) Let a_n be the number of rabbit pairs in the nth month. Clearly $a_1 = 1 = a_2$. In the nth month, each pair that is 2 or more months old (that is, a_{n-2} pairs) will produce a new pair to add to the a_{n-1} pairs already present. Thus, $a_n = a_{n-1} + a_{n-2}$, so that $\{a_n\} = \{f_n\}$, the Fibonacci sequence.

(b) $a_n = \dfrac{f_{n+1}}{f_n} \;\Rightarrow\; a_{n-1} = \dfrac{f_n}{f_{n-1}} = \dfrac{f_{n-1} + f_{n-2}}{f_{n-1}} = 1 + \dfrac{f_{n-2}}{f_{n-1}} = 1 + \dfrac{1}{f_{n-1}/f_{n-2}} = 1 + \dfrac{1}{a_{n-2}}$. If

$L = \lim\limits_{n \to \infty} a_n$, then $L = \lim\limits_{n \to \infty} a_{n-1}$ and $L = \lim\limits_{n \to \infty} a_{n-2}$, so L must satisfy $L = 1 + \dfrac{1}{L} \;\Rightarrow\;$

$L^2 - L - 1 = 0 \;\Rightarrow\; L = \dfrac{1+\sqrt{5}}{2}$ (since L must be positive).

67. (a)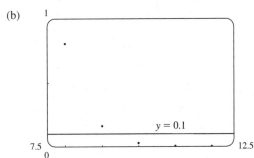

From the graph, it appears that the sequence $\left\{ \dfrac{n^5}{n!} \right\}$ converges to 0, that is,

$$\lim_{n \to \infty} \frac{n^5}{n!} = 0.$$

(b)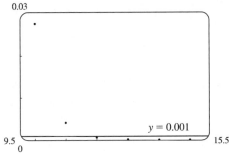

From the first graph, it seems that the smallest possible value of N corresponding to $\varepsilon = 0.1$ is 9, since $n^5/n! < 0.1$ whenever $n \ge 10$, but $9^5/9! > 0.1$. From the second graph, it seems that for $\varepsilon = 0.001$, the smallest possible value for N is 11.

69. If $\lim\limits_{n \to \infty} |a_n| = 0$ then $\lim\limits_{n \to \infty} -|a_n| = 0$, and since $-|a_n| \le a_n \le |a_n|$, we have that $\lim\limits_{n \to \infty} a_n = 0$ by the Squeeze Theorem.

71. (a) First we show that $a > a_1 > b_1 > b$.

$a_1 - b_1 = \dfrac{a+b}{2} - \sqrt{ab} = \dfrac{1}{2}\left(a - 2\sqrt{ab} + b\right) = \dfrac{1}{2}\left(\sqrt{a} - \sqrt{b}\right)^2 > 0$ (since $a > b$) $\;\Rightarrow\; a_1 > b_1$. Also

$a - a_1 = a - \dfrac{1}{2}(a + b) = \dfrac{1}{2}(a - b) > 0$ and $b - b_1 = b - \sqrt{ab} = \sqrt{b}\left(\sqrt{b} - \sqrt{a}\right) < 0$, so $a > a_1 > b_1 > b$.

In the same way we can show that $a_1 > a_2 > b_2 > b_1$ and so the given assertion is true for $n = 1$. Suppose it is true for $n = k$, that is, $a_k > a_{k+1} > b_{k+1} > b_k$. Then

$$a_{k+2} - b_{k+2} = \dfrac{1}{2}(a_{k+1} + b_{k+1}) - \sqrt{a_{k+1}b_{k+1}} = \dfrac{1}{2}\left(a_{k+1} - 2\sqrt{a_{k+1}b_{k+1}} + b_{k+1}\right)$$

$$= \dfrac{1}{2}\left(\sqrt{a_{k+1}} - \sqrt{b_{k+1}}\right)^2 > 0$$

$$a_{k+1} - a_{k+2} = a_{k+1} - \dfrac{1}{2}(a_{k+1} + b_{k+1}) = \dfrac{1}{2}(a_{k+1} - b_{k+1}) > 0$$

and $b_{k+1} - b_{k+2} = b_{k+1} - \sqrt{a_{k+1}b_{k+1}} = \sqrt{b_{k+1}}\left(\sqrt{b_{k+1}} - \sqrt{a_{k+1}}\right) < 0 \quad \Rightarrow$

$a_{k+1} > a_{k+2} > b_{k+2} > b_{k+1}$, so the assertion is true for $n = k + 1$. Thus, it is true for all n by mathematical induction.

(b) From part (a) we have $a > a_n > a_{n+1} > b_{n+1} > b_n > b$, which shows that both sequences, $\{a_n\}$ and $\{b_n\}$, are monotonic and bounded. So they are both convergent by the Monotonic Sequence Theorem.

(c) Let $\lim\limits_{n\to\infty} a_n = \alpha$ and $\lim\limits_{n\to\infty} b_n = \beta$. Then $\lim\limits_{n\to\infty} a_{n+1} = \lim\limits_{n\to\infty} \dfrac{a_n + b_n}{2} \quad \Rightarrow \quad \alpha = \dfrac{\alpha + \beta}{2} \quad \Rightarrow$

$2\alpha = \alpha + \beta \quad \Rightarrow \quad \alpha = \beta$.

73. (a) Suppose $\{p_n\}$ converges to p. Then $p_{n+1} = \dfrac{bp_n}{a + p_n} \quad \Rightarrow \quad \lim\limits_{n\to\infty} p_{n+1} = \dfrac{b \lim\limits_{n\to\infty} p_n}{a + \lim\limits_{n\to\infty} p_n} \quad \Rightarrow$

$p = \dfrac{bp}{a + p} \quad \Rightarrow \quad p^2 + ap = bp \quad \Rightarrow \quad p(p + a - b) = 0 \quad \Rightarrow \quad p = 0 \text{ or } p = b - a$.

(b) $p_{n+1} = \dfrac{bp_n}{a + p_n} = \dfrac{\dfrac{b}{a}p_n}{1 + \dfrac{p_n}{a}} < \dfrac{b}{a}p_n$ since $1 + \dfrac{p_n}{a} > 1$.

(c) By part (b), $p_1 < \left(\dfrac{b}{a}\right)p_0$, $p_2 < \left(\dfrac{b}{a}\right)p_1 < \left(\dfrac{b}{a}\right)^2 p_0$, $p_3 < \left(\dfrac{b}{a}\right)p_2 < \left(\dfrac{b}{a}\right)^3 p_0$, etc. In general,

$p_n < \left(\dfrac{b}{a}\right)^n p_0$, so $\lim\limits_{n\to\infty} p_n \leq \lim\limits_{n\to\infty} \left(\dfrac{b}{a}\right)^n \cdot p_0 = 0$ since $b < a$. [By result 8, $\lim\limits_{n\to\infty} r^n = 0$ if $-1 < r < 1$.

Here $r = \dfrac{b}{a} \in (0, 1)$.]

(d) Let $a < b$. We first show, by induction, that if $p_0 < b - a$, then $p_n < b - a$ and $p_{n+1} > p_n$.

For $n = 0$, we have $p_1 - p_0 = \dfrac{bp_0}{a + p_0} - p_0 = \dfrac{p_0(b - a - p_0)}{a + p_0} > 0$ since $p_0 < b - a$. So $p_1 > p_0$.

Now we suppose the assertion is true for $n = k$, that is, $p_k < b - a$ and $p_{k+1} > p_k$. Then

$b - a - p_{k+1} = b - a - \dfrac{bp_k}{a + p_k} = \dfrac{a(b - a) + bp_k - ap_k - bp_k}{a + p_k} = \dfrac{a(b - a - p_k)}{a + p_k} > 0$ because $p_k < b - a$.

So $p_{k+1} < b - a$. And $p_{k+2} - p_{k+1} = \dfrac{bp_{k+1}}{a + p_{k+1}} - p_{k+1} = \dfrac{p_{k+1}(b - a - p_{k+1})}{a + p_{k+1}} > 0$ since $p_{k+1} < b - a$.

Therefore, $p_{k+2} > p_{k+1}$. Thus, the assertion is true for $n = k + 1$. It is therefore true for all n by mathematical induction. A similar proof by induction shows that if $p_0 > b - a$, then $p_n > b - a$ and $\{p_n\}$ is decreasing.

In either case the sequence $\{p_n\}$ is bounded and monotonic, so it is convergent by the Monotonic Sequence Theorem. It then follows from part (a) that $\lim\limits_{n\to\infty} p_n = b - a$.

12.2 Series

1. (a) A sequence is an ordered list of numbers whereas a series is the *sum* of a list of numbers.

(b) A series is convergent if the sequence of partial sums is a convergent sequence. A series is divergent if it is not convergent.

3.

n	s_n
1	-2.40000
2	-1.92000
3	-2.01600
4	-1.99680
5	-2.00064
6	-1.99987
7	-2.00003
8	-1.99999
9	-2.00000
10	-2.00000

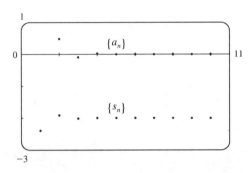

From the graph and the table, it seems that the series converges to -2. In fact, it is a geometric series with $a = -2.4$ and $r = -\frac{1}{5}$, so its sum is

$$\sum_{n=1}^{\infty} \frac{12}{(-5)^n} = \frac{-2.4}{1 - \left(-\frac{1}{5}\right)} = \frac{-2.4}{1.2} = -2.$$ Note that the dot corresponding to $n = 1$ is part of both $\{a_n\}$ and $\{s_n\}$.

TI-86 Note: To graph $\{a_n\}$ and $\{s_n\}$, set your calculator to Param mode and DrawDot mode. (DrawDot is under GRAPH, MORE, FORMT (F3).) Now under E(t) = make the assignments: xt1=t, yt1=12/(-5)^t, xt2=t, yt2=sum seq(yt1,t,1,t,1). (sum and seq are under LIST, OPS (F5), MORE.) Under WIND use 1,10,1,0,10,1,-3,1,1 to obtain a graph similar to the one above. Then use TRACE (F4) to see the values.

5.

n	s_n
1	1.55741
2	-0.62763
3	-0.77018
4	0.38764
5	-2.99287
6	-3.28388
7	-2.41243
8	-9.21214
9	-9.66446
10	-9.01610

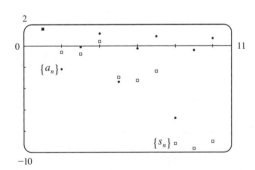

The series $\displaystyle\sum_{n=1}^{\infty} \tan n$ diverges, since its terms do not approach 0.

7.

n	s_n
1	0.64645
2	0.80755
3	0.87500
4	0.91056
5	0.93196
6	0.94601
7	0.95581
8	0.96296
9	0.96838
10	0.97259

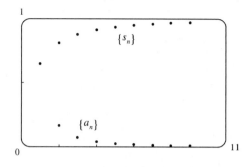

From the graph, it seems that the series converges to 1. To find the sum, we write

$$s_n = \sum_{i=1}^{n} \left(\frac{1}{i^{1.5}} - \frac{1}{(i+1)^{1.5}} \right)$$

$$= \left(1 - \frac{1}{2^{1.5}} \right) + \left(\frac{1}{2^{1.5}} - \frac{1}{3^{1.5}} \right) + \left(\frac{1}{3^{1.5}} - \frac{1}{4^{1.5}} \right) + \cdots + \left(\frac{1}{n^{1.5}} - \frac{1}{(n+1)^{1.5}} \right) = 1 - \frac{1}{(n+1)^{1.5}}$$

So the sum is $\lim_{n \to \infty} s_n = 1 - 0 = 1$.

9. (a) $\lim_{n \to \infty} a_n = \lim_{n \to \infty} \frac{2n}{3n+1} = \frac{2}{3}$, so the *sequence* $\{a_n\}$ is convergent by (12.1.1) [ET (11.1.1)].

(b) Since $\lim_{n \to \infty} a_n = \frac{2}{3} \neq 0$, the *series* $\sum_{n=1}^{\infty} a_n$ is divergent by the Test for Divergence (7).

11. $3 + 2 + \frac{4}{3} + \frac{8}{9} + \cdots$ is a geometric series with first term $a = 3$ and common ratio $r = \frac{2}{3}$. Since $|r| = \frac{2}{3} < 1$, the

series converges to $\frac{a}{1-r} = \frac{3}{1-2/3} = \frac{3}{1/3} = 9$.

13. $-2 + \frac{5}{2} - \frac{25}{8} + \frac{125}{32} - \cdots$ is a geometric series with $a = -2$ and $r = \frac{5/2}{-2} = -\frac{5}{4}$. Since $|r| = \frac{5}{4} > 1$, the series

diverges by (4).

15. $\sum_{n=1}^{\infty} 5 \left(\frac{2}{3} \right)^{n-1}$ is a geometric series with $a = 5$ and $r = \frac{2}{3}$. Since $|r| = \frac{2}{3} < 1$, the series converges to

$\frac{a}{1-r} = \frac{5}{1-2/3} = \frac{5}{1/3} = 15$.

17. $\sum_{n=1}^{\infty} \frac{(-3)^{n-1}}{4^n} = \frac{1}{4} \sum_{n=1}^{\infty} \left(-\frac{3}{4} \right)^{n-1}$. The latter series is geometric with $a = 1$ and $r = -\frac{3}{4}$. Since $|r| = \frac{3}{4} < 1$, it

converges to $\frac{1}{1-(-3/4)} = \frac{4}{7}$. Thus, the given series converges to $\left(\frac{1}{4} \right) \left(\frac{4}{7} \right) = \frac{1}{7}$.

19. $\sum_{n=0}^{\infty} \frac{\pi^n}{3^{n+1}} = \frac{1}{3} \sum_{n=0}^{\infty} \left(\frac{\pi}{3} \right)^n$ is a geometric series with ratio $r = \frac{\pi}{3}$. Since $|r| > 1$, the series diverges.

21. $\sum\limits_{n=1}^{\infty} \dfrac{n}{n+5}$ diverges since $\lim\limits_{n\to\infty} a_n = \lim\limits_{n\to\infty} \dfrac{n}{n+5} = 1 \neq 0$. [Use (7), the Test for Divergence.]

23. Using partial fractions, the partial sums are

$$s_n = \sum_{i=2}^{n} \frac{2}{(i-1)(i+1)} = \sum_{i=2}^{n} \left(\frac{1}{i-1} - \frac{1}{i+1} \right)$$

$$= \left(1 - \frac{1}{3}\right) + \left(\frac{1}{2} - \frac{1}{4}\right) + \left(\frac{1}{3} - \frac{1}{5}\right) + \cdots + \left(\frac{1}{n-3} - \frac{1}{n-1}\right) + \left(\frac{1}{n-2} - \frac{1}{n}\right)$$

This sum is a telescoping series and $s_n = 1 + \dfrac{1}{2} - \dfrac{1}{n-1} - \dfrac{1}{n}$.

Thus, $\sum\limits_{n=2}^{\infty} \dfrac{2}{n^2-1} = \lim\limits_{n\to\infty} \left(1 + \dfrac{1}{2} - \dfrac{1}{n-1} - \dfrac{1}{n}\right) = \dfrac{3}{2}$.

25. $\sum\limits_{k=2}^{\infty} \dfrac{k^2}{k^2-1}$ diverges by the Test for Divergence since $\lim\limits_{k\to\infty} a_k = \lim\limits_{k\to\infty} \dfrac{k^2}{k^2-1} = 1 \neq 0$.

27. Converges. $\sum\limits_{n=1}^{\infty} \dfrac{3^n + 2^n}{6^n} = \sum\limits_{n=1}^{\infty} \left(\dfrac{3^n}{6^n} + \dfrac{2^n}{6^n} \right) = \sum\limits_{n=1}^{\infty} \left[\left(\dfrac{1}{2}\right)^n + \left(\dfrac{1}{3}\right)^n \right] = \dfrac{1/2}{1-1/2} + \dfrac{1/3}{1-1/3} = 1 + \dfrac{1}{2} = \dfrac{3}{2}$

29. $\sum\limits_{n=1}^{\infty} \sqrt[n]{2} = 2 + \sqrt{2} + \sqrt[3]{2} + \sqrt[4]{2} + \cdots$ diverges by the Test for Divergence since

$\lim\limits_{n\to\infty} a_n = \lim\limits_{n\to\infty} \sqrt[n]{2} = \lim\limits_{n\to\infty} 2^{1/n} = 2^0 = 1 \neq 0$.

31. $\lim\limits_{n\to\infty} a_n = \lim\limits_{n\to\infty} \arctan n = \dfrac{\pi}{2} \neq 0$, so the series diverges by the Test for Divergence.

33. The first series is a telescoping sum:

$$\sum_{n=1}^{\infty} \frac{3}{n(n+3)} = \sum_{n=1}^{\infty} \left(\frac{1}{n} - \frac{1}{n+3} \right) = \sum_{n=1}^{\infty} \left(\frac{1}{n} - \frac{1}{n+1} + \frac{1}{n+1} - \frac{1}{n+2} + \frac{1}{n+2} - \frac{1}{n+3} \right)$$

$$= \sum_{n=1}^{\infty} \left(\frac{1}{n} - \frac{1}{n+1} \right) + \sum_{n=1}^{\infty} \left(\frac{1}{n+1} - \frac{1}{n+2} \right) + \sum_{n=1}^{\infty} \left(\frac{1}{n+2} - \frac{1}{n+3} \right)$$

$$= 1 + \frac{1}{2} + \frac{1}{3} = \frac{11}{6}$$

The second series is geometric with first term $\dfrac{5}{4}$ and ratio $\dfrac{1}{4}$: $\sum\limits_{n=1}^{\infty} \dfrac{5}{4^n} = \dfrac{5/4}{1-1/4} = \dfrac{5}{3}$. Thus,

$$\sum_{n=1}^{\infty} \left(\frac{3}{n(n+3)} + \frac{5}{4^n} \right) = \sum_{n=1}^{\infty} \frac{3}{n(n+3)} + \sum_{n=1}^{\infty} \frac{5}{4^n} \text{ [sum of two convergent series]} = \frac{11}{6} + \frac{5}{3} = \frac{7}{2}.$$

35. $0.\overline{2} = \dfrac{2}{10} + \dfrac{2}{10^2} + \cdots$ is a geometric series with $a = \dfrac{2}{10}$ and $r = \dfrac{1}{10}$. It converges to $\dfrac{a}{1-r} = \dfrac{2/10}{1-1/10} = \dfrac{2}{9}$.

37. $3.\overline{417} = 3 + \dfrac{417}{10^3} + \dfrac{417}{10^6} + \cdots = 3 + \dfrac{417/10^3}{1 - 1/10^3} = 3 + \dfrac{417}{999} = \dfrac{3414}{999} = \dfrac{1138}{333}$

39. $0.12\overline{3456} = \dfrac{123}{1000} + \dfrac{0.000456}{1 - 0.001} = \dfrac{123}{1000} + \dfrac{456}{999,000} = \dfrac{123,333}{999,000} = \dfrac{41,111}{333,000}$

41. $\displaystyle\sum_{n=1}^{\infty} \dfrac{x^n}{3^n} = \sum_{n=1}^{\infty} \left(\dfrac{x}{3}\right)^n$ is a geometric series with $r = \dfrac{x}{3}$, so the series converges $\Leftrightarrow$ $|r| < 1$ $\Leftrightarrow$ $\dfrac{|x|}{3} < 1$ $\Leftrightarrow$

$|x| < 3$; that is, $-3 < x < 3$. In that case, the sum of the series is $\dfrac{a}{1 - r} = \dfrac{x/3}{1 - x/3} = \dfrac{x/3}{1 - x/3} \cdot \dfrac{3}{3} = \dfrac{x}{3 - x}$.

43. $\sum_{n=0}^{\infty} 4^n x^n = \sum_{n=0}^{\infty} (4x)^n$ is a geometric series with $r = 4x$, so the series converges $\Leftrightarrow$ $|r| < 1$ $\Leftrightarrow$

$4|x| < 1$ $\Leftrightarrow$ $|x| < \frac{1}{4}$. In that case, the sum of the series is $\dfrac{1}{1 - 4x}$.

45. $\displaystyle\sum_{n=0}^{\infty} \dfrac{\cos^n x}{2^n}$ is a geometric series with first term 1 and ratio $r = \dfrac{\cos x}{2}$, so it converges $\Leftrightarrow$ $|r| < 1$. But

$|r| = \dfrac{|\cos x|}{2} \leq \dfrac{1}{2}$ for all x. Thus, the series converges for all real values of x and the sum of the series is

$\dfrac{1}{1 - (\cos x)/2} = \dfrac{2}{2 - \cos x}$.

47. After defining f, We use `convert(f,parfrac);` in Maple, `Apart` in Mathematica, or `Expand Rational`

and `Simplify` in Derive to find that the general term is $\dfrac{1}{(4n + 1)(4n - 3)} = -\dfrac{1/4}{4n + 1} + \dfrac{1/4}{4n - 3}$. So the

nth partial sum is

$$s_n = \sum_{k=1}^{n} \left(-\dfrac{1/4}{4k + 1} + \dfrac{1/4}{4k - 3}\right) = \dfrac{1}{4} \sum_{k=1}^{n} \left(\dfrac{1}{4k - 3} - \dfrac{1}{4k + 1}\right)$$

$$= \dfrac{1}{4}\left[\left(1 - \dfrac{1}{5}\right) + \left(\dfrac{1}{5} - \dfrac{1}{9}\right) + \left(\dfrac{1}{9} - \dfrac{1}{13}\right) + \cdots + \left(\dfrac{1}{4n - 3} - \dfrac{1}{4n + 1}\right)\right] = \dfrac{1}{4}\left(1 - \dfrac{1}{4n + 1}\right)$$

The series converges to $\lim\limits_{n \to \infty} s_n = \frac{1}{4}$. This can be confirmed by directly computing the sum using

`sum(f,1..infinity);` (in Maple), `Sum[f,{n,1,Infinity}]` (in Mathematica), or `Calculus Sum`

(from 1 to ∞) and `Simplify` (in Derive).

49. For $n = 1$, $a_1 = 0$ since $s_1 = 0$. For $n > 1$,

$$a_n = s_n - s_{n-1} = \dfrac{n - 1}{n + 1} - \dfrac{(n - 1) - 1}{(n - 1) + 1} = \dfrac{(n - 1)n - (n + 1)(n - 2)}{(n + 1)n} = \dfrac{2}{n(n + 1)}$$

Also, $\displaystyle\sum_{n=1}^{\infty} a_n = \lim_{n \to \infty} s_n = \lim_{n \to \infty} \dfrac{1 - 1/n}{1 + 1/n} = 1$.

51. (a) The first step in the chain occurs when the local government spends D dollars. The people who receive it spend

a fraction c of those D dollars, that is, Dc dollars. Those who receive the Dc dollars spend a fraction c of it, that

is, Dc^2 dollars. Continuing in this way, we see that the total spending after n transactions is

$$S_n = D + Dc + Dc^2 + \cdots + Dc^{n-1} = \frac{D(1 - c^n)}{1 - c} \text{ by (3).}$$

(b) $\lim_{n \to \infty} S_n = \lim_{n \to \infty} \frac{D(1 - c^n)}{1 - c} = \frac{D}{1 - c} \lim_{n \to \infty} (1 - c^n) = \frac{D}{1 - c} \text{ (since } 0 < c < 1 \Rightarrow \lim_{n \to \infty} c^n = 0)$

$= \frac{D}{s} \text{ (since } c + s = 1) = kD \text{ (since } k = 1/s)$

If $c = 0.8$, then $s = 1 - c = 0.2$ and the multiplier is $k = 1/s = 5$.

53. $\sum_{n=2}^{\infty} (1 + c)^{-n}$ is a geometric series with $a = (1 + c)^{-2}$ and $r = (1 + c)^{-1}$, so the series converges when

$\left|(1 + c)^{-1}\right| < 1 \Leftrightarrow |1 + c| > 1 \Leftrightarrow 1 + c > 1 \text{ or } 1 + c < -1 \Leftrightarrow c > 0 \text{ or } c < -2.$ We calculate

the sum of the series and set it equal to 2: $\dfrac{(1 + c)^{-2}}{1 - (1 + c)^{-1}} = 2 \Leftrightarrow \left(\dfrac{1}{1 + c}\right)^2 = 2 - 2\left(\dfrac{1}{1 + c}\right) \Leftrightarrow$

$1 = 2(1 + c)^2 - 2(1 + c) \Leftrightarrow 2c^2 + 2c - 1 = 0 \Leftrightarrow c = \dfrac{-2 \pm \sqrt{12}}{4} = \dfrac{\pm\sqrt{3} - 1}{2}.$ However, the negative root is

inadmissible because $-2 < \dfrac{-\sqrt{3} - 1}{2} < 0.$ So $c = \dfrac{\sqrt{3} - 1}{2}.$

55. Let d_n be the diameter of C_n. We draw lines from the centers of

the C_i to the center of D (or C), and using the Pythagorean

Theorem, we can write $1^2 + \left(1 - \frac{1}{2}d_1\right)^2 = \left(1 + \frac{1}{2}d_1\right)^2 \Leftrightarrow$

$1 = \left(1 + \frac{1}{2}d_1\right)^2 - \left(1 - \frac{1}{2}d_1\right)^2 = 2d_1 \text{ (difference of squares)}$

$\Rightarrow d_1 = \frac{1}{2}.$ Similarly,

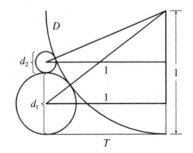

$1 = \left(1 + \frac{1}{2}d_2\right)^2 - \left(1 - d_1 - \frac{1}{2}d_2\right)^2 = 2d_2 + 2d_1 - d_1^2 - d_1 d_2$

$= (2 - d_1)(d_1 + d_2) \Leftrightarrow$

$d_2 = \dfrac{1}{2 - d_1} - d_1 = \dfrac{(1 - d_1)^2}{2 - d_1}, 1 = \left(1 + \frac{1}{2}d_3\right)^2 - \left(1 - d_1 - d_2 - \frac{1}{2}d_3\right)^2 \Leftrightarrow d_3 = \dfrac{[1 - (d_1 + d_2)]^2}{2 - (d_1 + d_2)},$ and

in general, $d_{n+1} = \dfrac{\left(1 - \sum_{i=1}^{n} d_i\right)^2}{2 - \sum_{i=1}^{n} d_i}.$ If we actually calculate d_2 and d_3 from the formulas above, we find that they

are $\dfrac{1}{6} = \dfrac{1}{2 \cdot 3}$ and $\dfrac{1}{12} = \dfrac{1}{3 \cdot 4}$ respectively, so we suspect that in general, $d_n = \dfrac{1}{n(n + 1)}.$ To prove this, we use

induction: Assume that for all $k \leq n$, $d_k = \dfrac{1}{k(k + 1)} = \dfrac{1}{k} - \dfrac{1}{k + 1}.$ Then

$$\sum_{i=1}^{n} d_i = 1 - \frac{1}{n+1} = \frac{n}{n+1} \quad \text{(telescoping sum). Substituting this into our formula for } d_{n+1}, \text{ we get}$$

$$d_{n+1} = \frac{\left[1 - \dfrac{n}{n+1}\right]^2}{2 - \left(\dfrac{n}{n+1}\right)} = \frac{\dfrac{1}{(n+1)^2}}{\dfrac{n+2}{n+1}} = \frac{1}{(n+1)(n+2)}, \text{ and the induction is complete.}$$

Now, we observe that the partial sums $\sum_{i=1}^{n} d_i$ of the diameters of the circles approach 1 as $n \to \infty$; that is,

$$\sum_{n=1}^{\infty} a_n = \sum_{n=1}^{\infty} \frac{1}{n(n+1)} = 1, \text{ which is what we wanted to prove.}$$

57. The series $1 - 1 + 1 - 1 + 1 - 1 + \cdots$ diverges (geometric series with $r = -1$) so we cannot say that

$0 = 1 - 1 + 1 - 1 + 1 - 1 + \cdots$.

59. $\sum_{n=1}^{\infty} ca_n = \lim\limits_{n \to \infty} \sum_{i=1}^{n} ca_i = \lim\limits_{n \to \infty} c \sum_{i=1}^{n} a_i = c \lim\limits_{n \to \infty} \sum_{i=1}^{n} a_i = c \sum_{n=1}^{\infty} a_n$, which exists by hypothesis.

61. Suppose on the contrary that $\sum(a_n + b_n)$ converges. Then $\sum(a_n + b_n)$ and $\sum a_n$ are convergent series. So by

Theorem 8, $\sum[(a_n + b_n) - a_n]$ would also be convergent. But $\sum[(a_n + b_n) - a_n] = \sum b_n$, a contradiction,

since $\sum b_n$ is given to be divergent.

63. The partial sums $\{s_n\}$ form an increasing sequence, since $s_n - s_{n-1} = a_n > 0$ for all n. Also, the sequence $\{s_n\}$

is bounded since $s_n \le 1000$ for all n. So by Theorem 12.1.11 [ET 11.1.11], the sequence of partial sums

converges, that is, the series $\sum a_n$ is convergent.

65. (a) At the first step, only the interval $\left(\frac{1}{3}, \frac{2}{3}\right)$ (length $\frac{1}{3}$) is removed. At the second step, we remove the intervals

$\left(\frac{1}{9}, \frac{2}{9}\right)$ and $\left(\frac{7}{9}, \frac{8}{9}\right)$, which have a total length of $2 \cdot \left(\frac{1}{3}\right)^2$. At the third step, we remove 2^2 intervals, each of

length $\left(\frac{1}{3}\right)^3$. In general, at the nth step we remove 2^{n-1} intervals, each of length $\left(\frac{1}{3}\right)^n$, for a length of

$2^{n-1} \cdot \left(\frac{1}{3}\right)^n = \frac{1}{3}\left(\frac{2}{3}\right)^{n-1}$. Thus, the total length of all removed intervals is $\sum\limits_{n=1}^{\infty} \frac{1}{3}\left(\frac{2}{3}\right)^{n-1} = \frac{1/3}{1 - 2/3} = 1$

(geometric series with $a = \frac{1}{3}$ and $r = \frac{2}{3}$). Notice that at the nth step, the leftmost interval that is removed is

$\left(\left(\frac{1}{3}\right)^n, \left(\frac{2}{3}\right)^n\right)$, so we never remove 0, and 0 is in the Cantor set. Also, the rightmost interval removed is

$\left(1 - \left(\frac{2}{3}\right)^n, 1 - \left(\frac{1}{3}\right)^n\right)$, so 1 is never removed. Some other numbers in the Cantor set are $\frac{1}{3}$, $\frac{2}{3}$, $\frac{1}{9}$, $\frac{2}{9}$, $\frac{7}{9}$, and $\frac{8}{9}$.

(b) The area removed at the first step is $\frac{1}{9}$; at the second step, $8 \cdot \left(\frac{1}{9}\right)^2$; at the third step, $(8)^2 \cdot \left(\frac{1}{9}\right)^3$. In general, the

area removed at the nth step is $(8)^{n-1}\left(\frac{1}{9}\right)^n = \frac{1}{9}\left(\frac{8}{9}\right)^{n-1}$, so the total area of all removed squares is

$$\sum_{n=1}^{\infty} \frac{1}{9}\left(\frac{8}{9}\right)^{n-1} = \frac{1/9}{1 - 8/9} = 1.$$

67. (a) For $\displaystyle\sum_{n=1}^{\infty} \frac{n}{(n+1)!}$, $s_1 = \dfrac{1}{1 \cdot 2} = \dfrac{1}{2}$, $s_2 = \dfrac{1}{2} + \dfrac{2}{1 \cdot 2 \cdot 3} = \dfrac{5}{6}$, $s_3 = \dfrac{5}{6} + \dfrac{3}{1 \cdot 2 \cdot 3 \cdot 4} = \dfrac{23}{24}$,

$s_4 = \dfrac{23}{24} + \dfrac{4}{1 \cdot 2 \cdot 3 \cdot 4 \cdot 5} = \dfrac{119}{120}$. The denominators are $(n+1)!$, so a guess would be $s_n = \dfrac{(n+1)! - 1}{(n+1)!}$.

(b) For $n = 1$, $s_1 = \dfrac{1}{2} = \dfrac{2! - 1}{2!}$, so the formula holds for $n = 1$. Assume $s_k = \dfrac{(k+1)! - 1}{(k+1)!}$. Then

$$s_{k+1} = \frac{(k+1)! - 1}{(k+1)!} + \frac{k+1}{(k+2)!} = \frac{(k+1)! - 1}{(k+1)!} + \frac{k+1}{(k+1)!(k+2)}$$

$$= \frac{(k+2)! - (k+2) + k + 1}{(k+2)!} = \frac{(k+2)! - 1}{(k+2)!}$$

Thus, the formula is true for $n = k + 1$. So by induction, the guess is correct.

(c) $\displaystyle\lim_{n \to \infty} s_n = \lim_{n \to \infty} \frac{(n+1)! - 1}{(n+1)!} = \lim_{n \to \infty} \left[1 - \frac{1}{(n+1)!} \right] = 1$ and so $\displaystyle\sum_{n=1}^{\infty} \frac{n}{(n+1)!} = 1$.

12.3 The Integral Test and Estimates of Sums $\qquad$ ET 11.3

1. The picture shows that $a_2 = \dfrac{1}{2^{1.3}} < \displaystyle\int_1^2 \frac{1}{x^{1.3}}\, dx$,

$a_3 = \dfrac{1}{3^{1.3}} < \displaystyle\int_2^3 \frac{1}{x^{1.3}}\, dx$, and so on, so $\displaystyle\sum_{n=2}^{\infty} \frac{1}{n^{1.3}} < \int_1^{\infty} \frac{1}{x^{1.3}}\, dx$. The

integral converges by (8.8.2) [ET (7.8.2)] with $p = 1.3 > 1$, so the series
converges.

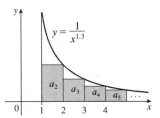

3. The function $f(x) = 1/x^4$ is continuous, positive, and decreasing on $[1, \infty)$, so the Integral Test applies.

$$\int_1^{\infty} \frac{1}{x^4}\, dx = \lim_{t \to \infty} \int_1^t x^{-4}\, dx = \lim_{t \to \infty} \left[\frac{x^{-3}}{-3} \right]_1^t = \lim_{t \to \infty} \left(-\frac{1}{3t^3} + \frac{1}{3} \right) = \frac{1}{3}.$$ Since this improper integral is

convergent, the series $\displaystyle\sum_{n=1}^{\infty} \frac{1}{n^4}$ is also convergent by the Integral Test.

5. The function $f(x) = 1/(3x + 1)$ is continuous, positive, and decreasing on $[1, \infty)$, so the Integral Test applies.

$$\int_1^{\infty} \frac{dx}{3x + 1} = \lim_{b \to \infty} \int_1^b \frac{dx}{3x + 1} = \lim_{b \to \infty} \left[\tfrac{1}{3} \ln(3x + 1) \right]_1^b = \lim_{b \to \infty} \left[\tfrac{1}{3} \ln(3b + 1) - \tfrac{1}{3} \ln 4 \right] = \infty$$

so the improper integral diverges, and so does the series $\sum_{n=1}^{\infty} 1/(3n + 1)$.

7. $f(x) = xe^{-x}$ is continuous and positive on $[1, \infty)$. $f'(x) = -xe^{-x} + e^{-x} = e^{-x}(1 - x) < 0$ for $x > 1$, so f is
decreasing on $[1, \infty)$. Thus, the Integral Test applies.

$$\int_1^{\infty} xe^{-x}\, dx = \lim_{b \to \infty} \int_1^b xe^{-x}\, dx = \lim_{b \to \infty} \left[-xe^{-x} - e^{-x} \right]_1^b \quad \text{(by parts)}$$

$$= \lim_{b \to \infty} \left[-be^{-b} - e^{-b} + e^{-1} + e^{-1} \right] = 2/e$$

since $\displaystyle\lim_{b \to \infty} be^{-b} = \lim_{b \to \infty} (b/e^b) \overset{\text{H}}{=} \lim_{b \to \infty} (1/e^b) = 0$ and $\displaystyle\lim_{b \to \infty} e^{-b} = 0$. Thus, $\sum_{n=1}^{\infty} ne^{-n}$ converges.

9. The series $\sum\limits_{n=1}^{\infty} \dfrac{1}{n^{0.85}}$ is a p-series with $p = 0.85 \leq 1$, so it diverges by (1). Therefore, the series $\sum\limits_{n=1}^{\infty} \dfrac{2}{n^{0.85}}$ must

also diverge, for if it converged, then $\sum\limits_{n=1}^{\infty} \dfrac{1}{n^{0.85}}$ would have to converge (by Theorem 8(i) in Section 12.2

[ET 11.2]).

11. $1 + \dfrac{1}{8} + \dfrac{1}{27} + \dfrac{1}{64} + \dfrac{1}{125} + \cdots = \sum\limits_{n=1}^{\infty} \dfrac{1}{n^3}$. This is a p-series with $p = 3 > 1$, so it converges by (1).

13. $\sum\limits_{n=1}^{\infty} \dfrac{5 - 2\sqrt{n}}{n^3} = 5 \sum\limits_{n=1}^{\infty} \dfrac{1}{n^3} - 2 \sum\limits_{n=1}^{\infty} \dfrac{1}{n^{5/2}}$ by Theorem 12.2.8 [ET 11.2.8], since $\sum\limits_{n=1}^{\infty} \dfrac{1}{n^3}$ and $\sum\limits_{n=1}^{\infty} \dfrac{1}{n^{5/2}}$ both

converge by (1) (with $p = 3 > 1$ and $p = \frac{5}{2} > 1$). Thus, $\sum\limits_{n=1}^{\infty} \dfrac{5 - 2\sqrt{n}}{n^3}$ converges.

15. The function $f(x) = \dfrac{1}{x^2 + 4}$ is continuous, positive, and decreasing on $[1, \infty)$, so we can apply the Integral Test.

$$\int_1^{\infty} \frac{1}{x^2 + 4} \, dx = \lim_{t \to \infty} \int_1^t \frac{1}{x^2 + 4} \, dx = \lim_{t \to \infty} \left[\frac{1}{2} \tan^{-1} \frac{x}{2} \right]_1^t = \frac{1}{2} \lim_{t \to \infty} \left[\tan^{-1} \left(\frac{t}{2} \right) - \tan^{-1} \left(\frac{1}{2} \right) \right]$$

$$= \frac{1}{2} \left[\frac{\pi}{2} - \tan^{-1} \left(\frac{1}{2} \right) \right]$$

Therefore, the series $\sum\limits_{n=1}^{\infty} \dfrac{1}{n^2 + 4}$ converges.

17. $f(x) = \dfrac{x}{x^2 + 1}$ is continuous and positive on $[1, \infty)$, and since

$$f'(x) = \frac{1 - x^2}{(x^2 + 1)^2} < 0 \text{ for } x > 1, \, f \text{ is also decreasing. Using the Integral Test,}$$

$$\int_1^{\infty} \frac{x}{x^2 + 1} \, dx = \lim_{t \to \infty} \int_1^t \frac{x}{x^2 + 1} \, dx = \lim_{t \to \infty} \left[\frac{\ln(x^2 + 1)}{2} \right]_1^t = \frac{1}{2} \lim_{t \to \infty} \left[\ln(t^2 + 1) - \ln 2 \right] = \infty, \text{ so the series}$$

diverges.

19. $f(x) = xe^{-x^2}$ is continuous and positive on $[1, \infty)$, and since $f'(x) = e^{-x^2} \left(1 - 2x^2 \right) < 0$ for

$x > 1$, f is decreasing as well. Thus, we can use the Integral Test.

$\int_1^{\infty} xe^{-x^2} \, dx = \lim\limits_{t \to \infty} \left[-\frac{1}{2} e^{-x^2} \right]_1^t = 0 - \left(-\frac{1}{2} e^{-1} \right) = 1/(2e)$. Since the integral converges, the series converges.

21. $f(x) = \dfrac{1}{x \ln x}$ is continuous and positive on $[2, \infty)$, and also decreasing since $f'(x) = -\dfrac{1 + \ln x}{x^2 (\ln x)^2} < 0$ for $x > 2$,

so we can use the Integral Test. $\int_2^{\infty} \dfrac{1}{x \ln x} \, dx = \lim\limits_{t \to \infty} \left[\ln(\ln x) \right]_2^t = \lim\limits_{t \to \infty} \left[\ln(\ln t) - \ln(\ln 2) \right] = \infty$, so the series

diverges.

23. The function $f(x) = \dfrac{1}{x^3 + x}$ is continuous, positive, and decreasing on $[1, \infty)$, so the Integral Test applies. We use

partial fractions to evaluate the integral:

$$\int_1^\infty \frac{1}{x^3 + x}\, dx = \lim_{t \to \infty} \int_1^t \left[\frac{1}{x} - \frac{x}{1 + x^2} \right] dx = \lim_{t \to \infty} \left[\ln x - \frac{1}{2} \ln(1 + x^2) \right]_1^t$$

$$= \lim_{t \to \infty} \left[\ln \frac{x}{\sqrt{1 + x^2}} \right]_1^t = \lim_{t \to \infty} \left(\ln \frac{t}{\sqrt{1 + t^2}} - \ln \frac{1}{\sqrt{2}} \right)$$

$$= \lim_{t \to \infty} \left(\ln \frac{1}{\sqrt{1 + 1/t^2}} + \frac{1}{2} \ln 2 \right) = \frac{1}{2} \ln 2$$

so the series $\displaystyle\sum_{n=1}^\infty \frac{1}{n^3 + n}$ converges.

25. We have already shown (in Exercise 21) that when $p = 1$ the series $\displaystyle\sum_{n=2}^\infty \frac{1}{n(\ln n)^p}$ diverges, so assume that $p \neq 1$.

$f(x) = \dfrac{1}{x(\ln x)^p}$ is continuous and positive on $[2, \infty)$, and $f'(x) = -\dfrac{p + \ln x}{x^2(\ln x)^{p+1}} < 0$ if $x > e^{-p}$, so that f is

eventually decreasing and we can use the Integral Test.

$$\int_2^\infty \frac{1}{x(\ln x)^p}\, dx = \lim_{t \to \infty} \left[\frac{(\ln x)^{1-p}}{1 - p} \right]_2^t \quad \text{(for } p \neq 1) \quad = \lim_{t \to \infty} \left[\frac{(\ln t)^{1-p}}{1 - p} \right] - \frac{(\ln 2)^{1-p}}{1 - p}$$

This limit exists whenever $1 - p < 0 \quad \Leftrightarrow \quad p > 1$, so the series converges for $p > 1$.

27. Clearly the series cannot converge if $p \geq -\frac{1}{2}$, because then $\displaystyle\lim_{n \to \infty} n(1 + n^2)^p \neq 0$. Also, if $p = -1$ the series

diverges (see Exercise 17). So assume $p < -\frac{1}{2}, p \neq -1$. Then $f(x) = x(1 + x^2)^p$ is continuous,

positive, and eventually decreasing on $[1, \infty)$, and we can use the Integral Test.

$$\int_1^\infty x(1 + x^2)^p\, dx = \lim_{t \to \infty} \left[\frac{1}{2} \cdot \frac{(1 + x^2)^{p+1}}{p + 1} \right]_1^t = \lim_{t \to \infty} \frac{1}{2} \cdot \frac{(1 + t^2)^{p+1}}{p + 1} - \frac{2^p}{p + 1}. \text{ This limit exists and is finite}$$

$\Leftrightarrow \quad p + 1 < 0 \quad \Leftrightarrow \quad p < -1$, so the series converges whenever $p < -1$.

29. Since this is a p-series with $p = x$, $\zeta(x)$ is defined when $x > 1$. Unless specified otherwise, the domain of a

function f is the set of numbers x such that the expression for $f(x)$ makes sense and defines a real number. So, in

the case of a series, it's the set of numbers x such that the series is convergent.

31. (a) $f(x) = \dfrac{1}{x^2}$ is positive and continuous and $f'(x) = -\dfrac{2}{x^3}$ is negative for $x > 0$, and so the Integral

Test applies. $\displaystyle\sum_{n=1}^\infty \frac{1}{n^2} \approx s_{10} = \frac{1}{1^2} + \frac{1}{2^2} + \frac{1}{3^2} + \cdots + \frac{1}{10^2} \approx 1.549768.$

$R_{10} \leq \displaystyle\int_{10}^\infty \frac{1}{x^2}\, dx = \lim_{t \to \infty} \left[\frac{-1}{x} \right]_{10}^t = \lim_{t \to \infty} \left(-\frac{1}{t} + \frac{1}{10} \right) = \frac{1}{10}$, so the error is at most 0.1.

(b) $s_{10} + \int_{11}^{\infty} \frac{1}{x^2}\, dx \leq s \leq s_{10} + \int_{10}^{\infty} \frac{1}{x^2}\, dx \quad \Rightarrow \quad s_{10} + \frac{1}{11} \leq s \leq s_{10} + \frac{1}{10} \quad \Rightarrow$

$1.549768 + 0.090909 = 1.640677 \leq s \leq 1.549768 + 0.1 = 1.649768$, so we get $s \approx 1.64522$ (the average of 1.640677 and 1.649768) with error ≤ 0.005 (the maximum of $1.649768 - 1.64522$ and $1.64522 - 1.640677$, rounded up).

(c) $R_n \leq \int_n^{\infty} \frac{1}{x^2}\, dx = \frac{1}{n}$. So $R_n < 0.001$ if $\frac{1}{n} < \frac{1}{1000} \quad \Leftrightarrow \quad n > 1000.$

33. $f(x) = x^{-3/2}$ is positive and continuous and $f'(x) = -\frac{3}{2} x^{-5/2}$ is negative for $x > 0$, so the Integral Test applies. From the end of Example 6, we see that the error is at most half the length of the interval. From (3), the interval is $\left(s_n + \int_{n+1}^{\infty} f(x)\, dx, \; s_n + \int_n^{\infty} f(x)\, dx \right)$, so its length is $\int_n^{\infty} f(x)\, dx - \int_{n+1}^{\infty} f(x)\, dx = \int_n^{n+1} f(x)\, dx$. Thus, we need n such that

$$0.01 > \frac{1}{2} \int_n^{n+1} x^{-3/2}\, dx = \frac{1}{2}\left[\frac{-2}{\sqrt{x}}\right]_n^{n+1} = \frac{1}{\sqrt{n}} - \frac{1}{\sqrt{n+1}}$$

$\Leftrightarrow \quad n > 13.08$ (use a graphing calculator to solve $1/\sqrt{x} - 1/\sqrt{x+1} < 0.01$). Again from the end of Example 6, we approximate s by the midpoint of this interval. In general, the midpoint is

$\frac{1}{2}\left[\left(s_n + \int_{n+1}^{\infty} f(x)\, dx\right) + \left(s_n + \int_n^{\infty} f(x)\, dx\right)\right] = s_n + \frac{1}{2}\left(\int_{n+1}^{\infty} f(x)\, dx + \int_n^{\infty} f(x)\, dx\right)$. So using $n = 14$,

we have $s \approx s_{14} + \frac{1}{2}\left(\int_{14}^{\infty} x^{-3/2}\, dx + \int_{15}^{\infty} x^{-3/2}\, dx\right) \approx 2.0872 + \frac{1}{\sqrt{14}} + \frac{1}{\sqrt{15}} \approx 2.6127 \approx 2.61$. Any larger

value of n will also work. For instance, $s \approx s_{30} + \frac{1}{\sqrt{30}} + \frac{1}{\sqrt{31}} \approx 2.6124$.

35. $\sum_{n=1}^{\infty} n^{-1.001} = \sum_{n=1}^{\infty} \frac{1}{n^{1.001}}$ is a convergent p-series with $p = 1.001 > 1$. Using (2), we get

$R_n \leq \int_n^{\infty} x^{-1.001}\, dx = \lim_{t \to \infty}\left[\frac{x^{-0.001}}{-0.001}\right]_n^t = -1000 \lim_{t \to \infty}\left[\frac{1}{x^{0.001}}\right]_n^t = -1000\left(-\frac{1}{n^{0.001}}\right) = \frac{1000}{n^{0.001}}$. We want

$R_n < 0.000\,000\,005 \quad \Leftrightarrow \quad \frac{1000}{n^{0.001}} < 5 \times 10^{-9} \quad \Leftrightarrow \quad n^{0.001} > \frac{1000}{5 \times 10^{-9}} \quad \Leftrightarrow$

$n > \left(2 \times 10^{11}\right)^{1000} = 2^{1000} \times 10^{11,000} \approx 1.07 \times 10^{301} \times 10^{11,000} = 1.07 \times 10^{11,301}.$

37. (a) From the figure, $a_2 + a_3 + \cdots + a_n \leq \int_1^n f(x)\, dx$, so with

$f(x) = \frac{1}{x}, \; \frac{1}{2} + \frac{1}{3} + \frac{1}{4} + \cdots + \frac{1}{n} \leq \int_1^n \frac{1}{x}\, dx = \ln n$. Thus,

$s_n = 1 + \frac{1}{2} + \frac{1}{3} + \frac{1}{4} + \cdots + \frac{1}{n} \leq 1 + \ln n.$

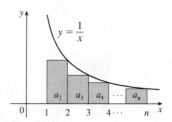

(b) By part (a), $s_{10^6} \leq 1 + \ln 10^6 \approx 14.82 < 15$ and $s_{10^9} \leq 1 + \ln 10^9 \approx 21.72 < 22.$

39. $b^{\ln n} = \left(e^{\ln b}\right)^{\ln n} = \left(e^{\ln n}\right)^{\ln b} = n^{\ln b} = \frac{1}{n^{-\ln b}}$. This is a p-series, which converges for all b such that $-\ln b > 1$

$\Leftrightarrow \quad \ln b < -1 \quad \Leftrightarrow \quad b < e^{-1} \quad \Leftrightarrow \quad b < 1/e$ [with $b > 0$].

12.4 The Comparison Tests

1. (a) We cannot say anything about $\sum a_n$. If $a_n > b_n$ for all n and $\sum b_n$ is convergent, then $\sum a_n$ could be convergent or divergent. (See the note after Example 2.)

(b) If $a_n < b_n$ for all n, then $\sum a_n$ is convergent. [This is part (i) of the Comparison Test.]

3. $\dfrac{1}{n^2 + n + 1} < \dfrac{1}{n^2}$ for all $n \geq 1$, so $\displaystyle\sum_{n=1}^{\infty} \dfrac{1}{n^2 + n + 1}$ converges by comparison with $\displaystyle\sum_{n=1}^{\infty} \dfrac{1}{n^2}$, which converges because it is a p-series with $p = 2 > 1$.

5. $\dfrac{5}{2 + 3^n} < \dfrac{5}{3^n}$ for all $n \geq 1$, so $\displaystyle\sum_{n=1}^{\infty} \dfrac{5}{2 + 3^n}$ converges by comparison with $\displaystyle\sum_{n=1}^{\infty} \dfrac{5}{3^n} = 5 \displaystyle\sum_{n=1}^{\infty} \dfrac{1}{3^n}$, which converges because $\displaystyle\sum_{n=1}^{\infty} \dfrac{1}{3^n}$ is a convergent geometric series with $r = \frac{1}{3}$ ($|r| < 1$).

7. $\dfrac{n+1}{n^2} > \dfrac{n}{n^2} = \dfrac{1}{n}$ for all $n \geq 1$, so $\displaystyle\sum_{n=1}^{\infty} \dfrac{n+1}{n^2}$ diverges by comparison with the harmonic series $\displaystyle\sum_{n=1}^{\infty} \dfrac{1}{n}$.

9. $\dfrac{\cos^2 n}{n^2 + 1} \leq \dfrac{1}{n^2 + 1} < \dfrac{1}{n^2}$, so the series $\displaystyle\sum_{n=1}^{\infty} \dfrac{\cos^2 n}{n^2 + 1}$ converges by comparison with the p-series $\displaystyle\sum_{n=1}^{\infty} \dfrac{1}{n^2}$ ($p = 2 > 1$).

11. If $a_n = \dfrac{n^2 + 1}{n^3 - 1}$ and $b_n = \dfrac{1}{n}$, then $\displaystyle\lim_{n \to \infty} \dfrac{a_n}{b_n} = \lim_{n \to \infty} \dfrac{n^3 + n}{n^3 - 1} = \lim_{n \to \infty} \dfrac{1 + 1/n^2}{1 - 1/n^3} = 1$, so $\displaystyle\sum_{n=2}^{\infty} \dfrac{n^2 + 1}{n^3 - 1}$ diverges by the Limit Comparison Test with the divergent (partial) harmonic series $\displaystyle\sum_{n=2}^{\infty} \dfrac{1}{n}$.

Or: Since $a_n = \dfrac{n^2 + 1}{n^3 - 1} > \dfrac{n^2 + 1}{n^3} > \dfrac{n^2}{n^3} = \dfrac{1}{n} = b_n$, we could use the Comparison Test.

13. $\dfrac{n-1}{n4^n}$ is positive for $n > 1$ and $\dfrac{n-1}{n4^n} < \dfrac{n}{n4^n} = \dfrac{1}{4^n}$, so $\displaystyle\sum_{n=1}^{\infty} \dfrac{n-1}{n4^n}$ converges by comparison with the convergent geometric series $\displaystyle\sum_{n=1}^{\infty} \left(\dfrac{1}{4}\right)^n$.

15. $\dfrac{2 + (-1)^n}{n\sqrt{n}} \leq \dfrac{3}{n\sqrt{n}}$, and $\displaystyle\sum_{n=1}^{\infty} \dfrac{3}{n\sqrt{n}}$ converges because it is a constant multiple of the convergent p-series $\displaystyle\sum_{n=1}^{\infty} \dfrac{1}{n\sqrt{n}}$ ($p = \frac{3}{2} > 1$), so the given series converges by the Comparison Test.

17. Use the Limit Comparison Test with $a_n = \dfrac{1}{\sqrt{n^2 + 1}}$ and $b_n = \dfrac{1}{n}$:

$$\lim_{n \to \infty} \dfrac{a_n}{b_n} = \lim_{n \to \infty} \dfrac{n}{\sqrt{n^2 + 1}} = \lim_{n \to \infty} \dfrac{1}{\sqrt{1 + (1/n^2)}} = 1 > 0.$$ Since the harmonic series $\displaystyle\sum_{n=1}^{\infty} \dfrac{1}{n}$ diverges, so does $\displaystyle\sum_{n=1}^{\infty} \dfrac{1}{\sqrt{n^2 + 1}}$.

19. $\dfrac{2^n}{1 + 3^n} < \dfrac{2^n}{3^n} = \left(\dfrac{2}{3}\right)^n$. $\displaystyle\sum_{n=1}^{\infty} \left(\dfrac{2}{3}\right)^n$ is a convergent geometric series ($|r| = \frac{2}{3} < 1$), so $\displaystyle\sum_{n=1}^{\infty} \dfrac{2^n}{1 + 3^n}$ converges by the Comparison Test.

21. Use the Limit Comparison Test with $a_n = \dfrac{1}{1 + \sqrt{n}}$ and $b_n = \dfrac{1}{\sqrt{n}}$: $\displaystyle\lim_{n\to\infty} \frac{a_n}{b_n} = \lim_{n\to\infty} \frac{\sqrt{n}}{1 + \sqrt{n}} = 1 > 0$. Since

$\displaystyle\sum_{n=1}^{\infty} \frac{1}{\sqrt{n}}$ is a divergent p-series $(p = \frac{1}{2} \le 1)$, $\displaystyle\sum_{n=1}^{\infty} \frac{1}{1 + \sqrt{n}}$ also diverges.

23. Use the Limit Comparison Test with $a_n = \dfrac{5 + 2n}{(1 + n^2)^2}$ and $b_n = \dfrac{1}{n^3}$:

$\displaystyle\lim_{n\to\infty} \frac{a_n}{b_n} = \lim_{n\to\infty} \frac{n^3 (5 + 2n)}{(1 + n^2)^2} = \lim_{n\to\infty} \frac{5n^3 + 2n^4}{(1 + n^2)^2} \cdot \frac{1/n^4}{1/(n^2)^2} = \lim_{n\to\infty} \frac{\frac{5}{n} + 2}{\left(\frac{1}{n^2} + 1\right)^2} = 2 > 0$. Since $\displaystyle\sum_{n=1}^{\infty} \frac{1}{n^3}$ is a

convergent p-series $(p = 3 > 1)$, the series $\displaystyle\sum_{n=1}^{\infty} \frac{5 + 2n}{(1 + n^2)^2}$ also converges.

25. If $a_n = \dfrac{1 + n + n^2}{\sqrt{1 + n^2 + n^6}}$ and $b_n = \dfrac{1}{n}$, then

$\displaystyle\lim_{n\to\infty} \frac{a_n}{b_n} = \lim_{n\to\infty} \frac{n + n^2 + n^3}{\sqrt{1 + n^2 + n^6}} = \lim_{n\to\infty} \frac{1/n^2 + 1/n + 1}{\sqrt{1/n^6 + 1/n^4 + 1}} = 1 > 0$, so $\displaystyle\sum_{n=1}^{\infty} \frac{1 + n + n^2}{\sqrt{1 + n^2 + n^6}}$ diverges by the

Limit Comparison Test with the divergent harmonic series $\displaystyle\sum_{n=1}^{\infty} \frac{1}{n}$.

27. Use the Limit Comparison Test with $a_n = \left(1 + \dfrac{1}{n}\right)^2 e^{-n}$ and $b_n = e^{-n}$: $\displaystyle\lim_{n\to\infty} \frac{a_n}{b_n} = \lim_{n\to\infty} \left(1 + \frac{1}{n}\right)^2 = 1 > 0$.

Since $\displaystyle\sum_{n=1}^{\infty} e^{-n} = \sum_{n=1}^{\infty} \frac{1}{e^n}$ is a convergent geometric series $\left(|r| = \frac{1}{e} < 1\right)$, the series $\displaystyle\sum_{n=1}^{\infty} \left(1 + \frac{1}{n}\right)^2 e^{-n}$ also
converges.

29. Clearly $n! = n(n - 1)(n - 2) \cdots (3)(2) \ge 2 \cdot 2 \cdot 2 \cdot \cdots \cdot 2 \cdot 2 = 2^{n-1}$, so $\dfrac{1}{n!} \le \dfrac{1}{2^{n-1}}$. $\displaystyle\sum_{n=1}^{\infty} \frac{1}{2^{n-1}}$ is a convergent

geometric series $\left(|r| = \frac{1}{2} < 1\right)$, so $\displaystyle\sum_{n=1}^{\infty} \frac{1}{n!}$ converges by the Comparison Test.

31. Use the Limit Comparison Test with $a_n = \sin\left(\dfrac{1}{n}\right)$ and $b_n = \dfrac{1}{n}$. Then $\sum a_n$ and $\sum b_n$ are series with positive

terms and $\displaystyle\lim_{n\to\infty} \frac{a_n}{b_n} = \lim_{n\to\infty} \frac{\sin(1/n)}{1/n} = \lim_{\theta\to 0} \frac{\sin\theta}{\theta} = 1 > 0$. Since $\sum_{n=1}^{\infty} b_n$ is the divergent harmonic series,

$\sum_{n=1}^{\infty} \sin(1/n)$ also diverges. (Note that we could also use l'Hospital's Rule to evaluate the limit:

$\displaystyle\lim_{x\to\infty} \frac{\sin(1/x)}{1/x} \overset{\text{H}}{=} \lim_{x\to\infty} \frac{\cos(1/x) \cdot (-1/x^2)}{-1/x^2} = \lim_{x\to\infty} \cos\frac{1}{x} = \cos 0 = 1$.)

33. $\displaystyle\sum_{n=1}^{10} \frac{1}{n^4 + n^2} = \frac{1}{2} + \frac{1}{20} + \frac{1}{90} + \cdots + \frac{1}{10,100} \approx 0.567975$. Now $\dfrac{1}{n^4 + n^2} < \dfrac{1}{n^4}$, so using the reasoning and

notation of Example 5, the error is $R_{10} \le T_{10} = \displaystyle\sum_{n=11}^{\infty} \frac{1}{n^4} \le \int_{10}^{\infty} \frac{dx}{x^4} = \lim_{t\to\infty} \left[-\frac{x^{-3}}{3}\right]_{10}^{t} = \frac{1}{3000} = 0.000\overline{3}$.

35. $\displaystyle\sum_{n=1}^{10} \frac{1}{1 + 2^n} = \frac{1}{3} + \frac{1}{5} + \frac{1}{9} + \cdots + \frac{1}{1025} \approx 0.76352$. Now $\dfrac{1}{1 + 2^n} < \dfrac{1}{2^n}$, so the error is

$R_{10} \le T_{10} = \displaystyle\sum_{n=11}^{\infty} \frac{1}{2^n} = \frac{1/2^{11}}{1 - 1/2}$ (geometric series) ≈ 0.00098.

37. Since $\dfrac{d_n}{10^n} \le \dfrac{9}{10^n}$ for each n, and since $\displaystyle\sum_{n=1}^{\infty} \dfrac{9}{10^n}$ is a convergent geometric series ($|r| = \frac{1}{10} < 1$),

$0.d_1 d_2 d_3 \ldots = \displaystyle\sum_{n=1}^{\infty} \dfrac{d_n}{10^n}$ will always converge by the Comparison Test.

39. Since $\sum a_n$ converges, $\displaystyle\lim_{n\to\infty} a_n = 0$, so there exists N such that $|a_n - 0| < 1$ for all $n > N$ $\Rightarrow$ $0 \le a_n < 1$

for all $n > N$ $\Rightarrow$ $0 \le a_n^2 \le a_n$. Since $\sum a_n$ converges, so does $\sum a_n^2$ by the Comparison Test.

41. (a) Since $\displaystyle\lim_{n\to\infty} \dfrac{a_n}{b_n} = \infty$, there is an integer N such that $\dfrac{a_n}{b_n} > 1$ whenever $n > N$. (Take $M = 1$ in

Definition 12.1.5 [ET 11.1.5].) Then $a_n > b_n$ whenever $n > N$ and since $\sum b_n$ is divergent, $\sum a_n$ is also

divergent by the Comparison Test.

(b) (i) If $a_n = \dfrac{1}{\ln n}$ and $b_n = \dfrac{1}{n}$ for $n \ge 2$, then

$\displaystyle\lim_{n\to\infty} \dfrac{a_n}{b_n} = \lim_{n\to\infty} \dfrac{n}{\ln n} = \lim_{x\to\infty} \dfrac{x}{\ln x} \overset{\text{H}}{=} \lim_{x\to\infty} \dfrac{1}{1/x} = \lim_{x\to\infty} x = \infty$, so by part (a), $\displaystyle\sum_{n=2}^{\infty} \dfrac{1}{\ln n}$ is divergent.

(ii) If $a_n = \dfrac{\ln n}{n}$ and $b_n = \dfrac{1}{n}$, then $\displaystyle\sum_{n=1}^{\infty} b_n$ is the divergent harmonic series and

$\displaystyle\lim_{n\to\infty} \dfrac{a_n}{b_n} = \lim_{n\to\infty} \ln n = \lim_{x\to\infty} \ln x = \infty$, so $\displaystyle\sum_{n=1}^{\infty} a_n$ diverges by part (a).

43. $\displaystyle\lim_{n\to\infty} n a_n = \lim_{n\to\infty} \dfrac{a_n}{1/n}$, so we apply the Limit Comparison Test with $b_n = \dfrac{1}{n}$. Since $\displaystyle\lim_{n\to\infty} n a_n > 0$ we know that

either both series converge or both series diverge, and we also know that $\displaystyle\sum_{n=0}^{\infty} \dfrac{1}{n}$ diverges (p-series with $p = 1$).

Therefore, $\sum a_n$ must be divergent.

45. Yes. Since $\sum a_n$ is a convergent series with positive terms, $\displaystyle\lim_{n\to\infty} a_n = 0$ by Theorem 12.2.6 [ET 11.2.6], and

$\sum b_n = \sum \sin(a_n)$ is a series with positive terms (for large enough n). We have

$\displaystyle\lim_{n\to\infty} \dfrac{b_n}{a_n} = \lim_{n\to\infty} \dfrac{\sin(a_n)}{a_n} = 1 > 0$ by Theorem 3.5.2 [ET 3.4.2]. Thus, $\sum b_n$ is also convergent by the Limit

Comparison Test.

12.5 Alternating Series   ET 11.5

1. (a) An alternating series is a series whose terms are alternately positive and negative.

(b) An alternating series $\sum_{n=1}^{\infty}(-1)^{n-1}b_n$ converges if $0 < b_{n+1} \le b_n$ for all n and $\displaystyle\lim_{n\to\infty} b_n = 0$. (This is the

Alternating Series Test.)

(c) The error involved in using the partial sum s_n as an approximation to the total sum s is the remainder

$R_n = s - s_n$ and the size of the error is smaller than b_{n+1}; that is, $|R_n| \le b_{n+1}$. (This is the Alternating Series

Estimation Theorem.)

3. $\dfrac{4}{7} - \dfrac{4}{8} + \dfrac{4}{9} - \dfrac{4}{10} + \dfrac{4}{11} - \cdots = \displaystyle\sum_{n=1}^{\infty}(-1)^{n-1}\dfrac{4}{n+6}$. Now $b_n = \dfrac{4}{n+6} > 0$, $\{b_n\}$ is decreasing, and

$\displaystyle\lim_{n\to\infty} b_n = 0$, so the series converges by the Alternating Series Test.

5. $b_n = \dfrac{1}{\sqrt{n}} > 0$, $\{b_n\}$ is decreasing, and $\lim\limits_{n\to\infty} b_n = 0$, so the series $\sum\limits_{n=1}^{\infty} \dfrac{(-1)^{n-1}}{\sqrt{n}}$ converges by the Alternating

Series Test.

7. $\sum\limits_{n=1}^{\infty} a_n = \sum\limits_{n=1}^{\infty} (-1)^n \dfrac{3n-1}{2n+1} = \sum\limits_{n=1}^{\infty} (-1)^n b_n$. Now $\lim\limits_{n\to\infty} b_n = \lim\limits_{n\to\infty} \dfrac{3 - 1/n}{2 + 1/n} = \dfrac{3}{2} \neq 0$. Since $\lim\limits_{n\to\infty} a_n \neq 0$

(in fact the limit does not exist), the series diverges by the Test for Divergence.

9. $b_n = \dfrac{1}{4n^2 + 1} > 0$, $\{b_n\}$ is decreasing, and $\lim\limits_{n\to\infty} b_n = 0$, so the series $\sum\limits_{n=1}^{\infty} \dfrac{(-1)^{n+1}}{4n^2 + 1}$ converges by the Alternating

Series Test.

11. $b_n = \dfrac{n^2}{n^3 + 4} > 0$ for $n \geq 1$. $\{b_n\}$ is decreasing for $n \geq 2$ since

$$\left(\frac{x^2}{x^3 + 4} \right)' = \frac{(x^3 + 4)(2x) - x^2(3x^2)}{(x^3 + 4)^2} = \frac{x(2x^3 + 8 - 3x^3)}{(x^3 + 4)^2} = \frac{x(8 - x^3)}{(x^3 + 4)^2} < 0 \text{ for } x > 2. \text{ Also,}$$

$$\lim_{n\to\infty} b_n = \lim_{n\to\infty} \frac{1/n}{1 + 4/n^3} = 0. \text{ Thus, the series } \sum_{n=1}^{\infty} (-1)^{n+1} \frac{n^2}{n^3 + 4} \text{ converges by the Alternating Series Test.}$$

13. $\sum\limits_{n=2}^{\infty} (-1)^n \dfrac{n}{\ln n}$. $\lim\limits_{n\to\infty} \dfrac{n}{\ln n} = \lim\limits_{x\to\infty} \dfrac{x}{\ln x} \overset{\text{H}}{=} \lim\limits_{x\to\infty} \dfrac{1}{1/x} = \infty$, so the series diverges by the Test for Divergence.

15. $\sum\limits_{n=1}^{\infty} \dfrac{\cos n\pi}{n^{3/4}} = \sum\limits_{n=1}^{\infty} \dfrac{(-1)^n}{n^{3/4}}$. $b_n = \dfrac{1}{n^{3/4}}$ is decreasing and positive and $\lim\limits_{n\to\infty} \dfrac{1}{n^{3/4}} = 0$, so the series converges by

the Alternating Series Test.

17. $\sum\limits_{n=1}^{\infty} (-1)^n \sin \dfrac{\pi}{n}$. $b_n = \sin \dfrac{\pi}{n} > 0$ for $n \geq 2$ and $\sin \dfrac{\pi}{n} \geq \sin \dfrac{\pi}{n+1}$, and $\lim\limits_{n\to\infty} \sin \dfrac{\pi}{n} = \sin 0 = 0$, so the series

converges by the Alternating Series Test.

19. $\dfrac{n^n}{n!} = \dfrac{n \cdot n \cdot \cdots \cdot n}{1 \cdot 2 \cdot \cdots \cdot n} \geq n \;\Rightarrow\; \lim\limits_{n\to\infty} \dfrac{n^n}{n!} = \infty \;\Rightarrow\; \lim\limits_{n\to\infty} \dfrac{(-1)^n n^n}{n!}$ does not exist. So the series diverges by

the Test for Divergence.

21.

n	a_n	s_n
1	1	1
2	−0.35355	0.64645
3	0.19245	0.83890
4	−0.125	0.71390
5	0.08944	0.80334
6	−0.06804	0.73530
7	0.05399	0.78929
8	−0.04419	0.74510
9	0.03704	0.78214
10	−0.03162	0.75051

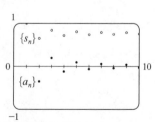

By the Alternating Series Estimation Theorem, the error in the

approximation $\sum\limits_{n=1}^{\infty} \dfrac{(-1)^{n-1}}{n^{3/2}} \approx 0.75051$ is

$|s - s_{10}| \leq b_{11} = 1/(11)^{3/2} \approx 0.0275$ (to four decimal places,

rounded up).

23. The series $\sum_{n=1}^{\infty} (-1)^{n-1} \dfrac{1}{n^2}$ satisfies (i) of the Alternating Series Test because $\dfrac{1}{(n+1)^2} < \dfrac{1}{n^2}$ and

(ii) $\lim_{n\to\infty} \dfrac{1}{n^2} = 0$, so the series is convergent. Now $b_{10} = \dfrac{1}{10^2} = 0.01$ and $b_{11} = \dfrac{1}{11^2} = \dfrac{1}{121} \approx 0.008 < 0.01$, so

by the Alternating Series Estimation Theorem, $n = 10$. (That is, since the 11th term is less than the desired error, we need to add the first 10 terms to get the sum to the desired accuracy.)

25. The series $\sum_{n=1}^{\infty} \dfrac{(-2)^n}{n!} = \sum_{n=1}^{\infty} (-1)^n \dfrac{2^n}{n!}$ satisfies (i) of the Alternating Series Test

because $b_{n+1} = \dfrac{2^{n+1}}{(n+1)!} = \dfrac{2 \cdot 2^n}{(n+1)n!} = \dfrac{2}{n+1} \cdot \dfrac{2^n}{n!} = \dfrac{2}{n+1} \cdot b_n \le b_n$ and (ii)

$\lim_{n\to\infty} \dfrac{2^n}{n!} = \dfrac{2}{n} \cdot \dfrac{2}{n-1} \cdot \ldots \cdot \dfrac{2}{2} \cdot \dfrac{2}{1} = 0$, so the series is convergent. Now $b_7 = 2^7/7! \approx 0.025 > 0.01$ and

$b_8 = 2^8/8! \approx 0.006 < 0.01$, so by the Alternating Series Estimation Theorem, $n = 7$. (That is, since the 8th term is less than the desired error, we need to add the first 7 terms to get the sum to the desired accuracy.)

27. $b_7 = \dfrac{1}{7^5} = \dfrac{1}{16,807} \approx 0.000\,059\,5$, so

$\sum_{n=1}^{\infty} \dfrac{(-1)^{n+1}}{n^5} \approx s_6 = \sum_{n=1}^{6} \dfrac{(-1)^{n+1}}{n^5} = 1 - \frac{1}{32} + \frac{1}{243} - \frac{1}{1024} + \frac{1}{3125} - \frac{1}{7776} \approx 0.972\,080$. Adding b_7 to s_6 does

not change the fourth decimal place of s_6, so the sum of the series, correct to four decimal places, is 0.9721.

29. $b_7 = \dfrac{7^2}{10^7} = 0.000\,004\,9$, so

$\sum_{n=1}^{\infty} \dfrac{(-1)^{n-1}n^2}{10^n} \approx s_6 = \sum_{n=1}^{6} \dfrac{(-1)^{n-1}n^2}{10^n} = \frac{1}{10} - \frac{4}{100} + \frac{9}{1000} - \frac{16}{10,000} + \frac{25}{100,000} - \frac{36}{1,000,000} = 0.067\,614$.

Adding b_7 to s_6 does not change the fourth decimal place of s_6, so the sum of the series, correct to four decimal places, is 0.0676.

31. $\sum_{n=1}^{\infty} \dfrac{(-1)^{n-1}}{n} = 1 - \dfrac{1}{2} + \dfrac{1}{3} - \dfrac{1}{4} + \cdots + \dfrac{1}{49} - \dfrac{1}{50} + \dfrac{1}{51} - \dfrac{1}{52} + \cdots$. The 50th partial sum of this series is an

underestimate, since $\sum_{n=1}^{\infty} \dfrac{(-1)^{n-1}}{n} = s_{50} + \left(\dfrac{1}{51} - \dfrac{1}{52} \right) + \left(\dfrac{1}{53} - \dfrac{1}{54} \right) + \cdots$, and the terms in parentheses are

all positive. The result can be seen geometrically in Figure 1.

33. Clearly $b_n = \dfrac{1}{n+p}$ is decreasing and eventually positive and $\lim_{n\to\infty} b_n = 0$ for any p. So the series converges (by

the Alternating Series Test) for any p for which every b_n is defined, that is, $n + p \ne 0$ for $n \ge 1$, or p is not a negative integer.

35. $\sum b_{2n} = \sum 1/(2n)^2$ clearly converges (by comparison with the p-series for $p = 2$). So suppose that

$\sum (-1)^{n-1} b_n$ converges. Then by Theorem 12.2.8(ii) [ET 11.2.8(ii)], so does

$\sum \left[(-1)^{n-1} b_n + b_n \right] = 2 \left(1 + \frac{1}{3} + \frac{1}{5} + \cdots \right) = 2\sum \dfrac{1}{2n-1}$. But this diverges by comparison with the

harmonic series, a contradiction. Therefore, $\sum (-1)^{n-1} b_n$ must diverge. The Alternating Series Test does not apply since $\{b_n\}$ is not decreasing.

12.6 Absolute Convergence and the Ratio and Root Tests   ET 11.6

1. (a) Since $\lim\limits_{n\to\infty}\left|\dfrac{a_{n+1}}{a_n}\right| = 8 > 1$, part (b) of the Ratio Test tells us that the series $\sum a_n$ is divergent.

   (b) Since $\lim\limits_{n\to\infty}\left|\dfrac{a_{n+1}}{a_n}\right| = 0.8 < 1$, part (a) of the Ratio Test tells us that the series $\sum a_n$ is absolutely convergent
   (and therefore convergent).

   (c) Since $\lim\limits_{n\to\infty}\left|\dfrac{a_{n+1}}{a_n}\right| = 1$, the Ratio Test fails and the series $\sum a_n$ might converge or it might diverge.

3. $\sum\limits_{n=0}^{\infty}\dfrac{(-10)^n}{n!}$. Using the Ratio Test, $\lim\limits_{n\to\infty}\left|\dfrac{a_{n+1}}{a_n}\right| = \lim\limits_{n\to\infty}\left|\dfrac{(-10)^{n+1}}{(n+1)!}\cdot\dfrac{n!}{(-10)^n}\right| = \lim\limits_{n\to\infty}\left|\dfrac{-10}{n+1}\right| = 0 < 1$, so the
series is absolutely convergent.

5. $\sum\limits_{n=1}^{\infty}\dfrac{(-1)^{n+1}}{\sqrt[4]{n}}$ converges by the Alternating Series Test, but $\sum\limits_{n=1}^{\infty}\dfrac{1}{\sqrt[4]{n}}$ is a divergent p-series $\left(p = \tfrac{1}{4} \le 1\right)$, so the
given series is conditionally convergent.

7. $\lim\limits_{n\to\infty}|a_n| = \lim\limits_{n\to\infty}\dfrac{n}{5+n} = \lim\limits_{n\to\infty}\dfrac{1}{5/n+1} = 1$, so $\lim\limits_{n\to\infty}a_n \ne 0$. Thus, the given series is divergent by the Test
for Divergence.

9. $\lim\limits_{n\to\infty}\left|\dfrac{a_{n+1}}{a_n}\right| = \lim\limits_{n\to\infty}\dfrac{1/(2n+2)!}{1/(2n)!} = \lim\limits_{n\to\infty}\dfrac{(2n)!}{(2n+2)!} = \lim\limits_{n\to\infty}\dfrac{(2n)!}{(2n+2)(2n+1)(2n)!}$

   $= \lim\limits_{n\to\infty}\dfrac{1}{(2n+2)(2n+1)} = 0 < 1$, so the series $\sum\limits_{n=1}^{\infty}\dfrac{1}{(2n)!}$ is absolutely convergent by the Ratio Test. Of course,
   absolute convergence is the same as convergence for this series, since all of its terms are positive.

11. Since $0 \le \dfrac{e^{1/n}}{n^3} \le \dfrac{e}{n^3} = e\left(\dfrac{1}{n^3}\right)$ and $\sum\limits_{n=1}^{\infty}\dfrac{1}{n^3}$ is a convergent p-series $(p = 3 > 1)$, $\sum\limits_{n=1}^{\infty}\dfrac{e^{1/n}}{n^3}$ converges, and so

   $\sum\limits_{n=1}^{\infty}\dfrac{(-1)^n e^{1/n}}{n^3}$ is absolutely convergent.

13. $\lim\limits_{n\to\infty}\left|\dfrac{a_{n+1}}{a_n}\right| = \lim\limits_{n\to\infty}\left[\dfrac{(n+1)\,3^{n+1}}{4^n}\cdot\dfrac{4^{n-1}}{n\cdot3^n}\right] = \lim\limits_{n\to\infty}\left(\dfrac{3}{4}\cdot\dfrac{n+1}{n}\right) = \dfrac{3}{4} < 1$, so the series $\sum\limits_{n=1}^{\infty}\dfrac{n(-3)^n}{4^{n-1}}$ is
absolutely convergent by the Ratio Test.

15. $\lim\limits_{n\to\infty}\left|\dfrac{a_{n+1}}{a_n}\right| = \lim\limits_{n\to\infty}\left[\dfrac{10^{n+1}}{(n+2)\,4^{2n+3}}\cdot\dfrac{(n+1)\,4^{2n+1}}{10^n}\right] = \lim\limits_{n\to\infty}\left(\dfrac{10}{4^2}\cdot\dfrac{n+1}{n+2}\right) = \dfrac{5}{8} < 1$, so the series

   $\sum\limits_{n=1}^{\infty}\dfrac{10^n}{(n+1)4^{2n+1}}$ is absolutely convergent by the Ratio Test. Since the terms of this series are positive, absolute
   convergence is the same as convergence.

17. $\displaystyle\sum_{n=2}^{\infty} \frac{(-1)^n}{\ln n}$ converges by the Alternating Series Test since $\displaystyle\lim_{n\to\infty} \frac{1}{\ln n} = 0$ and $\left\{\dfrac{1}{\ln n}\right\}$ is decreasing. Now

$\ln n < n$, so $\dfrac{1}{\ln n} > \dfrac{1}{n}$, and since $\displaystyle\sum_{n=2}^{\infty} \frac{1}{n}$ is the divergent (partial) harmonic series, $\displaystyle\sum_{n=2}^{\infty} \frac{1}{\ln n}$ diverges by the

Comparison Test. Thus, $\displaystyle\sum_{n=2}^{\infty} \frac{(-1)^n}{\ln n}$ is conditionally convergent.

19. $\dfrac{|\cos(n\pi/3)|}{n!} \le \dfrac{1}{n!}$ and $\displaystyle\sum_{n=1}^{\infty} \frac{1}{n!}$ converges (use the Ratio Test or the result of Exercise 12.4.29 [ET 11.4.29]), so the

series $\displaystyle\sum_{n=1}^{\infty} \frac{\cos(n\pi/3)}{n!}$ converges absolutely by the Comparison Test.

21. $\displaystyle\lim_{n\to\infty} \sqrt[n]{|a_n|} = \lim_{n\to\infty}\left(\frac{n^n}{3^{1+3n}}\right)^{1/n} = \lim_{n\to\infty} \frac{n}{\sqrt[n]{3}\cdot 3^3} = \infty$, so the series $\displaystyle\sum_{n=1}^{\infty} \frac{n^n}{3^{1+3n}}$ is divergent by the Root Test.

$Or:\ \displaystyle\lim_{n\to\infty}\left|\frac{a_{n+1}}{a_n}\right| = \lim_{n\to\infty}\left[\frac{(n+1)^{n+1}}{3^{4+3n}}\cdot\frac{3^{1+3n}}{n^n}\right] = \lim_{n\to\infty}\left[\frac{1}{3^3}\cdot\left(\frac{n+1}{n}\right)^n (n+1)\right]$

$\qquad = \dfrac{1}{27}\displaystyle\lim_{n\to\infty}\left(1+\frac{1}{n}\right)^n \lim_{n\to\infty}(n+1) = \tfrac{1}{27}e \lim_{n\to\infty}(n+1) = \infty,$

so the series is divergent by the Ratio Test.

23. $\displaystyle\lim_{n\to\infty}\sqrt[n]{|a_n|} = \lim_{n\to\infty}\frac{n^2+1}{2n^2+1} = \lim_{n\to\infty}\frac{1+1/n^2}{2+1/n^2} = \frac{1}{2} < 1$, so the series $\displaystyle\sum_{n=1}^{\infty}\left(\frac{n^2+1}{2n^2+1}\right)^n$ is absolutely

convergent by the Root Test.

25. Use the Ratio Test with the series $1 - \dfrac{1\cdot 3}{3!} + \dfrac{1\cdot 3\cdot 5}{5!} - \dfrac{1\cdot 3\cdot 5\cdot 7}{7!} + \cdots + (-1)^{n-1}\dfrac{1\cdot 3\cdot 5\cdots (2n-1)}{(2n-1)!} + \cdots$

$= \displaystyle\sum_{n=1}^{\infty} (-1)^{n-1}\frac{1\cdot 3\cdot 5\cdots (2n-1)}{(2n-1)!}.$

$\displaystyle\lim_{n\to\infty}\left|\frac{a_{n+1}}{a_n}\right| = \lim_{n\to\infty}\left|\frac{(-1)^n\cdot 1\cdot 3\cdot 5\cdots (2n-1)[2(n+1)-1]}{[2(n+1)-1]!}\cdot\frac{(2n-1)!}{(-1)^{n-1}\cdot 1\cdot 3\cdot 5\cdots (2n-1)}\right|$

$\qquad = \displaystyle\lim_{n\to\infty}\left|\frac{(-1)(2n+1)(2n-1)!}{(2n+1)(2n)(2n-1)!}\right|$

$\qquad = \displaystyle\lim_{n\to\infty}\frac{1}{2n} = 0 < 1,$

so the given series is absolutely convergent and therefore convergent.

27. $\displaystyle\sum_{n=1}^{\infty}\frac{2\cdot 4\cdot 6\cdots (2n)}{n!} = \sum_{n=1}^{\infty}\frac{(2\cdot 1)\cdot(2\cdot 2)\cdot(2\cdot 3)\cdots(2\cdot n)}{n!} = \sum_{n=1}^{\infty}\frac{2^n n!}{n!} = \sum_{n=1}^{\infty} 2^n$, which diverges by the

Test for Divergence since $\displaystyle\lim_{n\to\infty} 2^n = \infty$.

29. By the recursive definition, $\displaystyle\lim_{n\to\infty}\left|\frac{a_{n+1}}{a_n}\right| = \lim_{n\to\infty}\left|\frac{5n+1}{4n+3}\right| = \frac{5}{4} > 1$, so the series diverges by the Ratio Test.

31. (a) $\lim\limits_{n\to\infty}\left|\dfrac{1/(n+1)^3}{1/n^3}\right| = \lim\limits_{n\to\infty}\dfrac{n^3}{(n+1)^3} = \lim\limits_{n\to\infty}\dfrac{1}{(1+1/n)^3} = 1.$ Inconclusive.

(b) $\lim\limits_{n\to\infty}\left|\dfrac{(n+1)}{2^{n+1}}\cdot\dfrac{2^n}{n}\right| = \lim\limits_{n\to\infty}\dfrac{n+1}{2n} = \lim\limits_{n\to\infty}\left(\dfrac{1}{2}+\dfrac{1}{2n}\right) = \dfrac{1}{2}.$ Conclusive (convergent).

(c) $\lim\limits_{n\to\infty}\left|\dfrac{(-3)^n}{\sqrt{n+1}}\cdot\dfrac{\sqrt{n}}{(-3)^{n-1}}\right| = 3\lim\limits_{n\to\infty}\sqrt{\dfrac{n}{n+1}} = 3\lim\limits_{n\to\infty}\sqrt{\dfrac{1}{1+1/n}} = 3.$ Conclusive (divergent).

(d) $\lim\limits_{n\to\infty}\left|\dfrac{\sqrt{n+1}}{1+(n+1)^2}\cdot\dfrac{1+n^2}{\sqrt{n}}\right| = \lim\limits_{n\to\infty}\left[\sqrt{1+\dfrac{1}{n}}\cdot\dfrac{1/n^2+1}{1/n^2+(1+1/n)^2}\right] = 1.$ Inconclusive.

33. (a) $\lim\limits_{n\to\infty}\left|\dfrac{a_{n+1}}{a_n}\right| = \lim\limits_{n\to\infty}\left|\dfrac{x^{n+1}}{(n+1)!}\cdot\dfrac{n!}{x^n}\right| = \lim\limits_{n\to\infty}\left|\dfrac{x}{n+1}\right| = |x|\lim\limits_{n\to\infty}\dfrac{1}{n+1} = |x|\cdot 0 = 0 < 1,$ so by the Ratio

Test the series $\displaystyle\sum_{n=0}^{\infty}\dfrac{x^n}{n!}$ converges for all x.

(b) Since the series of part (a) always converges, we must have $\lim\limits_{n\to\infty}\dfrac{x^n}{n!} = 0$ by Theorem 12.2.6 [ET 11.2.6].

35. (a) $s_5 = \displaystyle\sum_{n=1}^{5}\dfrac{1}{n2^n} = \dfrac{1}{2}+\dfrac{1}{8}+\dfrac{1}{24}+\dfrac{1}{64}+\dfrac{1}{160} = \dfrac{661}{960} \approx 0.68854.$ Now the ratios

$r_n = \dfrac{a_{n+1}}{a_n} = \dfrac{n2^n}{(n+1)2^{n+1}} = \dfrac{n}{2(n+1)}$ form an increasing sequence, since

$r_{n+1}-r_n = \dfrac{n+1}{2(n+2)}-\dfrac{n}{2(n+1)} = \dfrac{(n+1)^2-n(n+2)}{2(n+1)(n+2)} = \dfrac{1}{2(n+1)(n+2)} > 0.$ So by Exercise 34(b),

the error in using s_5 is $R_5 \le \dfrac{a_6}{1-\lim\limits_{n\to\infty}r_n} = \dfrac{1/(6\cdot 2^6)}{1-1/2} = \dfrac{1}{192} \approx 0.00521.$

(b) The error in using s_n as an approximation to the sum is $R_n = \dfrac{a_{n+1}}{1-\frac{1}{2}} = \dfrac{2}{(n+1)2^{n+1}}.$ We want

$R_n < 0.00005 \iff \dfrac{1}{(n+1)2^n} < 0.00005 \iff (n+1)2^n > 20{,}000.$ To find such an n we can use trial

and error or a graph. We calculate $(11+1)2^{11} = 24{,}576,$ so $s_{11} = \displaystyle\sum_{n=1}^{11}\dfrac{1}{n2^n} \approx 0.693109$ is within 0.00005 of

the actual sum.

37. Summing the inequalities $-|a_i| \le a_i \le |a_i|$ for $i = 1, 2, \dots, n$, we get $-\sum_{i=1}^{n}|a_i| \le \sum_{i=1}^{n}a_i \le \sum_{i=1}^{n}|a_i|$

$\Rightarrow -\lim\limits_{n\to\infty}\sum_{i=1}^{n}|a_i| \le \lim\limits_{n\to\infty}\sum_{i=1}^{n}a_i \le \lim\limits_{n\to\infty}\sum_{i=1}^{n}|a_i| \Rightarrow -\sum_{n=1}^{\infty}|a_n| \le \sum_{n=1}^{\infty}a_n \le \sum_{n=1}^{\infty}|a_n| \Rightarrow$

$\left|\sum_{n=1}^{\infty}a_n\right| \le \sum_{n=1}^{\infty}|a_n|.$

39. (a) Since $\sum a_n$ is absolutely convergent, and since $|a_n^+| \le |a_n|$ and $|a_n^-| \le |a_n|$ (because a_n^+ and a_n^- each equal

either a_n or 0), we conclude by the Comparison Test that both $\sum a_n^+$ and $\sum a_n^-$ must be absolutely convergent.

(Or use Theorem 12.2.8 [ET 11.2.8].)

(b) We will show by contradiction that both $\sum a_n^+$ and $\sum a_n^-$ must diverge. For suppose that $\sum a_n^+$

converged. Then so would $\sum\left(a_n^+ - \frac{1}{2}a_n\right)$ by Theorem 12.2.8 [ET 11.2.8]. But

$\sum\left(a_n^+ - \frac{1}{2}a_n\right) = \sum\left[\frac{1}{2}\left(a_n+|a_n|\right) - \frac{1}{2}a_n\right] = \frac{1}{2}\sum|a_n|,$ which diverges because $\sum a_n$ is only conditionally

convergent. Hence, $\sum a_n^+$ can't converge. Similarly, neither can $\sum a_n^-$.

12.7 Strategy for Testing Series ET 11.7

1. $\lim\limits_{n\to\infty} a_n = \lim\limits_{n\to\infty} \dfrac{n^2-1}{n^2+1} = \lim\limits_{n\to\infty} \dfrac{1-1/n^2}{1+1/n} = 1 \neq 0$, so the series $\sum\limits_{n=1}^{\infty} \dfrac{n^2-1}{n^2+1}$ diverges by the Test for Divergence.

3. $\dfrac{1}{n^2+n} < \dfrac{1}{n^2}$ for all $n \geq 1$, so $\sum\limits_{n=1}^{\infty} \dfrac{1}{n^2+n}$ converges by the Comparison Test with $\sum\limits_{n=1}^{\infty} \dfrac{1}{n^2}$, a p-series that converges because $p = 2 > 1$.

5. $\lim\limits_{n\to\infty} \left| \dfrac{a_{n+1}}{a_n} \right| = \lim\limits_{n\to\infty} \left| \dfrac{(-3)^{n+2}}{2^{3(n+1)}} \cdot \dfrac{2^{3n}}{(-3)^{n+1}} \right| = \lim\limits_{n\to\infty} \left| \dfrac{-3 \cdot 2^{3n}}{2^{3n} \cdot 2^3} \right| = \lim\limits_{n\to\infty} \dfrac{3}{2^3} = \dfrac{3}{8} < 1$, so the series

$\sum\limits_{n=1}^{\infty} \dfrac{(-3)^{n+1}}{2^{3n}}$ is absolutely convergent by the Ratio Test.

7. Let $f(x) = \dfrac{1}{x\sqrt{\ln x}}$. Then f is positive, continuous, and decreasing on $[2, \infty)$, so we can apply the Integral Test.

Since $\displaystyle\int \dfrac{1}{x\sqrt{\ln x}}\,dx \quad \begin{bmatrix} u = \ln x, \\ du = dx/x \end{bmatrix} = \int u^{-1/2}\,du = 2u^{1/2} + C = 2\sqrt{\ln x} + C$, we find

$\displaystyle\int_2^\infty \dfrac{dx}{x\sqrt{\ln x}} = \lim\limits_{t\to\infty} \int_2^t \dfrac{dx}{x\sqrt{\ln x}} = \lim\limits_{t\to\infty} \left[2\sqrt{\ln x} \right]_2^t = \lim\limits_{t\to\infty} \left(2\sqrt{\ln t} - 2\sqrt{\ln 2} \right) = \infty$. Since the integral

diverges, the given series $\sum\limits_{n=2}^{\infty} \dfrac{1}{n\sqrt{\ln n}}$ diverges.

9. $\sum\limits_{k=1}^{\infty} k^2 e^{-k} = \sum\limits_{k=1}^{\infty} \dfrac{k^2}{e^k}$. Using the Ratio Test, we get

$\lim\limits_{k\to\infty} \left| \dfrac{a_{k+1}}{a_k} \right| = \lim\limits_{k\to\infty} \left| \dfrac{(k+1)^2}{e^{k+1}} \cdot \dfrac{e^k}{k^2} \right| = \lim\limits_{k\to\infty} \left[\left(\dfrac{k+1}{k} \right)^2 \cdot \dfrac{1}{e} \right] = 1^2 \cdot \dfrac{1}{e} = \dfrac{1}{e} < 1$, so the series converges.

11. $b_n = \dfrac{1}{n\ln n} > 0$ for $n \geq 2$, $\{b_n\}$ is decreasing, and $\lim\limits_{n\to\infty} b_n = 0$, so the given series $\sum\limits_{n=2}^{\infty} \dfrac{(-1)^{n+1}}{n\ln n}$ converges by the Alternating Series Test.

13. $\lim\limits_{n\to\infty} \left| \dfrac{a_{n+1}}{a_n} \right| = \lim\limits_{n\to\infty} \left| \dfrac{3^{n+1}(n+1)^2}{(n+1)!} \cdot \dfrac{n!}{3^n n^2} \right| = \lim\limits_{n\to\infty} \left[\dfrac{3(n+1)^2}{(n+1)n^2} \right] = 3 \lim\limits_{n\to\infty} \dfrac{n+1}{n^2} = 0 < 1$, so the series

$\sum\limits_{n=1}^{\infty} \dfrac{3^n n^2}{n!}$ converges by the Ratio Test.

15. $\lim\limits_{n\to\infty} \left| \dfrac{a_{n+1}}{a_n} \right| = \lim\limits_{n\to\infty} \left| \dfrac{(n+1)!}{2 \cdot 5 \cdot 8 \cdots (3n+2)[3(n+1)+2]} \cdot \dfrac{2 \cdot 5 \cdot 8 \cdots (3n+2)}{n!} \right|$

$\qquad = \lim\limits_{n\to\infty} \dfrac{n+1}{3n+5} = \dfrac{1}{3} < 1$

so the series $\sum\limits_{n=0}^{\infty} \dfrac{n!}{2 \cdot 5 \cdot 8 \cdots (3n+2)}$ converges by the Ratio Test.

17. $\lim\limits_{n\to\infty} 2^{1/n} = 2^0 = 1$, so $\lim\limits_{n\to\infty} (-1)^n 2^{1/n}$ does not exist and the series $\sum\limits_{n=1}^{\infty} (-1)^n 2^{1/n}$ diverges by the

Test for Divergence.

19. Let $f(x) = \dfrac{\ln x}{\sqrt{x}}$. Then $f'(x) = \dfrac{2 - \ln x}{2x^{3/2}} < 0$ when $\ln x > 2$ or $x > e^2$, so $\dfrac{\ln n}{\sqrt{n}}$ is decreasing for $n > e^2$.

By l'Hospital's Rule, $\lim\limits_{n\to\infty} \dfrac{\ln n}{\sqrt{n}} = \lim\limits_{n\to\infty} \dfrac{1/n}{1/(2\sqrt{n})} = \lim\limits_{n\to\infty} \dfrac{2}{\sqrt{n}} = 0$, so the series $\sum\limits_{n=1}^{\infty} (-1)^n \dfrac{\ln n}{\sqrt{n}}$ converges by the Alternating Series Test.

21. $\sum\limits_{n=1}^{\infty} \dfrac{(-2)^{2n}}{n^n} = \sum\limits_{n=1}^{\infty} \left(\dfrac{4}{n}\right)^n$. $\lim\limits_{n\to\infty} \sqrt[n]{|a_n|} = \lim\limits_{n\to\infty} \dfrac{4}{n} = 0 < 1$, so the given series is absolutely convergent by the Root Test.

23. Using the Limit Comparison Test with $a_n = \tan\left(\dfrac{1}{n}\right)$ and $b_n = \dfrac{1}{n}$, we have

$$\lim_{n\to\infty} \frac{a_n}{b_n} = \lim_{n\to\infty} \frac{\tan(1/n)}{1/n} = \lim_{x\to\infty} \frac{\tan(1/x)}{1/x} \overset{\text{H}}{=} \lim_{x\to\infty} \frac{\sec^2(1/x) \cdot (-1/x^2)}{-1/x^2} = \lim_{x\to\infty} \sec^2(1/x) = 1^2 = 1 > 0.$$

Since $\sum_{n=1}^{\infty} b_n$ is the divergent harmonic series, $\sum_{n=1}^{\infty} a_n$ is also divergent.

25. Use the Ratio Test. $\lim\limits_{n\to\infty} \left|\dfrac{a_{n+1}}{a_n}\right| = \lim\limits_{n\to\infty} \left|\dfrac{(n+1)!}{e^{(n+1)^2}} \cdot \dfrac{e^{n^2}}{n!}\right| = \lim\limits_{n\to\infty} \dfrac{(n+1)n! \cdot e^{n^2}}{e^{n^2+2n+1}n!} = \lim\limits_{n\to\infty} \dfrac{n+1}{e^{2n+1}} = 0 < 1$, so

$\sum\limits_{n=1}^{\infty} \dfrac{n!}{e^{n^2}}$ converges.

27. $\displaystyle\int_2^{\infty} \dfrac{\ln x}{x^2}\, dx = \lim\limits_{t\to\infty} \left[-\dfrac{\ln x}{x} - \dfrac{1}{x}\right]_1^t$ (using integration by parts) $\overset{\text{H}}{=} 1$. So $\sum\limits_{n=1}^{\infty} \dfrac{\ln n}{n^2}$ converges by the Integral Test,

and since $\dfrac{k \ln k}{(k+1)^3} < \dfrac{k \ln k}{k^3} = \dfrac{\ln k}{k^2}$, the given series $\sum\limits_{k=1}^{\infty} \dfrac{k \ln k}{(k+1)^3}$ converges by the Comparison Test.

29. $0 < \dfrac{\tan^{-1} n}{n^{3/2}} < \dfrac{\pi/2}{n^{3/2}}$. $\sum\limits_{n=1}^{\infty} \dfrac{\pi/2}{n^{3/2}} = \dfrac{\pi}{2} \sum\limits_{n=1}^{\infty} \dfrac{1}{n^{3/2}}$ which is a convergent p-series ($p = \tfrac{3}{2} > 1$), so

$\sum\limits_{n=1}^{\infty} \dfrac{\tan^{-1} n}{n^{3/2}}$ converges by the Comparison Test.

31. $\lim\limits_{k\to\infty} a_k = \lim\limits_{k\to\infty} \dfrac{5^k}{3^k + 4^k} = [\text{divide by } 4^k] \lim\limits_{k\to\infty} \dfrac{(5/4)^k}{(3/4)^k + 1} = \infty$ since $\lim\limits_{k\to\infty} \left(\dfrac{3}{4}\right)^k = 0$ and $\lim\limits_{k\to\infty} \left(\dfrac{5}{4}\right)^k = \infty$.

Thus, $\sum\limits_{k=1}^{\infty} \dfrac{5^k}{3^k + 4^k}$ diverges by the Test for Divergence.

33. Let $a_n = \dfrac{\sin(1/n)}{\sqrt{n}}$ and $b_n = \dfrac{1}{n\sqrt{n}}$. Then $\lim\limits_{n\to\infty} \dfrac{a_n}{b_n} = \lim\limits_{n\to\infty} \dfrac{\sin(1/n)}{1/n} = 1 > 0$, so $\sum\limits_{n=1}^{\infty} \dfrac{\sin(1/n)}{\sqrt{n}}$ converges by

limit comparison with the convergent p-series $\sum\limits_{n=1}^{\infty} \dfrac{1}{n^{3/2}}$ ($p = 3/2 > 1$).

35. $\lim\limits_{n\to\infty} \sqrt[n]{|a_n|} = \lim\limits_{n\to\infty} \left(\dfrac{n}{n+1}\right)^{n^2/n} = \lim\limits_{n\to\infty} \dfrac{1}{[(n+1)/n]^n} = \dfrac{1}{\lim\limits_{n\to\infty} (1+1/n)^n} = \dfrac{1}{e} < 1$, so the series

$\sum\limits_{n=1}^{\infty} \left(\dfrac{n}{n+1}\right)^{n^2}$ converges by the Root Test.

37. $\lim\limits_{n\to\infty} \sqrt[n]{|a_n|} = \lim\limits_{n\to\infty} \left(2^{1/n} - 1\right) = 1 - 1 = 0 < 1$, so the series $\sum\limits_{n=1}^{\infty} \left(\sqrt[n]{2} - 1\right)^n$ converges by the Root Test.

12.8 Power Series

ET 11.8

1. A power series is a series of the form $\sum_{n=0}^{\infty} c_n x^n = c_0 + c_1 x + c_2 x^2 + c_3 x^3 + \cdots$, where x is a variable and the c_n's are constants called the coefficients of the series.

More generally, a series of the form $\sum_{n=0}^{\infty} c_n (x - a)^n = c_0 + c_1 (x - a) + c_2 (x - a)^2 + \cdots$ is called a power series in $(x - a)$ or a power series centered at a or a power series about a, where a is a constant.

3. If $a_n = \dfrac{x^n}{\sqrt{n}}$, then $\lim\limits_{n \to \infty} \left| \dfrac{a_{n+1}}{a_n} \right| = \lim\limits_{n \to \infty} \left| \dfrac{x^{n+1}}{\sqrt{n+1}} \cdot \dfrac{\sqrt{n}}{x} \right| = \lim\limits_{n \to \infty} \left| \dfrac{x}{\sqrt{n+1}/\sqrt{n}} \right| = \lim\limits_{n \to \infty} \dfrac{|x|}{\sqrt{1 + 1/n}} = |x|.$

By the Ratio Test, the series $\sum\limits_{n=1}^{\infty} \dfrac{x^n}{\sqrt{n}}$ converges when $|x| < 1$, so the radius of convergence $R = 1$. Now we'll

check the endpoints, that is, $x = \pm 1$. When $x = 1$, the series $\sum\limits_{n=1}^{\infty} \dfrac{1}{\sqrt{n}}$ diverges because it is a p-series with

$p = \frac{1}{2} \le 1$. When $x = -1$, the series $\sum\limits_{n=1}^{\infty} \dfrac{(-1)^n}{\sqrt{n}}$ converges by the Alternating Series Test. Thus, the interval of

convergence is $I = [-1, 1)$.

5. If $a_n = \dfrac{(-1)^{n-1} x^n}{n^3}$, then $\lim\limits_{n \to \infty} \left| \dfrac{a_{n+1}}{a_n} \right| = \lim\limits_{n \to \infty} \left| \dfrac{(-1)^n x^{n+1}}{(n+1)^3} \cdot \dfrac{n^3}{(-1)^{n-1} x^n} \right| = \lim\limits_{n \to \infty} \left| \dfrac{(-1) x n^3}{(n+1)^3} \right|$

$= \lim\limits_{n \to \infty} \left[\left(\dfrac{n}{n+1} \right)^3 |x| \right] = 1^3 \cdot |x| = |x|.$ By the Ratio Test, the series $\sum\limits_{n=1}^{\infty} \dfrac{(-1)^{n-1} x^n}{n^3}$ converges when $|x| < 1$,

so the radius of convergence $R = 1$. Now we'll check the endpoints, that is, $x = \pm 1$. When $x = 1$, the series

$\sum\limits_{n=1}^{\infty} \dfrac{(-1)^{n-1}}{n^3}$ converges by the Alternating Series Test. When $x = -1$, the series

$\sum\limits_{n=1}^{\infty} \dfrac{(-1)^{n-1}(-1)^n}{n^3} = -\sum\limits_{n=1}^{\infty} \dfrac{1}{n^3}$ converges because it is a constant multiple of a convergent p-series $(p = 3 > 1)$.

Thus, the interval of convergence is $I = [-1, 1]$.

7. If $a_n = \dfrac{x^n}{n!}$, then $\lim\limits_{n \to \infty} \left| \dfrac{a_{n+1}}{a_n} \right| = \lim\limits_{n \to \infty} \left| \dfrac{x^{n+1}}{(n+1)!} \cdot \dfrac{n!}{x^n} \right| = \lim\limits_{n \to \infty} \left| \dfrac{x}{n+1} \right| = |x| \lim\limits_{n \to \infty} \dfrac{1}{n+1} = |x| \cdot 0 = 0 < 1$ for

all real x. So, by the Ratio Test, $R = \infty$, and $I = (-\infty, \infty)$.

9. $\lim\limits_{n \to \infty} \left| \dfrac{a_{n+1}}{a_n} \right| = \lim\limits_{n \to \infty} \dfrac{(n+1) 4^{n+1} |x|^{n+1}}{n 4^n |x|^n} = \lim\limits_{n \to \infty} \left(1 + \dfrac{1}{n} \right) 4 |x| = 4 |x|.$ Now $4 |x| < 1 \iff |x| < \frac{1}{4}$, so by

the Ratio Test, $R = \frac{1}{4}$. When $x = \frac{1}{4}$, we get the divergent series $\sum_{n=1}^{\infty} (-1)^n n$, and when $x = -\frac{1}{4}$, we get the

divergent series $\sum_{n=1}^{\infty} n$. Thus, $I = \left(-\frac{1}{4}, \frac{1}{4} \right)$.

11. $a_n = \dfrac{(-2)^n x^n}{\sqrt[4]{n}}$, so $\lim\limits_{n \to \infty} \left| \dfrac{a_{n+1}}{a_n} \right| = \lim\limits_{n \to \infty} \dfrac{2^{n+1} |x|^{n+1}}{\sqrt[4]{n+1}} \cdot \dfrac{\sqrt[4]{n}}{2^n |x|^n} = \lim\limits_{n \to \infty} 2 |x| \sqrt[4]{\dfrac{n}{n+1}} = 2 |x|,$ so by the

Ratio Test, the series converges when $2 |x| < 1 \iff |x| < \frac{1}{2}$, so $R = \frac{1}{2}$. When $x = -\frac{1}{2}$, we get the divergent

p-series $\sum\limits_{n=1}^{\infty} \dfrac{1}{\sqrt[4]{n}} \left(p = \frac{1}{4} \le 1 \right)$. When $x = \frac{1}{2}$, we get the series $\sum\limits_{n=1}^{\infty} \dfrac{(-1)^n}{\sqrt[4]{n}}$, which converges by the Alternating

Series Test. Thus, $I = \left(-\frac{1}{2}, \frac{1}{2} \right].$

13. If $a_n = (-1)^n \dfrac{x^n}{4^n \ln n}$, then

$$\lim_{n\to\infty} \left| \frac{a_{n+1}}{a_n} \right| = \lim_{n\to\infty} \left| \frac{x^{n+1}}{4^{n+1}\ln(n+1)} \cdot \frac{4^n \ln n}{x^n} \right| = \frac{|x|}{4} \lim_{n\to\infty} \frac{\ln n}{\ln(n+1)} = \frac{|x|}{4} \cdot 1 \ \text{(by l'Hospital's Rule)} = \frac{|x|}{4}.$$

By the Ratio Test, the series converges when $\dfrac{|x|}{4} < 1 \quad \Leftrightarrow \quad |x| < 4$, so $R = 4$. When $x = -4$,

$$\sum_{n=2}^{\infty} (-1)^n \frac{x^n}{4^n \ln n} = \sum_{n=2}^{\infty} \frac{[(-1)(-4)]^n}{4^n \ln n} = \sum_{n=2}^{\infty} \frac{1}{\ln n}. \ \text{Since } \ln n < n \text{ for } n \geq 2, \ \frac{1}{\ln n} > \frac{1}{n} \text{ and } \sum_{n=2}^{\infty} \frac{1}{n} \text{ is the}$$

divergent harmonic series (without the $n = 1$ term), $\displaystyle\sum_{n=2}^{\infty} \frac{1}{\ln n}$ is divergent by the Comparison Test. When $x = 4$,

$$\sum_{n=2}^{\infty} (-1)^n \frac{x^n}{4^n \ln n} = \sum_{n=2}^{\infty} (-1)^n \frac{1}{\ln n}, \ \text{which converges by the Alternating Series Test. Thus, } I = (-4, 4].$$

15. If $a_n = \sqrt{n}\,(x-1)^n$, then $\displaystyle\lim_{n\to\infty} \left| \frac{a_{n+1}}{a_n} \right| = \lim_{n\to\infty} \frac{\sqrt{n+1}\,|x-1|^{n+1}}{\sqrt{n}\,|x-1|^n} = \lim_{n\to\infty} \sqrt{1 + \frac{1}{n}}\,|x-1| = |x-1|$. By

the Ratio Test, the series converges when $|x-1| < 1$ [so $R = 1$] $\Leftrightarrow -1 < x - 1 < 1 \Leftrightarrow 0 < x < 2$.
When $x = 0$, the series becomes $\sum_{n=0}^{\infty} (-1)^n \sqrt{n}$, which diverges by the Test for Divergence. When $x = 2$, the
series becomes $\sum_{n=0}^{\infty} \sqrt{n}$, which also diverges by the Test for Divergence. Thus, $I = (0, 2)$.

17. If $a_n = (-1)^n \dfrac{(x+2)^n}{n2^n}$, then

$$\lim_{n\to\infty} \left| \frac{a_{n+1}}{a_n} \right| = \lim_{n\to\infty} \left[\frac{|x+2|^{n+1}}{(n+1)2^{n+1}} \cdot \frac{n2^n}{|x+2|^n} \right] = \lim_{n\to\infty} \frac{n}{n+1} \cdot \frac{|x+2|}{2} = \frac{|x+2|}{2}. \ \text{By the Ratio Test, the}$$

series converges when $\dfrac{|x+2|}{2} < 1 \Leftrightarrow |x+2| < 2$ [so $R = 2$] $\Leftrightarrow -2 < x + 2 < 2 \Leftrightarrow -4 < x < 0$.

When $x = -4$, the series becomes $\displaystyle\sum_{n=1}^{\infty} (-1)^n \frac{(-2)^n}{n2^n} = \sum_{n=1}^{\infty} \frac{2^n}{n2^n} = \sum_{n=1}^{\infty} \frac{1}{n}$, which is the divergent harmonic series.

When $x = 0$, the series is $\displaystyle\sum_{n=1}^{\infty} \frac{(-1)^n}{n}$, the alternating harmonic series, which converges by the Alternating Series

Test. Thus, $I = (-4, 0]$.

19. If $a_n = \dfrac{(x-2)^n}{n^n}$, then $\displaystyle\lim_{n\to\infty} \sqrt[n]{|a_n|} = \lim_{n\to\infty} \frac{|x-2|}{n} = 0$, so the series converges for all x (by the Root Test).
$R = \infty$ and $I = (-\infty, \infty)$.

21. $a_n = \dfrac{n}{b^n}(x-a)^n$, where $b > 0$.

$$\lim_{n\to\infty} \left| \frac{a_{n+1}}{a_n} \right| = \lim_{n\to\infty} \frac{(n+1)\,|x-a|^{n+1}}{b^{n+1}} \cdot \frac{b^n}{n\,|x-a|^n} = \lim_{n\to\infty} \left(1 + \frac{1}{n}\right) \frac{|x-a|}{b} = \frac{|x-a|}{b}.$$

By the Ratio Test, the series converges when $\dfrac{|x-a|}{b} < 1 \Leftrightarrow |x-a| < b$ [so $R = b$] $\Leftrightarrow$

$-b < x - a < b \Leftrightarrow a - b < x < a + b$. When $|x-a| = b$, $\displaystyle\lim_{n\to\infty} |a_n| = \lim_{n\to\infty} n = \infty$, so the series diverges.
Thus, $I = (a-b, a+b)$.

23. If $a_n = n!(2x-1)^n$, then $\displaystyle\lim_{n\to\infty} \left| \frac{a_{n+1}}{a_n} \right| = \lim_{n\to\infty} \left| \frac{(n+1)!(2x-1)^{n+1}}{n!(2x-1)^n} \right| = \lim_{n\to\infty} (n+1)\,|2x-1| \to \infty$

as $n \to \infty$ for all $x \neq \frac{1}{2}$. Since the series diverges for all $x \neq \frac{1}{2}$, $R = 0$ and $I = \left\{ \frac{1}{2} \right\}$.

25. $\lim\limits_{n\to\infty} \left|\dfrac{a_{n+1}}{a_n}\right| = \lim\limits_{n\to\infty}\left[\dfrac{|4x+1|^{n+1}}{(n+1)^2}\cdot\dfrac{n^2}{|4x+1|^n}\right] = \lim\limits_{n\to\infty}\dfrac{|4x+1|}{(1+1/n)^2} = |4x+1|$, so by the Ratio Test, the

series converges when $|4x+1| < 1 \Leftrightarrow -1 < 4x+1 < 1 \Leftrightarrow -2 < 4x < 0 \Leftrightarrow -\frac{1}{2} < x < 0$, so

$R = \frac{1}{4}$. When $x = -\frac{1}{2}$, the series becomes $\sum\limits_{n=1}^{\infty}\dfrac{(-1)^n}{n^2}$, which converges by the Alternating Series Test. When

$x = 0$, the series becomes $\sum\limits_{n=1}^{\infty}\dfrac{1}{n^2}$, a convergent p-series ($p = 2 > 1$). $I = \left[-\frac{1}{2}, 0\right]$.

27. If $a_n = \dfrac{x^n}{(\ln n)^n}$, then $\lim\limits_{n\to\infty}\sqrt[n]{|a_n|} = \lim\limits_{n\to\infty}\dfrac{|x|}{\ln n} = 0 < 1$ for all x, so $R = \infty$ and $I = (-\infty, \infty)$ by the

Root Test.

29. (a) We are given that the power series $\sum_{n=0}^{\infty} c_n x^n$ is convergent for $x = 4$. So by Theorem 3, it must converge for

at least $-4 < x \le 4$. In particular, it converges when $x = -2$; that is, $\sum_{n=0}^{\infty} c_n(-2)^n$ is convergent.

(b) It does not follow that $\sum_{n=0}^{\infty} c_n(-4)^n$ is necessarily convergent. [See the comments after Theorem 3 about

convergence at the endpoint of an interval. An example is $c_n = (-1)^n/(n4^n)$.]

31. If $a_n = \dfrac{(n!)^k}{(kn)!}x^n$, then

$$\lim\limits_{n\to\infty}\left|\dfrac{a_{n+1}}{a_n}\right| = \lim\limits_{n\to\infty}\dfrac{[(n+1)!]^k (kn)!}{(n!)^k [k(n+1)]!}|x| = \lim\limits_{n\to\infty}\dfrac{(n+1)^k}{(kn+k)(kn+k-1)\cdots(kn+2)(kn+1)}|x|$$

$$= \lim\limits_{n\to\infty}\left[\dfrac{(n+1)}{(kn+1)}\dfrac{(n+1)}{(kn+2)}\cdots\dfrac{(n+1)}{(kn+k)}\right]|x|$$

$$= \lim\limits_{n\to\infty}\left[\dfrac{n+1}{kn+1}\right]\lim\limits_{n\to\infty}\left[\dfrac{n+1}{kn+2}\right]\cdots\lim\limits_{n\to\infty}\left[\dfrac{n+1}{kn+k}\right]|x| = \left(\dfrac{1}{k}\right)^k|x| < 1 \Leftrightarrow$$

$|x| < k^k$ for convergence, and the radius of convergence is $R = k^k$.

33. (a) If $a_n = \dfrac{(-1)^n x^{2n+1}}{n!(n+1)! 2^{2n+1}}$, then

$$\lim\limits_{n\to\infty}\left|\dfrac{a_{n+1}}{a_n}\right| = \lim\limits_{n\to\infty}\left|\dfrac{x^{2n+3}}{(n+1)!(n+2)! 2^{2n+3}}\cdot\dfrac{n!(n+1)! 2^{2n+1}}{x^{2n+1}}\right| = \left(\dfrac{x}{2}\right)^2\lim\limits_{n\to\infty}\dfrac{1}{(n+1)(n+2)} = 0$$ for

all x. So $J_1(x)$ converges for all x and its domain is $(-\infty, \infty)$.

(b), (c) The initial terms of $J_1(x)$ up to $n = 5$ are

$a_0 = \dfrac{x}{2}$, $a_1 = -\dfrac{x^3}{16}$, $a_2 = \dfrac{x^5}{384}$, $a_3 = -\dfrac{x^7}{18,432}$,

$a_4 = \dfrac{x^9}{1,474,560}$, and $a_5 = -\dfrac{x^{11}}{176,947,200}$. The

partial sums seem to approximate $J_1(x)$ well near

the origin, but as $|x|$ increases, we need to take a

large number of terms to get a good approximation.

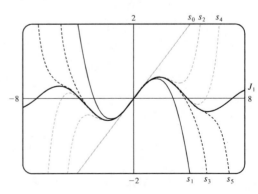

35. $s_{2n-1} = 1 + 2x + x^2 + 2x^3 + x^4 + 2x^5 + \cdots + x^{2n-2} + 2x^{2n-1}$

$\qquad = 1(1 + 2x) + x^2(1 + 2x) + x^4(1 + 2x) + \cdots + x^{2n-2}(1 + 2x)$

$\qquad = (1 + 2x)(1 + x^2 + x^4 + \cdots + x^{2n-2})$

$\qquad = (1 + 2x)\dfrac{1 - x^{2n}}{1 - x^2}$ [by (12.2.3 [ET 11.2.3]) with $r = x^2$]

$\qquad \to \dfrac{1 + 2x}{1 - x^2}$ as $n \to \infty$ [by (12.2.4 [ET 11.2.4])], when $|x| < 1$. Also $s_{2n} = s_{2n-1} + x^{2n} \to \dfrac{1 + 2x}{1 - x^2}$ since

$x^{2n} \to 0$ for $|x| < 1$. Therefore, $s_n \to \dfrac{1 + 2x}{1 - x^2}$ since s_{2n} and s_{2n-1} both approach $\dfrac{1 + 2x}{1 - x^2}$ as $n \to \infty$. Thus, the

interval of convergence is $(-1, 1)$ and $f(x) = \dfrac{1 + 2x}{1 - x^2}$.

37. We use the Root Test on the series $\sum c_n x^n$. We need $\lim\limits_{n \to \infty} \sqrt[n]{|c_n x^n|} = |x| \lim\limits_{n \to \infty} \sqrt[n]{|c_n|} = c\,|x| < 1$ for

convergence, or $|x| < 1/c$, so $R = 1/c$.

39. For $2 < x < 3$, $\sum c_n x^n$ diverges and $\sum d_n x^n$ converges. By Exercise 12.2.61 [ET 11.2.61], $\sum(c_n + d_n)\,x^n$

diverges. Since both series converge for $|x| < 2$, the radius of convergence of $\sum(c_n + d_n)\,x^n$ is 2.

12.9 Representations of Functions as Power Series ET 11.9

1. If $f(x) = \sum\limits_{n=0}^{\infty} c_n x^n$ has radius of convergence 10, then $f'(x) = \sum\limits_{n=1}^{\infty} n c_n x^{n-1}$ also has radius of convergence 10

by Theorem 2.

3. Our goal is to write the function in the form $\dfrac{1}{1 - r}$, and then use Equation (1) to represent the function as a sum of a

power series. $f(x) = \dfrac{1}{1 + x} = \dfrac{1}{1 - (-x)} = \sum\limits_{n=0}^{\infty} (-x)^n = \sum\limits_{n=0}^{\infty} (-1)^n x^n$ with $|-x| < 1 \iff |x| < 1$, so $R = 1$

and $I = (-1, 1)$.

5. Replacing x with x^3 in (1) gives $f(x) = \dfrac{1}{1 - x^3} = \sum\limits_{n=0}^{\infty} (x^3)^n = \sum\limits_{n=0}^{\infty} x^{3n}$. The series converges when $|x^3| < 1$

$\iff |x|^3 < 1 \iff |x| < \sqrt[3]{1} \iff |x| < 1$. Thus, $R = 1$ and $I = (-1, 1)$.

7. $f(x) = \dfrac{1}{x - 5} = -\dfrac{1}{5}\left(\dfrac{1}{1 - x/5}\right) = -\dfrac{1}{5} \sum\limits_{n=0}^{\infty} \left(\dfrac{x}{5}\right)^n$ or equivalently, $-\sum\limits_{n=0}^{\infty} \dfrac{1}{5^{n+1}} x^n$. The series converges when

$\left|\dfrac{x}{5}\right| < 1$; that is, when $|x| < 5$, so $I = (-5, 5)$.

9. $f(x) = \dfrac{x}{9 + x^2} = \dfrac{x}{9}\left[\dfrac{1}{1 + (x/3)^2}\right] = \dfrac{x}{9}\left[\dfrac{1}{1 - \{-(x/3)^2\}}\right] = \dfrac{x}{9} \sum\limits_{n=0}^{\infty} \left[-\left(\dfrac{x}{3}\right)^2\right]^n$

$= \dfrac{x}{9} \sum\limits_{n=0}^{\infty} (-1)^n \dfrac{x^{2n}}{9^n} = \sum\limits_{n=0}^{\infty} (-1)^n \dfrac{x^{2n+1}}{9^{n+1}}$. The geometric series $\sum\limits_{n=0}^{\infty} \left[-\left(\dfrac{x}{3}\right)^2\right]^n$ converges when

$\left|-\left(\dfrac{x}{3}\right)^2\right| < 1 \iff \dfrac{|x^2|}{9} < 1 \iff |x|^2 < 9 \iff |x| < 3$, so $R = 3$ and $I = (-3, 3)$.

11. $f(x) = \dfrac{3}{x^2 + x - 2} = \dfrac{3}{(x + 2)(x - 1)} = \dfrac{A}{x + 2} + \dfrac{B}{x - 1}$ ⟹ $3 = A(x - 1) + B(x + 2)$. Taking $x = -2$, we

get $A = -1$. Taking $x = 1$, we get $B = 1$. Thus,

$$\frac{3}{x^2 + x - 2} = \frac{1}{x - 1} - \frac{1}{x + 2} = -\frac{1}{1 - x} - \frac{1}{2}\frac{1}{1 + x/2} = -\sum_{n=0}^{\infty} x^n - \frac{1}{2}\sum_{n=0}^{\infty}\left(-\frac{x}{2}\right)^n$$

$$= \sum_{n=0}^{\infty}\left[-1 - \frac{1}{2}\left(-\frac{1}{2}\right)^n\right]x^n = \sum_{n=0}^{\infty}\left[-1 + \left(-\frac{1}{2}\right)^{n+1}\right]x^n = \sum_{n=0}^{\infty}\left[\frac{(-1)^{n+1}}{2^{n+1}} - 1\right]x^n$$

We represented the given function as the sum of two geometric series; the first converges for $x \in (-1, 1)$ and the

second converges for $x \in (-2, 2)$. Thus, the sum converges for $x \in (-1, 1) = I$.

13. (a) $f(x) = \dfrac{1}{(1 + x)^2} = \dfrac{d}{dx}\left(\dfrac{-1}{1 + x}\right) = -\dfrac{d}{dx}\left[\displaystyle\sum_{n=0}^{\infty}(-1)^n x^n\right]$ [from Exercise 3]

$$= \sum_{n=1}^{\infty}(-1)^{n+1}nx^{n-1} \text{ [from Theorem 2(i)] } = \sum_{n=0}^{\infty}(-1)^n(n + 1)x^n \text{ with } R = 1.$$

In the last step, note that we *decreased* the initial value of the summation variable n by 1, and then *increased*

each occurrence of n in the term by 1 [also note that $(-1)^{n+2} = (-1)^n$].

(b) $f(x) = \dfrac{1}{(1 + x)^3} = -\dfrac{1}{2}\dfrac{d}{dx}\left[\dfrac{1}{(1 + x)^2}\right] = -\dfrac{1}{2}\dfrac{d}{dx}\left[\displaystyle\sum_{n=0}^{\infty}(-1)^n(n + 1)x^n\right]$ [from part (a)]

$$= -\frac{1}{2}\sum_{n=1}^{\infty}(-1)^n(n + 1)nx^{n-1} = \frac{1}{2}\sum_{n=0}^{\infty}(-1)^n(n + 2)(n + 1)x^n \text{ with } R = 1.$$

(c) $f(x) = \dfrac{x^2}{(1 + x)^3} = x^2 \cdot \dfrac{1}{(1 + x)^3} = x^2 \cdot \dfrac{1}{2}\displaystyle\sum_{n=0}^{\infty}(-1)^n(n + 2)(n + 1)x^n$ [from part (b)]

$$= \frac{1}{2}\sum_{n=0}^{\infty}(-1)^n(n + 2)(n + 1)x^{n+2}. \text{ To write the power series with } x^n \text{ rather than } x^{n+2},$$

we will *decrease* each occurrence of n in the term by 2 and *increase* the initial value of the summation variable

by 2. This gives us $\dfrac{1}{2}\displaystyle\sum_{n=2}^{\infty}(-1)^n(n)(n - 1)x^n$.

15. $f(x) = \ln(5 - x) = -\displaystyle\int\frac{dx}{5 - x} = -\frac{1}{5}\int\frac{dx}{1 - x/5}$

$$= -\frac{1}{5}\int\left[\sum_{n=0}^{\infty}\left(\frac{x}{5}\right)^n\right]dx = C - \frac{1}{5}\sum_{n=0}^{\infty}\frac{x^{n+1}}{5^n(n + 1)} = C - \sum_{n=1}^{\infty}\frac{x^n}{n5^n}$$

Putting $x = 0$, we get $C = \ln 5$. The series converges for $|x/5| < 1$ ⟺ $|x| < 5$, so $R = 5$.

17. $\dfrac{1}{2 - x} = \dfrac{1}{2(1 - x/2)} = \dfrac{1}{2}\displaystyle\sum_{n=0}^{\infty}\left(\frac{x}{2}\right)^n = \sum_{n=0}^{\infty}\frac{1}{2^{n+1}}x^n$ for $\left|\dfrac{x}{2}\right| < 1$ ⟺ $|x| < 2$. Now

$$\frac{1}{(x - 2)^2} = \frac{d}{dx}\left(\frac{1}{2 - x}\right) = \frac{d}{dx}\left(\sum_{n=0}^{\infty}\frac{1}{2^{n+1}}x^n\right) = \sum_{n=1}^{\infty}\frac{n}{2^{n+1}}x^{n-1} = \sum_{n=0}^{\infty}\frac{n + 1}{2^{n+2}}x^n. \text{ So}$$

$$f(x) = \frac{x^3}{(x - 2)^2} = x^3\sum_{n=0}^{\infty}\frac{n + 1}{2^{n+2}}x^n = \sum_{n=0}^{\infty}\frac{n + 1}{2^{n+2}}x^{n+3} \text{ or } \sum_{n=3}^{\infty}\frac{n - 2}{2^{n-1}}x^n \text{ for } |x| < 2. \text{ Thus, } R = 2 \text{ and}$$

$I = (-2, 2)$.

19. $f(x) = \ln(3 + x) = \displaystyle\int \frac{dx}{3 + x} = \frac{1}{3}\int \frac{dx}{1 + x/3} = \frac{1}{3}\int \frac{dx}{1 - (-x/3)} = \frac{1}{3}\int \sum_{n=0}^{\infty}\left(-\frac{x}{3}\right)^n dx$

$= C + \dfrac{1}{3}\displaystyle\sum_{n=0}^{\infty} \dfrac{(-1)^n}{(n+1)3^n}x^{n+1} = \ln 3 + \dfrac{1}{3}\sum_{n=1}^{\infty}\dfrac{(-1)^{n-1}}{n3^{n-1}}x^n \ \ [C = f(0) = \ln 3]$

$= \ln 3 + \displaystyle\sum_{n=1}^{\infty}\dfrac{(-1)^{n-1}}{n3^n}x^n$. The series converges when $|-x/3| < 1 \ \Leftrightarrow \ |x| < 3$, so $R = 3$.

The terms of the series are $a_0 = \ln 3, a_1 = \dfrac{x}{3}, a_2 = -\dfrac{x^2}{18}, a_3 = \dfrac{x^3}{81}, a_4 = -\dfrac{x^4}{324}, a_5 = \dfrac{x^5}{1215}, \ldots\ .$

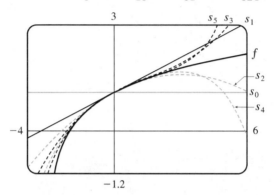

As n increases, $s_n(x)$ approximates f better on the interval of convergence, which is $(-3, 3)$.

21. $f(x) = \ln\left(\dfrac{1+x}{1-x}\right) = \ln(1+x) - \ln(1-x) = \displaystyle\int \dfrac{dx}{1+x} + \int \dfrac{dx}{1-x}$

$= \displaystyle\int \dfrac{dx}{1-(-x)} + \int \dfrac{dx}{1-x} = \int \left[\sum_{n=0}^{\infty}(-1)^n x^n + \sum_{n=0}^{\infty}x^n\right]dx$

$= \displaystyle\int \left[(1 - x + x^2 - x^3 + x^4 - \cdots) + (1 + x + x^2 + x^3 + x^4 + \cdots)\right]dx$

$= \displaystyle\int \left(2 + 2x^2 + 2x^4 + \cdots\right)dx = \int \sum_{n=0}^{\infty} 2x^{2n}\,dx = C + \sum_{n=0}^{\infty}\dfrac{2x^{2n+1}}{2n+1}$

But $f(0) = \ln\frac{1}{1} = 0$, so $C = 0$ and we have $f(x) = \displaystyle\sum_{n=0}^{\infty}\dfrac{2x^{2n+1}}{2n+1}$ with $R = 1$. If $x = \pm 1$, then

$f(x) = \pm 2\displaystyle\sum_{n=0}^{\infty}\dfrac{1}{2n+1}$, which both diverge by the Limit Comparison Test with $b_n = \dfrac{1}{n}$.

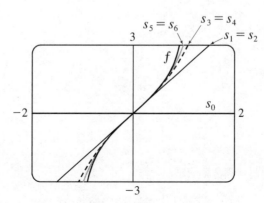

As n increases, $s_n(x)$ approximates f better on the interval of convergence, which is $(-1, 1)$.

23. $\dfrac{t}{1-t^8} = t \cdot \dfrac{1}{1-t^8} = t \displaystyle\sum_{n=0}^{\infty}(t^8)^n = \displaystyle\sum_{n=0}^{\infty} t^{8n+1} \quad\Rightarrow\quad \displaystyle\int \dfrac{t}{1-t^8}\,dt = C + \displaystyle\sum_{n=0}^{\infty}\dfrac{t^{8n+2}}{8n+2}$. The series for $\dfrac{1}{1-t^8}$

converges when $\left|t^8\right| < 1 \quad\Leftrightarrow\quad |t| < 1$, so $R = 1$ for that series and also the series for $t/(1-t^8)$. By Theorem

2, the series for $\displaystyle\int \dfrac{t}{1-t^8}\,dt$ also has $R = 1$.

25. By Example 7, $\tan^{-1}x = \displaystyle\sum_{n=0}^{\infty}(-1)^n \dfrac{x^{2n+1}}{2n+1}$ with $R = 1$, so

$x - \tan^{-1}x = x - \left(x - \dfrac{x^3}{3} + \dfrac{x^5}{5} - \dfrac{x^7}{7} + \cdots \right) = \dfrac{x^3}{3} - \dfrac{x^5}{5} + \dfrac{x^7}{7} - \cdots = \displaystyle\sum_{n=1}^{\infty}(-1)^{n+1}\dfrac{x^{2n+1}}{2n+1}$

and $\dfrac{x - \tan^{-1}x}{x^3} = \displaystyle\sum_{n=1}^{\infty}(-1)^{n+1}\dfrac{x^{2n-2}}{2n+1}$, so

$\displaystyle\int \dfrac{x - \tan^{-1}x}{x^3}\,dx = C + \displaystyle\sum_{n=1}^{\infty}(-1)^{n+1}\dfrac{x^{2n-1}}{(2n+1)(2n-1)} = C + \displaystyle\sum_{n=1}^{\infty}(-1)^{n+1}\dfrac{x^{2n-1}}{4n^2-1}$. By Theorem 2, $R = 1$.

27. $\dfrac{1}{1+x^5} = \dfrac{1}{1-(-x^5)} = \displaystyle\sum_{n=0}^{\infty}\left(-x^5\right)^n = \displaystyle\sum_{n=0}^{\infty}(-1)^n x^{5n} \quad\Rightarrow$

$\displaystyle\int \dfrac{1}{1+x^5}\,dx = \displaystyle\int \displaystyle\sum_{n=0}^{\infty}(-1)^n x^{5n}\,dx = C + \displaystyle\sum_{n=0}^{\infty}(-1)^n \dfrac{x^{5n+1}}{5n+1}$. Thus,

$I = \displaystyle\int_0^{0.2} \dfrac{1}{1+x^5}\,dx = \left[x - \dfrac{x^6}{6} + \dfrac{x^{11}}{11} - \cdots \right]_0^{0.2} = 0.2 - \dfrac{(0.2)^6}{6} + \dfrac{(0.2)^{11}}{11} - \cdots$. The series is alternating, so

if we use the first two terms, the error is at most $(0.2)^{11}/11 \approx 1.9 \times 10^{-9}$. So $I \approx 0.2 - (0.2)^6/6 \approx 0.199989$ to

six decimal places.

29. We substitute x^4 for x in Example 7, and find that

$$\int x^2 \tan^{-1}\left(x^4\right)dx = \int x^2 \displaystyle\sum_{n=0}^{\infty}(-1)^n \dfrac{\left(x^4\right)^{2n+1}}{2n+1}\,dx$$

$$= \int \displaystyle\sum_{n=0}^{\infty}(-1)^n \dfrac{x^{8n+6}}{2n+1}\,dx = C + \displaystyle\sum_{n=0}^{\infty}(-1)^n \dfrac{x^{8n+7}}{(2n+1)(8n+7)}$$

So $\displaystyle\int_0^{1/3} x^2 \tan^{-1}\left(x^4\right)dx = \left[\dfrac{x^7}{7} - \dfrac{x^{15}}{45} + \cdots \right]_0^{1/3} = \dfrac{1}{7 \cdot 3^7} - \dfrac{1}{45 \cdot 3^{15}} + \cdots$. The series is alternating,

so if we use only one term, the error is at most $1/\left(45 \cdot 3^{15}\right) \approx 1.5 \times 10^{-9}$. So

$\int_0^{1/3} x^2 \tan^{-1}\left(x^4\right)dx \approx 1/\left(7 \cdot 3^7\right) \approx 0.000065$ to six decimal places.

31. Using the result of Example 6, $\ln(1 - x) = -\sum\limits_{n=1}^{\infty} \dfrac{x^n}{n}$, with $x = -0.1$, we have

$$\ln 1.1 = \ln[1 - (-0.1)] = 0.1 - \frac{0.01}{2} + \frac{0.001}{3} - \frac{0.0001}{4} + \frac{0.00001}{5} - \cdots. \text{ The series is alternating, so if}$$

we use only the first four terms, the error is at most $\dfrac{0.00001}{5} = 0.000002$. So

$$\ln 1.1 \approx 0.1 - \frac{0.01}{2} + \frac{0.001}{3} - \frac{0.0001}{4} \approx 0.09531.$$

33. (a) $J_0(x) = \sum\limits_{n=0}^{\infty} \dfrac{(-1)^n \, x^{2n}}{2^{2n}(n!)^2}$, $J_0'(x) = \sum\limits_{n=1}^{\infty} \dfrac{(-1)^n \, 2nx^{2n-1}}{2^{2n}(n!)^2}$, and $J_0''(x) = \sum\limits_{n=1}^{\infty} \dfrac{(-1)^n \, 2n(2n-1)x^{2n-2}}{2^{2n}(n!)^2}$, so

$$x^2 J_0''(x) + x J_0'(x) + x^2 J_0(x) = \sum_{n=1}^{\infty} \frac{(-1)^n \, 2n(2n-1)x^{2n}}{2^{2n}(n!)^2} + \sum_{n=1}^{\infty} \frac{(-1)^n \, 2nx^{2n}}{2^{2n}(n!)^2} + \sum_{n=0}^{\infty} \frac{(-1)^n \, x^{2n+2}}{2^{2n}(n!)^2}$$

$$= \sum_{n=1}^{\infty} \frac{(-1)^n \, 2n(2n-1)x^{2n}}{2^{2n}(n!)^2} + \sum_{n=1}^{\infty} \frac{(-1)^n \, 2nx^{2n}}{2^{2n}(n!)^2} + \sum_{n=1}^{\infty} \frac{(-1)^{n-1} \, x^{2n}}{2^{2n-2}\left[(n-1)!\right]^2}$$

$$= \sum_{n=1}^{\infty} \frac{(-1)^n \, 2n(2n-1)x^{2n}}{2^{2n}(n!)^2} + \sum_{n=1}^{\infty} \frac{(-1)^n \, 2nx^{2n}}{2^{2n}(n!)^2} + \sum_{n=1}^{\infty} \frac{(-1)^n(-1)^{-1} 2^2 n^2 x^{2n}}{2^{2n}(n!)^2}$$

$$= \sum_{n=1}^{\infty} (-1)^n \left[\frac{2n(2n-1) + 2n - 2^2 n^2}{2^{2n}(n!)^2}\right] x^{2n} = \sum_{n=1}^{\infty} (-1)^n \left[\frac{4n^2 - 2n + 2n - 4n^2}{2^{2n}(n!)^2}\right] x^{2n} = 0$$

(b) $\displaystyle\int_0^1 J_0(x)\,dx = \int_0^1 \left[\sum_{n=0}^{\infty} \frac{(-1)^n \, x^{2n}}{2^{2n}(n!)^2}\right] dx = \int_0^1 \left(1 - \frac{x^2}{4} + \frac{x^4}{64} - \frac{x^6}{2304} + \cdots\right) dx$

$$= \left[x - \frac{x^3}{3 \cdot 4} + \frac{x^5}{5 \cdot 64} - \frac{x^7}{7 \cdot 2304} + \cdots\right]_0^1 = 1 - \frac{1}{12} + \frac{1}{320} - \frac{1}{16{,}128} + \cdots$$

Since $\frac{1}{16{,}128} \approx 0.000062$, it follows from The Alternating Series Estimation Theorem that, correct to three

decimal places, $\int_0^1 J_0(x)\,dx \approx 1 - \frac{1}{12} + \frac{1}{320} \approx 0.920$.

35. (a) $f(x) = \sum\limits_{n=0}^{\infty} \dfrac{x^n}{n!} \;\Rightarrow\; f'(x) = \sum\limits_{n=1}^{\infty} \dfrac{nx^{n-1}}{n!} = \sum\limits_{n=1}^{\infty} \dfrac{x^{n-1}}{(n-1)!} = \sum\limits_{n=0}^{\infty} \dfrac{x^n}{n!} = f(x)$

(b) By Theorem 10.4.2 [ET 9.4.2], the only solution to the differential equation $df(x)/dx = f(x)$ is $f(x) = Ke^x$, but $f(0) = 1$, so $K = 1$ and $f(x) = e^x$.

Or: We could solve the equation $df(x)/dx = f(x)$ as a separable differential equation.

37. If $a_n = \dfrac{x^n}{n^2}$, then by the Ratio Test, $\lim\limits_{n \to \infty}\left|\dfrac{a_{n+1}}{a_n}\right| = \lim\limits_{n \to \infty}\left|\dfrac{x^{n+1}}{(n+1)^2} \cdot \dfrac{n^2}{x^n}\right| = |x|\lim\limits_{n \to \infty}\left(\dfrac{n}{n+1}\right)^2 = |x| < 1$ for

convergence, so $R = 1$. When $x = \pm 1$, $\sum\limits_{n=1}^{\infty}\left|\dfrac{x^n}{n^2}\right| = \sum\limits_{n=1}^{\infty}\dfrac{1}{n^2}$ which is a convergent p-series ($p = 2 > 1$), so the

interval of convergence for f is $[-1, 1]$. By Theorem 2, the radii of convergence of f' and f'' are both 1, so we need

only check the endpoints. $f(x) = \sum\limits_{n=1}^{\infty}\dfrac{x^n}{n^2}$ $\Rightarrow$ $f'(x) = \sum\limits_{n=1}^{\infty}\dfrac{nx^{n-1}}{n^2} = \sum\limits_{n=0}^{\infty}\dfrac{x^n}{n+1}$, and this series diverges for

$x = 1$ (harmonic series) and converges for $x = -1$ (Alternating Series Test), so the interval of convergence

is $[-1, 1)$. $f''(x) = \sum\limits_{n=1}^{\infty}\dfrac{nx^{n-1}}{n+1}$ diverges at both 1 and -1 (Test for Divergence) since $\lim\limits_{n \to \infty}\dfrac{n}{n+1} = 1 \neq 0$, so its

interval of convergence is $(-1, 1)$.

39. By Example 7, $\tan^{-1} x = \sum\limits_{n=0}^{\infty}(-1)^n\dfrac{x^{2n+1}}{2n+1}$ for $|x| < 1$. In particular, for $x = \dfrac{1}{\sqrt{3}}$, we have

$$\dfrac{\pi}{6} = \tan^{-1}\left(\dfrac{1}{\sqrt{3}}\right) = \sum\limits_{n=0}^{\infty}(-1)^n\dfrac{\left(1/\sqrt{3}\right)^{2n+1}}{2n+1} = \sum\limits_{n=0}^{\infty}(-1)^n\left(\dfrac{1}{3}\right)^n\dfrac{1}{\sqrt{3}}\dfrac{1}{2n+1}, \text{ so}$$

$$\pi = \dfrac{6}{\sqrt{3}}\sum\limits_{n=0}^{\infty}\dfrac{(-1)^n}{(2n+1)3^n} = 2\sqrt{3}\sum\limits_{n=0}^{\infty}\dfrac{(-1)^n}{(2n+1)3^n}.$$

12.10 Taylor and Maclaurin Series   ET 11.10

1. Using Theorem 5 with $\sum\limits_{n=0}^{\infty} b_n(x-5)^n$, $b_n = \dfrac{f^{(n)}(a)}{n!}$, so $b_8 = \dfrac{f^{(8)}(5)}{8!}$.

3.

n	$f^{(n)}(x)$	$f^{(n)}(0)$
0	$\cos x$	1
1	$-\sin x$	0
2	$-\cos x$	-1
3	$\sin x$	0
4	$\cos x$	1
⋮	⋮	⋮

We use Equation 7 with $f(x) = \cos x$.

$$\cos x = f(0) + f'(0)x + \dfrac{f''(0)}{2!}x^2 + \dfrac{f^{(3)}(0)}{3!}x^3 + \dfrac{f^{(4)}(0)}{4!}x^4 + \cdots$$

$$= 1 - \dfrac{x^2}{2!} + \dfrac{x^4}{4!} - \cdots = \sum\limits_{n=0}^{\infty}\dfrac{(-1)^n x^{2n}}{(2n)!}$$

If $a_n = \dfrac{(-1)^n x^{2n}}{(2n)!}$, then

$$\lim\limits_{n \to \infty}\left|\dfrac{a_{n+1}}{a_n}\right| = \lim\limits_{n \to \infty}\left|\dfrac{x^{2n+2}}{(2n+2)!} \cdot \dfrac{(2n)!}{x^{2n}}\right| = x^2\lim\limits_{n \to \infty}\dfrac{1}{(2n+2)(2n+1)} = 0 < 1 \text{ for all } x.$$

So $R = \infty$ (Ratio Test).

5.

n	$f^{(n)}(x)$	$f^{(n)}(0)$
0	$(1+x)^{-3}$	1
1	$-3(1+x)^{-4}$	-3
2	$12(1+x)^{-5}$	12
3	$-60(1+x)^{-6}$	-60
4	$360(1+x)^{-7}$	360
$\vdots$	$\vdots$	$\vdots$

$$(1+x)^{-3} = f(0) + f'(0)x + \frac{f''(0)}{2!}x^2 + \frac{f'''(0)}{3!}x^3 + \frac{f^{(4)}(0)}{4!}x^4 + \cdots$$

$$= 1 - 3x + \frac{4 \cdot 3}{2!}x^2 - \frac{5 \cdot 4 \cdot 3}{3!}x^3 + \frac{6 \cdot 5 \cdot 4 \cdot 3}{4!}x^4 - \cdots$$

$$= 1 - 3x + \frac{4 \cdot 3 \cdot 2}{2 \cdot 2!}x^2 - \frac{5 \cdot 4 \cdot 3 \cdot 2}{2 \cdot 3!}x^3 + \frac{6 \cdot 5 \cdot 4 \cdot 3 \cdot 2}{2 \cdot 4!}x^4 - \cdots$$

$$= \sum_{n=0}^{\infty} \frac{(-1)^n (n+2)! \, x^n}{2(n!)} = \sum_{n=0}^{\infty} \frac{(-1)^n (n+2)(n+1)x^n}{2}$$

$$\lim_{n \to \infty} \left| \frac{a_{n+1}}{a_n} \right| = \lim_{n \to \infty} \left| \frac{(n+3)(n+2)x^{n+1}}{2} \cdot \frac{2}{(n+2)(n+1)x^n} \right| = |x| \lim_{n \to \infty} \frac{n+3}{n+1} = |x| < 1 \text{ for convergence,}$$

so $R = 1$ (Ratio Test).

7.

n	$f^{(n)}(x)$	$f^{(n)}(0)$
0	e^{5x}	1
1	$5e^{5x}$	5
2	$5^2 e^{5x}$	25
3	$5^3 e^{5x}$	125
4	$5^4 e^{5x}$	625
$\vdots$	$\vdots$	$\vdots$

$$e^{5x} = \sum_{n=0}^{\infty} \frac{f^{(n)}(0)}{n!}x^n = \sum_{n=0}^{\infty} \frac{5^n}{n!}x^n.$$

$$\lim_{n \to \infty} \left| \frac{a_{n+1}}{a_n} \right| = \lim_{n \to \infty} \left[\frac{5^{n+1} |x|^{n+1}}{(n+1)!} \cdot \frac{n!}{5^n |x|^n} \right]$$

$$= \lim_{n \to \infty} \frac{5 |x|}{n+1} = 0 < 1 \text{ for all } x, \text{ so } R = \infty.$$

9.

n	$f^{(n)}(x)$	$f^{(n)}(0)$
0	$\sinh x$	0
1	$\cosh x$	1
2	$\sinh x$	0
3	$\cosh x$	1
4	$\sinh x$	0
$\vdots$	$\vdots$	$\vdots$

$$f^{(n)}(0) = \begin{cases} 0 & \text{if } n \text{ is even} \\ 1 & \text{if } n \text{ is odd} \end{cases} \text{ so } \sinh x = \sum_{n=0}^{\infty} \frac{x^{2n+1}}{(2n+1)!}.$$

Use the Ratio Test to find R. If $a_n = \dfrac{x^{2n+1}}{(2n+1)!}$, then

$$\lim_{n \to \infty} \left| \frac{a_{n+1}}{a_n} \right| = \lim_{n \to \infty} \left| \frac{x^{2n+3}}{(2n+3)!} \cdot \frac{(2n+1)!}{x^{2n+1}} \right|$$

$$= x^2 \cdot \lim_{n \to \infty} \frac{1}{(2n+3)(2n+2)} = 0 < 1$$

for all x, so $R = \infty$.

11.

n	$f^{(n)}(x)$	$f^{(n)}(2)$
0	$1 + x + x^2$	7
1	$1 + 2x$	5
2	2	2
3	0	0
4	0	0
$\vdots$	$\vdots$	$\vdots$

$$f(x) = 7 + 5(x-2) + \frac{2}{2!}(x-2)^2 + \sum_{n=3}^{\infty} \frac{0}{n!}(x-2)^n$$
$$= 7 + 5(x-2) + (x-2)^2$$

Since $a_n = 0$ for large n, $R = \infty$.

13. Clearly, $f^{(n)}(x) = e^x$, so $f^{(n)}(3) = e^3$ and $e^x = \sum_{n=0}^{\infty} \frac{e^3}{n!}(x-3)^n$. If $a_n = \frac{e^3}{n!}(x-3)^n$, then

$$\lim_{n\to\infty}\left|\frac{a_{n+1}}{a_n}\right| = \lim_{n\to\infty}\left|\frac{e^3(x-3)^{n+1}}{(n+1)!}\cdot\frac{n!}{e^3(x-3)^n}\right| = \lim_{n\to\infty}\frac{|x-3|}{n+1} = 0 < 1 \text{ for all } x, \text{ so } R = \infty.$$

15.

n	$f^{(n)}(x)$	$f^{(n)}(\pi)$
0	$\cos x$	-1
1	$-\sin x$	0
2	$-\cos x$	1
3	$\sin x$	0
4	$\cos x$	-1
$\vdots$	$\vdots$	$\vdots$

$$\cos x = \sum_{k=0}^{\infty}\frac{f^{(k)}(\pi)}{k!}(x-\pi)^k = -1 + \frac{(x-\pi)^2}{2!} - \frac{(x-\pi)^4}{4!} + \frac{(x-\pi)^6}{6!} - \cdots = \sum_{n=0}^{\infty}(-1)^{n+1}\frac{(x-\pi)^{2n}}{(2n)!}.$$

$$\lim_{n\to\infty}\left|\frac{a_{n+1}}{a_n}\right| = \lim_{n\to\infty}\left[\frac{|x-\pi|^{2n+2}}{(2n+2)!}\cdot\frac{(2n)!}{|x-\pi|^{2n}}\right] = \lim_{n\to\infty}\frac{|x-\pi|^2}{(2n+2)(2n+1)} = 0 < 1 \text{ for all } x, \text{ so } R = \infty.$$

17.

n	$f^{(n)}(x)$	$f^{(n)}(9)$
0	$x^{-1/2}$	$\frac{1}{3}$
1	$-\frac{1}{2}x^{-3/2}$	$-\frac{1}{2}\cdot\frac{1}{3^3}$
2	$\frac{3}{4}x^{-5/2}$	$-\frac{1}{2}\cdot\left(-\frac{3}{2}\right)\cdot\frac{1}{3^5}$
3	$-\frac{15}{8}x^{-7/2}$	$-\frac{1}{2}\cdot\left(-\frac{3}{2}\right)\cdot\left(-\frac{5}{2}\right)\cdot\frac{1}{3^7}$
$\vdots$	$\vdots$	$\vdots$

$$\frac{1}{\sqrt{x}} = \frac{1}{3} - \frac{1}{2\cdot 3^3}(x-9) + \frac{3}{2^2\cdot 3^5}\frac{(x-9)^2}{2!} - \frac{3\cdot 5}{2^3\cdot 3^7}\frac{(x-9)^3}{3!} + \cdots$$

$$= \sum_{n=0}^{\infty}(-1)^n\frac{1\cdot 3\cdot 5\cdot\ \cdots\ \cdot(2n-1)}{2^n\cdot 3^{2n+1}\cdot n!}(x-9)^n.$$

$$\lim_{n\to\infty}\left|\frac{a_{n+1}}{a_n}\right| = \lim_{n\to\infty}\left[\frac{1\cdot3\cdot5\cdot\ \cdots\ \cdot(2n-1)[2(n+1)-1]\,|x-9|^{n+1}}{2^{n+1}\cdot3^{[2(n+1)+1]}\cdot(n+1)!}\cdot\frac{2^n\cdot3^{2n+1}\cdot n!}{1\cdot3\cdot5\cdot\ \cdots\ \cdot(2n-1)\,|x-9|^n}\right]$$

$$= \lim_{n\to\infty}\left[\frac{(2n+1)\,|x-9|}{2\cdot3^2(n+1)}\right] = \frac{1}{9}\,|x-9| < 1$$

for convergence, so $|x-9| < 9$ and $R = 9$.

19. If $f(x) = \cos x$, then $f^{(n+1)}(x) = \pm\sin x$ or $\pm\cos x$. In each case, $\left|f^{(n+1)}(x)\right| \le 1$, so by Formula 9 with $a = 0$

and $M = 1$, $|R_n(x)| \le \dfrac{1}{(n+1)!}\,|x|^{n+1}$. Thus, $|R_n(x)| \to 0$ as $n \to \infty$ by Equation 10. So $\lim\limits_{n\to\infty} R_n(x) = 0$ and,

by Theorem 8, the series in Exercise 3 represents $\cos x$ for all x.

21. If $f(x) = \sinh x$, then for all n, $f^{(n+1)}(x) = \cosh x$ or $\sinh x$. Since $|\sinh x| < |\cosh x| = \cosh x$ for all x, we

have $\left|f^{(n+1)}(x)\right| \le \cosh x$ for all n. If d is any positive number and $|x| \le d$, then $\left|f^{(n+1)}(x)\right| \le \cosh x \le \cosh d$,

so by Formula 9 with $a = 0$ and $M = \cosh d$, we have $|R_n(x)| \le \dfrac{\cosh d}{(n+1)!}\,|x|^{n+1}$. It follows that $|R_n(x)| \to 0$

as $n \to \infty$ for $|x| \le d$ (by Equation 10). But d was an arbitrary positive number. So by Theorem 8, the series

represents $\sinh x$ for all x.

23. $\cos x = \displaystyle\sum_{n=0}^{\infty}(-1)^n\frac{x^{2n}}{(2n)!} \quad\Rightarrow\quad f(x) = \cos(\pi x) = \displaystyle\sum_{n=0}^{\infty}\frac{(-1)^n(\pi x)^{2n}}{(2n)!} = \displaystyle\sum_{n=0}^{\infty}\frac{(-1)^n\pi^{2n}x^{2n}}{(2n)!},\ R = \infty$

25. $\tan^{-1}x = \displaystyle\sum_{n=0}^{\infty}(-1)^n\frac{x^{2n+1}}{2n+1} \quad\Rightarrow\quad f(x) = x\tan^{-1}x = x\displaystyle\sum_{n=0}^{\infty}(-1)^n\frac{x^{2n+1}}{2n+1} = \displaystyle\sum_{n=0}^{\infty}(-1)^n\frac{x^{2n+2}}{2n+1},\ R = 1$

27. $e^x = \displaystyle\sum_{n=0}^{\infty}\frac{x^n}{n!} \quad\Rightarrow\quad f(x) = x^2e^{-x} = x^2\displaystyle\sum_{n=0}^{\infty}\frac{(-x)^n}{n!} = \displaystyle\sum_{n=0}^{\infty}\frac{(-1)^n\,x^{n+2}}{n!},\ R = \infty$

29. $\sin^2 x = \dfrac{1}{2}(1 - \cos 2x) = \dfrac{1}{2}\left[1 - \displaystyle\sum_{n=0}^{\infty}\frac{(-1)^n(2x)^{2n}}{(2n)!}\right] = \dfrac{1}{2}\left[1 - 1 - \displaystyle\sum_{n=1}^{\infty}\frac{(-1)^n(2x)^{2n}}{(2n)!}\right]$

$$= \displaystyle\sum_{n=1}^{\infty}\frac{(-1)^{n+1}2^{2n-1}x^{2n}}{(2n)!},\ R = \infty$$

31. $\dfrac{\sin x}{x} = \dfrac{1}{x}\displaystyle\sum_{n=0}^{\infty}\frac{(-1)^n x^{2n+1}}{(2n+1)!} = \displaystyle\sum_{n=0}^{\infty}\frac{(-1)^n x^{2n}}{(2n+1)!}$ and this series also gives the required value at $x = 0$ (namely 1);

$R = \infty$.

33.

n	$f^{(n)}(x)$	$f^{(n)}(0)$
0	$(1+x)^{1/2}$	1
1	$\frac{1}{2}(1+x)^{-1/2}$	$\frac{1}{2}$
2	$-\frac{1}{4}(1+x)^{-3/2}$	$-\frac{1}{4}$
3	$\frac{3}{8}(1+x)^{-5/2}$	$\frac{3}{8}$
4	$-\frac{15}{16}(1+x)^{-7/2}$	$-\frac{15}{16}$
$\vdots$	$\vdots$	$\vdots$

So $f^{(n)}(0) = \dfrac{(-1)^{n-1}\,1\cdot 3\cdot 5\cdot\,\cdots\,\cdot(2n-3)}{2^n}$ for $n \geq 2$, and

$$\sqrt{1+x} = 1 + \frac{x}{2} + \sum_{n=2}^{\infty}\frac{(-1)^{n-1}1\cdot 3\cdot 5\cdot\,\cdots\,\cdot(2n-3)}{2^n n!}x^n. \text{ If } a_n = \frac{(-1)^{n-1}1\cdot 3\cdot 5\cdot\,\cdots\,\cdot(2n-3)}{2^n n!}x^n,$$

then $\displaystyle\lim_{n\to\infty}\left|\frac{a_{n+1}}{a_n}\right| = \lim_{n\to\infty}\left|\frac{1\cdot 3\cdot 5\cdot\,\cdots\,\cdot(2n-3)(2n-1)x^{n+1}}{2^{n+1}(n+1)!}\cdot\frac{2^n n!}{1\cdot 3\cdot 5\cdot\,\cdots\,\cdot(2n-3)x^n}\right|$

$$= \frac{|x|}{2}\lim_{n\to\infty}\frac{2n-1}{n+1} = \frac{|x|}{2}\cdot 2 = |x| < 1 \text{ for convergence, so } R = 1.$$

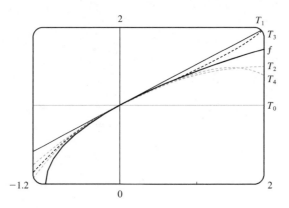

Notice that, as n increases, $T_n(x)$ becomes a better approximation to $f(x)$ for $-1 < x < 1$.

35. $\cos x = \displaystyle\sum_{n=0}^{\infty}(-1)^n\frac{x^{2n}}{(2n)!} \ \Rightarrow\ f(x) = \cos(x^2) = \sum_{n=0}^{\infty}\frac{(-1)^n\left(x^2\right)^{2n}}{(2n)!} = \sum_{n=0}^{\infty}\frac{(-1)^n x^{4n}}{(2n)!}, \ R = \infty$

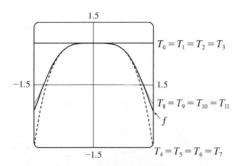

Notice that, as n increases, $T_n(x)$

becomes a better approximation to $f(x)$.

37. $e^x = \sum\limits_{n=0}^{\infty} \dfrac{x^n}{n!}$, so

$$e^{-0.2} = \sum_{n=0}^{\infty} \frac{(-0.2)^n}{n!} = 1 - 0.2 + \frac{1}{2!}(0.2)^2 - \frac{1}{3!}(0.2)^3 + \frac{1}{4!}(0.2)^4 - \frac{1}{5!}(0.2)^5 + \frac{1}{6!}(0.2)^6 - \cdots. \text{ But}$$

$\dfrac{1}{6!}(0.2)^6 = 8.\overline{8} \times 10^{-8}$, so by the Alternating Series Estimation Theorem, $e^{-0.2} \approx \sum\limits_{n=0}^{5} \dfrac{(-0.2)^n}{n!} \approx 0.81873$,

correct to five decimal places.

39. $\cos x \overset{(16)}{=} \sum\limits_{n=0}^{\infty} (-1)^n \dfrac{x^{2n}}{(2n)!} \quad \Rightarrow \quad \cos(x^3) = \sum\limits_{n=0}^{\infty} (-1)^n \dfrac{(x^3)^{2n}}{(2n)!} = \sum\limits_{n=0}^{\infty} (-1)^n \dfrac{x^{6n}}{(2n)!} \quad \Rightarrow$

$x \cos(x^3) = \sum\limits_{n=0}^{\infty} (-1)^n \dfrac{x^{6n+1}}{(2n)!} \quad \Rightarrow \quad \displaystyle\int x\cos(x^3)\, dx = C + \sum\limits_{n=0}^{\infty} (-1)^n \dfrac{x^{6n+2}}{(6n+2)(2n)!}$, with $R = \infty$.

41. Using the series from Exercise 33 and substituting x^3 for x, we get

$$\int \sqrt{x^3+1}\, dx = \int \left[1 + \frac{x^3}{2} + \sum_{n=2}^{\infty} \frac{(-1)^{n-1} 1 \cdot 3 \cdot 5 \cdot \cdots \cdot (2n-3)}{2^n n!} x^{3n} \right] dx$$

$$= C + x + \frac{x^4}{8} + \sum_{n=2}^{\infty} \frac{(-1)^{n-1} 1 \cdot 3 \cdot 5 \cdot \cdots \cdot (2n-3)}{2^n n! (3n+1)} x^{3n+1}$$

43. By Exercise 39, $\displaystyle\int x\cos(x^3)\, dx = C + \sum\limits_{n=0}^{\infty} (-1)^n \dfrac{x^{6n+2}}{(6n+2)(2n)!}$, so $\displaystyle\int_0^1 x\cos(x^3)\, dx$

$$= \left[\sum_{n=0}^{\infty} (-1)^n \frac{x^{6n+2}}{(6n+2)(2n)!} \right]_0^1 = \sum_{n=0}^{\infty} \frac{(-1)^n}{(6n+2)(2n)!} = \frac{1}{2} - \frac{1}{8 \cdot 2!} + \frac{1}{14 \cdot 4!} - \frac{1}{20 \cdot 6!} + \cdots, \text{ but}$$

$\dfrac{1}{20 \cdot 6!} = \dfrac{1}{14{,}400} \approx 0.000\,069$, so $\displaystyle\int_0^1 x\cos(x^3)\, dx \approx \frac{1}{2} - \frac{1}{16} + \frac{1}{336} \approx 0.440$ (correct to three decimal places)

by the Alternating Series Estimation Theorem.

45. We first find a series representation for $f(x) = (1+x)^{-1/2}$, and then substitute.

n	$f^{(n)}(x)$	$f^{(n)}(0)$
0	$(1+x)^{-1/2}$	1
1	$-\frac{1}{2}(1+x)^{-3/2}$	$-\frac{1}{2}$
2	$\frac{3}{4}(1+x)^{-5/2}$	$\frac{3}{4}$
3	$-\frac{15}{8}(1+x)^{-7/2}$	$-\frac{15}{8}$
$\vdots$	$\vdots$	$\vdots$

$$\frac{1}{\sqrt{1+x}} = 1 - \frac{x}{2} + \frac{3}{4}\left(\frac{x^2}{2!}\right) - \frac{15}{8}\left(\frac{x^3}{3!}\right) + \cdots \quad \Rightarrow \quad \frac{1}{\sqrt{1+x^3}} = 1 - \frac{1}{2}x^3 + \frac{3}{8}x^6 - \frac{5}{16}x^9 + \cdots \quad \Rightarrow$$

$$\int_0^{0.1} \frac{dx}{\sqrt{1+x^3}} = \left[x - \frac{1}{8}x^4 + \frac{3}{56}x^7 - \frac{1}{32}x^{10} + \cdots \right]_0^{0.1} \approx (0.1) - \frac{1}{8}(0.1)^4, \text{ by the Alternating Series}$$

Estimation Theorem, since $\frac{3}{56}(0.1)^7 \approx 0.000\,000\,005\,4 < 10^{-8}$, which is the maximum desired error. Therefore,

$$\int_0^{0.1} \frac{dx}{\sqrt{1+x^3}} \approx 0.099\,987\,50.$$

47. $\displaystyle\lim_{x\to 0} \frac{x - \tan^{-1} x}{x^3} = \lim_{x\to 0} \frac{x - \left(x - \frac{1}{3}x^3 + \frac{1}{5}x^5 - \frac{1}{7}x^7 + \cdots\right)}{x^3} = \lim_{x\to 0} \frac{\frac{1}{3}x^3 - \frac{1}{5}x^5 + \frac{1}{7}x^7 - \cdots}{x^3}$

$$= \lim_{x\to 0} \left(\frac{1}{3} - \frac{1}{5}x^2 + \frac{1}{7}x^4 - \cdots\right) = \frac{1}{3}$$

since power series are continuous functions.

49. $\displaystyle\lim_{x\to 0} \frac{\sin x - x + \frac{1}{6}x^3}{x^5} = \lim_{x\to 0} \frac{\left(x - \frac{1}{3!}x^3 + \frac{1}{5!}x^5 - \frac{1}{7!}x^7 + \cdots\right) - x + \frac{1}{6}x^3}{x^5}$

$$= \lim_{x\to 0} \frac{\frac{1}{5!}x^5 - \frac{1}{7!}x^7 + \cdots}{x^5} = \lim_{x\to 0} \left(\frac{1}{5!} - \frac{x^2}{7!} + \frac{x^4}{9!} - \cdots\right) = \frac{1}{5!} = \frac{1}{120}$$

since power series are continuous functions.

51. As in Example 8(a), we have $e^{-x^2} = 1 - \dfrac{x^2}{1!} + \dfrac{x^4}{2!} - \dfrac{x^6}{3!} + \cdots$ and we know that $\cos x = 1 - \dfrac{x^2}{2!} + \dfrac{x^4}{4!} - \cdots$

from Equation 16. Therefore, $e^{-x^2}\cos x = \left(1 - x^2 + \frac{1}{2}x^4 - \cdots\right)\left(1 - \frac{1}{2}x^2 + \frac{1}{24}x^4 - \cdots\right)$. Writing only the

terms with degree ≤ 4, we get $e^{-x^2}\cos x = 1 - \frac{1}{2}x^2 + \frac{1}{24}x^4 - x^2 + \frac{1}{2}x^4 + \frac{1}{2}x^4 + \cdots = 1 - \frac{3}{2}x^2 + \frac{25}{24}x^4 + \cdots$.

53.

$$
\begin{array}{r}
1 + \frac{1}{6}x^2 + \frac{7}{360}x^4 + \cdots \\
x - \frac{1}{6}x^3 + \frac{1}{120}x^5 - \cdots \overline{\smash{\big)}\, x } \\
\underline{x - \frac{1}{6}x^3 + \frac{1}{120}x^5 - \cdots} \\
\frac{1}{6}x^3 - \frac{1}{120}x^5 + \cdots \\
\underline{\frac{1}{6}x^3 - \frac{1}{36}x^5 + \cdots} \\
\frac{7}{360}x^5 + \cdots \\
\underline{\frac{7}{360}x^5 + \cdots} \\
\cdots
\end{array}
$$

$\dfrac{x}{\sin x} \overset{(15)}{=} \dfrac{x}{x - \frac{1}{6}x^3 + \frac{1}{120}x^5 - \cdots}$. From the long division above, $\dfrac{x}{\sin x} = 1 + \frac{1}{6}x^2 + \frac{7}{360}x^4 + \cdots$.

55. $\displaystyle\sum_{n=0}^{\infty} (-1)^n \frac{x^{4n}}{n!} = \sum_{n=0}^{\infty} \frac{\left(-x^4\right)^n}{n!} = e^{-x^4}$, by (11).

57. $\displaystyle\sum_{n=0}^{\infty} \frac{(-1)^n \pi^{2n+1}}{4^{2n+1}(2n+1)!} = \sum_{n=0}^{\infty} \frac{(-1)^n \left(\frac{\pi}{4}\right)^{2n+1}}{(2n+1)!} = \sin\frac{\pi}{4} = \frac{1}{\sqrt{2}}$, by (15).

59. $3 + \dfrac{9}{2!} + \dfrac{27}{3!} + \dfrac{81}{4!} + \cdots = \dfrac{3^1}{1!} + \dfrac{3^2}{2!} + \dfrac{3^3}{3!} + \dfrac{3^4}{4!} + \cdots = \displaystyle\sum_{n=1}^{\infty} \dfrac{3^n}{n!} = \displaystyle\sum_{n=0}^{\infty} \dfrac{3^n}{n!} - 1 = e^3 - 1$, by (11).

61. Assume that $|f'''(x)| \le M$, so $f'''(x) \le M$ for $a \le x \le a + d$. Now $\int_a^x f'''(t)\,dt \le \int_a^x M\,dt \quad\Rightarrow$

$f''(x) - f''(a) \le M(x - a) \quad\Rightarrow\quad f''(x) \le f''(a) + M(x - a)$. Thus, $\int_a^x f''(t)\,dt \le \int_a^x [f''(a) + M(t - a)]\,dt$

$\Rightarrow\quad f'(x) - f'(a) \le f''(a)(x - a) + \tfrac{1}{2}M(x - a)^2 \quad\Rightarrow\quad f'(x) \le f'(a) + f''(a)(x - a) + \tfrac{1}{2}M(x - a)^2 \quad\Rightarrow$

$\int_a^x f'(t)\,dt \le \int_a^x \left[f'(a) + f''(a)(t - a) + \tfrac{1}{2}M(t - a)^2 \right] dt \quad\Rightarrow$

$f(x) - f(a) \le f'(a)(x - a) + \tfrac{1}{2}f''(a)(x - a)^2 + \tfrac{1}{6}M(x - a)^3$. So

$f(x) - f(a) - f'(a)(x - a) - \tfrac{1}{2}f''(a)(x - a)^2 \le \tfrac{1}{6}M(x - a)^3$. But

$R_2(x) = f(x) - T_2(x) = f(x) - f(a) - f'(a)(x - a) - \tfrac{1}{2}f''(a)(x - a)^2$, so $R_2(x) \le \tfrac{1}{6}M(x - a)^3$.

A similar argument using $f'''(x) \ge -M$ shows that $R_2(x) \ge -\tfrac{1}{6}M(x - a)^3$. So $|R_2(x_2)| \le \tfrac{1}{6}M|x - a|^3$.

Although we have assumed that $x > a$, a similar calculation shows that this inequality is also true if $x < a$.

12.11 The Binomial Series   ET 11.11

1. The general binomial series in (2) is

$$(1 + x)^k = \sum_{n=0}^{\infty} \binom{k}{n} x^n = 1 + kx + \frac{k(k - 1)}{2!}x^2 + \frac{k(k - 1)(k - 2)}{3!}x^3 + \cdots.$$

$$(1 + x)^{1/2} = \sum_{n=0}^{\infty} \binom{\tfrac{1}{2}}{n} x^n = 1 + \left(\tfrac{1}{2}\right)x + \frac{\left(\tfrac{1}{2}\right)\left(-\tfrac{1}{2}\right)}{2!}x^2 + \frac{\left(\tfrac{1}{2}\right)\left(-\tfrac{1}{2}\right)\left(-\tfrac{3}{2}\right)}{3!}x^3 + \cdots$$

$$= 1 + \frac{x}{2} - \frac{x^2}{2^2 \cdot 2!} + \frac{1 \cdot 3 \cdot x^3}{2^3 \cdot 3!} - \frac{1 \cdot 3 \cdot 5 \cdot x^4}{2^4 \cdot 4!} + \cdots$$

$$= 1 + \frac{x}{2} + \sum_{n=2}^{\infty} \frac{(-1)^{n-1}\, 1 \cdot 3 \cdot 5 \cdots\cdots (2n - 3)x^n}{2^n \cdot n!} \quad \text{for } |x| < 1, \text{ so } R = 1$$

3. $\dfrac{1}{(2 + x)^3} = \dfrac{1}{[2(1 + x/2)]^3} = \dfrac{1}{8}\left(1 + \dfrac{x}{2}\right)^{-3} = \dfrac{1}{8}\displaystyle\sum_{n=0}^{\infty} \binom{-3}{n}\left(\dfrac{x}{2}\right)^n$. The binomial coefficient is

$$\binom{-3}{n} = \frac{(-3)(-4)(-5) \cdots\cdots (-3 - n + 1)}{n!} = \frac{(-3)(-4)(-5) \cdots\cdots [-(n + 2)]}{n!}$$

$$= \frac{(-1)^n \cdot 2 \cdot 3 \cdot 4 \cdot 5 \cdots\cdots (n + 1)(n + 2)}{2 \cdot n!} = \frac{(-1)^n(n + 1)(n + 2)}{2}$$

Thus, $\dfrac{1}{(2 + x)^3} = \dfrac{1}{8}\displaystyle\sum_{n=0}^{\infty} \dfrac{(-1)^n(n + 1)(n + 2)}{2}\dfrac{x^n}{2^n} = \displaystyle\sum_{n=0}^{\infty} \dfrac{(-1)^n(n + 1)(n + 2)x^n}{2^{n+4}}$ for $\left|\dfrac{x}{2}\right| < 1 \quad\Leftrightarrow$

$|x| < 2$, so $R = 2$.

5. $\sqrt[4]{1 - 8x} = (1 - 8x)^{1/4} = \sum\limits_{n=0}^{\infty} \binom{\frac{1}{4}}{n} (-8x)^n$

$$= 1 + \tfrac{1}{4}(-8x) + \frac{\frac{1}{4}\left(-\frac{3}{4}\right)}{2!}(-8x)^2 + \frac{\left(\frac{1}{4}\right)\left(-\frac{3}{4}\right)\left(-\frac{7}{4}\right)}{3!}(-8x)^3 + \cdots$$

$$= 1 - 2x + \sum_{n=2}^{\infty} \frac{(-1)^n (-1)^{n-1} \cdot 3 \cdot 7 \cdots \cdot (4n - 5) \, 8^n}{4^n \cdot n!} x^n$$

$$= 1 - 2x - \sum_{n=2}^{\infty} \frac{3 \cdot 7 \cdots \cdot (4n - 5) \, 2^n}{n!} x^n$$

and $|-8x| < 1 \quad \Leftrightarrow \quad |x| < \tfrac{1}{8}$, so $R = \tfrac{1}{8}$.

7. We must write the binomial in the form $(1+$ expression$)$, so we'll factor out a 4.

$$\frac{x}{\sqrt{4 + x^2}} = \frac{x}{\sqrt{4(1 + x^2/4)}} = \frac{x}{2\sqrt{1 + x^2/4}} = \frac{x}{2}\left(1 + \frac{x^2}{4}\right)^{-1/2} = \frac{x}{2} \sum_{n=0}^{\infty} \binom{-\frac{1}{2}}{n}\left(\frac{x^2}{4}\right)^n$$

$$= \frac{x}{2}\left[1 + \left(-\tfrac{1}{2}\right)\frac{x^2}{4} + \frac{\left(-\frac{1}{2}\right)\left(-\frac{3}{2}\right)}{2!}\left(\frac{x^2}{4}\right)^2 + \frac{\left(-\frac{1}{2}\right)\left(-\frac{3}{2}\right)\left(-\frac{5}{2}\right)}{3!}\left(\frac{x^2}{4}\right)^3 + \cdots\right]$$

$$= \frac{x}{2} + \frac{x}{2} \sum_{n=1}^{\infty} (-1)^n \frac{1 \cdot 3 \cdot 5 \cdots \cdot (2n - 1)}{2^n \cdot 4^n \cdot n!} x^{2n}$$

$$= \frac{x}{2} + \sum_{n=1}^{\infty} (-1)^n \frac{1 \cdot 3 \cdot 5 \cdots \cdot (2n - 1)}{n! \, 2^{3n+1}} x^{2n+1} \text{ and } \frac{x^2}{4} < 1 \quad \Leftrightarrow \quad \frac{|x|}{2} < 1 \quad \Leftrightarrow$$

$|x| < 2$, so $R = 2$.

9. $(1 + 2x)^{3/4} = 1 + \frac{3}{4}(2x) + \frac{\left(\frac{3}{4}\right)\left(-\frac{1}{4}\right)}{2!}(2x)^2 + \frac{\left(\frac{3}{4}\right)\left(-\frac{1}{4}\right)\left(-\frac{5}{4}\right)}{3!}(2x)^3 + \cdots$

$$= 1 + \frac{3}{2}x + 3 \sum_{n=2}^{\infty} (-1)^{n+1} \frac{1 \cdot 5 \cdot 9 \cdots \cdot (4n - 7)}{4^n \cdot n!} \cdot 2^n x^n$$

$$= 1 + \frac{3}{2}x + 3 \sum_{n=2}^{\infty} (-1)^{n+1} \frac{1 \cdot 5 \cdot 9 \cdots \cdot (4n - 7)}{2^n \cdot n!} x^n \text{ and } |2x| < 1 \quad \Leftrightarrow \quad |x| < \frac{1}{2}, \text{ so } R = \frac{1}{2}.$$

The three Taylor polynomials are $T_1(x) = 1 + \dfrac{3}{2}x$, $T_2(x) = 1 + \dfrac{3}{2}x - \dfrac{3}{8}x^2$, and

$T_3(x) = 1 + \dfrac{3}{2}x - \dfrac{3}{8}x^2 + \dfrac{5}{16}x^3.$

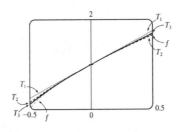

11. (a) $1/\sqrt{1 - x^2} = \left[1 + (-x^2)\right]^{-1/2}$

$$= 1 + \left(-\tfrac{1}{2}\right)\left(-x^2\right) + \frac{\left(-\tfrac{1}{2}\right)\left(-\tfrac{3}{2}\right)}{2!}\left(-x^2\right)^2 + \frac{\left(-\tfrac{1}{2}\right)\left(-\tfrac{3}{2}\right)\left(-\tfrac{5}{2}\right)}{3!}\left(-x^2\right)^3 + \cdots$$

$$= 1 + \sum_{n=1}^{\infty} \frac{1 \cdot 3 \cdot 5 \cdot \cdots \cdot (2n - 1)}{2^n \cdot n!} x^{2n}$$

(b) $\sin^{-1} x = \displaystyle\int \frac{1}{\sqrt{1 - x^2}}\, dx = C + x + \sum_{n=1}^{\infty} \frac{1 \cdot 3 \cdot 5 \cdot \cdots \cdot (2n - 1)}{(2n + 1)2^n \cdot n!} x^{2n+1}$

$$= x + \sum_{n=1}^{\infty} \frac{1 \cdot 3 \cdot 5 \cdot \cdots \cdot (2n - 1)}{(2n + 1)2^n \cdot n!} x^{2n+1} \quad \text{since } 0 = \sin^{-1} 0 = C.$$

13. (a) $\sqrt[3]{1 + x} = (1 + x)^{1/3} = \displaystyle\sum_{n=0}^{\infty} \binom{\tfrac{1}{3}}{n} x^n$

$$= 1 + \frac{1}{3}x + \frac{\left(\tfrac{1}{3}\right)\left(-\tfrac{2}{3}\right)}{2!}x^2 + \frac{\left(\tfrac{1}{3}\right)\left(-\tfrac{2}{3}\right)\left(-\tfrac{5}{3}\right)}{3!}x^3 + \cdots$$

$$= 1 + \frac{x}{3} + \sum_{n=2}^{\infty}(-1)^{n+1}\frac{2 \cdot 5 \cdot 8 \cdot \cdots \cdot (3n - 4)}{3^n \cdot n!}x^n$$

(b) $\sqrt[3]{1 + x} = 1 + \dfrac{1}{3}x - \dfrac{1}{9}x^2 + \dfrac{5}{81}x^3 - \cdots$. $\sqrt[3]{1.01} = \sqrt[3]{1 + 0.01}$, so let $x = 0.01$. The sum of the first two

terms is then $1 + \dfrac{1}{3}(0.01) \approx 1.0033$. The third term is $\dfrac{1}{9}(0.01)^2 \approx 0.000\,01$, which does not affect the fourth

decimal place of the sum, so we have $\sqrt[3]{1.01} \approx 1.0033$.

15. (a) $[1 + (-x)]^{-2} = 1 + (-2)(-x) + \dfrac{(-2)(-3)}{2!}(-x)^2 + \dfrac{(-2)(-3)(-4)}{3!}(-x)^3 + \cdots$

$$= 1 + 2x + 3x^2 + 4x^3 + \cdots = \sum_{n=0}^{\infty}(n + 1)x^n,$$

so $\dfrac{x}{(1 - x)^2} = x \displaystyle\sum_{n=0}^{\infty}(n + 1)x^n = \sum_{n=0}^{\infty}(n + 1)x^{n+1} = \sum_{n=1}^{\infty} n x^n.$

(b) With $x = \tfrac{1}{2}$ in part (a), we have $\displaystyle\sum_{n=1}^{\infty} n\left(\tfrac{1}{2}\right)^n = \sum_{n=1}^{\infty} \frac{n}{2^n} = \frac{\tfrac{1}{2}}{\left(1 - \tfrac{1}{2}\right)^2} = \frac{\tfrac{1}{2}}{\tfrac{1}{4}} = 2.$

17. (a) $\left(1 + x^2\right)^{1/2} = 1 + \left(\tfrac{1}{2}\right)x^2 + \dfrac{\left(\tfrac{1}{2}\right)\left(-\tfrac{1}{2}\right)}{2!}\left(x^2\right)^2 + \dfrac{\left(\tfrac{1}{2}\right)\left(-\tfrac{1}{2}\right)\left(-\tfrac{3}{2}\right)}{3!}\left(x^2\right)^3 + \cdots$

$$= 1 + \frac{x^2}{2} + \sum_{n=2}^{\infty} \frac{(-1)^{n-1}\, 1 \cdot 3 \cdot 5 \cdot \cdots \cdot (2n - 3)}{2^n \cdot n!} x^{2n}$$

(b) The coefficient of x^{10} (corresponding to $n = 5$) in the above Maclaurin series is $\dfrac{f^{(10)}(0)}{10!}$, so

$$\frac{f^{(10)}(0)}{10!} = \frac{(-1)^4 \cdot 1 \cdot 3 \cdot 5 \cdot 7}{2^5 \cdot 5!} \quad \Rightarrow \quad f^{(10)}(0) = 10!\left(\frac{1 \cdot 3 \cdot 5 \cdot 7}{2^5 \cdot 5!}\right) = 99{,}225.$$

19. (a) $g(x) = \sum\limits_{n=0}^{\infty} \binom{k}{n} x^n \;\; \Rightarrow \;\; g'(x) = \sum\limits_{n=1}^{\infty} \binom{k}{n} n x^{n-1}$, so

$$(1+x)g'(x) = (1+x) \sum_{n=1}^{\infty} \binom{k}{n} n x^{n-1} = \sum_{n=1}^{\infty} \binom{k}{n} n x^{n-1} + \sum_{n=1}^{\infty} \binom{k}{n} n x^{n}$$

$$= \sum_{n=0}^{\infty} \binom{k}{n+1} (n+1) x^n + \sum_{n=0}^{\infty} \binom{k}{n} n x^n \qquad \begin{bmatrix} \text{Replace } n \text{ with } n+1 \\ \text{in the first series} \end{bmatrix}$$

$$= \sum_{n=0}^{\infty} (n+1) \frac{k(k-1)(k-2)\cdots(k-n+1)(k-n)}{(n+1)!} x^n$$

$$+ \sum_{n=0}^{\infty} \left[(n) \frac{k(k-1)(k-2)\cdots(k-n+1)}{n!} \right] x^n$$

$$= \sum_{n=0}^{\infty} \frac{(n+1)k(k-1)(k-2)\cdots(k-n+1)}{(n+1)!} \left[(k-n) + n \right] x^n$$

$$= k \sum_{n=0}^{\infty} \frac{k(k-1)(k-2)\cdots(k-n+1)}{n!} x^n = k \sum_{n=0}^{\infty} \binom{k}{n} x^n = kg(x)$$

Thus, $g'(x) = \dfrac{kg(x)}{1+x}$.

(b) $h(x) = (1+x)^{-k} g(x) \;\; \Rightarrow$

$h'(x) = -k(1+x)^{-k-1} g(x) + (1+x)^{-k} g'(x)$ [Product Rule]

$\qquad = -k(1+x)^{-k-1} g(x) + (1+x)^{-k} \dfrac{kg(x)}{1+x}$ [from part (a)]

$\qquad = -k(1+x)^{-k-1} g(x) + k(1+x)^{-k-1} g(x) = 0$

(c) From part (b) we see that $h(x)$ must be constant for $x \in (-1,1)$, so $h(x) = h(0) = 1$ for $x \in (-1,1)$.

Thus, $h(x) = 1 = (1+x)^{-k} g(x) \;\; \Leftrightarrow \;\; g(x) = (1+x)^k$ for $x \in (-1,1)$.

12.12 Applications of Taylor Polynomials

ET 11.12

1. (a)

n	$f^{(n)}(x)$	$f^{(n)}(0)$	$T_n(x)$
0	$\cos x$	1	1
1	$-\sin x$	0	1
2	$-\cos x$	-1	$1 - \frac{1}{2}x^2$
3	$\sin x$	0	$1 - \frac{1}{2}x^2$
4	$\cos x$	1	$1 - \frac{1}{2}x^2 + \frac{1}{24}x^4$
5	$-\sin x$	0	$1 - \frac{1}{2}x^2 + \frac{1}{24}x^4$
6	$-\cos x$	-1	$1 - \frac{1}{2}x^2 + \frac{1}{24}x^4 - \frac{1}{720}x^6$

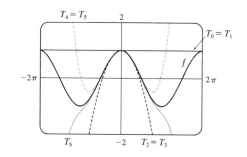

(b)

x	f	$T_0 = T_1$	$T_2 = T_3$	$T_4 = T_5$	T_6
$\frac{\pi}{4}$	0.7071	1	0.6916	0.7074	0.7071
$\frac{\pi}{2}$	0	1	-0.2337	0.0200	-0.0009
π	-1	1	-3.9348	0.1239	-1.2114

(c) As n increases, $T_n(x)$ is a good approximation to $f(x)$ on a larger and larger interval.

3.

n	$f^{(n)}(x)$	$f^{(n)}(1)$
0	$\ln x$	0
1	$1/x$	1
2	$-1/x^2$	-1
3	$2/x^3$	2
4	$-6/x^4$	-6

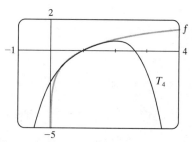

$$T_4(x) = \sum_{n=0}^{4} \frac{f^{(n)}(1)}{n!}(x-1)^n = 0 + (x-1) - \tfrac{1}{2}(x-1)^2 + \tfrac{1}{3}(x-1)^3 - \tfrac{1}{4}(x-1)^4$$

5.

n	$f^{(n)}(x)$	$f^{(n)}\left(\frac{\pi}{6}\right)$
0	$\sin x$	$\frac{1}{2}$
1	$\cos x$	$\frac{\sqrt{3}}{2}$
2	$-\sin x$	$-\frac{1}{2}$
3	$-\cos x$	$-\frac{\sqrt{3}}{2}$

$$T_3(x) = \sum_{n=0}^{3} \frac{f^{(n)}\left(\frac{\pi}{6}\right)}{n!}\left(x - \tfrac{\pi}{6}\right)^n = \tfrac{1}{2} + \tfrac{\sqrt{3}}{2}\left(x - \tfrac{\pi}{6}\right) - \tfrac{1}{4}\left(x - \tfrac{\pi}{6}\right)^2 - \tfrac{\sqrt{3}}{12}\left(x - \tfrac{\pi}{6}\right)^3$$

7.

n	$f^{(n)}(x)$	$f^{(n)}(0)$
0	$\arcsin x$	0
1	$1/\sqrt{1-x^2}$	1
2	$x/(1-x^2)^{3/2}$	0
3	$(2x^2+1)/(1-x^2)^{5/2}$	1

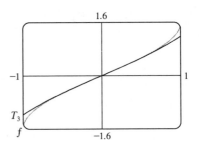

$$T_3(x) = \sum_{n=0}^{3} \frac{f^{(n)}(0)}{n!}x^n = x + \frac{x^3}{6}$$

9.

n	$f^{(n)}(x)$	$f^{(n)}(0)$
0	xe^{-2x}	0
1	$(1-2x)e^{-2x}$	1
2	$4(x-1)e^{-2x}$	-4
3	$4(3-2x)e^{-2x}$	12

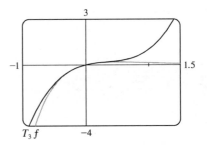

$$T_3(x) = \sum_{n=0}^{3} \frac{f^{(n)}(0)}{n!}x^n = \frac{0}{1}\cdot 1 + \frac{1}{1}x^1 + \frac{-4}{2}x^2 + \frac{12}{6}x^3 = x - 2x^2 + 2x^3$$

11. In Maple, we can find the Taylor polynomials by the following method: first define `f:=sec(x);` and then set

`T2:=convert(taylor(f,x=0,3),polynom);`, `T4:=convert(taylor(f,x=0,5),polynom);`,

etc. (The third argument in the `taylor` function is one more than the degree of the desired polynomial). We must

convert to the type `polynom` because the output of the

`taylor` function contains an error term which we do not

want. In Mathematica, we use

`Tn:=Normal[Series[f,{x,0,n}]]`, with n=2, 4,

etc. Note that in Mathematica, the "degree" argument is the

same as the degree of the desired polynomial. In Derive,

author $\sec x$, then enter `Calculus,Taylor,8,0;` and

then simplify the expression. The eighth Taylor polynomial is

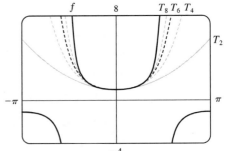

$$T_8(x) = 1 + \tfrac{1}{2}x^2 + \tfrac{5}{24}x^4 + \tfrac{61}{720}x^6 + \tfrac{277}{8064}x^8.$$

13.

n	$f^{(n)}(x)$	$f^{(n)}(4)$
0	$\sqrt{x}$	2
1	$\frac{1}{2}x^{-1/2}$	$\frac{1}{4}$
2	$-\frac{1}{4}x^{-3/2}$	$-\frac{1}{32}$
3	$\frac{3}{8}x^{-5/2}$	

(a) $f(x) = \sqrt{x} \approx T_2(x) = 2 + \frac{1}{4}(x-4) - \frac{1/32}{2!}(x-4)^2 = 2 + \frac{1}{4}(x-4) - \frac{1}{64}(x-4)^2$

(b) $|R_2(x)| \leq \dfrac{M}{3!}|x-4|^3$, where $|f'''(x)| \leq M$. Now $4 \leq x \leq 4.2 \;\Rightarrow\; |x-4| \leq 0.2 \;\Rightarrow\;$
$|x-4|^3 \leq 0.008$. Since $f'''(x)$ is decreasing on $[4, 4.2]$, we can take $M = |f'''(4)| = \frac{3}{8}4^{-5/2} = \frac{3}{256}$, so
$|R_2(x)| \leq \frac{3/256}{6}(0.008) = \frac{0.008}{512} = 0.000015625$.

(c) From the graph of $|R_2(x)| = |\sqrt{x} - T_2(x)|$, it seems

that the error is less than 1.52×10^{-5} on $[4, 4.2]$.

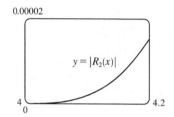

15.

n	$f^{(n)}(x)$	$f^{(n)}(1)$
0	$x^{2/3}$	1
1	$\frac{2}{3}x^{-1/3}$	$\frac{2}{3}$
2	$-\frac{2}{9}x^{-4/3}$	$-\frac{2}{9}$
3	$\frac{8}{27}x^{-7/3}$	$\frac{8}{27}$
4	$-\frac{56}{81}x^{-10/3}$	

(a) $f(x) = x^{2/3} \approx T_3(x) = 1 + \frac{2}{3}(x-1) - \frac{2/9}{2!}(x-1)^2 + \frac{8/27}{3!}(x-1)^3$

$\qquad = 1 + \frac{2}{3}(x-1) - \frac{1}{9}(x-1)^2 + \frac{4}{81}(x-1)^3$

(b) $|R_3(x)| \leq \frac{M}{4!}|x-1|^4$, where $\left|f^{(4)}(x)\right| \leq M$. Now $0.8 \leq x \leq 1.2 \;\Rightarrow\; |x-1| \leq 0.2 \;\Rightarrow\;$
$|x-1|^4 \leq 0.0016$. Since $\left|f^{(4)}(x)\right|$ is decreasing on $[0.8, 1.2]$, we can take $M = \left|f^{(4)}(0.8)\right| = \frac{56}{81}(0.8)^{-10/3}$,
so $|R_3(x)| \leq \dfrac{\frac{56}{81}(0.8)^{-10/3}}{24}(0.0016) \approx 0.000\,096\,97$.

(c) From the graph of $|R_3(x)| = \left|x^{2/3} - T_3(x)\right|$, it seems

that the error is less than $0.000\,053\,3$ on $[0.8, 1.2]$.

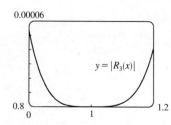

17.

n	$f^{(n)}(x)$	$f^{(n)}(0)$
0	$\tan x$	0
1	$\sec^2 x$	1
2	$2\sec^2 x \tan x$	0
3	$4\sec^2 x \tan^2 x + 2\sec^4 x$	2
4	$8\sec^2 x \tan^3 x + 16\sec^4 x \tan x$	

(a) $f(x) = \tan x \approx T_3(x) = x + \frac{1}{3}x^3$

(b) $|R_3(x)| \le \dfrac{M}{4!}\,|x|^4$, where $\left|f^{(4)}(x)\right| \le M$. Now

$\quad 0 \le x \le \frac{\pi}{6} \;\Rightarrow\; x^4 \le \left(\frac{\pi}{6}\right)^4$, and letting $x = \frac{\pi}{6}$

$\quad \left[\text{since } f^{(4)} \text{ is increasing on } \left(0, \frac{\pi}{6}\right)\right]$ gives

$\quad |R_3(x)| \le \dfrac{8\left(\frac{2}{\sqrt{3}}\right)^2 \left(\frac{1}{\sqrt{3}}\right)^3 + 16\left(\frac{2}{\sqrt{3}}\right)^4 \left(\frac{1}{\sqrt{3}}\right)}{4!}\left(\frac{\pi}{6}\right)^4$

$\quad = \frac{4\sqrt{3}}{9}\left(\frac{\pi}{6}\right)^4 \approx 0.057859$

(c)

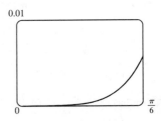

From the graph of

$|R_3(x)| = |\tan x - T_3(3)|$, it seems that

the error is less than 0.006 on $[0, \pi/6]$.

19.

n	$f^{(n)}(x)$	$f^{(n)}(0)$
0	e^{x^2}	1
1	$e^{x^2}(2x)$	0
2	$e^{x^2}(2 + 4x^2)$	2
3	$e^{x^2}(12x + 8x^3)$	0
4	$e^{x^2}(12 + 48x^2 + 16x^4)$	

(a) $f(x) = e^{x^2} \approx T_3(x) = 1 + \frac{2}{2!}x^2 = 1 + x^2$

(b) $|R_3(x)| \le \dfrac{M}{4!}\,|x|^4$, where $\left|f^{(4)}(x)\right| \le M$.

Now $0 \le x \le 0.1 \;\Rightarrow\; x^4 \le (0.1)^4$, and

letting $x = 0.1$ gives

$|R_3(x)| \le \dfrac{e^{0.01}\,(12 + 0.48 + 0.0016)}{24}(0.1)^4 \approx$

0.00006.

(c)

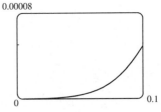

From the graph of

$|R_3(x)| = \left|e^{x^2} - (1 + x^2)\right|$, it appears that

the error is less than 0.000051 on $[0, 0.1]$.

21.

n	$f^{(n)}(x)$	$f^{(n)}(0)$
0	$x \sin x$	0
1	$\sin x + x \cos x$	0
2	$2 \cos x - x \sin x$	2
3	$-3 \sin x - x \cos x$	0
4	$-4 \cos x + x \sin x$	-4
5	$5 \sin x + x \cos x$	

(a) $f(x) = x \sin x \approx T_4(x) = \frac{2}{2!}(x-0)^2 + \frac{-4}{4!}(x-0)^4 = x^2 - \frac{1}{6}x^4$

(b) $|R_4(x)| \leq \frac{M}{5!}|x|^5$, where $\left| f^{(5)}(x) \right| \leq M$. Now $-1 \leq x \leq 1 \;\Rightarrow\; |x| \leq 1$, and a graph of $f^{(5)}(x)$ shows that

$\left| f^{(5)}(x) \right| \leq 5$ for $-1 \leq x \leq 1$. Thus, we can take $M = 5$ and get $|R_4(x)| \leq \frac{5}{5!} \cdot 1^5 = \dfrac{1}{24} = 0.041\overline{6}$.

(c) From the graph of $|R_4(x)| = |x \sin x - T_4(x)|$, it

seems that the error is less than 0.0082 on $[-1, 1]$.

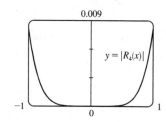

23. From Exercise 5, $\sin x = \frac{1}{2} + \frac{\sqrt{3}}{2}\left(x - \frac{\pi}{6}\right) - \frac{1}{4}\left(x - \frac{\pi}{6}\right)^2 - \frac{\sqrt{3}}{12}\left(x - \frac{\pi}{6}\right)^3 + R_3(x)$, where $|R_3(x)| \leq \dfrac{M}{4!}\left|x - \frac{\pi}{6}\right|^4$

with $\left| f^{(4)}(x) \right| = |\sin x| \leq M = 1$. Now $x = 35° = (30° + 5°) = \left(\frac{\pi}{6} + \frac{\pi}{36}\right)$ radians, so the error is

$\left|R_3\left(\frac{\pi}{36}\right)\right| \leq \dfrac{\left(\frac{\pi}{36}\right)^4}{4!} < 0.000003.$ Therefore, to five decimal places,

$\sin 35° \approx \frac{1}{2} + \frac{\sqrt{3}}{2}\left(\frac{\pi}{36}\right) - \frac{1}{4}\left(\frac{\pi}{36}\right)^2 - \frac{\sqrt{3}}{12}\left(\frac{\pi}{36}\right)^3 \approx 0.57358.$

25. All derivatives of e^x are e^x, so $|R_n(x)| \leq \dfrac{e^x}{(n+1)!}|x|^{n+1}$, where $0 < x < 0.1$. Letting $x = 0.1$,

$R_n(0.1) \leq \dfrac{e^{0.1}}{(n+1)!}(0.1)^{n+1} < 0.00001$, and by trial and error we find that $n = 3$ satisfies this inequality since

$R_3(0.1) < 0.0000046$. Thus, by adding the four terms of the Maclaurin series for e^x corresponding to $n = 0, 1, 2,$

and 3, we can estimate $e^{0.1}$ to within 0.00001. (In fact, this sum is $1.1051\overline{6}$ and $e^{0.1} \approx 1.10517$.)

27. $\sin x = x - \frac{1}{3!}x^3 + \frac{1}{5!}x^5 - \cdots$. By the Alternating

Series Estimation Theorem, the error in the

approximation $\sin x = x - \frac{1}{3!}x^3$ is less than

$\left|\frac{1}{5!}x^5\right| < 0.01 \quad \Leftrightarrow \quad \left|x^5\right| < 120(0.01) \quad \Leftrightarrow$

$|x| < (1.2)^{1/5} \approx 1.037$. The curves $y = x - \frac{1}{6}x^3$ and

$y = \sin x - 0.01$ intersect at $x \approx 1.043$, so the graph

confirms our estimate. Since both the sine function and

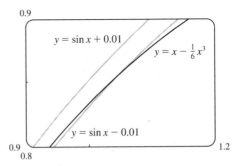

the given approximation are odd functions, we need to check the estimate only for $x > 0$. Thus, the desired range of

values for x is $-1.037 < x < 1.037$.

29. Let $s(t)$ be the position function of the car, and for convenience set $s(0) = 0$. The velocity of the car is

$v(t) = s'(t)$ and the acceleration is $a(t) = s''(t)$, so the second degree Taylor polynomial is

$T_2(t) = s(0) + v(0)t + \frac{a(0)}{2}t^2 = 20t + t^2$. We estimate the distance travelled during the next second to be

$s(1) \approx T_2(1) = 20 + 1 = 21$ m. The function $T_2(t)$ would not be accurate over a full minute, since the car could

not possibly maintain an acceleration of 2 m/s^2 for that long (if it did, its final speed would be

140 m/s ≈ 313 mi/h!)

31. $E = \dfrac{q}{D^2} - \dfrac{q}{(D+d)^2} = \dfrac{q}{D^2} - \dfrac{q}{D^2(1+d/D)^2} = \dfrac{q}{D^2}\left[1 - \left(1 + \dfrac{d}{D}\right)^{-2}\right]$.

We use the Binomial Series to expand $(1 + d/D)^{-2}$:

$$E = \frac{q}{D^2}\left[1 - \left(1 - 2\left(\frac{d}{D}\right) + \frac{2\cdot 3}{2!}\left(\frac{d}{D}\right)^2 - \frac{2\cdot 3\cdot 4}{3!}\left(\frac{d}{D}\right)^3 + \cdots\right)\right]$$

$$= \frac{q}{D^2}\left[2\left(\frac{d}{D}\right) - 3\left(\frac{d}{D}\right)^2 + 4\left(\frac{d}{D}\right)^3 - \cdots\right] \approx \frac{q}{D^2}\cdot 2\left(\frac{d}{D}\right) = 2qd\cdot\frac{1}{D^3}$$

when D is much larger than d; that is, when P is far away from the dipole.

33. (a) If the water is deep, then $2\pi d/L$ is large, and we know that $\tanh x \to 1$ as $x \to \infty$. So we can approximate

$\tanh(2\pi d/L) \approx 1$, and so $v^2 \approx gL/(2\pi) \quad \Leftrightarrow \quad v \approx \sqrt{gL/(2\pi)}$.

(b) From the table, the first term in the Maclaurin

series of $\tanh x$ is x, so if the water is shallow,

we can approximate $\tanh \dfrac{2\pi d}{L} \approx \dfrac{2\pi d}{L}$, and so

$v^2 \approx \dfrac{gL}{2\pi}\cdot\dfrac{2\pi d}{L} \quad \Leftrightarrow \quad v \approx \sqrt{gd}$.

n	$f^{(n)}(x)$	$f^{(n)}(0)$
0	$\tanh x$	0
1	$\text{sech}^2 x$	1
2	$-2\,\text{sech}^2 x \tanh x$	0
3	$2\,\text{sech}^2 x\,(3\tanh^2 x - 1)$	-2

(c) Since $\tanh x$ is an odd function, its Maclaurin series is alternating, so the error in the approximation

$$\tanh \frac{2\pi d}{L} \approx \frac{2\pi d}{L} \text{ is less than the first neglected term, which is } \frac{|f'''(0)|}{3!}\left(\frac{2\pi d}{L}\right)^3 = \frac{1}{3}\left(\frac{2\pi d}{L}\right)^3.$$

If $L > 10d$, then $\dfrac{1}{3}\left(\dfrac{2\pi d}{L}\right)^3 < \dfrac{1}{3}\left(2\pi \cdot \dfrac{1}{10}\right)^3 = \dfrac{\pi^3}{375}$, so the error in the approximation $v^2 = gd$ is less

than $\dfrac{gL}{2\pi} \cdot \dfrac{\pi^3}{375} \approx 0.0132gL$.

35. (a) L is the length of the arc subtended by the angle θ, so $L = R\theta$ $\Rightarrow$

$\theta = L/R$. Now $\sec\theta = (R + C)/R$ $\Rightarrow$ $R\sec\theta = R + C$ $\Rightarrow$

$C = R\sec\theta - R = R\sec(L/R) - R$.

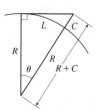

(b) From Exercise 11, $\sec x \approx T_4(x) = 1 + \frac{1}{2}x^2 + \frac{5}{24}x^4$. By part (a),

$$C \approx R\left[1 + \frac{1}{2}\left(\frac{L}{R}\right)^2 + \frac{5}{24}\left(\frac{L}{R}\right)^4\right] - R = R + \frac{1}{2}R\cdot\frac{L^2}{R^2} + \frac{5}{24}R\cdot\frac{L^4}{R^4} - R = \frac{L^2}{2R} + \frac{5L^4}{24R^3}.$$

(c) Taking $L = 100$ km and $R = 6370$ km, the formula in part (a) says that

$C = R\sec(L/R) - R = 6370\sec(100/6370) - 6370 \approx 0.785\,009\,965\,44$ km

The formula in part (b) says that

$$C \approx \frac{L^2}{2R} + \frac{5L^4}{24R^3} = \frac{100^2}{2\cdot6370} + \frac{5\cdot100^4}{24\cdot6370^3} \approx 0.785\,009\,957\,36 \text{ km}.$$

The difference between these two results is only $0.000\,000\,008\,08$ km, or $0.000\,008\,08$ m!

37. Using $f(x) = T_n(x) + R_n(x)$ with $n = 1$ and $x = r$, we have $f(r) = T_1(r) + R_1(r)$, where T_1 is the first-degree

Taylor polynomial of f at a. Because $a = x_n$, $f(r) = f(x_n) + f'(x_n)(r - x_n) + R_1(r)$. But r is a root of f, so

$f(r) = 0$ and we have $0 = f(x_n) + f'(x_n)(r - x_n) + R_1(r)$. Taking the first two terms to the left side gives us

$f'(x_n)(x_n - r) - f(x_n) = R_1(r)$. Dividing by $f'(x_n)$, we get $x_n - r - \dfrac{f(x_n)}{f'(x_n)} = \dfrac{R_1(r)}{f'(x_n)}$. By the formula for

Newton's method, the left side of the preceding equation is $x_{n+1} - r$, so $|x_{n+1} - r| = \left|\dfrac{R_1(r)}{f'(x_n)}\right|$. Taylor's

Inequality gives us $|R_1(r)| \leq \dfrac{|f''(r)|}{2!}|r - x_n|^2$. Combining this inequality with the facts $|f''(x)| \leq M$ and

$|f'(x)| \geq K$ gives us $|x_{n+1} - r| \leq \dfrac{M}{2K}|x_n - r|^2$.

12 Review

—————————————— CONCEPT CHECK ——————————————

1. (a) See Definition 12.1.1 [ET 11.1.1].

  (b) See Definition 12.2.2 [ET 11.2.2].

  (c) The terms of the sequence $\{a_n\}$ approach 3 as n becomes large.

  (d) By adding sufficiently many terms of the series, we can make the partial sums as close to 3 as we like.

2. (a) See Definition 12.1.10 [ET 11.1.10].

  (b) A sequence is monotonic if it is either increasing or decreasing.

  (c) By Theorem 12.1.11 [ET 11.1.11], every bounded, monotonic sequence is convergent.

3. (a) See (4) in Section 12.2 [ET 11.2].

  (b) The p-series $\displaystyle\sum_{n=1}^{\infty} \frac{1}{n^p}$ is convergent if $p > 1$.

4. If $\sum a_n = 3$, then $\displaystyle\lim_{n \to \infty} a_n = 0$ and $\displaystyle\lim_{n \to \infty} s_n = 3$.

5. (a) See the Test for Divergence on page 754 [ET page 718].

  (b) See the Integral Test on page 760 [ET page 724].

  (c) See the Comparison Test on page 767 [ET page 731].

  (d) See the Limit Comparison Test on page 768 [ET page 732].

  (e) See the Alternating Series Test on page 772 [ET page 736].

  (f) See the Ratio Test on page 778 [ET page 742].

  (g) See the Root Test on page 780 [ET page 744].

6. (a) A series $\sum a_n$ is called *absolutely convergent* if the series of absolute values $\sum |a_n|$ is convergent.

  (b) If a series $\sum a_n$ is absolutely convergent, then it is convergent.

  (c) A series $\sum a_n$ is called *conditionally convergent* if it is convergent but not absolutely convergent.

7. (a) Use (3) in Section 12.3 [ET 11.3].

  (b) See Example 5 in Section 12.4 [ET 11.4].

  (c) By adding terms until you reach the desired accuracy given by the Alternating Series Estimation Theorem on page 774 [ET page 738].

8. (a) $\sum_{n=0}^{\infty} c_n(x - a)^n$

  (b) Given the power series $\sum_{n=0}^{\infty} c_n(x - a)^n$, the radius of convergence is:

    (i) 0 if the series converges only when $x = a$

    (ii) ∞ if the series converges for all x, or

    (iii) a positive number R such that the series converges if $|x - a| < R$ and diverges if $|x - a| > R$.

  (c) The interval of convergence of a power series is the interval that consists of all values of x for which the series converges. Corresponding to the cases in part (b), the interval of convergence is: (i) the single point $\{a\}$, (ii) all real numbers, that is, the real number line $(-\infty, \infty)$, or (iii) an interval with endpoints $a - R$ and $a + R$ which can contain neither, either, or both of the endpoints. In this case, we must test the series for convergence at each endpoint to determine the interval of convergence.

9. (a), (b) See Theorem 12.9.2 [ET 11.9.2].

10. (a) $T_n(x) = \sum\limits_{i=0}^{n} \dfrac{f^{(i)}(a)}{i!}(x-a)^i$

(b) $\sum\limits_{n=0}^{\infty} \dfrac{f^{(n)}(a)}{n!}(x-a)^n$

(c) $\sum\limits_{n=0}^{\infty} \dfrac{f^{(n)}(0)}{n!}x^n$ [$a=0$ in part (b)]

(d) See Theorem 12.10.8 [ET 11.10.8].

(e) See Taylor's Inequality ((12.10.9) [ET (11.10.9)]).

11. (a) – (e) See the table on page 803 [ET page 767].

12. See the Binomial Series ((12.11.2) [ET (11.11.2)]) for the expansion. The radius of convergence for the binomial series is 1.

──────────────── TRUE-FALSE QUIZ ────────────────

1. False. See Note 2 after Theorem 12.2.6 [ET 11.2.6].

3. True. If $\lim\limits_{n\to\infty} a_n = L$, then given any $\varepsilon > 0$, we can find a positive integer N such that $|a_n - L| < \varepsilon$ whenever $n > N$. If $n > N$, then $2n + 1 > N$ and $|a_{2n+1} - L| < \varepsilon$. Thus, $\lim\limits_{n\to\infty} a_{2n+1} = L$.

5. False. For example, take $c_n = (-1)^n / (n6^n)$.

7. False, since $\lim\limits_{n\to\infty} \left| \dfrac{a_{n+1}}{a_n} \right| = \lim\limits_{n\to\infty} \left| \dfrac{1}{(n+1)^3} \cdot \dfrac{n^3}{1} \right| = \lim\limits_{n\to\infty} \left| \dfrac{n^3}{(n+1)^3} \cdot \dfrac{1/n^3}{1/n^3} \right| = \lim\limits_{n\to\infty} \dfrac{1}{(1+1/n)^3} = 1.$

9. False. See the note after Example 2 in Section 12.4 [ET 11.4].

11. True. See (8) in Section 12.1 [ET 11.1].

13. True. By Theorem 12.10.5 [ET 11.10.5] the coefficient of x^3 is $\dfrac{f'''(0)}{3!} = \dfrac{1}{3}$ $\Rightarrow$ $f'''(0) = 2$.

Or: Use Theorem 12.9.2 [ET 11.9.2] to differentiate f three times.

15. False. For example, let $a_n = b_n = (-1)^n$. Then $\{a_n\}$ and $\{b_n\}$ are divergent, but $a_n b_n = 1$, so $\{a_n b_n\}$ is convergent.

17. True by Theorem 12.6.3 [ET 11.6.3]. [$\sum (-1)^n a_n$ is absolutely convergent and hence convergent.]

──────────────── EXERCISES ────────────────

1. $\left\{ \dfrac{2+n^3}{1+2n^3} \right\}$ converges since $\lim\limits_{n\to\infty} \dfrac{2+n^3}{1+2n^3} = \lim\limits_{n\to\infty} \dfrac{2/n^3 + 1}{1/n^3 + 2} = \dfrac{1}{2}$.

3. $\lim\limits_{n\to\infty} a_n = \lim\limits_{n\to\infty} \dfrac{n^3}{1+n^2} = \lim\limits_{n\to\infty} \dfrac{n}{1/n^2 + 1} = \infty$, so the sequence diverges.

5. $|a_n| = \left| \dfrac{n\sin n}{n^2+1} \right| \le \dfrac{n}{n^2+1} < \dfrac{1}{n}$, so $|a_n| \to 0$ as $n \to \infty$. Thus, $\lim\limits_{n\to\infty} a_n = 0$. The sequence $\{a_n\}$ is convergent.

7. $\left\{\left(1 + \dfrac{3}{n}\right)^{4n}\right\}$ is convergent. Let $y = \left(1 + \dfrac{3}{x}\right)^{4x}$. Then

$$\lim_{x \to \infty} \ln y = \lim_{x \to \infty} 4x \ln (1 + 3/x) = \lim_{x \to \infty} \frac{\ln (1 + 3/x)}{1/(4x)} \overset{\text{H}}{=} \lim_{x \to \infty} \frac{\dfrac{1}{1 + 3/x}\left(-\dfrac{3}{x^2}\right)}{-1/(4x^2)} = \lim_{x \to \infty} \frac{12}{1 + 3/x} = 12$$

so $\displaystyle\lim_{x \to \infty} y = \lim_{n \to \infty} \left(1 + \frac{3}{n}\right)^{4n} = e^{12}$.

Or: Use Exercise 7.7.54 [ET 4.4.54].

9. We use induction, hypothesizing that $a_{n-1} < a_n < 2$. Note first that $1 < a_2 = \frac{1}{3}(1 + 4) = \frac{5}{3} < 2$, so the
hypothesis holds for $n = 2$. Now assume that $a_{k-1} < a_k < 2$. Then
$a_k = \frac{1}{3}(a_{k-1} + 4) < \frac{1}{3}(a_k + 4) < \frac{1}{3}(2 + 4) = 2$. So $a_k < a_{k+1} < 2$, and the induction is complete. To find the
limit of the sequence, we note that $L = \lim_{n \to \infty} a_n = \lim_{n \to \infty} a_{n+1} \implies L = \frac{1}{3}(L + 4) \implies L = 2$.

11. $\dfrac{n}{n^3 + 1} < \dfrac{n}{n^3} = \dfrac{1}{n^2}$, so $\displaystyle\sum_{n=1}^{\infty} \dfrac{n}{n^3 + 1}$ converges by the Comparison Test with the convergent p-series

$\displaystyle\sum_{n=1}^{\infty} \dfrac{1}{n^2}$ $(p = 2 > 1)$.

13. $\displaystyle\lim_{n \to \infty} \left| \dfrac{a_{n+1}}{a_n} \right| = \lim_{n \to \infty} \left[\dfrac{(n+1)^3}{5^{n+1}} \cdot \dfrac{5^n}{n^3} \right] = \lim_{n \to \infty} \left(1 + \dfrac{1}{n}\right)^3 \cdot \dfrac{1}{5} = \dfrac{1}{5} < 1$, so $\displaystyle\sum_{n=1}^{\infty} \dfrac{n^3}{5^n}$ converges by the Ratio Test.

15. Let $f(x) = \dfrac{1}{x\sqrt{\ln x}}$. Then f is continuous, positive, and decreasing on $[2, \infty)$, so the Integral Test applies.

$$\int_2^{\infty} f(x)\, dx = \lim_{t \to \infty} \int_2^t \frac{1}{x\sqrt{\ln x}}\, dx \quad \begin{bmatrix} u = \ln x, \\ du = \dfrac{1}{x}\, dx \end{bmatrix} = \lim_{t \to \infty} \int_{\ln 2}^{\ln t} u^{-1/2}\, du = \lim_{t \to \infty} \left[2\sqrt{u} \right]_{\ln 2}^{\ln t}$$

$$= \lim_{t \to \infty} (2\sqrt{\ln t} - 2\sqrt{\ln 2}) = \infty, \text{ so the series } \sum_{n=2}^{\infty} \frac{1}{n\sqrt{\ln n}} \text{ diverges.}$$

17. $|a_n| = \left| \dfrac{\cos 3n}{1 + (1.2)^n} \right| \le \dfrac{1}{1 + (1.2)^n} < \dfrac{1}{(1.2)^n} = \left(\dfrac{5}{6}\right)^n$, so $\displaystyle\sum_{n=1}^{\infty} |a_n|$ converges by comparison with the convergent

geometric series $\displaystyle\sum_{n=1}^{\infty} \left(\dfrac{5}{6}\right)^n$ $\left(r = \frac{5}{6} < 1\right)$. It follows that $\displaystyle\sum_{n=1}^{\infty} a_n$ converges (by Theorem 3 in Section 12.6

[ET 11.6]).

19. $\displaystyle\lim_{n \to \infty} \left| \dfrac{a_{n+1}}{a_n} \right| = \lim_{n \to \infty} \dfrac{1 \cdot 3 \cdot 5 \cdot \cdots \cdot (2n-1)(2n+1)}{5^{n+1}(n+1)!} \cdot \dfrac{5^n n!}{1 \cdot 3 \cdot 5 \cdot \cdots \cdot (2n-1)} = \lim_{n \to \infty} \dfrac{2n+1}{5(n+1)} = \dfrac{2}{5} < 1$, so
the series converges by the Ratio Test.

21. $b_n = \dfrac{\sqrt{n}}{n+1} > 0$, $\{b_n\}$ is decreasing, and $\displaystyle\lim_{n \to \infty} b_n = 0$, so the series $\displaystyle\sum_{n=1}^{\infty} (-1)^{n-1} \dfrac{\sqrt{n}}{n+1}$ converges by the
Alternating Series Test.

23. Consider the series of absolute values: $\sum_{n=1}^{\infty} n^{-1/3}$ is a p-series with $p = \frac{1}{3} \le 1$ and is therefore divergent. But if
we apply the Alternating Series Test, we see that $b_n = \dfrac{1}{\sqrt[3]{n}} > 0$, $\{b_n\}$ is decreasing, and $\displaystyle\lim_{n \to \infty} b_n = 0$, so the series

$\displaystyle\sum_{n=1}^{\infty} (-1)^{n-1} n^{-1/3}$ converges. Thus, $\displaystyle\sum_{n=1}^{\infty} (-1)^{n-1} n^{-1/3}$ is conditionally convergent.

25. $\left| \dfrac{a_{n+1}}{a_n} \right| = \left| \dfrac{(-1)^{n+1}(n+2)\,3^{n+1}}{2^{2n+3}} \cdot \dfrac{2^{2n+1}}{(-1)^n(n+1)\,3^n} \right| = \dfrac{n+2}{n+1} \cdot \dfrac{3}{4} = \dfrac{1+(2/n)}{1+(1/n)} \cdot \dfrac{3}{4} \to \dfrac{3}{4} < 1$ as $n \to \infty$, so

by the Ratio Test, $\displaystyle\sum_{n=1}^{\infty} \dfrac{(-1)^n(n+1)\,3^n}{2^{2n+1}}$ is absolutely convergent.

27. $\dfrac{2^{2n+1}}{5^n} = \dfrac{2^{2n} \cdot 2^1}{5^n} = \dfrac{(2^2)^n \cdot 2}{5^n} = 2\left(\dfrac{4}{5}\right)^n$, so $\displaystyle\sum_{n=1}^{\infty} \dfrac{2^{2n+1}}{5^n} = 2\sum_{n=1}^{\infty} \left(\dfrac{4}{5}\right)^n$ is a geometric series with $a = \dfrac{8}{5}$ and

$r = \dfrac{4}{5}$. Since $|r| = \dfrac{4}{5} < 1$, the series converges to $\dfrac{a}{1-r} = \dfrac{8/5}{1-4/5} = \dfrac{8/5}{1/5} = 8$.

29. $\displaystyle\sum_{n=1}^{\infty}\left[\tan^{-1}(n+1) - \tan^{-1}n\right] = \lim_{n\to\infty} s_n = \lim_{n\to\infty}\left[(\tan^{-1}2 - \tan^{-1}1) + (\tan^{-1}3 - \tan^{-1}2) + \cdots\right.$

$$\left. + \,(\tan^{-1}(n+1) - \tan^{-1}n)\right]$$

$$= \lim_{n\to\infty}\left[\tan^{-1}(n+1) - \tan^{-1}1\right] = \tfrac{\pi}{2} - \tfrac{\pi}{4} = \tfrac{\pi}{4}$$

31. $1 - e + \dfrac{e^2}{2!} - \dfrac{e^3}{3!} + \dfrac{e^4}{4!} - \cdots = \displaystyle\sum_{n=0}^{\infty}(-1)^n\dfrac{e^n}{n!} = \sum_{n=0}^{\infty}\dfrac{(-e)^n}{n!} = e^{-e}$ since $e^x = \displaystyle\sum_{n=0}^{\infty}\dfrac{x^n}{n!}$ for all x.

33. $\cosh x = \dfrac{1}{2}(e^x + e^{-x}) = \dfrac{1}{2}\left(\displaystyle\sum_{n=0}^{\infty}\dfrac{x^n}{n!} + \sum_{n=0}^{\infty}\dfrac{(-x)^n}{n!}\right)$

$$= \dfrac{1}{2}\left[\left(1 + x + \dfrac{x^2}{2!} + \dfrac{x^3}{3!} + \dfrac{x^4}{4!} + \cdots\right) + \left(1 - x + \dfrac{x^2}{2!} - \dfrac{x^3}{3!} + \dfrac{x^4}{4!} - \cdots\right)\right]$$

$$= \dfrac{1}{2}\left(2 + 2\cdot\dfrac{x^2}{2!} + 2\cdot\dfrac{x^4}{4!} + \cdots\right)$$

$$= 1 + \dfrac{1}{2}x^2 + \displaystyle\sum_{n=2}^{\infty}\dfrac{x^{2n}}{(2n)!}$$

$$\geq 1 + \dfrac{1}{2}x^2 \quad\text{for all } x$$

35. $\displaystyle\sum_{n=1}^{\infty}\dfrac{(-1)^{n+1}}{n^5} = 1 - \dfrac{1}{32} + \dfrac{1}{243} - \dfrac{1}{1024} + \dfrac{1}{3125} - \dfrac{1}{7776} + \dfrac{1}{16,807} - \dfrac{1}{32,768} + \cdots.$

Since $b_8 = \dfrac{1}{8^5} = \dfrac{1}{32,768} < 0.000031$, $\displaystyle\sum_{n=1}^{\infty}\dfrac{(-1)^{n+1}}{n^5} \approx \sum_{n=1}^{7}\dfrac{(-1)^{n+1}}{n^5} \approx 0.9721.$

37. $\displaystyle\sum_{n=1}^{\infty}\dfrac{1}{2+5^n} \approx \sum_{n=1}^{8}\dfrac{1}{2+5^n} \approx 0.18976224.$ To estimate the error, note that $\dfrac{1}{2+5^n} < \dfrac{1}{5^n}$, so the remainder term is

$R_8 = \displaystyle\sum_{n=9}^{\infty}\dfrac{1}{2+5^n} < \sum_{n=9}^{\infty}\dfrac{1}{5^n} = \dfrac{1/5^9}{1-1/5} = 6.4 \times 10^{-7}$ (geometric series with $a = \tfrac{1}{5^9}$ and $r = \tfrac{1}{5}$).

39. Use the Limit Comparison Test. $\displaystyle\lim_{n\to\infty}\left|\dfrac{\left(\frac{n+1}{n}\right)a_n}{a_n}\right| = \lim_{n\to\infty}\dfrac{n+1}{n} = \lim_{n\to\infty}\left(1 + \dfrac{1}{n}\right) = 1 > 0.$

Since $\sum|a_n|$ is convergent, so is $\sum\left|\left(\dfrac{n+1}{n}\right)a_n\right|$, by the Limit Comparison Test.

41. $\lim\limits_{n\to\infty}\left|\dfrac{a_{n+1}}{a_n}\right| = \lim\limits_{n\to\infty}\left[\dfrac{|x+2|^{n+1}}{(n+1)\,4^{n+1}}\cdot\dfrac{n4^n}{|x+2|^n}\right] = \lim\limits_{n\to\infty}\left[\dfrac{n}{n+1}\dfrac{|x+2|}{4}\right] = \dfrac{|x+2|}{4} < 1 \quad\Leftrightarrow\quad |x+2| < 4,$

so $R = 4$. $|x+2| < 4 \quad\Leftrightarrow\quad -4 < x+2 < 4 \quad\Leftrightarrow\quad -6 < x < 2$. If $x = -6$, then the series $\sum\limits_{n=1}^{\infty}\dfrac{(x+2)^n}{n4^n}$

becomes $\sum\limits_{n=1}^{\infty}\dfrac{(-4)^n}{n4^n} = \sum\limits_{n=1}^{\infty}\dfrac{(-1)^n}{n}$, the alternating harmonic series, which converges by the Alternating Series

Test. When $x = 2$, the series becomes the harmonic series $\sum\limits_{n=1}^{\infty}\dfrac{1}{n}$, which diverges. Thus, $I = [-6, 2)$.

43. $\lim\limits_{n\to\infty}\left|\dfrac{a_{n+1}}{a_n}\right| = \lim\limits_{n\to\infty}\left|\dfrac{2^{n+1}\,(x-3)^{n+1}}{\sqrt{n+4}}\cdot\dfrac{\sqrt{n+3}}{2^n\,(x-3)^n}\right| = 2\,|x-3|\,\lim\limits_{n\to\infty}\sqrt{\dfrac{n+3}{n+4}} = 2\,|x-3| < 1 \quad\Leftrightarrow$

$|x-3| < \frac{1}{2}$, so $R = \frac{1}{2}$. $|x-3| < \frac{1}{2} \quad\Leftrightarrow\quad -\frac{1}{2} < x-3 < \frac{1}{2} \quad\Leftrightarrow\quad \frac{5}{2} < x < \frac{7}{2}$. For $x = \frac{7}{2}$, the series

$\sum\limits_{n=1}^{\infty}\dfrac{2^n\,(x-3)^n}{\sqrt{n+3}}$ becomes $\sum\limits_{n=0}^{\infty}\dfrac{1}{\sqrt{n+3}} = \sum\limits_{n=3}^{\infty}\dfrac{1}{n^{1/2}}$, which diverges $(p = \frac{1}{2} \le 1)$, but for $x = \frac{5}{2}$, we get

$\sum\limits_{n=0}^{\infty}\dfrac{(-1)^n}{\sqrt{n+3}}$, which is a convergent alternating series, so $I = \left[\frac{5}{2}, \frac{7}{2}\right)$.

45.

n	$f^{(n)}(x)$	$f^{(n)}\!\left(\frac{\pi}{6}\right)$
0	$\sin x$	$\frac{1}{2}$
1	$\cos x$	$\frac{\sqrt{3}}{2}$
2	$-\sin x$	$-\frac{1}{2}$
3	$-\cos x$	$-\frac{\sqrt{3}}{2}$
4	$\sin x$	$\frac{1}{2}$
$\vdots$	$\vdots$	$\vdots$

$\sin x = f\!\left(\dfrac{\pi}{6}\right) + f'\!\left(\dfrac{\pi}{6}\right)\!\left(x - \dfrac{\pi}{6}\right) + \dfrac{f''\!\left(\frac{\pi}{6}\right)}{2!}\left(x - \dfrac{\pi}{6}\right)^2 + \dfrac{f^{(3)}\!\left(\frac{\pi}{6}\right)}{3!}\left(x - \dfrac{\pi}{6}\right)^3 + \dfrac{f^{(4)}\!\left(\frac{\pi}{6}\right)}{4!}\left(x - \dfrac{\pi}{6}\right)^4 + \cdots$

$= \dfrac{1}{2}\left[1 - \dfrac{1}{2!}\left(x - \dfrac{\pi}{6}\right)^2 + \dfrac{1}{4!}\left(x - \dfrac{\pi}{6}\right)^4 - \cdots\right] + \dfrac{\sqrt{3}}{2}\left[\left(x - \dfrac{\pi}{6}\right) - \dfrac{1}{3!}\left(x - \dfrac{\pi}{6}\right)^3 + \cdots\right]$

$= \dfrac{1}{2}\sum\limits_{n=0}^{\infty}(-1)^n\dfrac{1}{(2n)!}\left(x - \dfrac{\pi}{6}\right)^{2n} + \dfrac{\sqrt{3}}{2}\sum\limits_{n=0}^{\infty}(-1)^n\dfrac{1}{(2n+1)!}\left(x - \dfrac{\pi}{6}\right)^{2n+1}$

47. $\dfrac{1}{1+x} = \dfrac{1}{1-(-x)} = \sum\limits_{n=0}^{\infty}(-x)^n = \sum\limits_{n=0}^{\infty}(-1)^n\,x^n$ for $|x| < 1 \quad\Rightarrow\quad \dfrac{x^2}{1+x} = \sum\limits_{n=0}^{\infty}(-1)^n\,x^{n+2}$ with $R = 1$.

49. $\dfrac{1}{1-x} = \sum\limits_{n=0}^{\infty}x^n$ for $|x| < 1 \quad\Rightarrow\quad \ln(1-x) = -\displaystyle\int\dfrac{dx}{1-x} = -\int\sum\limits_{n=0}^{\infty}x^n\,dx = C - \sum\limits_{n=0}^{\infty}\dfrac{x^{n+1}}{n+1}.$

$\ln(1-0) = C - 0 \quad\Rightarrow\quad C = 0 \quad\Rightarrow\quad \ln(1-x) = -\sum\limits_{n=0}^{\infty}\dfrac{x^{n+1}}{n+1} = \sum\limits_{n=1}^{\infty}\dfrac{-x^n}{n}$ with $R = 1$.

51. $\sin x = \sum\limits_{n=0}^{\infty} \dfrac{(-1)^n x^{2n+1}}{(2n+1)!} \quad \Rightarrow \quad \sin\left(x^4\right) = \sum\limits_{n=0}^{\infty} \dfrac{(-1)^n \left(x^4\right)^{2n+1}}{(2n+1)!} = \sum\limits_{n=0}^{\infty} \dfrac{(-1)^n x^{8n+4}}{(2n+1)!}$ for all x, so the radius of

convergence is ∞.

53. $f(x) = 1 \,/\, \sqrt[4]{16 - x} = 1/\left(\sqrt[4]{16}\,\sqrt[4]{1 - \tfrac{1}{16}x}\right) = \tfrac{1}{2}\left(1 - \tfrac{1}{16}x\right)^{-1/4}$

$$= \frac{1}{2}\left[1 + \left(-\frac{1}{4}\right)\left(-\frac{x}{16}\right) + \frac{\left(-\frac{1}{4}\right)\left(-\frac{5}{4}\right)}{2!}\left(-\frac{x}{16}\right)^2 + \frac{\left(-\frac{1}{4}\right)\left(-\frac{5}{4}\right)\left(-\frac{9}{4}\right)}{3!}\left(-\frac{x}{16}\right)^3 + \cdots\right]$$

$$= \frac{1}{2} + \sum_{n=1}^{\infty} \frac{1 \cdot 5 \cdot 9 \cdots \cdots (4n-3)}{2 \cdot 4^n \cdot n! \cdot 16^n}x^n = \frac{1}{2} + \sum_{n=1}^{\infty} \frac{1 \cdot 5 \cdot 9 \cdots \cdots (4n-3)}{2^{6n+1}\, n!}x^n$$

for $\left|-\dfrac{x}{16}\right| < 1 \quad \Leftrightarrow \quad |x| < 16$, so $R = 16$.

55. $e^x = \sum\limits_{n=0}^{\infty} \dfrac{x^n}{n!}$, so $\dfrac{e^x}{x} = \dfrac{1}{x} + \sum\limits_{n=1}^{\infty} \dfrac{x^{n-1}}{n!}$ and $\displaystyle\int \dfrac{e^x}{x}\,dx = C + \ln|x| + \sum\limits_{n=1}^{\infty} \dfrac{x^n}{n \cdot n!}$.

57. (a)

n	$f^{(n)}(x)$	$f^{(n)}(1)$
0	$x^{1/2}$	1
1	$\frac{1}{2}x^{-1/2}$	$\frac{1}{2}$
2	$-\frac{1}{4}x^{-3/2}$	$-\frac{1}{4}$
3	$\frac{3}{8}x^{-5/2}$	$\frac{3}{8}$
4	$-\frac{15}{16}x^{-7/2}$	$-\frac{15}{16}$
$\vdots$	$\vdots$	$\vdots$

$$\sqrt{x} \approx T_3(x) = 1 + \frac{1/2}{1!}(x-1) - \frac{1/4}{2!}(x-1)^2 + \frac{3/8}{3!}(x-1)^3$$

$$= 1 + \tfrac{1}{2}(x-1) - \tfrac{1}{8}(x-1)^2 + \tfrac{1}{16}(x-1)^3$$

(b)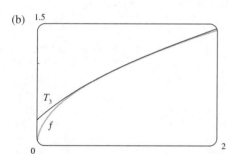

(c) $|R_3(x)| \leq \dfrac{M}{4!}|x-1|^4$, where $\left|f^{(4)}(x)\right| \leq M$ with

$f^{(4)}(x) = -\frac{15}{16}x^{-7/2}$. Now $0.9 \leq x \leq 1.1 \;\Rightarrow$

$-0.1 \leq x - 1 \leq 0.1 \quad \Rightarrow \quad (x-1)^4 \leq (0.1)^4$,

and letting $x = 0.9$ gives $M = \dfrac{15}{16(0.9)^{7/2}}$, so

$|R_3(x)| \leq \dfrac{15}{16(0.9)^{7/2}4!}(0.1)^4 \approx 0.000005648.$

(d)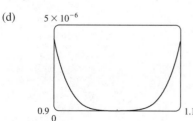

From the graph of $|R_3(x)| = |\sqrt{x} - T_3(x)|$, it appears

that the error is less than 5×10^{-6} on $[0.9, 1.1]$.

59. $\sin x = \sum\limits_{n=0}^{\infty} (-1)^n \dfrac{x^{2n+1}}{(2n+1)!} = x - \dfrac{x^3}{3!} + \dfrac{x^5}{5!} - \dfrac{x^7}{7!} + \cdots$, so $\sin x - x = -\dfrac{x^3}{3!} + \dfrac{x^5}{5!} - \dfrac{x^7}{7!} + \cdots$ and

$\dfrac{\sin x - x}{x^3} = -\dfrac{1}{3!} + \dfrac{x^2}{5!} - \dfrac{x^4}{7!} + \cdots$. Thus, $\lim\limits_{x\to 0} \dfrac{\sin x - x}{x^3} = \lim\limits_{x\to 0} \left(-\dfrac{1}{6} + \dfrac{x^2}{120} - \dfrac{x^4}{5040} + \cdots \right) = -\dfrac{1}{6}$.

61. $f(x) = \sum_{n=0}^{\infty} c_n x^n \quad \Rightarrow \quad f(-x) = \sum_{n=0}^{\infty} c_n (-x)^n = \sum_{n=0}^{\infty} (-1)^n c_n x^n$

(a) If f is an odd function, then $f(-x) = -f(x) \quad \Rightarrow \quad \sum_{n=0}^{\infty} (-1)^n c_n x^n = \sum_{n=0}^{\infty} -c_n x^n$. The coefficients of

any power series are uniquely determined (by Theorem 12.10.5 [ET 11.10.5]), so $(-1)^n c_n = -c_n$. If n is

even, then $(-1)^n = 1$, so $c_n = -c_n \quad \Rightarrow \quad 2c_n = 0 \quad \Rightarrow \quad c_n = 0$. Thus, all even coefficients are 0, that is,

$c_0 = c_2 = c_4 = \cdots = 0$.

(b) If f is even, then $f(-x) = f(x) \quad \Rightarrow \quad \sum_{n=0}^{\infty} (-1)^n c_n x^n = \sum_{n=0}^{\infty} c_n x^n \quad \Rightarrow \quad (-1)^n c_n = c_n$.

If n is odd, then $(-1)^n = -1$, so $-c_n = c_n \quad \Rightarrow \quad 2c_n = 0 \quad \Rightarrow \quad c_n = 0$. Thus, all odd coefficients are 0,

that is, $c_1 = c_3 = c_5 = \cdots = 0$.

☐ PROBLEMS PLUS

1. It would be far too much work to compute 15 derivatives of f. The key idea is to remember that $f^{(n)}(0)$ occurs in the coefficient of x^n in the Maclaurin series of f. We start with the Maclaurin series for sin:

$$\sin x = x - \frac{x^3}{3!} + \frac{x^5}{5!} - \cdots. \text{ Then } \sin(x^3) = x^3 - \frac{x^9}{3!} + \frac{x^{15}}{5!} - \cdots, \text{ and so the coefficient of } x^{15} \text{ is}$$

$$\frac{f^{(15)}(0)}{15!} = \frac{1}{5!}. \text{ Therefore, } f^{(15)}(0) = \frac{15!}{5!} = 6 \cdot 7 \cdot 8 \cdot 9 \cdot 10 \cdot 11 \cdot 12 \cdot 13 \cdot 14 \cdot 15 = 10{,}897{,}286{,}400.$$

3. (a) From Formula 14a in Appendix D, with $x = y = \theta$, we get $\tan 2\theta = \dfrac{2\tan\theta}{1 - \tan^2\theta}$, so $\cot 2\theta = \dfrac{1 - \tan^2\theta}{2\tan\theta}$ $\Rightarrow$

$$2\cot 2\theta = \frac{1 - \tan^2\theta}{\tan\theta} = \cot\theta - \tan\theta. \text{ Replacing } \theta \text{ by } \tfrac{1}{2}x, \text{ we get } 2\cot x = \cot\tfrac{1}{2}x - \tan\tfrac{1}{2}x, \text{ or}$$

$$\tan\tfrac{1}{2}x = \cot\tfrac{1}{2}x - 2\cot x$$

(b) From part (a) with $\dfrac{x}{2^{n-1}}$ in place of x, $\tan\dfrac{x}{2^n} = \cot\dfrac{x}{2^n} - 2\cot\dfrac{x}{2^{n-1}}$, so the nth partial sum of

$\displaystyle\sum_{n=1}^{\infty} \frac{1}{2^n}\tan\frac{x}{2^n}$ is

$$s_n = \frac{\tan(x/2)}{2} + \frac{\tan(x/4)}{4} + \frac{\tan(x/8)}{8} + \cdots + \frac{\tan(x/2^n)}{2^n}$$

$$= \left[\frac{\cot(x/2)}{2} - \cot x\right] + \left[\frac{\cot(x/4)}{4} - \frac{\cot(x/2)}{2}\right] + \left[\frac{\cot(x/8)}{8} - \frac{\cot(x/4)}{4}\right] + \cdots$$

$$+ \left[\frac{\cot(x/2^n)}{2^n} - \frac{\cot(x/2^{n-1})}{2^{n-1}}\right] = -\cot x + \frac{\cot(x/2^n)}{2^n} \quad \text{(telescoping sum)}$$

Now $\dfrac{\cot(x/2^n)}{2^n} = \dfrac{\cos(x/2^n)}{2^n\sin(x/2^n)} = \dfrac{\cos(x/2^n)}{x} \cdot \dfrac{x/2^n}{\sin(x/2^n)} \to \dfrac{1}{x} \cdot 1 = \dfrac{1}{x}$ as $n \to \infty$ since $x/2^n \to 0$

for $x \neq 0$. Therefore, if $x \neq 0$ and $x \neq k\pi$ where k is any integer, then

$$\sum_{n=1}^{\infty}\frac{1}{2^n}\tan\frac{x}{2^n} = \lim_{n\to\infty} s_n = \lim_{n\to\infty}\left(-\cot x + \frac{1}{2^n}\cot\frac{x}{2^n}\right) = -\cot x + \frac{1}{x}$$

If $x = 0$, then all terms in the series are 0, so the sum is 0.

5. (a) At each stage, each side is replaced by four shorter sides, each of length $\frac{1}{3}$ of the side length at the preceding stage. Writing s_0 and ℓ_0 for the number of sides and the length of the side of the initial triangle, we generate the table at right. In general, we have $s_n = 3 \cdot 4^n$ and $\ell_n = \left(\frac{1}{3}\right)^n$, so the length of the perimeter at the nth stage of construction is $p_n = s_n\ell_n = 3 \cdot 4^n \cdot \left(\frac{1}{3}\right)^n = 3 \cdot \left(\frac{4}{3}\right)^n$.

$s_0 = 3$	$\ell_0 = 1$
$s_1 = 3 \cdot 4$	$\ell_1 = 1/3$
$s_2 = 3 \cdot 4^2$	$\ell_2 = 1/3^2$
$s_3 = 3 \cdot 4^3$	$\ell_3 = 1/3^3$
$\cdots$	$\cdots$

(b) $p_n = \dfrac{4^n}{3^{n-1}} = 4\left(\dfrac{4}{3}\right)^{n-1}$. Since $\frac{4}{3} > 1$, $p_n \to \infty$ as $n \to \infty$.

(c) The area of each of the small triangles added at a given stage is one-ninth of the area of the triangle added at the preceding stage. Let a be the area of the original triangle. Then the area a_n of each of the small triangles added at stage n is $a_n = a \cdot \dfrac{1}{9^n} = \dfrac{a}{9^n}$. Since a small triangle is added to each side at every stage, it follows that the total area A_n added to the figure at the nth stage is

$$A_n = s_{n-1} \cdot a_n = 3 \cdot 4^{n-1} \cdot \frac{a}{9^n} = a \cdot \frac{4^{n-1}}{3^{2n-1}}. \quad \text{Then the total area enclosed by the snowflake curve is}$$

$$A = a + A_1 + A_2 + A_3 + \cdots = a + a \cdot \frac{1}{3} + a \cdot \frac{4}{3^3} + a \cdot \frac{4^2}{3^5} + a \cdot \frac{4^3}{3^7} + \cdots . \quad \text{After the first term, this is a}$$

geometric series with common ratio $\frac{4}{9}$, so $A = a + \dfrac{a/3}{1 - \frac{4}{9}} = a + \dfrac{a}{3} \cdot \dfrac{9}{5} = \dfrac{8a}{5}$. But the area of the original

equilateral triangle with side 1 is $a = \frac{1}{2} \cdot 1 \cdot \sin\frac{\pi}{3} = \frac{\sqrt{3}}{4}$. So the area enclosed by the snowflake curve is

$\frac{8}{5} \cdot \frac{\sqrt{3}}{4} = \frac{2\sqrt{3}}{5}$.

7. (a) Let $a = \arctan x$ and $b = \arctan y$. Then, from Formula 14b in Appendix D,

$$\tan(a - b) = \frac{\tan a - \tan b}{1 + \tan a \tan b} = \frac{\tan(\arctan x) - \tan(\arctan y)}{1 + \tan(\arctan x)\tan(\arctan y)} = \frac{x - y}{1 + xy}.$$

Now $\arctan x - \arctan y = a - b = \arctan(\tan(a - b)) = \arctan\dfrac{x - y}{1 + xy}$ since $-\frac{\pi}{2} < a - b < \frac{\pi}{2}$.

(b) From part (a) we have

$$\arctan\tfrac{120}{119} - \arctan\tfrac{1}{239} = \arctan\frac{\frac{120}{119} - \frac{1}{239}}{1 + \frac{120}{119} \cdot \frac{1}{239}} = \arctan\frac{\frac{28,561}{28,441}}{\frac{28,561}{28,441}} = \arctan 1 = \tfrac{\pi}{4}$$

(c) Replacing y by $-y$ in the formula of part (a), we get $\arctan x + \arctan y = \arctan\dfrac{x + y}{1 - xy}$. So

$$4\arctan\tfrac{1}{5} = 2\left(\arctan\tfrac{1}{5} + \arctan\tfrac{1}{5}\right) = 2\arctan\frac{\frac{1}{5} + \frac{1}{5}}{1 - \frac{1}{5} \cdot \frac{1}{5}} = 2\arctan\tfrac{5}{12} = \arctan\tfrac{5}{12} + \arctan\tfrac{5}{12}$$

$$= \arctan\frac{\frac{5}{12} + \frac{5}{12}}{1 - \frac{5}{12} \cdot \frac{5}{12}} = \arctan\tfrac{120}{119}$$

Thus, from part (b), we have $4\arctan\tfrac{1}{5} - \arctan\tfrac{1}{239} = \arctan\tfrac{120}{119} - \arctan\tfrac{1}{239} = \tfrac{\pi}{4}$.

(d) From Example 7 in Section 12.9 [ET 11.9] we have $\arctan x = x - \dfrac{x^3}{3} + \dfrac{x^5}{5} - \dfrac{x^7}{7} + \dfrac{x^9}{9} - \dfrac{x^{11}}{11} + \cdots$, so

$$\arctan\frac{1}{5} = \frac{1}{5} - \frac{1}{3 \cdot 5^3} + \frac{1}{5 \cdot 5^5} - \frac{1}{7 \cdot 5^7} + \frac{1}{9 \cdot 5^9} - \frac{1}{11 \cdot 5^{11}} + \cdots$$

This is an alternating series and the size of the terms decreases to 0, so by the Alternating Series Estimation Theorem, the sum lies between s_5 and s_6, that is, $0.197395560 < \arctan\frac{1}{5} < 0.197395562$.

(e) From the series in part (d) we get $\arctan \frac{1}{239} = \frac{1}{239} - \frac{1}{3 \cdot 239^3} + \frac{1}{5 \cdot 239^5} - \cdots$. The third term is less than

2.6×10^{-13}, so by the Alternating Series Estimation Theorem, we have, to nine decimal places,

$\arctan \frac{1}{239} \approx s_2 \approx 0.004184076$. Thus, $0.004184075 < \arctan \frac{1}{239} < 0.004184077$.

(f) From part (c) we have $\pi = 16 \arctan \frac{1}{5} - 4 \arctan \frac{1}{239}$, so from parts (d) and (e) we have

$16(0.197395560) - 4(0.004184077) < \pi < 16(0.197395562) - 4(0.004184075) \Rightarrow$

$3.141592652 < \pi < 3.141592692$. So, to 7 decimal places, $\pi \approx 3.1415927$.

9. We start with the geometric series $\sum\limits_{n=0}^{\infty} x^n = \frac{1}{1-x}$, $|x| < 1$, and differentiate:

$\sum\limits_{n=1}^{\infty} n x^{n-1} = \frac{d}{dx} \left(\sum\limits_{n=0}^{\infty} x^n \right) = \frac{d}{dx} \left(\frac{1}{1-x} \right) = \frac{1}{(1-x)^2}$ for $|x| < 1 \Rightarrow$

$\sum\limits_{n=1}^{\infty} n x^n = x \sum\limits_{n=1}^{\infty} n x^{n-1} = \frac{x}{(1-x)^2}$ for $|x| < 1$. Differentiate again:

$\sum\limits_{n=1}^{\infty} n^2 x^{n-1} = \frac{d}{dx} \frac{x}{(1-x)^2} = \frac{(1-x)^2 - x \cdot 2(1-x)(-1)}{(1-x)^4} = \frac{x+1}{(1-x)^3} \Rightarrow \sum\limits_{n=1}^{\infty} n^2 x^n = \frac{x^2+x}{(1-x)^3} \Rightarrow$

$\sum\limits_{n=1}^{\infty} n^3 x^{n-1} = \frac{d}{dx} \frac{x^2+x}{(1-x)^3} = \frac{(1-x)^3(2x+1) - (x^2+x)3(1-x)^2(-1)}{(1-x)^6} = \frac{x^2+4x+1}{(1-x)^4} \Rightarrow$

$\sum\limits_{n=1}^{\infty} n^3 x^n = \frac{x^3+4x^2+x}{(1-x)^4}$, $|x| < 1$. The radius of convergence is 1 because that is the radius of convergence for

the geometric series we started with. If $x = \pm 1$, the series is $\sum n^3 (\pm 1)^n$, which diverges by the Test For

Divergence, so the interval of convergence is $(-1, 1)$.

11. $\ln \left(1 - \frac{1}{n^2} \right) = \ln \left(\frac{n^2 - 1}{n^2} \right) = \ln \frac{(n+1)(n-1)}{n^2} = \ln[(n+1)(n-1)] - \ln n^2$

$= \ln(n+1) + \ln(n-1) - 2 \ln n$

$= \ln(n-1) - \ln n - \ln n + \ln(n+1)$

$= \ln \frac{n-1}{n} - [\ln n - \ln(n+1)] = \ln \frac{n-1}{n} - \ln \frac{n}{n+1}$.

Let $s_k = \sum\limits_{n=2}^{k} \ln \left(1 - \frac{1}{n^2} \right) = \sum\limits_{n=2}^{k} \left(\ln \frac{n-1}{n} - \ln \frac{n}{n+1} \right)$ for $k \geq 2$. Then

$s_k = \left(\ln \frac{1}{2} - \ln \frac{2}{3} \right) + \left(\ln \frac{2}{3} - \ln \frac{3}{4} \right) + \cdots + \left(\ln \frac{k-1}{k} - \ln \frac{k}{k+1} \right) = \ln \frac{1}{2} - \ln \frac{k}{k+1}$, so

$\sum\limits_{n=2}^{\infty} \ln \left(1 - \frac{1}{n^2} \right) = \lim\limits_{k \to \infty} s_k = \lim\limits_{k \to \infty} \left(\ln \frac{1}{2} - \ln \frac{k}{k+1} \right) = \ln \frac{1}{2} - \ln 1 = \ln 1 - \ln 2 - \ln 1 = -\ln 2$.

13. $u = 1 + \dfrac{x^3}{3!} + \dfrac{x^6}{6!} + \dfrac{x^9}{9!} + \cdots, v = x + \dfrac{x^4}{4!} + \dfrac{x^7}{7!} + \dfrac{x^{10}}{10!} + \cdots, w = \dfrac{x^2}{2!} + \dfrac{x^5}{5!} + \dfrac{x^8}{8!} + \cdots.$

Use the Ratio Test to show that the series for u, v, and w have positive radii of convergence (∞ in each case), so

Theorem 12.9.2 [ET 11.9.2] applies, and hence, we may differentiate each of these series:

$$\frac{du}{dx} = \frac{3x^2}{3!} + \frac{6x^5}{6!} + \frac{9x^8}{9!} + \cdots = \frac{x^2}{2!} + \frac{x^5}{5!} + \frac{x^8}{8!} + \cdots = w$$

Similarly, $\dfrac{dv}{dx} = 1 + \dfrac{x^3}{3!} + \dfrac{x^6}{6!} + \dfrac{x^9}{9!} + \cdots = u$, and $\dfrac{dw}{dx} = x + \dfrac{x^4}{4!} + \dfrac{x^7}{7!} + \dfrac{x^{10}}{10!} + \cdots = v$.

So $u' = w$, $v' = u$, and $w' = v$. Now differentiate the left hand side of the desired equation:

$$\frac{d}{dx}\left(u^3 + v^3 + w^3 - 3uvw\right) = 3u^2u' + 3v^2v' + 3w^2w' - 3(u'vw + uv'w + uvw')$$

$$= 3u^2w + 3v^2u + 3w^2v - 3(vw^2 + u^2w + uv^2) = 0 \quad \Rightarrow$$

$u^3 + v^3 + w^3 - 3uvw = C$. To find the value of the constant C, we put $x = 0$ in the last equation and get

$1^3 + 0^3 + 0^3 - 3(1 \cdot 0 \cdot 0) = C \quad \Rightarrow \quad C = 1$, so $u^3 + v^3 + w^3 - 3uvw = 1$.

15. If L is the length of a side of the equilateral triangle, then the area is $A = \frac{1}{2}L \cdot \frac{\sqrt{3}}{2}L = \frac{\sqrt{3}}{4}L^2$ and so $L^2 = \frac{4}{\sqrt{3}}A$.

Let r be the radius of one of the circles. When there are n rows of circles, the figure shows that

$$L = \sqrt{3}r + r + (n-2)(2r) + r + \sqrt{3}r = r\left(2n - 2 + 2\sqrt{3}\right), \text{ so } r = \frac{L}{2\left(n + \sqrt{3} - 1\right)}.$$

The number of circles is $1 + 2 + \cdots + n = \dfrac{n(n+1)}{2}$, and so the total area of the circles is

$$A_n = \frac{n(n+1)}{2}\pi r^2 = \frac{n(n+1)}{2}\pi\frac{L^2}{4\left(n + \sqrt{3} - 1\right)^2} = \frac{n(n+1)}{2}\pi\frac{4A/\sqrt{3}}{4\left(n + \sqrt{3} - 1\right)^2}$$

$$= \frac{n(n+1)}{\left(n + \sqrt{3} - 1\right)^2}\frac{\pi A}{2\sqrt{3}} \quad \Rightarrow$$

$$\frac{A_n}{A} = \frac{n(n+1)}{\left(n + \sqrt{3} - 1\right)^2}\frac{\pi}{2\sqrt{3}}$$

$$= \frac{1 + 1/n}{\left[1 + (\sqrt{3} - 1)/n\right]^2}\frac{\pi}{2\sqrt{3}} \to \frac{\pi}{2\sqrt{3}} \text{ as } n \to \infty$$

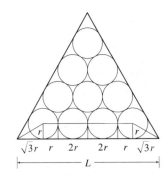

$\sqrt{3}r \quad r \quad 2r \quad 2r \quad r \quad \sqrt{3}r$

L

17. As in Section 12.9 [ET 11.9] we have to integrate the function x^x by integrating series.

Writing $x^x = (e^{\ln x})^x = e^{x \ln x}$ and using the Maclaurin series for e^x, we have

$$x^x = (e^{\ln x})^x = e^{x \ln x} = \sum_{n=0}^{\infty} \frac{(x \ln x)^n}{n!} = \sum_{n=0}^{\infty} \frac{x^n (\ln x)^n}{n!}.$$

As with power series, we can integrate this series term-by-term:

$$\int_0^1 x^x \, dx = \sum_{n=0}^{\infty} \int_0^1 \frac{x^n (\ln x)^n}{n!} \, dx = \sum_{n=0}^{\infty} \frac{1}{n!} \int_0^1 x^n (\ln x)^n \, dx$$

We integrate by parts with $u = (\ln x)^n$, $dv = x^n \, dx$, so $du = \dfrac{n(\ln x)^{n-1}}{x} \, dx$ and $v = \dfrac{x^{n+1}}{n+1}$:

$$\int_0^1 x^n (\ln x)^n \, dx = \lim_{t \to 0^+} \int_t^1 x^n (\ln x)^n \, dx = \lim_{t \to 0^+} \left[\frac{x^{n+1}}{n+1} (\ln x)^n \right]_t^1 - \lim_{t \to 0^+} \int_t^1 \frac{n}{n+1} x^n (\ln x)^{n-1} \, dx$$

$$= 0 - \frac{n}{n+1} \int_0^1 x^n (\ln x)^{n-1} \, dx$$

(where l'Hospital's Rule was used to help evaluate the first limit).

Further integration by parts gives $\displaystyle\int_0^1 x^n (\ln x)^k \, dx = -\frac{k}{n+1} \int_0^1 x^n (\ln x)^{k-1} \, dx$ and, combining

these steps, we get $\displaystyle\int_0^1 x^n (\ln x)^n \, dx = \frac{(-1)^n \, n!}{(n+1)^n} \int_0^1 x^n \, dx = \frac{(-1)^n \, n!}{(n+1)^{n+1}} \quad \Rightarrow$

$$\int_0^1 x^x \, dx = \sum_{n=0}^{\infty} \frac{1}{n!} \int_0^1 x^n (\ln x)^n \, dx = \sum_{n=0}^{\infty} \frac{1}{n!} \frac{(-1)^n \, n!}{(n+1)^{n+1}} = \sum_{n=0}^{\infty} \frac{(-1)^n}{(n+1)^{n+1}} = \sum_{n=1}^{\infty} \frac{(-1)^{n-1}}{n^n}.$$

19. Let $f(x) = \sum_{m=0}^{\infty} c_m x^m$ and $g(x) = e^{f(x)} = \sum_{n=0}^{\infty} d_n x^n$. Then $g'(x) = \sum_{n=0}^{\infty} n d_n x^{n-1}$, so $n d_n$ occurs as

the coefficient of x^{n-1}. But also

$$g'(x) = e^{f(x)} f'(x) = \left(\sum_{n=0}^{\infty} d_n x^n \right) \left(\sum_{m=1}^{\infty} m c_m x^{m-1} \right)$$

$$= (d_0 + d_1 x + d_2 x^2 + \cdots + d_{n-1} x^{n-1} + \cdots)(c_1 + 2 c_2 x + 3 c_3 x^2 + \cdots + n c_n x^{n-1} + \cdots)$$

so the coefficient of x^{n-1} is $c_1 d_{n-1} + 2 c_2 d_{n-2} + 3 c_3 d_{n-3} + \cdots + n c_n d_0 = \sum_{i=1}^{n} i c_i d_{n-i}$. Therefore,

$n d_n = \sum_{i=1}^{n} i c_i d_{n-i}$.

21. Call the series S. We group the terms according to the number of digits in their denominators:

$$S = \underbrace{\left(\tfrac{1}{1} + \tfrac{1}{2} + \cdots + \tfrac{1}{8} + \tfrac{1}{9} \right)}_{g_1} + \underbrace{\left(\tfrac{1}{11} + \cdots + \tfrac{1}{99} \right)}_{g_2} + \underbrace{\left(\tfrac{1}{111} + \cdots + \tfrac{1}{999} \right)}_{g_3} + \cdots$$

Now in the group g_n, since we have 9 choices for each of the n digits in the denominator, there are 9^n terms.

Furthermore, each term in g_n is less than $\frac{1}{10^{n-1}}$ [except for the first term in g_1]. So $g_n < 9^n \cdot \frac{1}{10^{n-1}} = 9 \left(\frac{9}{10} \right)^{n-1}$.

Now $\sum_{n=1}^{\infty} 9 \left(\frac{9}{10} \right)^{n-1}$ is a geometric series with $a = 9$ and $r = \frac{9}{10} < 1$. Therefore, by the Comparison Test,

$S = \sum_{n=1}^{\infty} g_n < \sum_{n=1}^{\infty} 9 \left(\frac{9}{10} \right)^{n-1} = \frac{9}{1 - 9/10} = 90.$

13 □ VECTORS AND THE GEOMETRY OF SPACE □ ET 12

13.1 Three-Dimensional Coordinate Systems

1. We start at the origin, which has coordinates $(0, 0, 0)$. First we move 4 units along the positive x-axis, affecting only the x-coordinate, bringing us to the point $(4, 0, 0)$. We then move 3 units straight downward, in the negative z-direction. Thus only the z-coordinate is affected, and we arrive at $(4, 0, -3)$.

3. The distance from a point to the xz-plane is the absolute value of the y-coordinate of the point. $Q(-5, -1, 4)$ has the y-coordinate with the smallest absolute value, so Q is the point closest to the xz-plane. $R(0, 3, 8)$ must lie in the yz-plane since the distance from R to the yz-plane, given by the x-coordinate of R, is 0.

5. The equation $x + y = 2$ represents the set of all points in $\mathbb{R}^3$ whose x- and y-coordinates have a sum of 2, or equivalently where $y = 2 - x$. This is the set $\{(x, 2 - x, z) \mid x \in \mathbb{R}, z \in \mathbb{R}\}$ which is a vertical plane that intersects the xy-plane in the line $y = 2 - x$, $z = 0$.

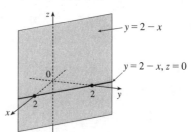

7. We first find the lengths of the sides of the triangle by using the distance formula between pairs of vertices:

$$|PQ| = \sqrt{[1 - (-2)]^2 + (2 - 4)^2 + (-1 - 0)^2} = \sqrt{9 + 4 + 1} = \sqrt{14}$$

$$|QR| = \sqrt{(-1 - 1)^2 + (1 - 2)^2 + [2 - (-1)]^2} = \sqrt{4 + 1 + 9} = \sqrt{14}$$

$$|PR| = \sqrt{[-1 - (-2)]^2 + (1 - 4)^2 + (2 - 0)^2} = \sqrt{1 + 9 + 4} = \sqrt{14}$$

Since all three sides have the same length, PQR is an equilateral triangle.

9. (a) First we find the distances between points:

$$|AB| = \sqrt{(7 - 5)^2 + (9 - 1)^2 + (-1 - 3)^2} = \sqrt{84} = 2\sqrt{21}$$

$$|BC| = \sqrt{(1 - 7)^2 + (-15 - 9)^2 + [11 - (-1)]^2} = \sqrt{756} = 6\sqrt{21}$$

$$|AC| = \sqrt{(1 - 5)^2 + (-15 - 1)^2 + (11 - 3)^2} = \sqrt{336} = 4\sqrt{21}$$

In order for the points to lie on a straight line, the sum of the two shortest distances must be equal to the longest distance. Since $|AB| + |AC| = |BC|$, the three points lie on a straight line.

(b) The distances between points are

$$|KL| = \sqrt{(1 - 0)^2 + (2 - 3)^2 + [-2 - (-4)]^2} = \sqrt{6}$$

$$|LM| = \sqrt{(3 - 1)^2 + (0 - 2)^2 + [1 - (-2)]^2} = \sqrt{17}$$

$$|KM| = \sqrt{(3 - 0)^2 + (0 - 3)^2 + [1 - (-4)]^2} = \sqrt{43}$$

Since $\sqrt{6} + \sqrt{17} \neq \sqrt{43}$, the three points do not lie on a straight line.

11. An equation of the sphere with center $(1, -4, 3)$ and radius 5 is $(x - 1)^2 + [y - (-4)]^2 + (z - 3)^2 = 5^2$ or $(x - 1)^2 + (y + 4)^2 + (z - 3)^2 = 25$. The intersection of this sphere with the xz-plane is the set of points on the sphere whose y-coordinate is 0. Putting $y = 0$ into the equation, we have $(x - 1)^2 + 4^2 + (z - 3)^2 = 25$, $y = 0$ or $(x - 1)^2 + (z - 3)^2 = 9$, $y = 0$, which represents a circle in the xz-plane with center $(1, 0, 3)$ and radius 3.

13. The radius of the sphere is the distance between $(4, 3, -1)$ and $(3, 8, 1)$:

$r = \sqrt{(3 - 4)^2 + (8 - 3)^2 + [1 - (-1)]^2} = \sqrt{30}$. Thus, an equation of the sphere is

$(x - 3)^2 + (y - 8)^2 + (z - 1)^2 = 30$.

15. Completing squares in the equation $x^2 + y^2 + z^2 - 6x + 4y - 2z = 11$ gives

$(x^2 - 6x + 9) + (y^2 + 4y + 4) + (z^2 - 2z + 1) = 11 + 9 + 4 + 1 \quad \Rightarrow$

$(x - 3)^2 + (y + 2)^2 + (z - 1)^2 = 25$ which we recognize as an equation of a sphere with center $(3, -2, 1)$ and radius 5.

17. Completing squares in the equation gives $\left(x^2 - x + \frac{1}{4}\right) + \left(y^2 - y + \frac{1}{4}\right) + \left(z^2 - z + \frac{1}{4}\right) = \frac{1}{4} + \frac{1}{4} + \frac{1}{4} \quad \Rightarrow$

$\left(x - \frac{1}{2}\right)^2 + \left(y - \frac{1}{2}\right)^2 + \left(z - \frac{1}{2}\right)^2 = \frac{3}{4}$ which we recognize as an equation of a sphere with center $\left(\frac{1}{2}, \frac{1}{2}, \frac{1}{2}\right)$ and

radius $\sqrt{\frac{3}{4}} = \frac{\sqrt{3}}{2}$.

19. (a) If the midpoint of the line segment from $P_1(x_1, y_1, z_1)$ to $P_2(x_2, y_2, z_2)$ is

$Q = \left(\dfrac{x_1 + x_2}{2}, \dfrac{y_1 + y_2}{2}, \dfrac{z_1 + z_2}{2}\right)$, then the distances $|P_1Q|$ and $|QP_2|$ are equal, and each is half of $|P_1P_2|$.

We verify that this is the case:

$$|P_1P_2| = \sqrt{(x_2 - x_1)^2 + (y_2 - y_1)^2 + (z_2 - z_1)^2}$$

$$|P_1Q| = \sqrt{\left[\tfrac{1}{2}(x_1 + x_2) - x_1\right]^2 + \left[\tfrac{1}{2}(y_1 + y_2) - y_1\right]^2 + \left[\tfrac{1}{2}(z_1 + z_2) - z_1\right]^2}$$

$$= \sqrt{\left(\tfrac{1}{2}x_2 - \tfrac{1}{2}x_1\right)^2 + \left(\tfrac{1}{2}y_2 - \tfrac{1}{2}y_1\right)^2 + \left(\tfrac{1}{2}z_2 - \tfrac{1}{2}z_1\right)^2}$$

$$= \sqrt{\left(\tfrac{1}{2}\right)^2\left[(x_2 - x_1)^2 + (y_2 - y_1)^2 + (z_2 - z_1)^2\right]}$$

$$= \tfrac{1}{2}\sqrt{(x_2 - x_1)^2 + (y_2 - y_1)^2 + (z_2 - z_1)^2}$$

$$= \tfrac{1}{2}|P_1P_2|$$

$$|QP_2| = \sqrt{\left[x_2 - \tfrac{1}{2}(x_1 + x_2)\right]^2 + \left[y_2 - \tfrac{1}{2}(y_1 + y_2)\right]^2 + \left[z_2 - \tfrac{1}{2}(z_1 + z_2)\right]^2}$$

$$= \sqrt{\left(\tfrac{1}{2}x_2 - \tfrac{1}{2}x_1\right)^2 + \left(\tfrac{1}{2}y_2 - \tfrac{1}{2}y_1\right)^2 + \left(\tfrac{1}{2}z_2 - \tfrac{1}{2}z_1\right)^2}$$

$$= \sqrt{\left(\tfrac{1}{2}\right)^2\left[(x_2 - x_1)^2 + (y_2 - y_1)^2 + (z_2 - z_1)^2\right]}$$

$$= \tfrac{1}{2}\sqrt{(x_2 - x_1)^2 + (y_2 - y_1)^2 + (z_2 - z_1)^2}$$

$$= \tfrac{1}{2}|P_1P_2|$$

So Q is indeed the midpoint of P_1P_2.

(b) By part (a), the midpoints of sides AB, BC and CA are $P_1\left(-\frac{1}{2}, 1, 4\right)$, $P_2\left(1, \frac{1}{2}, 5\right)$ and $P_3\left(\frac{5}{2}, \frac{3}{2}, 4\right)$. (Recall that a median of a triangle is a line segment from a vertex to the midpoint of the opposite side.) Then the lengths of the medians are:

$$|AP_2| = \sqrt{0^2 + \left(\tfrac{1}{2} - 2\right)^2 + (5 - 3)^2} = \sqrt{\tfrac{9}{4} + 4} = \sqrt{\tfrac{25}{4}} = \tfrac{5}{2}$$

$$|BP_3| = \sqrt{\left(\tfrac{5}{2} + 2\right)^2 + \left(\tfrac{3}{2}\right)^2 + (4 - 5)^2} = \sqrt{\tfrac{81}{4} + \tfrac{9}{4} + 1} = \sqrt{\tfrac{94}{4}} = \tfrac{1}{2}\sqrt{94}$$

$$|CP_1| = \sqrt{\left(-\tfrac{1}{2} - 4\right)^2 + (1 - 1)^2 + (4 - 5)^2} = \sqrt{\tfrac{81}{4} + 1} = \tfrac{1}{2}\sqrt{85}$$

21. (a) Since the sphere touches the xy-plane, its radius is the distance from its center, $(2, -3, 6)$, to the xy-plane, namely 6. Therefore $r = 6$ and an equation of the sphere is $(x - 2)^2 + (y + 3)^2 + (z - 6)^2 = 6^2 = 36$.

(b) The radius of this sphere is the distance from its center $(2, -3, 6)$ to the yz-plane, which is 2. Therefore, an equation is $(x - 2)^2 + (y + 3)^2 + (z - 6)^2 = 4$.

(c) Here the radius is the distance from the center $(2, -3, 6)$ to the xz-plane, which is 3. Therefore, an equation is $(x - 2)^2 + (y + 3)^2 + (z - 6)^2 = 9$.

23. The equation $y = -4$ represents a plane parallel to the xz-plane and 4 units to the left of it.

25. The inequality $x > 3$ represents a half-space consisting of all points in front of the plane $x = 3$.

27. The inequality $0 \leq z \leq 6$ represents all points on or between the horizontal planes $z = 0$ (the xy-plane) and $z = 6$.

29. The inequality $x^2 + y^2 + z^2 > 1$ is equivalent to $\sqrt{x^2 + y^2 + z^2} > 1$, so the region consists of those points whose distance from the origin is greater than 1. This is the set of all points outside the sphere with radius 1 and center $(0, 0, 0)$.

31. Completing the square in z gives $x^2 + y^2 + (z^2 - 2z + 1) < 3 + 1$ or $x^2 + y^2 + (z - 1)^2 < 4$, which is equivalent to $\sqrt{x^2 + y^2 + (z - 1)^2} < 2$. Thus the region consists of those points whose distance from the point $(0, 0, 1)$ is less than 2. This is the set of all points inside the sphere with radius 2 and center $(0, 0, 1)$.

33. Here $x^2 + z^2 \leq 9$ or equivalently $\sqrt{x^2 + z^2} \leq 3$ which describes the set of all points in $\mathbb{R}^3$ whose distance from the y-axis is at most 3. Thus, the inequality represents the region consisting of all points on or inside a circular cylinder of radius 3 with axis the y-axis.

35. This describes all points with negative y-coordinates, that is, $y < 0$.

37. This describes a region all of whose points have a distance to the origin which is greater than r, but smaller than R. So inequalities describing the region are $r < \sqrt{x^2 + y^2 + z^2} < R$, or $r^2 < x^2 + y^2 + z^2 < R^2$.

39. (a) To find the x- and y-coordinates of the point P, we project it

onto L_2 and project the resulting point Q onto the x- and

y-axes. To find the z-coordinate, we project P onto either the

xz-plane or the yz-plane (using our knowledge of its x- or

y-coordinate) and then project the resulting point onto the

z-axis. (Or, we could draw a line parallel to QO from P to

the z-axis.) The coordinates of P are $(2, 1, 4)$.

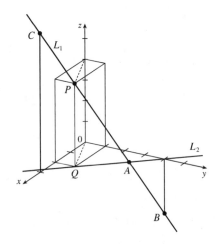

(b) A is the intersection of L_1 and L_2, B is directly below

the y-intercept of L_2, and C is directly above the x-intercept

of L_2.

41. We need to find a set of points $\{P(x, y, z) \mid |AP| = |BP|\}$.

$$\sqrt{(x+1)^2 + (y-5)^2 + (z-3)^2} = \sqrt{(x-6)^2 + (y-2)^2 + (z+2)^2} \Rightarrow$$
$$(x+1)^2 + (y-5) + (z-3)^2 = (x-6)^2 + (y-2)^2 + (z+2)^2 \Rightarrow$$
$$x^2 + 2x + 1 + y^2 - 10y + 25 + z^2 - 6z + 9 = x^2 - 12x + 36 + y^2 - 4y + 4 + z^2 + 4z + 4 \Rightarrow$$
$14x - 6y - 10z = 9$. Thus the set of points is a plane perpendicular to the line segment joining A and B (since this

plane must contain the perpendicular bisector of the line segment AB).

13.2 Vectors   ET 12.2

1. (a) The cost of a theater ticket is a scalar, because it has only magnitude.

   (b) The current in a river is a vector, because it has both magnitude (the speed of the current) and direction at any
       given location.

   (c) If we assume that the initial path is linear, the initial flight path from Houston to Dallas is a vector, because it has
       both magnitude (distance) and direction.

   (d) The population of the world is a scalar, because it has only magnitude.

3. Vectors are equal when they share the same length and direction (but not necessarily location). Using the symmetry
   of the parallelogram as a guide, we see that $\overrightarrow{AB} = \overrightarrow{DC}$, $\overrightarrow{DA} = \overrightarrow{CB}$, $\overrightarrow{DE} = \overrightarrow{EB}$, and $\overrightarrow{EA} = \overrightarrow{CE}$.

5. (a)   (b)

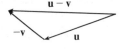

(c)   (d)

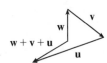

7. $\mathbf{a} = \langle -2 - 2, 1 - 3 \rangle = \langle -4, -2 \rangle$

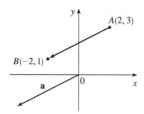

9. $\mathbf{a} = \langle -3 - (-1), 4 - (-1) \rangle = \langle -2, 5 \rangle$

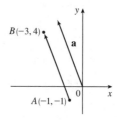

11. $\mathbf{a} = \langle 2 - 0, 3 - 3, -1 - 1 \rangle = \langle 2, 0, -2 \rangle$

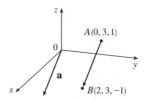

13. $\langle 3, -1 \rangle + \langle -2, 4 \rangle = \langle 3 + (-2), -1 + 4 \rangle$
$$= \langle 1, 3 \rangle$$

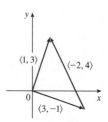

15. $\langle 0, 1, 2 \rangle + \langle 0, 0, -3 \rangle = \langle 0 + 0, 1 + 0, 2 + (-3) \rangle$
$$= \langle 0, 1, -1 \rangle$$

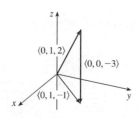

17. $|\mathbf{a}| = \sqrt{(-4)^2 + 3^2} = \sqrt{25} = 5$

$\mathbf{a} + \mathbf{b} = \langle -4 + 6, 3 + 2 \rangle = \langle 2, 5 \rangle$

$\mathbf{a} - \mathbf{b} = \langle -4 - 6, 3 - 2 \rangle = \langle -10, 1 \rangle$

$2\mathbf{a} = \langle 2(-4), 2(3) \rangle = \langle -8, 6 \rangle$

$3\mathbf{a} + 4\mathbf{b} = \langle -12, 9 \rangle + \langle 24, 8 \rangle = \langle 12, 17 \rangle$

19. $|\mathbf{a}| = \sqrt{6^2 + 2^2 + 3^2} = \sqrt{49} = 7$

$\mathbf{a} + \mathbf{b} = \langle 6 + (-1), 2 + 5, 3 + (-2) \rangle = \langle 5, 7, 1 \rangle$

$\mathbf{a} - \mathbf{b} = \langle 6 - (-1), 2 - 5, 3 - (-2) \rangle$

$\quad = \langle 7, -3, 5 \rangle$

$2\mathbf{a} = \langle 2(6), 2(2), 2(3) \rangle = \langle 12, 4, 6 \rangle$

$3\mathbf{a} + 4\mathbf{b} = \langle 18, 6, 9 \rangle + \langle -4, 20, -8 \rangle$

$\quad = \langle 14, 26, 1 \rangle$

21. $|\mathbf{a}| = \sqrt{1^2 + (-2)^2 + 1^2} = \sqrt{6}$

$\mathbf{a} + \mathbf{b} = (\mathbf{i} - 2\mathbf{j} + \mathbf{k}) + (\mathbf{j} + 2\mathbf{k}) = \mathbf{i} - \mathbf{j} + 3\mathbf{k}$

$\mathbf{a} - \mathbf{b} = (\mathbf{i} - 2\mathbf{j} + \mathbf{k}) - (\mathbf{j} + 2\mathbf{k}) = \mathbf{i} - 3\mathbf{j} - \mathbf{k}$

$2\mathbf{a} = 2(\mathbf{i} - 2\mathbf{j} + \mathbf{k}) = 2\mathbf{i} - 4\mathbf{j} + 2\mathbf{k}$

$3\mathbf{a} + 4\mathbf{b} = 3(\mathbf{i} - 2\mathbf{j} + \mathbf{k}) + 4(\mathbf{j} + 2\mathbf{k})$

$\quad = 3\mathbf{i} - 6\mathbf{j} + 3\mathbf{k} + 4\mathbf{j} + 8\mathbf{k}$

$\quad = 3\mathbf{i} - 2\mathbf{j} + 11\mathbf{k}$

23. $|\langle 9, -5 \rangle| = \sqrt{9^2 + (-5)^2} = \sqrt{106}$, so $\mathbf{u} = \dfrac{1}{\sqrt{106}} \langle 9, -5 \rangle = \left\langle \dfrac{9}{\sqrt{106}}, \dfrac{-5}{\sqrt{106}} \right\rangle.$

25. The vector $8\,\mathbf{i} - \mathbf{j} + 4\,\mathbf{k}$ has length $|8\,\mathbf{i} - \mathbf{j} + 4\,\mathbf{k}| = \sqrt{8^2 + (-1)^2 + 4^2} = \sqrt{81} = 9$, so by Equation 4 the unit

vector with the same direction is $\frac{1}{9}(8\,\mathbf{i} - \mathbf{j} + 4\,\mathbf{k}) = \frac{8}{9}\,\mathbf{i} - \frac{1}{9}\,\mathbf{j} + \frac{4}{9}\,\mathbf{k}$.

27. From the figure, we see that the x-component of $\mathbf{v}$ is

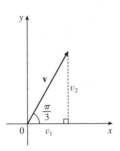

$v_1 = |\mathbf{v}|\cos(\pi/3) = 4 \cdot \frac{1}{2} = 2$ and the y-component is

$v_2 = |\mathbf{v}|\sin(\pi/3) = 4 \cdot \frac{\sqrt{3}}{2} = 2\sqrt{3}$. Thus

$\mathbf{v} = \langle v_1, v_2 \rangle = \langle 2, 2\sqrt{3} \rangle$.

29. $|\mathbf{F}_1| = 10$ lb and $|\mathbf{F}_2| = 12$ lb.

$\mathbf{F}_1 = -|\mathbf{F}_1|\cos 45^\circ\,\mathbf{i} + |\mathbf{F}_1|\sin 45^\circ\,\mathbf{j} = -10\cos 45^\circ\,\mathbf{i} + 10\sin 45^\circ\,\mathbf{j}$

$\quad = -5\sqrt{2}\,\mathbf{i} + 5\sqrt{2}\,\mathbf{j}$

$\mathbf{F}_2 = |\mathbf{F}_2|\cos 30^\circ\,\mathbf{i} + |\mathbf{F}_2|\sin 30^\circ\,\mathbf{j} = 12\cos 30^\circ\,\mathbf{i} + 12\sin 30^\circ\,\mathbf{j} = 6\sqrt{3}\,\mathbf{i} + 6\,\mathbf{j}$

$\mathbf{F} = \mathbf{F}_1 + \mathbf{F}_2 = \left(6\sqrt{3} - 5\sqrt{2}\right)\mathbf{i} + \left(6 + 5\sqrt{2}\right)\mathbf{j} \approx 3.32\,\mathbf{i} + 13.07\,\mathbf{j}$

$|\mathbf{F}| \approx \sqrt{(3.32)^2 + (13.07)^2} \approx 13.5$ lb. $\tan\theta = \dfrac{6 + 5\sqrt{2}}{6\sqrt{3} - 5\sqrt{2}} \;\Rightarrow\; \theta = \tan^{-1}\dfrac{6 + 5\sqrt{2}}{6\sqrt{3} - 5\sqrt{2}} \approx 76^\circ$.

31. With respect to the water's surface, the woman's velocity is the vector sum of the velocity of the ship with respect to

the water, and the woman's velocity with respect to the ship. If we let north be the positive y-direction, then

$\mathbf{v} = \langle 0, 22 \rangle + \langle -3, 0 \rangle = \langle -3, 22 \rangle$. The woman's speed is $|\mathbf{v}| = \sqrt{9 + 484} \approx 22.2$ mi/h. The vector $\mathbf{v}$ makes an

angle θ with the east, where $\theta = \tan^{-1}\left(\frac{22}{-3}\right) \approx 98^\circ$. Therefore, the woman's direction is about

$\mathrm{N}(98 - 90)^\circ\,\mathrm{W} = \mathrm{N}8^\circ\,\mathrm{W}$.

33. Let $\mathbf{T}_1$ and $\mathbf{T}_2$ represent the tension vectors in each side of the

clothesline as shown in the figure. $\mathbf{T}_1$ and $\mathbf{T}_2$ have equal vertical

components and opposite horizontal components, so we can write

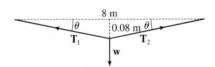

$\mathbf{T}_1 = -a\,\mathbf{i} + b\,\mathbf{j}$ and $\mathbf{T}_2 = a\,\mathbf{i} + b\,\mathbf{j}$ $(a, b > 0)$. By similar triangles, $\dfrac{b}{a} = \dfrac{0.08}{4} \;\Rightarrow\; a = 50b$. The force due to

gravity acting on the shirt has magnitude $0.8g \approx (0.8)(9.8) = 7.84$ N, hence we have $\mathbf{w} = -7.84\,\mathbf{j}$. The resultant

$\mathbf{T}_1 + \mathbf{T}_2$ of the tensile forces counterbalances $\mathbf{w}$, so $\mathbf{T}_1 + \mathbf{T}_2 = -\mathbf{w} \;\Rightarrow\; (-a\,\mathbf{i} + b\,\mathbf{j}) + (a\,\mathbf{i} + b\,\mathbf{j}) = 7.84\,\mathbf{j}$

$\Rightarrow\; (-50b\,\mathbf{i} + b\,\mathbf{j}) + (50b\,\mathbf{i} + b\,\mathbf{j}) = 2b\,\mathbf{j} = 7.84\,\mathbf{j} \;\Rightarrow\; b = \frac{7.84}{2} = 3.92$ and $a = 50b = 196$. Thus the tensions

are $\mathbf{T}_1 = -a\,\mathbf{i} + b\,\mathbf{j} = -196\,\mathbf{i} + 3.92\,\mathbf{j}$ and $\mathbf{T}_2 = a\,\mathbf{i} + b\,\mathbf{j} = 196\,\mathbf{i} + 3.92\,\mathbf{j}$.

Alternatively, we can find the value of θ and proceed as in Example 7.

35. By the Triangle Law, $\overrightarrow{AB} + \overrightarrow{BC} = \overrightarrow{AC}$. Then $\overrightarrow{AB} + \overrightarrow{BC} + \overrightarrow{CA} = \overrightarrow{AC} + \overrightarrow{CA}$, but

$\overrightarrow{AC} + \overrightarrow{CA} = \overrightarrow{AC} + \left(-\overrightarrow{AC}\right) = \mathbf{0}$. So $\overrightarrow{AB} + \overrightarrow{BC} + \overrightarrow{CA} = \mathbf{0}$.

37. (a), (b)

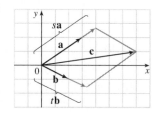

(c) From the sketch, we estimate that $s \approx 1.3$ and $t \approx 1.6$.

(d) $\mathbf{c} = s\,\mathbf{a} + t\,\mathbf{b}$ $\Leftrightarrow$ $7 = 3s + 2t$ and $1 = 2s - t$.

Solving these equations gives $s = \frac{9}{7}$ and $t = \frac{11}{7}$.

39. $|\mathbf{r} - \mathbf{r}_0|$ is the distance between the points (x, y, z) and (x_0, y_0, z_0), so the set of points is a sphere with radius 1 and center (x_0, y_0, z_0).

Alternate method: $|\mathbf{r} - \mathbf{r}_0| = 1$ $\Leftrightarrow$ $\sqrt{(x - x_0)^2 + (y - y_0)^2 + (z - z_0)^2} = 1$ $\Leftrightarrow$

$(x - x_0)^2 + (y - y_0)^2 + (z - z_0)^2 = 1$, which is the equation of a sphere with radius 1 and center (x_0, y_0, z_0).

41. $\mathbf{a} + (\mathbf{b} + \mathbf{c}) = \langle a_1, a_2 \rangle + (\langle b_1, b_2 \rangle + \langle c_1, c_2 \rangle) = \langle a_1, a_2 \rangle + \langle b_1 + c_1, b_2 + c_2 \rangle$

$\quad = \langle a_1 + b_1 + c_1, a_2 + b_2 + c_2 \rangle = \langle (a_1 + b_1) + c_1, (a_2 + b_2) + c_2 \rangle$

$\quad = \langle a_1 + b_1, a_2 + b_2 \rangle + \langle c_1, c_2 \rangle = (\langle a_1, a_2 \rangle + \langle b_1, b_2 \rangle) + \langle c_1, c_2 \rangle$

$\quad = (\mathbf{a} + \mathbf{b}) + \mathbf{c}$

43. Consider triangle ABC, where D and E are the midpoints of AB and BC. We know that $\overrightarrow{AB} + \overrightarrow{BC} = \overrightarrow{AC}$ (1)

and $\overrightarrow{DB} + \overrightarrow{BE} = \overrightarrow{DE}$ (2). However, $\overrightarrow{DB} = \frac{1}{2}\overrightarrow{AB}$, and $\overrightarrow{BE} = \frac{1}{2}\overrightarrow{BC}$. Substituting these expressions for $\overrightarrow{DB}$ and

$\overrightarrow{BE}$ into (2) gives $\frac{1}{2}\overrightarrow{AB} + \frac{1}{2}\overrightarrow{BC} = \overrightarrow{DE}$. Comparing this with (1) gives $\overrightarrow{DE} = \frac{1}{2}\overrightarrow{AC}$. Therefore $\overrightarrow{AC}$ and $\overrightarrow{DE}$ are

parallel and $\left| \overrightarrow{DE} \right| = \frac{1}{2} \left| \overrightarrow{AC} \right|$.

13.3 The Dot Product   ET 12.3

1. (a) $\mathbf{a} \cdot \mathbf{b}$ is a scalar, and the dot product is defined only for vectors, so $(\mathbf{a} \cdot \mathbf{b}) \cdot \mathbf{c}$ has no meaning.

(b) $(\mathbf{a} \cdot \mathbf{b})\,\mathbf{c}$ is a scalar multiple of a vector, so it does have meaning.

(c) Both $|\mathbf{a}|$ and $\mathbf{b} \cdot \mathbf{c}$ are scalars, so $|\mathbf{a}|\,(\mathbf{b} \cdot \mathbf{c})$ is an ordinary product of real numbers, and has meaning.

(d) Both $\mathbf{a}$ and $\mathbf{b} + \mathbf{c}$ are vectors, so the dot product $\mathbf{a} \cdot (\mathbf{b} + \mathbf{c})$ has meaning.

(e) $\mathbf{a} \cdot \mathbf{b}$ is a scalar, but $\mathbf{c}$ is a vector, and so the two quantities cannot be added and this expression has no meaning.

(f) $|\mathbf{a}|$ is a scalar, and the dot product is defined only for vectors, so $|\mathbf{a}| \cdot (\mathbf{b} + \mathbf{c})$ has no meaning.

3. $\mathbf{a} \cdot \mathbf{b} = \langle 4, -1 \rangle \cdot \langle 3, 6 \rangle = (4)(3) + (-1)(6) = 6$

5. $\mathbf{a} \cdot \mathbf{b} = \langle 5, 0, -2 \rangle \cdot \langle 3, -1, 10 \rangle = (5)(3) + (0)(-1) + (-2)(10) = -5$

7. $\mathbf{a} \cdot \mathbf{b} = (\mathbf{i} - 2\,\mathbf{j} + 3\,\mathbf{k}) \cdot (5\,\mathbf{i} + 9\,\mathbf{k}) = (1)(5) + (-2)(0) + (3)(9) = 32$

9. Use Theorem 3: $\mathbf{a} \cdot \mathbf{b} = |\mathbf{a}|\,|\mathbf{b}| \cos\theta = (12)(15)\cos\frac{\pi}{6} = 180 \cdot \frac{\sqrt{3}}{2} = 90\sqrt{3} \approx 155.9$

11. $\mathbf{u}$, $\mathbf{v}$, and $\mathbf{w}$ are all unit vectors, so the triangle is an equilateral triangle. Thus the angle between $\mathbf{u}$ and $\mathbf{v}$ is $60\,°$ and

$\mathbf{u} \cdot \mathbf{v} = |\mathbf{u}|\,|\mathbf{v}| \cos 60\,° = (1)(1)\left(\frac{1}{2}\right) = \frac{1}{2}$. If $\mathbf{w}$ is moved so it has the same initial point as $\mathbf{u}$, we can see that the

angle between them is $120\,°$ and we have $\mathbf{u} \cdot \mathbf{w} = |\mathbf{u}|\,|\mathbf{w}| \cos 120\,° = (1)(1)\left(-\frac{1}{2}\right) = -\frac{1}{2}$.

13. (a) $\mathbf{i} \cdot \mathbf{j} = \langle 1, 0, 0 \rangle \cdot \langle 0, 1, 0 \rangle = (1)(0) + (0)(1) + (0)(0) = 0$. Similarly $\mathbf{j} \cdot \mathbf{k} = (0)(0) + (1)(0) + (0)(1) = 0$ and
$\mathbf{k} \cdot \mathbf{i} = (0)(1) + (0)(0) + (1)(0) = 0$.

Another method: Because $\mathbf{i}$, $\mathbf{j}$, and $\mathbf{k}$ are mutually perpendicular, the cosine factor in each dot product
(see Theorem 3) is $\cos \frac{\pi}{2} = 0$.

(b) By Property 1 of the dot product, $\mathbf{i} \cdot \mathbf{i} = |\mathbf{i}|^2 = 1^2 = 1$ since $\mathbf{i}$ is a unit vector. Similarly, $\mathbf{j} \cdot \mathbf{j} = |\mathbf{j}|^2 = 1$ and
$\mathbf{k} \cdot \mathbf{k} = |\mathbf{k}|^2 = 1$.

15. $|\mathbf{a}| = \sqrt{3^2 + 4^2} = 5$, $|\mathbf{b}| = \sqrt{5^2 + 12^2} = 13$, and $\mathbf{a} \cdot \mathbf{b} = (3)(5) + (4)(12) = 63$. Using Corollary 6, we have
$\cos \theta = \dfrac{\mathbf{a} \cdot \mathbf{b}}{|\mathbf{a}| \, |\mathbf{b}|} = \dfrac{63}{5 \cdot 13} = \dfrac{63}{65}$. So the angle between $\mathbf{a}$ and $\mathbf{b}$ is $\theta = \cos^{-1} \left(\frac{63}{65} \right) \approx 14°$.

17. $|\mathbf{a}| = \sqrt{1^2 + 2^2 + 3^2} = \sqrt{14}$, $|\mathbf{b}| = \sqrt{4^2 + 0^2 + (-1)^2} = \sqrt{17}$, and $\mathbf{a} \cdot \mathbf{b} = (1)(4) + (2)(0) + (3)(-1) = 1$.
Then $\cos \theta = \dfrac{\mathbf{a} \cdot \mathbf{b}}{|\mathbf{a}| \, |\mathbf{b}|} = \dfrac{1}{\sqrt{14} \cdot \sqrt{17}} = \dfrac{1}{\sqrt{238}}$ and the angle between $\mathbf{a}$ and $\mathbf{b}$ is $\theta = \cos^{-1} \left(\frac{1}{\sqrt{238}} \right) \approx 86°$.

19. $|\mathbf{a}| = \sqrt{0^2 + 1^2 + 1^2} = \sqrt{2}$, $|\mathbf{b}| = \sqrt{1^2 + 2^2 + (-3)^2} = \sqrt{14}$, and $\mathbf{a} \cdot \mathbf{b} = (0)(1) + (1)(2) + (1)(-3) = -1$.
Then $\cos \theta = \dfrac{\mathbf{a} \cdot \mathbf{b}}{|\mathbf{a}| \, |\mathbf{b}|} = \dfrac{-1}{\sqrt{2} \cdot \sqrt{14}} = \dfrac{-1}{2\sqrt{7}}$ and $\theta = \cos^{-1} \left(-\dfrac{1}{2\sqrt{7}} \right) \approx 101°$.

21. Let a, b, and c be the angles at vertices A, B, and C respectively.

Then a is the angle between vectors $\overrightarrow{AB}$ and $\overrightarrow{AC}$, b is the angle

between vectors $\overrightarrow{BA}$ and $\overrightarrow{BC}$, and c is the angle between vectors

$\overrightarrow{CA}$ and $\overrightarrow{CB}$.

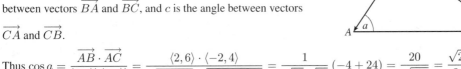

Thus $\cos a = \dfrac{\overrightarrow{AB} \cdot \overrightarrow{AC}}{\left| \overrightarrow{AB} \right| \left| \overrightarrow{AC} \right|} = \dfrac{\langle 2, 6 \rangle \cdot \langle -2, 4 \rangle}{\sqrt{2^2 + 6^2} \sqrt{(-2)^2 + 4^2}} = \dfrac{1}{\sqrt{40} \sqrt{20}} (-4 + 24) = \dfrac{20}{\sqrt{800}} = \dfrac{\sqrt{2}}{2}$

and $a = \cos^{-1} \left(\dfrac{\sqrt{2}}{2} \right) = 45°$. Similarly,

$\cos b = \dfrac{\overrightarrow{BA} \cdot \overrightarrow{BC}}{\left| \overrightarrow{BA} \right| \left| \overrightarrow{BC} \right|} = \dfrac{\langle -2, -6 \rangle \cdot \langle -4, -2 \rangle}{\sqrt{4 + 36} \sqrt{16 + 4}} = \dfrac{1}{\sqrt{40} \sqrt{20}} (8 + 12) = \dfrac{20}{\sqrt{800}} = \dfrac{\sqrt{2}}{2}$ so

$b = \cos^{-1} \left(\dfrac{\sqrt{2}}{2} \right) = 45°$ and $c = 180° - (45° + 45°) = 90°$.

Alternate solution: Apply the Law of Cosines three times as follows: $\cos a = \dfrac{\left| \overrightarrow{BC} \right|^2 - \left| \overrightarrow{AB} \right|^2 - \left| \overrightarrow{AC} \right|^2}{2 \left| \overrightarrow{AB} \right| \left| \overrightarrow{AC} \right|}$,

$\cos b = \dfrac{\left| \overrightarrow{AC} \right|^2 - \left| \overrightarrow{AB} \right|^2 - \left| \overrightarrow{BC} \right|^2}{2 \left| \overrightarrow{AB} \right| \left| \overrightarrow{BC} \right|}$, and $\cos c = \dfrac{\left| \overrightarrow{AB} \right|^2 - \left| \overrightarrow{AC} \right|^2 - \left| \overrightarrow{BC} \right|^2}{2 \left| \overrightarrow{AC} \right| \left| \overrightarrow{BC} \right|}$.

23. (a) $\mathbf{a} \cdot \mathbf{b} = (-5)(6) + (3)(-8) + (7)(2) = -40 \neq 0$, so $\mathbf{a}$ and $\mathbf{b}$ are not orthogonal. Also, since $\mathbf{a}$ is not a scalar
multiple of $\mathbf{b}$, $\mathbf{a}$ and $\mathbf{b}$ are not parallel.

(b) $\mathbf{a} \cdot \mathbf{b} = (4)(-3) + (6)(2) = 0$, so $\mathbf{a}$ and $\mathbf{b}$ are orthogonal (and not parallel).

(c) $\mathbf{a} \cdot \mathbf{b} = (-1)(3) + (2)(4) + (5)(-1) = 0$, so $\mathbf{a}$ and $\mathbf{b}$ are orthogonal (and not parallel).

(d) Because $\mathbf{a} = -\frac{2}{3} \mathbf{b}$, $\mathbf{a}$ and $\mathbf{b}$ are parallel.

25. $\overrightarrow{QP} = \langle -1, -3, 2 \rangle$, $\overrightarrow{QR} = \langle 4, -2, -1 \rangle$, and $\overrightarrow{QP} \cdot \overrightarrow{QR} = -4 + 6 - 2 = 0$. Thus $\overrightarrow{QP}$ and $\overrightarrow{QR}$ are orthogonal, so the angle of the triangle at vertex Q is a right angle.

27. Let $\mathbf{a} = a_1\,\mathbf{i} + a_2\,\mathbf{j} + a_3\,\mathbf{k}$ be a vector orthogonal to both $\mathbf{i} + \mathbf{j}$ and $\mathbf{i} + \mathbf{k}$. Then $\mathbf{a} \cdot (\mathbf{i} + \mathbf{j}) = 0$ ⟺ $a_1 + a_2 = 0$ and $\mathbf{a} \cdot (\mathbf{i} + \mathbf{k}) = 0$ ⟺ $a_1 + a_3 = 0$, so $a_1 = -a_2 = -a_3$. Furthermore $\mathbf{a}$ is to be a unit vector, so $1 = a_1^2 + a_2^2 + a_3^2 = 3a_1^2$ implies $a_1 = \pm\frac{1}{\sqrt{3}}$. Thus $\mathbf{a} = \frac{1}{\sqrt{3}}\,\mathbf{i} - \frac{1}{\sqrt{3}}\,\mathbf{j} - \frac{1}{\sqrt{3}}\,\mathbf{k}$ and $\mathbf{a} = -\frac{1}{\sqrt{3}}\,\mathbf{i} + \frac{1}{\sqrt{3}}\,\mathbf{j} + \frac{1}{\sqrt{3}}\,\mathbf{k}$ are two such unit vectors.

29. Since $|\langle 3, 4, 5 \rangle| = \sqrt{9 + 16 + 25} = \sqrt{50} = 5\sqrt{2}$, using Equations 8 and 9 we have $\cos\alpha = \frac{3}{5\sqrt{2}}$, $\cos\beta = \frac{4}{5\sqrt{2}}$, and $\cos\gamma = \frac{5}{5\sqrt{2}} = \frac{1}{\sqrt{2}}$. The direction angles are given by $\alpha = \cos^{-1}\left(\frac{3}{5\sqrt{2}}\right) \approx 65°$, $\beta = \cos^{-1}\left(\frac{4}{5\sqrt{2}}\right) \approx 56°$, and $\gamma = \cos^{-1}\left(\frac{1}{\sqrt{2}}\right) = 45°$.

31. Since $|2\,\mathbf{i} + 3\,\mathbf{j} - 6\,\mathbf{k}| = \sqrt{4 + 9 + 36} = \sqrt{49} = 7$, Equations 8 and 9 give $\cos\alpha = \frac{2}{7}$, $\cos\beta = \frac{3}{7}$, and $\cos\gamma = \frac{-6}{7}$, while $\alpha = \cos^{-1}\left(\frac{2}{7}\right) \approx 73°$, $\beta = \cos^{-1}\left(\frac{3}{7}\right) \approx 65°$, and $\gamma = \cos^{-1}\left(-\frac{6}{7}\right) \approx 149°$.

33. $|\langle c, c, c \rangle| = \sqrt{c^2 + c^2 + c^2} = \sqrt{3}c$ (since $c > 0$), so $\cos\alpha = \cos\beta = \cos\gamma = \frac{c}{\sqrt{3}c} = \frac{1}{\sqrt{3}}$ and $\alpha = \beta = \gamma = \cos^{-1}\left(\frac{1}{\sqrt{3}}\right) \approx 55°$.

35. $|\mathbf{a}| = \sqrt{3^2 + (-4)^2} = 5$. The scalar projection of $\mathbf{b}$ onto $\mathbf{a}$ is $\text{comp}_\mathbf{a}\,\mathbf{b} = \frac{\mathbf{a} \cdot \mathbf{b}}{|\mathbf{a}|} = \frac{3 \cdot 5 + (-4) \cdot 0}{5} = 3$ and the vector projection of $\mathbf{b}$ onto $\mathbf{a}$ is $\text{proj}_\mathbf{a}\,\mathbf{b} = \left(\frac{\mathbf{a} \cdot \mathbf{b}}{|\mathbf{a}|}\right)\frac{\mathbf{a}}{|\mathbf{a}|} = 3 \cdot \frac{1}{5}\langle 3, -4 \rangle = \langle \frac{9}{5}, -\frac{12}{5} \rangle$.

37. $|\mathbf{a}| = \sqrt{16 + 4 + 0} = 2\sqrt{5}$ so the scalar projection of $\mathbf{b}$ onto $\mathbf{a}$ is $\text{comp}_\mathbf{a}\,\mathbf{b} = \frac{\mathbf{a} \cdot \mathbf{b}}{|\mathbf{a}|} = \frac{1}{2\sqrt{5}}(4 + 2 + 0) = \frac{3}{\sqrt{5}}$. The vector projection of $\mathbf{b}$ onto $\mathbf{a}$ is $\text{proj}_\mathbf{a}\,\mathbf{b} = \frac{3}{\sqrt{5}}\frac{\mathbf{a}}{|\mathbf{a}|} = \frac{3}{\sqrt{5}} \cdot \frac{1}{2\sqrt{5}}\langle 4, 2, 0 \rangle = \frac{1}{5}\langle 6, 3, 0 \rangle = \langle \frac{6}{5}, \frac{3}{5}, 0 \rangle$.

39. $|\mathbf{a}| = \sqrt{1 + 0 + 1} = \sqrt{2}$ so the scalar projection of $\mathbf{b}$ onto $\mathbf{a}$ is $\text{comp}_\mathbf{a}\,\mathbf{b} = \frac{\mathbf{a} \cdot \mathbf{b}}{|\mathbf{a}|} = \frac{1}{\sqrt{2}}(1 + 0 + 0) = \frac{1}{\sqrt{2}}$ while the vector projection of $\mathbf{b}$ onto $\mathbf{a}$ is $\text{proj}_\mathbf{a}\,\mathbf{b} = \frac{1}{\sqrt{2}}\frac{\mathbf{a}}{|\mathbf{a}|} = \frac{1}{\sqrt{2}} \cdot \frac{1}{\sqrt{2}}(\mathbf{i} + \mathbf{k}) = \frac{1}{2}(\mathbf{i} + \mathbf{k})$.

41. $(\text{orth}_\mathbf{a}\,\mathbf{b}) \cdot \mathbf{a} = (\mathbf{b} - \text{proj}_\mathbf{a}\,\mathbf{b}) \cdot \mathbf{a} = \mathbf{b} \cdot \mathbf{a} - (\text{proj}_\mathbf{a}\,\mathbf{b}) \cdot \mathbf{a} = \mathbf{b} \cdot \mathbf{a} - \frac{\mathbf{a} \cdot \mathbf{b}}{|\mathbf{a}|^2}\,\mathbf{a} \cdot \mathbf{a}$

$$= \mathbf{b} \cdot \mathbf{a} - \frac{\mathbf{a} \cdot \mathbf{b}}{|\mathbf{a}|^2}\,|\mathbf{a}|^2 = \mathbf{b} \cdot \mathbf{a} - \mathbf{a} \cdot \mathbf{b} = 0$$

So they are orthogonal by (7).

43. $\text{comp}_\mathbf{a}\,\mathbf{b} = \frac{\mathbf{a} \cdot \mathbf{b}}{|\mathbf{a}|} = 2$ ⟺ $\mathbf{a} \cdot \mathbf{b} = 2\,|\mathbf{a}| = 2\sqrt{10}$. If $\mathbf{b} = \langle b_1, b_2, b_3 \rangle$, then we need $3b_1 + 0b_2 - 1b_3 = 2\sqrt{10}$. One possible solution is obtained by taking $b_1 = 0$, $b_2 = 0$, $b_3 = -2\sqrt{10}$. In general, $\mathbf{b} = \langle s, t, 3s - 2\sqrt{10} \rangle$, $s, t \in \mathbb{R}$.

45. Here $\mathbf{D} = (4 - 2)\,\mathbf{i} + (9 - 3)\,\mathbf{j} + (15 - 0)\,\mathbf{k} = 2\,\mathbf{i} + 6\,\mathbf{j} + 15\,\mathbf{k}$ so by Equation 12 we have $W = \mathbf{F} \cdot \mathbf{D} = 20 + 108 - 90 = 38$ joules.

47. $W = |\mathbf{F}|\,|\mathbf{D}|\cos\theta = (25)(10)\cos 20\,°$

≈ 235 ft-lb

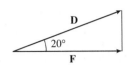

49. First note that $\mathbf{n} = \langle a, b\rangle$ is perpendicular to the line, because if $Q_1 = (a_1, b_1)$ and $Q_2 = (a_2, b_2)$ lie on the line,

then $\mathbf{n}\cdot\overrightarrow{Q_1Q_2} = aa_2 - aa_1 + bb_2 - bb_1 = 0$, since $aa_2 + bb_2 = -c = aa_1 + bb_1$ from the equation of the line.

Let $P_2 = (x_2, y_2)$ lie on the line. Then the distance from P_1 to the line is the absolute value of the scalar projection

of $\overrightarrow{P_1P_2}$ onto $\mathbf{n}$. $\text{comp}_{\mathbf{n}}\left(\overrightarrow{P_1P_2}\right) = \dfrac{|\mathbf{n}\cdot\langle x_2 - x_1, y_2 - y_1\rangle|}{|\mathbf{n}|} = \dfrac{|ax_2 - ax_1 + by_2 - by_1|}{\sqrt{a^2 + b^2}} = \dfrac{|ax_1 + by_1 + c|}{\sqrt{a^2 + b^2}}$

since $ax_2 + by_2 = -c$. The required distance is $\dfrac{|3\cdot-2 + -4\cdot 3 + 5|}{\sqrt{3^2 + 4^2}} = \dfrac{13}{5}$.

51. For convenience, consider the unit cube positioned so that its back left corner is at the origin, and its edges lie along

the coordinate axes. The diagonal of the cube that begins at the origin and ends at $(1, 1, 1)$ has vector representation

$\langle 1, 1, 1\rangle$. The angle θ between this vector and the vector of the edge which also begins at the origin and runs along

the x-axis [that is, $\langle 1, 0, 0\rangle$] is given by $\cos\theta = \dfrac{\langle 1, 1, 1\rangle\cdot\langle 1, 0, 0\rangle}{|\langle 1, 1, 1\rangle|\,|\langle 1, 0, 0\rangle|} = \dfrac{1}{\sqrt{3}} \quad\Rightarrow\quad \theta = \cos^{-1}\left(\dfrac{1}{\sqrt{3}}\right) \approx 55\,°$.

53. Consider the H-C-H combination consisting of the sole carbon atom and the two hydrogen atoms that are at $(1, 0, 0)$

and $(0, 1, 0)$ (or any H-C-H combination, for that matter). Vector representations of the line segments emanating

from the carbon atom and extending to these two hydrogen atoms are $\langle 1 - \frac{1}{2}, 0 - \frac{1}{2}, 0 - \frac{1}{2}\rangle = \langle\frac{1}{2}, -\frac{1}{2}, -\frac{1}{2}\rangle$ and

$\langle 0 - \frac{1}{2}, 1 - \frac{1}{2}, 0 - \frac{1}{2}\rangle = \langle -\frac{1}{2}, \frac{1}{2}, -\frac{1}{2}\rangle$. The bond angle, θ, is therefore given by

$\cos\theta = \dfrac{\langle\frac{1}{2}, -\frac{1}{2}, -\frac{1}{2}\rangle\cdot\langle -\frac{1}{2}, \frac{1}{2}, -\frac{1}{2}\rangle}{|\langle\frac{1}{2}, -\frac{1}{2}, -\frac{1}{2}\rangle|\,|\langle -\frac{1}{2}, \frac{1}{2}, -\frac{1}{2}\rangle|} = \dfrac{-\frac{1}{4} - \frac{1}{4} + \frac{1}{4}}{\sqrt{\frac{3}{4}}\sqrt{\frac{3}{4}}} = -\dfrac{1}{3} \quad\Rightarrow\quad \theta = \cos^{-1}\left(-\dfrac{1}{3}\right) \approx 109.5\,°$.

55. Let $\mathbf{a} = \langle a_1, a_2, a_3\rangle$ and $= \langle b_1, b_2, b_3\rangle$.

Property 2: $\mathbf{a}\cdot\mathbf{b} = \langle a_1, a_2, a_3\rangle\cdot\langle b_1, b_2, b_3\rangle = a_1b_1 + a_2b_2 + a_3b_3$

$= b_1a_1 + b_2a_2 + b_3a_3 = \langle b_1, b_2, b_3\rangle\cdot\langle a_1, a_2, a_3\rangle = \mathbf{b}\cdot\mathbf{a}$

Property 4: $(c\,\mathbf{a})\cdot\mathbf{b} = \langle ca_1, ca_2, ca_3\rangle\cdot\langle b_1, b_2, b_3\rangle = (ca_1)b_1 + (ca_2)b_2 + (ca_3)b_3$

$= c\,(a_1b_1 + a_2b_2 + a_3b_3) = c\,(\mathbf{a}\cdot\mathbf{b}) = a_1(cb_1) + a_2(cb_2) + a_3(cb_3)$

$= \langle a_1, a_2, a_3\rangle\cdot\langle cb_1, cb_2, cb_3\rangle = \mathbf{a}\cdot(c\,\mathbf{b})$

Property 5: $\mathbf{0}\cdot\mathbf{a} = \langle 0, 0, 0\rangle\cdot\langle a_1, a_2, a_3\rangle = (0)(a_1) + (0)(a_2) + (0)(a_3) = 0$

57. $|\mathbf{a}\cdot\mathbf{b}| = |\,|\mathbf{a}|\,|\mathbf{b}|\cos\theta| = |\mathbf{a}|\,|\mathbf{b}|\,|\cos\theta|$. Since $|\cos\theta|\le 1$, $|\mathbf{a}\cdot\mathbf{b}| = |\mathbf{a}|\,|\mathbf{b}|\,|\cos\theta| \le |\mathbf{a}|\,|\mathbf{b}|$.

Note: We have equality in the case of $\cos\theta = \pm 1$, so $\theta = 0$ or $\theta = \pi$, thus equality when $\mathbf{a}$ and $\mathbf{b}$ are parallel.

59. (a)

The Parallelogram Law states that the sum of the squares of the lengths of the diagonals of a parallelogram equals the sum of the squares of its (four) sides.

(b) $|\mathbf{a} + \mathbf{b}|^2 = (\mathbf{a} + \mathbf{b})\cdot(\mathbf{a} + \mathbf{b}) = |\mathbf{a}|^2 + 2(\mathbf{a}\cdot\mathbf{b}) + |\mathbf{b}|^2$ and

$|\mathbf{a} - \mathbf{b}|^2 = (\mathbf{a} - \mathbf{b})\cdot(\mathbf{a} - \mathbf{b}) = |\mathbf{a}|^2 - 2(\mathbf{a}\cdot\mathbf{b}) + |\mathbf{b}|^2$.

Adding these two equations gives $|\mathbf{a} + \mathbf{b}|^2 + |\mathbf{a} - \mathbf{b}|^2 = 2\,|\mathbf{a}|^2 + 2\,|\mathbf{b}|^2$.

13.4 The Cross Product

ET 12.4

1. $\mathbf{a} \times \mathbf{b} = \begin{vmatrix} \mathbf{i} & \mathbf{j} & \mathbf{k} \\ 1 & 2 & 0 \\ 0 & 3 & 1 \end{vmatrix} = \begin{vmatrix} 2 & 0 \\ 3 & 1 \end{vmatrix} \mathbf{i} - \begin{vmatrix} 1 & 0 \\ 0 & 1 \end{vmatrix} \mathbf{j} + \begin{vmatrix} 1 & 2 \\ 0 & 3 \end{vmatrix} \mathbf{k} = (2 - 0)\mathbf{i} - (1 - 0)\mathbf{j} + (3 - 0)\mathbf{k} = 2\mathbf{i} - \mathbf{j} + 3\mathbf{k}$

Now $(\mathbf{a} \times \mathbf{b}) \cdot \mathbf{a} = \langle 2, -1, 3 \rangle \cdot \langle 1, 2, 0 \rangle = 2 - 2 + 0 = 0$ and $(\mathbf{a} \times \mathbf{b}) \cdot \mathbf{b} = \langle 2, -1, 3 \rangle \cdot \langle 0, 3, 1 \rangle = 0 - 3 + 3 = 0$,
so $\mathbf{a} \times \mathbf{b}$ is orthogonal to both $\mathbf{a}$ and $\mathbf{b}$.

3. $\mathbf{a} \times \mathbf{b} = \begin{vmatrix} \mathbf{i} & \mathbf{j} & \mathbf{k} \\ 2 & 1 & -1 \\ 0 & 1 & 2 \end{vmatrix} = \begin{vmatrix} 1 & -1 \\ 1 & 2 \end{vmatrix} \mathbf{i} - \begin{vmatrix} 2 & -1 \\ 0 & 2 \end{vmatrix} \mathbf{j} + \begin{vmatrix} 2 & 1 \\ 0 & 1 \end{vmatrix} \mathbf{k}$

$= [2 - (-1)]\mathbf{i} - (4 - 0)\mathbf{j} + (2 - 0)\mathbf{k} = 3\mathbf{i} - 4\mathbf{j} + 2\mathbf{k}$

Now $(\mathbf{a} \times \mathbf{b}) \cdot \mathbf{a} = (3\mathbf{i} - 4\mathbf{j} + 2\mathbf{k}) \cdot (2\mathbf{i} + \mathbf{j} - \mathbf{k}) = 6 - 4 - 2 = 0$ and

$(\mathbf{a} \times \mathbf{b}) \cdot \mathbf{b} = (3\mathbf{i} - 4\mathbf{j} + 2\mathbf{k}) \cdot (\mathbf{j} + 2\mathbf{k}) = 0 - 4 + 4 = 0$, so $\mathbf{a} \times \mathbf{b}$ is orthogonal to both $\mathbf{a}$ and $\mathbf{b}$.

5. $\mathbf{a} \times \mathbf{b} = \begin{vmatrix} \mathbf{i} & \mathbf{j} & \mathbf{k} \\ 3 & 2 & 4 \\ 1 & -2 & -3 \end{vmatrix} = \begin{vmatrix} 2 & 4 \\ -2 & -3 \end{vmatrix} \mathbf{i} - \begin{vmatrix} 3 & 4 \\ 1 & -3 \end{vmatrix} \mathbf{j} + \begin{vmatrix} 3 & 2 \\ 1 & -2 \end{vmatrix} \mathbf{k}$

$= [-6 - (-8)]\mathbf{i} - (-9 - 4)\mathbf{j} + (-6 - 2)\mathbf{k} = 2\mathbf{i} + 13\mathbf{j} - 8\mathbf{k}$

Since $(\mathbf{a} \times \mathbf{b}) \cdot \mathbf{a} = (2\mathbf{i} + 13\mathbf{j} - 8\mathbf{k}) \cdot (3\mathbf{i} + 2\mathbf{j} + 4\mathbf{k}) = 6 + 26 - 32 = 0$, $\mathbf{a} \times \mathbf{b}$ is orthogonal to $\mathbf{a}$.
Since $(\mathbf{a} \times \mathbf{b}) \cdot \mathbf{b} = (2\mathbf{i} + 13\mathbf{j} - 8\mathbf{k}) \cdot (\mathbf{i} - 2\mathbf{j} - 3\mathbf{k}) = 2 - 26 + 24 = 0$, $\mathbf{a} \times \mathbf{b}$ is orthogonal to $\mathbf{b}$.

7. $\mathbf{a} \times \mathbf{b} = \begin{vmatrix} \mathbf{i} & \mathbf{j} & \mathbf{k} \\ t & t^2 & t^3 \\ 1 & 2t & 3t^2 \end{vmatrix} = \begin{vmatrix} t^2 & t^3 \\ 2t & 3t^2 \end{vmatrix} \mathbf{i} - \begin{vmatrix} t & t^3 \\ 1 & 3t^2 \end{vmatrix} \mathbf{j} + \begin{vmatrix} t & t^2 \\ 1 & 2t \end{vmatrix} \mathbf{k}$

$= (3t^4 - 2t^4)\mathbf{i} - (3t^3 - t^3)\mathbf{j} + (2t^2 - t^2)\mathbf{k} = t^4\mathbf{i} - 2t^3\mathbf{j} + t^2\mathbf{k}$

Since $(\mathbf{a} \times \mathbf{b}) \cdot \mathbf{a} = \langle t^4, -2t^3, t^2 \rangle \cdot \langle t, t^2, t^3 \rangle = t^5 - 2t^5 + t^5 = 0$, $\mathbf{a} \times \mathbf{b}$ is orthogonal to $\mathbf{a}$.
Since $(\mathbf{a} \times \mathbf{b}) \cdot \mathbf{b} = \langle t^4, -2t^3, t^2 \rangle \cdot \langle 1, 2t, 3t^2 \rangle = t^4 - 4t^4 + 3t^4 = 0$, $\mathbf{a} \times \mathbf{b}$ is orthogonal to $\mathbf{b}$.

9. (a) Since $\mathbf{b} \times \mathbf{c}$ is a vector, the dot product $\mathbf{a} \cdot (\mathbf{b} \times \mathbf{c})$ is meaningful and is a scalar.

(b) $\mathbf{b} \cdot \mathbf{c}$ is a scalar, so $\mathbf{a} \times (\mathbf{b} \cdot \mathbf{c})$ is meaningless, as the cross product is defined only for two *vectors*.

(c) Since $\mathbf{b} \times \mathbf{c}$ is a vector, the cross product $\mathbf{a} \times (\mathbf{b} \times \mathbf{c})$ is meaningful and results in another vector.

(d) $\mathbf{a} \cdot \mathbf{b}$ is a scalar, so the cross product $(\mathbf{a} \cdot \mathbf{b}) \times \mathbf{c}$ is meaningless.

(e) Since $(\mathbf{a} \cdot \mathbf{b})$ and $(\mathbf{c} \cdot \mathbf{d})$ are both scalars, the cross product $(\mathbf{a} \cdot \mathbf{b}) \times (\mathbf{c} \cdot \mathbf{d})$ is meaningless.

(f) $\mathbf{a} \times \mathbf{b}$ and $\mathbf{c} \times \mathbf{d}$ are both vectors, so the dot product $(\mathbf{a} \times \mathbf{b}) \cdot (\mathbf{c} \times \mathbf{d})$ is meaningful and is a scalar.

11. If we sketch $\mathbf{u}$ and $\mathbf{v}$ starting from the same initial point,

we see that the angle between them is $30\,°$. Using Theorem 6,

we have

$$|\mathbf{u} \times \mathbf{v}| = |\mathbf{u}|\,|\mathbf{v}|\sin 30\,° = (6)(8)\left(\tfrac{1}{2}\right) = 24$$

By the right-hand rule, $\mathbf{u} \times \mathbf{v}$ is directed into the page.

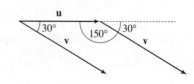

13. $\mathbf{a} \times \mathbf{b} = \begin{vmatrix} \mathbf{i} & \mathbf{j} & \mathbf{k} \\ 1 & 2 & 1 \\ 0 & 1 & 3 \end{vmatrix} = \begin{vmatrix} 2 & 1 \\ 1 & 3 \end{vmatrix} \mathbf{i} - \begin{vmatrix} 1 & 1 \\ 0 & 3 \end{vmatrix} \mathbf{j} + \begin{vmatrix} 1 & 2 \\ 0 & 1 \end{vmatrix} \mathbf{k} = (6-1)\mathbf{i} - (3-0)\mathbf{j} + (1-0)\mathbf{k} = 5\mathbf{i} - 3\mathbf{j} + \mathbf{k}$

$\mathbf{b} \times \mathbf{a} = \begin{vmatrix} \mathbf{i} & \mathbf{j} & \mathbf{k} \\ 0 & 1 & 3 \\ 1 & 2 & 1 \end{vmatrix} = \begin{vmatrix} 1 & 3 \\ 2 & 1 \end{vmatrix} \mathbf{i} - \begin{vmatrix} 0 & 3 \\ 1 & 1 \end{vmatrix} \mathbf{j} + \begin{vmatrix} 0 & 1 \\ 1 & 2 \end{vmatrix} \mathbf{k} = (1-6)\mathbf{i} - (0-3)\mathbf{j} + (0-1)\mathbf{k} = -5\mathbf{i} + 3\mathbf{j} - \mathbf{k}$

Notice $\mathbf{a} \times \mathbf{b} = -\mathbf{b} \times \mathbf{a}$ here, as we know is always true by Theorem 8.

15. We know that the cross product of two vectors is orthogonal to both. So we calculate

$\langle 1, -1, 1 \rangle \times \langle 0, 4, 4 \rangle = \begin{vmatrix} \mathbf{i} & \mathbf{j} & \mathbf{k} \\ 1 & -1 & 1 \\ 0 & 4 & 4 \end{vmatrix} = \begin{vmatrix} -1 & 1 \\ 4 & 4 \end{vmatrix} \mathbf{i} - \begin{vmatrix} 1 & 1 \\ 0 & 4 \end{vmatrix} \mathbf{j} + \begin{vmatrix} 1 & -1 \\ 0 & 4 \end{vmatrix} \mathbf{k} = -8\mathbf{i} - 4\mathbf{j} + 4\mathbf{k}$.

So two unit vectors orthogonal to both are $\pm \dfrac{\langle -8, -4, 4 \rangle}{\sqrt{64 + 16 + 16}} = \pm \dfrac{\langle -8, -4, 4 \rangle}{4\sqrt{6}}$, that is, $\left\langle -\dfrac{2}{\sqrt{6}}, -\dfrac{1}{\sqrt{6}}, \dfrac{1}{\sqrt{6}} \right\rangle$

and $\left\langle \dfrac{2}{\sqrt{6}}, \dfrac{1}{\sqrt{6}}, -\dfrac{1}{\sqrt{6}} \right\rangle$.

17. Let $\mathbf{a} = \langle a_1, a_2, a_3 \rangle$. Then

$\mathbf{0} \times \mathbf{a} = \begin{vmatrix} \mathbf{i} & \mathbf{j} & \mathbf{k} \\ 0 & 0 & 0 \\ a_1 & a_2 & a_3 \end{vmatrix} = \begin{vmatrix} 0 & 0 \\ a_2 & a_3 \end{vmatrix} \mathbf{i} - \begin{vmatrix} 0 & 0 \\ a_1 & a_3 \end{vmatrix} \mathbf{j} + \begin{vmatrix} 0 & 0 \\ a_1 & a_2 \end{vmatrix} \mathbf{k} = \mathbf{0}$,

$\mathbf{a} \times \mathbf{0} = \begin{vmatrix} \mathbf{i} & \mathbf{j} & \mathbf{k} \\ a_1 & a_2 & a_3 \\ 0 & 0 & 0 \end{vmatrix} = \begin{vmatrix} a_2 & a_3 \\ 0 & 0 \end{vmatrix} \mathbf{i} - \begin{vmatrix} a_1 & a_3 \\ 0 & 0 \end{vmatrix} \mathbf{j} + \begin{vmatrix} a_1 & a_2 \\ 0 & 0 \end{vmatrix} \mathbf{k} = \mathbf{0}$.

19. $\mathbf{a} \times \mathbf{b} = \langle a_2 b_3 - a_3 b_2, a_3 b_1 - a_1 b_3, a_1 b_2 - a_2 b_1 \rangle$

$\qquad = \langle (-1)(b_2 a_3 - b_3 a_2), (-1)(b_3 a_1 - b_1 a_3), (-1)(b_1 a_2 - b_2 a_1) \rangle$

$\qquad = - \langle b_2 a_3 - b_3 a_2, b_3 a_1 - b_1 a_3, b_1 a_2 - b_2 a_1 \rangle = -\mathbf{b} \times \mathbf{a}$

21. $\mathbf{a} \times (\mathbf{b} + \mathbf{c}) = \mathbf{a} \times \langle b_1 + c_1, b_2 + c_2, b_3 + c_3 \rangle$

$\qquad = \langle a_2(b_3 + c_3) - a_3(b_2 + c_2), a_3(b_1 + c_1) - a_1(b_3 + c_3), a_1(b_2 + c_2) - a_2(b_1 + c_1) \rangle$

$\qquad = \langle a_2 b_3 + a_2 c_3 - a_3 b_2 - a_3 c_2, a_3 b_1 + a_3 c_1 - a_1 b_3 - a_1 c_3, a_1 b_2 + a_1 c_2 - a_2 b_1 - a_2 c_1 \rangle$

$\qquad = \langle (a_2 b_3 - a_3 b_2) + (a_2 c_3 - a_3 c_2), (a_3 b_1 - a_1 b_3) + (a_3 c_1 - a_1 c_3),$

$\qquad\qquad\qquad\qquad\qquad\qquad (a_1 b_2 - a_2 b_1) + (a_1 c_2 - a_2 c_1) \rangle$

$\qquad = \langle a_2 b_3 - a_3 b_2, a_3 b_1 - a_1 b_3, a_1 b_2 - a_2 b_1 \rangle + \langle a_2 c_3 - a_3 c_2, a_3 c_1 - a_1 c_3, a_1 c_2 - a_2 c_1 \rangle$

$\qquad = (\mathbf{a} \times \mathbf{b}) + (\mathbf{a} \times \mathbf{c})$

23. By plotting the vertices, we can see that the parallelogram is determined

by the vectors $\overrightarrow{AB} = \langle 2, 3 \rangle$ and $\overrightarrow{AD} = \langle 4, -2 \rangle$. We know that the area

of the parallelogram determined by two vectors is equal to the length of

the cross product of these vectors. In order to compute the cross product,

we consider the vector $\overrightarrow{AB}$ as the three-dimensional vector $\langle 2, 3, 0 \rangle$

(and similarly for $\overrightarrow{AD}$), and then the area of parallelogram $ABCD$ is

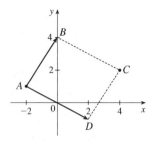

$$\left| \overrightarrow{AB} \times \overrightarrow{AD} \right| = \begin{vmatrix} \mathbf{i} & \mathbf{j} & \mathbf{k} \\ 2 & 3 & 0 \\ 4 & -2 & 0 \end{vmatrix} = |(0)\,\mathbf{i} - (0)\,\mathbf{j} + (-4 - 12)\,\mathbf{k}| = |-16\,\mathbf{k}| = 16$$

25. (a) Because the plane through P, Q, and R contains the vectors $\overrightarrow{PQ}$ and $\overrightarrow{PR}$, a vector orthogonal to both of these

vectors (such as their cross product) is also orthogonal to the plane. Here $\overrightarrow{PQ} = \langle -1, 2, 0 \rangle$ and

$\overrightarrow{PR} = \langle -1, 0, 3 \rangle$, so

$$\overrightarrow{PQ} \times \overrightarrow{PR} = \langle (2)(3) - (0)(0), (0)(-1) - (-1)(3), (-1)(0) - (2)(-1) \rangle = \langle 6, 3, 2 \rangle$$

Therefore, $\langle 6, 3, 2 \rangle$ (or any scalar multiple thereof) is orthogonal to the plane through P, Q, and R.

(b) Note that the area of the triangle determined by P, Q, and R is equal to half of the area of the parallelogram

determined by the three points. From part (a), the area of the parallelogram is

$$\left| \overrightarrow{PQ} \times \overrightarrow{PR} \right| = |\langle 6, 3, 2 \rangle| = \sqrt{36 + 9 + 4} = 7, \text{ so the area of the triangle is } \tfrac{1}{2}(7) = \tfrac{7}{2}.$$

27. (a) $\overrightarrow{PQ} = \langle 4, 3, -2 \rangle$ and $\overrightarrow{PR} = \langle 5, 5, 1 \rangle$, so a vector orthogonal to the plane through P, Q, and R is

$$\overrightarrow{PQ} \times \overrightarrow{PR} = \langle (3)(1) - (-2)(5), (-2)(5) - (4)(1), (4)(5) - (3)(5) \rangle = \langle 13, -14, 5 \rangle$$

(or any scalar mutiple thereof).

(b) The area of the parallelogram determined by $\overrightarrow{PQ}$ and $\overrightarrow{PR}$ is

$$\left| \overrightarrow{PQ} \times \overrightarrow{PR} \right| = |\langle 13, -14, 5 \rangle| = \sqrt{13^2 + (-14)^2 + 5^2} = \sqrt{390}, \text{ so the area of triangle } PQR \text{ is } \tfrac{1}{2}\sqrt{390}.$$

29. We know that the volume of the parallelepiped determined by $\mathbf{a}$, $\mathbf{b}$, and $\mathbf{c}$ is the magnitude of their scalar triple

product, which is

$$\mathbf{a} \cdot (\mathbf{b} \times \mathbf{c}) = \begin{vmatrix} 6 & 3 & -1 \\ 0 & 1 & 2 \\ 4 & -2 & 5 \end{vmatrix} = 6 \begin{vmatrix} 1 & 2 \\ -2 & 5 \end{vmatrix} - 3 \begin{vmatrix} 0 & 2 \\ 4 & 5 \end{vmatrix} + (-1) \begin{vmatrix} 0 & 1 \\ 4 & -2 \end{vmatrix}$$

$$= 6(5 + 4) - 3(0 - 8) - (0 - 4) = 82$$

Thus the volume of the parallelepiped is 82 cubic units.

31. $\mathbf{a} = \overrightarrow{PQ} = \langle 2, 1, 1 \rangle$, $\mathbf{b} = \overrightarrow{PR} = \langle 1, -1, 2 \rangle$, and $\mathbf{c} = \overrightarrow{PS} = \langle 0, -2, 3 \rangle$.

$$\mathbf{a} \cdot (\mathbf{b} \times \mathbf{c}) = \begin{vmatrix} 2 & 1 & 1 \\ 1 & -1 & 2 \\ 0 & -2 & 3 \end{vmatrix} = 2 \begin{vmatrix} -1 & 2 \\ -2 & 3 \end{vmatrix} - 1 \begin{vmatrix} 1 & 2 \\ 0 & 3 \end{vmatrix} + 1 \begin{vmatrix} 1 & -1 \\ 0 & -2 \end{vmatrix} = 2 - 3 - 2 = -3,$$

so the volume of the parallelepiped is 3 cubic units.

33. $\mathbf{a} \cdot (\mathbf{b} \times \mathbf{c}) = \begin{vmatrix} 2 & 3 & 1 \\ 1 & -1 & 0 \\ 7 & 3 & 2 \end{vmatrix} = 2 \begin{vmatrix} -1 & 0 \\ 3 & 2 \end{vmatrix} - 3 \begin{vmatrix} 1 & 0 \\ 7 & 2 \end{vmatrix} + 1 \begin{vmatrix} 1 & -1 \\ 7 & 3 \end{vmatrix} = -4 - 6 + 10 = 0$, which says that the

volume of the parallelepiped determined by $\mathbf{a}$, $\mathbf{b}$ and $\mathbf{c}$ is 0, and thus these three vectors are coplanar.

35. The magnitude of the torque is

$|\boldsymbol{\tau}| = |\mathbf{r} \times \mathbf{F}| = |\mathbf{r}|\,|\mathbf{F}| \sin \theta = (0.18 \text{ m})(60 \text{ N}) \sin(70 + 10)^\circ = 10.8 \sin 80^\circ \approx 10.6 \text{ J}$.

37. Using the notation of the text, $\mathbf{r} = \langle 0, 0.3, 0 \rangle$ and $\mathbf{F}$ has direction $\langle 0, 3, -4 \rangle$. The angle θ between them can be

determined by $\cos \theta = \dfrac{\langle 0, 0.3, 0 \rangle \cdot \langle 0, 3, -4 \rangle}{|\langle 0, 0.3, 0 \rangle|\,|\langle 0, 3, -4 \rangle|} \;\Rightarrow\; \cos \theta = \dfrac{0.9}{(0.3)(5)} \;\Rightarrow\; \cos \theta = 0.6 \;\Rightarrow\; \theta \approx 53.1^\circ$.

Then $|\boldsymbol{\tau}| = |\mathbf{r}|\,|\mathbf{F}| \sin \theta \;\Rightarrow\; 100 = 0.3\,|\mathbf{F}| \sin 53.1^\circ \;\Rightarrow\; |\mathbf{F}| \approx 417 \text{ N}$.

39. (a)

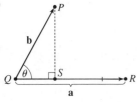

The distance between a point and a line is the length of the

perpendicular from the point to the line, here $\left|\overrightarrow{PS}\right| = d$. But referring

to triangle PQS, $d = \left|\overrightarrow{PS}\right| = \left|\overrightarrow{QP}\right| \sin \theta = |\mathbf{b}| \sin \theta$. But θ is the

angle between $\overrightarrow{QP} = \mathbf{b}$ and $\overrightarrow{QR} = \mathbf{a}$. Thus by Theorem 6,

$\sin \theta = \dfrac{|\mathbf{a} \times \mathbf{b}|}{|\mathbf{a}|\,|\mathbf{b}|}$ and so $d = |\mathbf{b}| \sin \theta = \dfrac{|\mathbf{b}|\,|\mathbf{a} \times \mathbf{b}|}{|\mathbf{a}|\,|\mathbf{b}|} = \dfrac{|\mathbf{a} \times \mathbf{b}|}{|\mathbf{a}|}$.

(b) $\mathbf{a} = \overrightarrow{QR} = \langle -1, -2, -1 \rangle$ and $\mathbf{b} = \overrightarrow{QP} = \langle 1, -5, -7 \rangle$. Then

$\mathbf{a} \times \mathbf{b} = \langle (-2)(-7) - (-1)(-5), (-1)(1) - (-1)(-7), (-1)(-5) - (-2)(1) \rangle = \langle 9, -8, 7 \rangle$. Thus the

distance is $d = \dfrac{|\mathbf{a} \times \mathbf{b}|}{|\mathbf{a}|} = \dfrac{1}{\sqrt{6}} \sqrt{81 + 64 + 49} = \sqrt{\dfrac{194}{6}} = \sqrt{\dfrac{97}{3}}$.

41. $(\mathbf{a} - \mathbf{b}) \times (\mathbf{a} + \mathbf{b}) = (\mathbf{a} - \mathbf{b}) \times \mathbf{a} + (\mathbf{a} - \mathbf{b}) \times \mathbf{b}$ by Theorem 8 #3

$= \mathbf{a} \times \mathbf{a} + (-\mathbf{b}) \times \mathbf{a} + \mathbf{a} \times \mathbf{b} + (-\mathbf{b}) \times \mathbf{b}$ by Theorem 8 #4

$= (\mathbf{a} \times \mathbf{a}) - (\mathbf{b} \times \mathbf{a}) + (\mathbf{a} \times \mathbf{b}) - (\mathbf{b} \times \mathbf{b})$ by Theorem 8 #2 (with $c = -1$)

$= \mathbf{0} - (\mathbf{b} \times \mathbf{a}) + (\mathbf{a} \times \mathbf{b}) - \mathbf{0}$ by Example 2

$= (\mathbf{a} \times \mathbf{b}) + (\mathbf{a} \times \mathbf{b})$ by Theorem 8 #1

$= 2(\mathbf{a} \times \mathbf{b})$

43. $\mathbf{a} \times (\mathbf{b} \times \mathbf{c}) + \mathbf{b} \times (\mathbf{c} \times \mathbf{a}) + \mathbf{c} \times (\mathbf{a} \times \mathbf{b})$

$= [(\mathbf{a} \cdot \mathbf{c})\mathbf{b} - (\mathbf{a} \cdot \mathbf{b})\mathbf{c}] + [(\mathbf{b} \cdot \mathbf{a})\mathbf{c} - (\mathbf{b} \cdot \mathbf{c})\mathbf{a}] + [(\mathbf{c} \cdot \mathbf{b})\mathbf{a} - (\mathbf{c} \cdot \mathbf{a})\mathbf{b}]$ by Exercise 42

$= (\mathbf{a} \cdot \mathbf{c})\mathbf{b} - (\mathbf{a} \cdot \mathbf{b})\mathbf{c} + (\mathbf{a} \cdot \mathbf{b})\mathbf{c} - (\mathbf{b} \cdot \mathbf{c})\mathbf{a} + (\mathbf{b} \cdot \mathbf{c})\mathbf{a} - (\mathbf{a} \cdot \mathbf{c})\mathbf{b} = 0$

45. (a) No. If $\mathbf{a} \cdot \mathbf{b} = \mathbf{a} \cdot \mathbf{c}$, then $\mathbf{a} \cdot (\mathbf{b} - \mathbf{c}) = 0$, so $\mathbf{a}$ is perpendicular to $\mathbf{b} - \mathbf{c}$, which can happen if $\mathbf{b} \neq \mathbf{c}$.

For example, let $\mathbf{a} = \langle 1, 1, 1 \rangle$, $\mathbf{b} = \langle 1, 0, 0 \rangle$ and $\mathbf{c} = \langle 0, 1, 0 \rangle$.

(b) No. If $\mathbf{a} \times \mathbf{b} = \mathbf{a} \times \mathbf{c}$ then $\mathbf{a} \times (\mathbf{b} - \mathbf{c}) = \mathbf{0}$, which implies that $\mathbf{a}$ is parallel to $\mathbf{b} - \mathbf{c}$, which of course can

happen if $\mathbf{b} \neq \mathbf{c}$.

(c) Yes. Since $\mathbf{a} \cdot \mathbf{c} = \mathbf{a} \cdot \mathbf{b}$, $\mathbf{a}$ is perpendicular to $\mathbf{b} - \mathbf{c}$, by part (a). From part (b), $\mathbf{a}$ is also parallel to $\mathbf{b} - \mathbf{c}$.

Thus since $\mathbf{a} \neq \mathbf{0}$ but is both parallel and perpendicular to $\mathbf{b} - \mathbf{c}$, we have $\mathbf{b} - \mathbf{c} = \mathbf{0}$, so $\mathbf{b} = \mathbf{c}$.

13.5 Equations of Lines and Planes $\qquad$ ET 12.5

1. (a) True; each of the first two lines has a direction vector parallel to the direction vector of the third line, so these vectors are each scalar multiples of the third direction vector. Then the first two direction vectors are also scalar multiples of each other, so these vectors, and hence the two lines, are parallel.

   (b) False; for example, the x- and y-axes are both perpendicular to the z-axis, yet the x- and y-axes are not parallel.

   (c) True; each of the first two planes has a normal vector parallel to the normal vector of the third plane, so these two normal vectors are parallel to each other and the planes are parallel.

   (d) False; for example, the xy- and yz-planes are not parallel, yet they are both perpendicular to the xz-plane.

   (e) False; the x- and y-axes are not parallel, yet they are both parallel to the plane $z = 1$.

   (f) True; if each line is perpendicular to a plane, then the lines' direction vectors are both parallel to a normal vector for the plane. Thus, the direction vectors are parallel to each other and the lines are parallel.

   (g) False; the planes $y = 1$ and $z = 1$ are not parallel, yet they are both parallel to the x-axis.

   (h) True; if each plane is perpendicular to a line, then any normal vector for each plane is parallel to a direction vector for the line. Thus, the normal vectors are parallel to each other and the planes are parallel.

   (i) True; see Figure 9 and the accompanying discussion.

   (j) False; they can be skew, as in Example 3.

   (k) True. Consider any normal vector for the plane and any direction vector for the line. If the normal vector is perpendicular to the direction vector, the line and plane are parallel. Otherwise, the vectors meet at an angle θ, $0° \leq \theta < 90°$, and the line will intersect the plane at an angle $90° - \theta$.

3. For this line, we have $\mathbf{r}_0 = -2\,\mathbf{i} + 4\,\mathbf{j} + 10\,\mathbf{k}$ and $\mathbf{v} = 3\,\mathbf{i} + \mathbf{j} - 8\,\mathbf{k}$, so a vector equation is $\mathbf{r} = \mathbf{r}_0 + t\,\mathbf{v} = (-2\,\mathbf{i} + 4\,\mathbf{j} + 10\,\mathbf{k}) + t(3\,\mathbf{i} + \mathbf{j} - 8\,\mathbf{k}) = (-2 + 3t)\,\mathbf{i} + (4 + t)\,\mathbf{j} + (10 - 8t)\,\mathbf{k}$ and parametric equations are $x = -2 + 3t$, $y = 4 + t$, $z = 10 - 8t$.

5. A line perpendicular to the given plane has the same direction as a normal vector to the plane, such as $\mathbf{n} = \langle 1, 3, 1 \rangle$. So $\mathbf{r}_0 = \mathbf{i} + 6\,\mathbf{k}$, and we can take $\mathbf{v} = \mathbf{i} + 3\,\mathbf{j} + \mathbf{k}$. Then a vector equation is $\mathbf{r} = (\mathbf{i} + 6\,\mathbf{k}) + t(\mathbf{i} + 3\,\mathbf{j} + \mathbf{k}) = (1 + t)\,\mathbf{i} + 3t\,\mathbf{j} + (6 + t)\,\mathbf{k}$, and parametric equations are $x = 1 + t$, $y = 3t$, $z = 6 + t$.

7. The vector $\mathbf{v} = \langle -4 - 1, 3 - 3, 0 - 2 \rangle = \langle -5, 0, -2 \rangle$ is parallel to the line. Letting $P_0 = (1, 3, 2)$, parametric equations are $x = 1 - 5t$, $y = 3 + 0t = 3$, $z = 2 - 2t$, while symmetric equations are $\dfrac{x - 1}{-5} = \dfrac{z - 2}{-2}$, $y = 3$.

   Notice here that the direction number $b = 0$, so rather than writing $\dfrac{y - 3}{0}$ in the symmetric equation we must write the equation $y = 3$ separately.

9. $\mathbf{v} = \left\langle 2 - 0, 1 - \frac{1}{2}, -3 - 1 \right\rangle = \left\langle 2, \frac{1}{2}, -4 \right\rangle$, and letting $P_0 = (2, 1, -3)$, parametric equations are $x = 2 + 2t$, $y = 1 + \frac{1}{2}t$, $z = -3 - 4t$, while symmetric equations are $\dfrac{x - 2}{2} = \dfrac{y - 1}{1/2} = \dfrac{z + 3}{-4}$ or $\dfrac{x - 2}{2} = 2y - 2 = \dfrac{z + 3}{-4}$.

11. The line has direction $\mathbf{v} = \langle 1, 2, 1 \rangle$. Letting $P_0 = (1, -1, 1)$, parametric equations are $x = 1 + t$, $y = -1 + 2t$, $z = 1 + t$ and symmetric equations are $x - 1 = \dfrac{y + 1}{2} = z - 1$.

13. Direction vectors of the lines are $\mathbf{v}_1 = \langle -2 - (-4), 0 - (-6), -3 - 1 \rangle = \langle 2, 6, -4 \rangle$ and $\mathbf{v}_2 = \langle 5 - 10, 3 - 18, 14 - 4 \rangle = \langle -5, -15, 10 \rangle$, and since $\mathbf{v}_2 = -\frac{5}{2}\mathbf{v}_1$, the direction vectors and thus the lines are parallel.

15. (a) A direction vector of the line with parametric equations $x = 1 + 2t$, $y = 3t$, $z = 5 - 7t$ is $\mathbf{v} = \langle 2, 3, -7 \rangle$ and
the desired parallel line must also have $\mathbf{v}$ as a direction vector. Here $P_0 = (0, 2, -1)$, so symmetric equations
for the line are $\dfrac{x}{2} = \dfrac{y - 2}{3} = \dfrac{z + 1}{-7}$.

(b) The line intersects the xy-plane when $z = 0$, so we need $\dfrac{x}{2} = \dfrac{y - 2}{3} = \dfrac{1}{-7}$ or $x = -\frac{2}{7}$, $y = \frac{11}{7}$. Thus the

point of intersection with the xy-plane is $\left(-\frac{2}{7}, \frac{11}{7}, 0\right)$. Similarly for the yz-plane, we need $x = 0$ $\Leftrightarrow$

$0 = \dfrac{y - 2}{3} = \dfrac{z + 1}{-7}$ $\Leftrightarrow$ $y = 2$, $z = -1$. Thus the line intersects the yz-plane at $(0, 2, -1)$. For the

xz-plane, we need $y = 0$ $\Leftrightarrow$ $\dfrac{x}{2} = -\dfrac{2}{3} = \dfrac{z + 1}{-7}$ $\Leftrightarrow$ $x = -\frac{4}{3}$, $z = \frac{11}{3}$. So the line intersects the xz-plane

at $\left(-\frac{4}{3}, 0, \frac{11}{3}\right)$.

17. From Equation 4, the line segment from $\mathbf{r}_0 = 2\,\mathbf{i} - \mathbf{j} + 4\,\mathbf{k}$ to $\mathbf{r}_1 = 4\,\mathbf{i} + 6\,\mathbf{j} + \mathbf{k}$ is
$\mathbf{r}(t) = (1 - t)\,\mathbf{r}_0 + t\,\mathbf{r}_1 = (1 - t)(2\,\mathbf{i} - \mathbf{j} + 4\,\mathbf{k}) + t(4\,\mathbf{i} + 6\,\mathbf{j} + \mathbf{k}) = (2\,\mathbf{i} - \mathbf{j} + 4\,\mathbf{k}) + t(2\,\mathbf{i} + 7\,\mathbf{j} - 3\,\mathbf{k})$,
$0 \le t \le 1$.

19. Since the direction vectors are $\mathbf{v}_1 = \langle -6, 9, -3 \rangle$ and $\mathbf{v}_2 = \langle 2, -3, 1 \rangle$, we have $\mathbf{v}_1 = -3\mathbf{v}_2$ so the lines are
parallel.

21. Since the direction vectors $\langle 1, 2, 3 \rangle$ and $\langle -4, -3, 2 \rangle$ are not scalar multiples of each other, the lines are not parallel,
so we check to see if the lines intersect. The parametric equations of the lines are L_1: $x = t$, $y = 1 + 2t$,
$z = 2 + 3t$ and L_2: $x = 3 - 4s$, $y = 2 - 3s$, $z = 1 + 2s$. For the lines to intersect, we must be able to find one
value of t and one value of s that produce the same point from the respective parametric equations. Thus we need to
satisfy the following three equations: $t = 3 - 4s$, $1 + 2t = 2 - 3s$, $2 + 3t = 1 + 2s$. Solving the first two equations
we get $t = -1$, $s = 1$ and checking, we see that these values don't satisfy the third equation. Thus the lines aren't
parallel and don't intersect, so they must be skew lines.

23. Since the plane is perpendicular to the vector $\langle -2, 1, 5 \rangle$, we can take $\langle -2, 1, 5 \rangle$ as a normal vector to the plane.
$(6, 3, 2)$ is a point on the plane, so setting $a = -2$, $b = 1$, $c = 5$ and $x_0 = 6$, $y_0 = 3$, $z_0 = 2$ in Equation 7 gives
$-2(x - 6) + 1(y - 3) + 5(z - 2) = 0$ or $-2x + y + 5z = 1$ to be an equation of the plane.

25. $\mathbf{i} + \mathbf{j} - \mathbf{k} = \langle 1, 1, -1 \rangle$ is a normal vector to the plane and $(1, -1, 1)$ is a point on the plane, so setting $a = 1$, $b = 1$,
$c = -1$, $x_0 = 1$, $y_0 = -1$, $z_0 = 1$ in Equation 7 gives $1\,(x - 1) + 1[y - (-1)] - 1(z - 1) = 0$ or
$x + y - z = -1$ to be an equation of the plane.

27. Since the two planes are parallel, they will have the same normal vectors. So we can take $\mathbf{n} = \langle 2, -1, 3 \rangle$, and an
equation of the plane is $2(x - 0) - 1(y - 0) + 3(z - 0) = 0$ or $2x - y + 3z = 0$.

29. Since the two planes are parallel, they will have the same normal vectors. So we can take $\mathbf{n} = \langle 3, 0, -7 \rangle$, and an
equation of the plane is $3(x - 4) + 0[y - (-2)] - 7(z - 3) = 0$ or $3x - 7z = -9$.

31. Here the vectors $\mathbf{a} = \langle 1 - 0, 0 - 1, 1 - 1 \rangle = \langle 1, -1, 0 \rangle$ and $\mathbf{b} = \langle 1 - 0, 1 - 1, 0 - 1 \rangle = \langle 1, 0, -1 \rangle$ lie in the
plane, so $\mathbf{a} \times \mathbf{b}$ is a normal vector to the plane. Thus, we can take $\mathbf{n} = \mathbf{a} \times \mathbf{b} = \langle 1 - 0, 0 + 1, 0 + 1 \rangle = \langle 1, 1, 1 \rangle$.
If P_0 is the point $(0, 1, 1)$, an equation of the plane is $1(x - 0) + 1(y - 1) + 1(z - 1) = 0$ or $x + y + z = 2$.

33. Here the vectors $\mathbf{a} = \langle 8 - 3, 2 - (-1), 4 - 2 \rangle = \langle 5, 3, 2 \rangle$ and
$\mathbf{b} = \langle -1 - 3, -2 - (-1), -3 - 2 \rangle = \langle -4, -1, -5 \rangle$ lie in the plane, so a normal vector to the plane is
$\mathbf{n} = \mathbf{a} \times \mathbf{b} = \langle -15 + 2, -8 + 25, -5 + 12 \rangle = \langle -13, 17, 7 \rangle$ and an equation of the plane is
$-13(x - 3) + 17[y - (-1)] + 7(z - 2) = 0$ or $-13x + 17y + 7z = -42$.

35. If we first find two nonparallel vectors in the plane, their cross product will be a normal vector to the plane. Since the given line lies in the plane, its direction vector $\mathbf{a} = \langle -2, 5, 4 \rangle$ is one vector in the plane. We can verify that the given point $(6, 0, -2)$ does not lie on this line, so to find another nonparallel vector $\mathbf{b}$ which lies in the plane, we can pick any point on the line and find a vector connecting the points. If we put $t = 0$, we see that $(4, 3, 7)$ is on the line, so $\mathbf{b} = \langle 6 - 4, 0 - 3, -2 - 7 \rangle = \langle 2, -3, -9 \rangle$ and $\mathbf{n} = \mathbf{a} \times \mathbf{b} = \langle -45 + 12, 8 - 18, 6 - 10 \rangle = \langle -33, -10, -4 \rangle$. Thus, an equation of the plane is $-33(x - 6) - 10(y - 0) - 4[z - (-2)] = 0$ or $33x + 10y + 4z = 190$.

37. A direction vector for the line of intersection is $\mathbf{a} = \mathbf{n}_1 \times \mathbf{n}_2 = \langle 1, 1, -1 \rangle \times \langle 2, -1, 3 \rangle = \langle 2, -5, -3 \rangle$, and $\mathbf{a}$ is parallel to the desired plane. Another vector parallel to the plane is the vector connecting any point on the line of intersection to the given point $(-1, 2, 1)$ in the plane. Setting $x = 0$, the equations of the planes reduce to $y - z = 2$ and $-y + 3z = 1$ with simultaneous solution $y = \frac{7}{2}$ and $z = \frac{3}{2}$. So a point on the line is $\left(0, \frac{7}{2}, \frac{3}{2}\right)$ and another vector parallel to the plane is $\left\langle -1, -\frac{3}{2}, -\frac{1}{2} \right\rangle$. Then a normal vector to the plane is $\mathbf{n} = \langle 2, -5, -3 \rangle \times \left\langle -1, -\frac{3}{2}, -\frac{1}{2} \right\rangle = \langle -2, 4, -8 \rangle$ and an equation of the plane is $-2(x + 1) + 4(y - 2) - 8(z - 1) = 0$ or $x - 2y + 4z = -1$.

39. Substitute the parametric equations of the line into the equation of the plane: $(3 - t) - (2 + t) + 2(5t) = 9$ ⇒ $8t = 8$ ⇒ $t = 1$. Therefore, the point of intersection of the line and the plane is given by $x = 3 - 1 = 2$, $y = 2 + 1 = 3$, and $z = 5(1) = 5$, that is, the point $(2, 3, 5)$.

41. Parametric equations for the line are $x = t$, $y = 1 + t$, $z = \frac{1}{2}t$ and substituting into the equation of the plane gives $4(t) - (1 + t) + 3\left(\frac{1}{2}t\right) = 8$ ⇒ $\frac{9}{2}t = 9$ ⇒ $t = 2$. Thus $x = 2$, $y = 1 + 2 = 3$, $z = \frac{1}{2}(2) = 1$ and the point of intersection is $(2, 3, 1)$.

43. Setting $x = 0$, we see that $(0, 1, 0)$ satisfies the equations of both planes, so that they do in fact have a line of intersection. $\mathbf{v} = \mathbf{n}_1 \times \mathbf{n}_2 = \langle 1, 1, 1 \rangle \times \langle 1, 0, 1 \rangle = \langle 1, 0, -1 \rangle$ is the direction of this line. Therefore, direction numbers of the intersecting line are $1, 0, -1$.

45. Normal vectors for the planes are $\mathbf{n}_1 = \langle 1, 4, -3 \rangle$ and $\mathbf{n}_2 = \langle -3, 6, 7 \rangle$, so the normals (and thus the planes) aren't parallel. But $\mathbf{n}_1 \cdot \mathbf{n}_2 = -3 + 24 - 21 = 0$, so the normals (and thus the planes) are perpendicular.

47. Normal vectors for the planes are $\mathbf{n}_1 = \langle 1, 1, 1 \rangle$ and $\mathbf{n}_2 = \langle 1, -1, 1 \rangle$. The normals are not parallel, so neither are the planes. Furthermore, $\mathbf{n}_1 \cdot \mathbf{n}_2 = 1 - 1 + 1 = 1 \neq 0$, so the planes aren't perpendicular. The angle between them is given by $\cos \theta = \dfrac{\mathbf{n}_1 \cdot \mathbf{n}_2}{|\mathbf{n}_1|\,|\mathbf{n}_2|} = \dfrac{1}{\sqrt{3}\,\sqrt{3}} = \dfrac{1}{3}$ ⇒ $\theta = \cos^{-1}\left(\frac{1}{3}\right) \approx 70.5\,°$.

49. The normals are $\mathbf{n}_1 = \langle 1, -4, 2 \rangle$ and $\mathbf{n}_2 = \langle 2, -8, 4 \rangle$. Since $\mathbf{n}_2 = 2\mathbf{n}_1$, the normals (and thus the planes) are parallel.

51. (a) To find a point on the line of intersection, set one of the variables equal to a constant, say $z = 0$. (This will only work if the line of intersection crosses the xy-plane; otherwise, try setting x or y equal to 0.) Then the equations of the planes reduce to $x + y = 2$ and $3x - 4y = 6$. Solving these two equations gives $x = 2$, $y = 0$. So a point on the line of intersection is $(2, 0, 0)$. The direction of the line is $\mathbf{v} = \mathbf{n}_1 \times \mathbf{n}_2 = \langle 5 - 4, -3 - 5, -4 - 3 \rangle = \langle 1, -8, -7 \rangle$, and symmetric equations for the line are

$$x - 2 = \frac{y}{-8} = \frac{z}{-7}.$$

(b) The angle between the planes satisfies $\cos \theta = \dfrac{\mathbf{n}_1 \cdot \mathbf{n}_2}{|\mathbf{n}_1|\,|\mathbf{n}_2|} = \dfrac{3 - 4 - 5}{\sqrt{3}\,\sqrt{50}} = -\dfrac{\sqrt{6}}{5}$. Therefore $\theta = \cos^{-1}\left(-\frac{\sqrt{6}}{5}\right) \approx 119°$ (or $61°$).

53. Setting $x = 0$, the equations of the two planes become $z = y$ and $5y + z = -1$, which intersect at $y = -\frac{1}{6}$ and

$z = -\frac{1}{6}$. Thus we can choose $(x_0, y_0, z_0) = \left(0, -\frac{1}{6}, -\frac{1}{6}\right)$. The vector giving the direction of this intersecting line,

$\mathbf{v}$, is perpendicular to the normal vectors of both planes. So $\mathbf{v} = \mathbf{n_1} \times \mathbf{n_2} = \langle 2, -5, -1 \rangle \times \langle 1, 1, -1 \rangle = \langle 6, 1, 7 \rangle$.

Therefore, by Equations 2, parametric equations for this line are $x = 6t$, $y = -\frac{1}{6} + t$, $z = -\frac{1}{6} + 7t$.

55. The plane contains all perpendicular bisectors of the line segment joining $(1, 1, 0)$ and $(0, 1, 1)$. All of these

bisectors pass through the midpoint of this segment $\left(\frac{1}{2}, \frac{1+1}{2}, \frac{1}{2}\right) = \left(\frac{1}{2}, 1, \frac{1}{2}\right)$. The direction of this line segment

$\langle 1 - 0, 1 - 1, 0 - 1 \rangle = \langle 1, 0, -1 \rangle$ is perpendicular to the plane so that we can choose this to be $\mathbf{n}$. Therefore the

equation of the plane is $1\left(x - \frac{1}{2}\right) + 0(y - 1) - 1\left(z - \frac{1}{2}\right) = 0 \iff x = z$.

57. The plane contains the points $(a, 0, 0)$, $(0, b, 0)$ and $(0, 0, c)$. Thus the vectors $\mathbf{a} = \langle -a, b, 0 \rangle$ and $\mathbf{b} = \langle -a, 0, c \rangle$

lie in the plane, and $\mathbf{n} = \mathbf{a} \times \mathbf{b} = \langle bc - 0, 0 + ac, 0 + ab \rangle = \langle bc, ac, ab \rangle$ is a normal vector to the plane. The

equation of the plane is therefore $bcx + acy + abz = abc + 0 + 0$ or $bcx + acy + abz = abc$. Notice that if $a \neq 0$,

$b \neq 0$ and $c \neq 0$ then we can rewrite the equation as $\dfrac{x}{a} + \dfrac{y}{b} + \dfrac{z}{c} = 1$. This is a good equation to remember!

59. Two vectors which are perpendicular to the required line are the normal of the given plane, $\langle 1, 1, 1 \rangle$, and a direction

vector for the given line, $\langle 1, -1, 2 \rangle$. So a direction vector for the required line is

$\langle 1, 1, 1 \rangle \times \langle 1, -1, 2 \rangle = \langle 3, -1, -2 \rangle$. Thus L is given by $\langle x, y, z \rangle = \langle 0, 1, 2 \rangle + t\langle 3, -1, -2 \rangle$, or in parametric

form, $x = 3t$, $y = 1 - t$, $z = 2 - 2t$.

61. Let P_i have normal vector $\mathbf{n}_i$. Then $\mathbf{n}_1 = \langle 4, -2, 6 \rangle$, $\mathbf{n}_2 = \langle 4, -2, -2 \rangle$, $\mathbf{n}_3 = \langle -6, 3, -9 \rangle$, $\mathbf{n}_4 = \langle 2, -1, -1 \rangle$.

Now $\mathbf{n}_1 = -\frac{2}{3}\mathbf{n}_3$, so $\mathbf{n}_1$ and $\mathbf{n}_3$ are parallel, and hence P_1 and P_3 are parallel; similarly P_2 and P_4 are parallel

because $\mathbf{n}_2 = 2\mathbf{n}_4$. However, $\mathbf{n}_1$ and $\mathbf{n}_2$ are not parallel. $\left(0, 0, \frac{1}{2}\right)$ lies on P_1, but not on P_3, so they are not the

same plane, but both P_2 and P_4 contain the point $(0, 0, -3)$, so these two planes are identical.

63. Let $Q = (2, 2, 0)$ and $R = (3, -1, 5)$, points on the line corresponding to $t = 0$ and $t = 1$. Let

$P = (1, 2, 3)$. Then $\mathbf{a} = \overrightarrow{QR} = \langle 1, -3, 5 \rangle$, $\mathbf{b} = \overrightarrow{QP} = \langle -1, 0, 3 \rangle$. The distance is

$$d = \frac{|\mathbf{a} \times \mathbf{b}|}{|\mathbf{a}|} = \frac{|\langle 1, -3, 5 \rangle \times \langle -1, 0, 3 \rangle|}{|\langle 1, -3, 5 \rangle|} = \frac{|\langle -9, -8, -3 \rangle|}{|\langle 1, -3, 5 \rangle|} = \frac{\sqrt{9^2 + 8^2 + 3^2}}{\sqrt{1^2 + 3^2 + 5^2}} = \frac{\sqrt{154}}{\sqrt{35}} = \sqrt{\frac{22}{5}}.$$

65. By Equation 9, the distance is $D = \dfrac{1}{\sqrt{1 + 4 + 4}} \left[(1)(2) + (-2)(8) + (-2)(5) - 1\right] = \dfrac{25}{3}$.

67. Put $y = z = 0$ in the equation of the first plane to get the point $(-1, 0, 0)$ on the plane. Because the planes are

parallel, the distance D between them is the distance from $(-1, 0, 0)$ to the second plane. By Equation 9,

$$D = \frac{|3(-1) + 6(0) - 3(0) - 4|}{\sqrt{3^2 + 6^2 + (-3)^2}} = \frac{7}{3\sqrt{6}} \text{ or } \frac{7\sqrt{6}}{18}.$$

69. The distance between two parallel planes is the same as the distance between a point on one of the planes and the

other plane. Let $P_0 = (x_0, y_0, z_0)$ be a point on the plane given by $ax + by + cz + d_1 = 0$. Then

$ax_0 + by_0 + cz_0 + d_1 = 0$ and the distance between P_0 and the plane given by $ax + by + cz + d_2 = 0$ is, from

Equation 9, $D = \dfrac{|ax_0 + by_0 + cz_0 + d_2|}{\sqrt{a^2 + b^2 + c^2}} = \dfrac{|-d_1 + d_2|}{\sqrt{a^2 + b^2 + c^2}} = \dfrac{|d_1 - d_2|}{\sqrt{a^2 + b^2 + c^2}}$.

71. L_1: $x = y = z$ $\Rightarrow$ $x = y$ (1). L_2: $x + 1 = y/2 = z/3$ $\Rightarrow$ $x + 1 = y/2$ (2). The solution of (1) and (2) is $x = y = -2$. However, when $x = -2$, $x = z$ $\Rightarrow$ $z = -2$, but $x + 1 = z/3$ $\Rightarrow$ $z = -3$, a contradiction. Hence the lines do not intersect. For L_1, $\mathbf{v}_1 = \langle 1, 1, 1 \rangle$, and for L_2, $\mathbf{v}_2 = \langle 1, 2, 3 \rangle$, so the lines are not parallel. Thus the lines are skew lines. If two lines are skew, they can be viewed as lying in two parallel planes and so the distance between the skew lines would be the same as the distance between these parallel planes. The common normal vector to the planes must be perpendicular to both $\langle 1, 1, 1 \rangle$ and $\langle 1, 2, 3 \rangle$, the direction vectors of the two lines. So set $\mathbf{n} = \langle 1, 1, 1 \rangle \times \langle 1, 2, 3 \rangle = \langle 3 - 2, -3 + 1, 2 - 1 \rangle = \langle 1, -2, 1 \rangle$. From above, we know that $(-2, -2, -2)$ and $(-2, -2, -3)$ are points of L_1 and L_2 respectively. So in the notation of Equation 8, $1(-2) - 2(-2) + 1(-2) + d_1 = 0$ $\Rightarrow$ $d_1 = 0$ and $1(-2) - 2(-2) + 1(-3) + d_2 = 0$ $\Rightarrow$ $d_2 = 1$.

By Exercise 69, the distance between these two skew lines is $D = \dfrac{|0 - 1|}{\sqrt{1 + 4 + 1}} = \dfrac{1}{\sqrt{6}}$.

Alternate solution (without reference to planes): A vector which is perpendicular to both of the lines is $\mathbf{n} = \langle 1, 1, 1 \rangle \times \langle 1, 2, 3 \rangle = \langle 1, -2, 1 \rangle$. Pick any point on each of the lines, say $(-2, -2, -2)$ and $(-2, -2, -3)$, and form the vector $\mathbf{b} = \langle 0, 0, 1 \rangle$ connecting the two points. The distance between the two skew lines is the absolute value of the scalar projection of $\mathbf{b}$ along $\mathbf{n}$, that is, $D = \dfrac{|\mathbf{n} \cdot \mathbf{b}|}{|\mathbf{n}|} = \dfrac{|1 \cdot 0 - 2 \cdot 0 + 1 \cdot 1|}{\sqrt{1 + 4 + 1}} = \dfrac{1}{\sqrt{6}}$.

73. If $a \neq 0$, then $ax + by + cz + d = 0$ $\Rightarrow$ $a(x + d/a) + b(y - 0) + c(z - 0) = 0$ which by (7) is the scalar equation of the plane through the point $(-d/a, 0, 0)$ with normal vector $\langle a, b, c \rangle$. Similarly, if $b \neq 0$ (or if $c \neq 0$) the equation of the plane can be rewritten as $a(x - 0) + b(y + d/b) + c(z - 0) = 0$ [or as $a(x - 0) + b(y - 0) + c(z + d/c) = 0$] which by (7) is the scalar equation of a plane through the point $(0, -d/b, 0)$ [or the point $(0, 0, -d/c)$] with normal vector $\langle a, b, c \rangle$.

13.6 Cylinders and Quadric Surfaces   ET 12.6

1. (a) In $\mathbb{R}^2$, the equation $y = x^2$ represents a parabola.

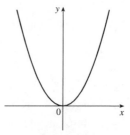

(b) In $\mathbb{R}^3$, the equation $y = x^2$ doesn't involve z, so any horizontal plane with equation $z = k$ intersects the graph in a curve with equation $y = x^2$. Thus, the surface is a parabolic cylinder, made up of infinitely many shifted copies of the same parabola. The rulings are parallel to the z-axis.

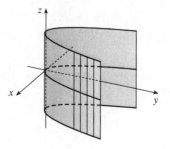

(c) In $\mathbb{R}^3$, the equation $z = y^2$ also represents a parabolic cylinder. Since x doesn't appear, the graph is formed by moving the parabola $z = y^2$ in the direction of the x-axis. Thus, the rulings of the cylinder are parallel to the x-axis.

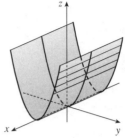

3. Since x is missing from the equation, the vertical traces $y^2 + 4z^2 = 4$, $x = k$, are copies of the same ellipse in the plane $x = k$. Thus, the surface $y^2 + 4z^2 = 4$ is an elliptic cylinder with rulings parallel to the x-axis.

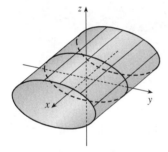

5. Since z is missing, each horizontal trace $x = y^2$, $z = k$, is a copy of the same parabola in the plane $z = k$. Thus, the surface $x - y^2 = 0$ is a parabolic cylinder with rulings parallel to the z-axis.

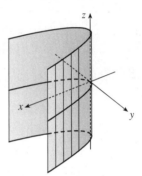

7. Since y is missing, each vertical trace $z = \cos x$, $y = k$ is a copy of a cosine curve in the plane $y = k$. Thus, the surface $z = \cos x$ is a cylindrical surface with rulings parallel to the y-axis.

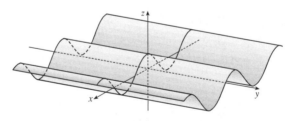

9. (a) The traces of $x^2 + y^2 - z^2 = 1$ in $x = k$ are $y^2 - z^2 = 1 - k^2$, a family of hyperbolas. (Note that the hyperbolas are oriented differently for $-1 < k < 1$ than for $k < -1$ or $k > 1$.) The traces in $y = k$ are $x^2 - z^2 = 1 - k^2$, a similar family of hyperbolas. The traces in $z = k$ are $x^2 + y^2 = 1 + k^2$, a family of circles. For $k = 0$, the trace in the xy-plane, the circle is of radius 1. As $|k|$ increases, so does the radius of the circle. This behavior, combined with the hyperbolic vertical traces, gives the graph of the hyperboloid of one sheet in Table 1.

(b) The shape of the surface is unchanged, but the hyperboloid is rotated so that its axis is the y-axis. Traces in $y = k$ are circles, while traces in $x = k$ and $z = k$ are hyperbolas.

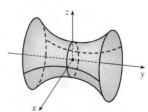

(c) Completing the square in y gives $x^2 + (y+1)^2 - z^2 = 1$. The surface is a hyperboloid identical to the one in part (a) but shifted one unit in the negative y-direction.

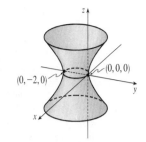

11. Traces: $x = k$, $9y^2 + 36z^2 = 36 - 4k^2$, an ellipse for $|k| < 3$; $y = k$, $4x^2 + 36z^2 = 36 - 9k^2$, an ellipse for $|k| < 2$; $z = k$, $4x^2 + 9y^2 = 36(1 - k^2)$, an ellipse for $|k| < 1$. Thus the surface is an ellipsoid with center at the origin and axes along the x-, y- and z-axes.

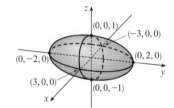

13. Traces: $x = k$, $y^2 = k^2 + z^2$ or $y^2 - z^2 = k^2$, a hyperbola for $k \neq 0$ and two intersecting lines for $k = 0$; $y = k$, $x^2 + z^2 = k^2$, a circle for $k \neq 0$; $z = k$, $y^2 = x^2 + k^2$ or $y^2 - x^2 = k^2$, a hyperbola for $k \neq 0$ and two intersecting lines for $k = 0$. Thus the surface is a cone (right circular) with axis the y-axis and vertex the origin.

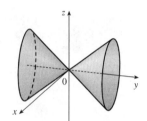

15. Traces: $x = k$, $4y^2 - z^2 = 4 + k^2$, a hyperbola; $y = k$, $x^2 + z^2 = 4k^2 - 4$, a circle for $|k| > 1$; $z = k$, $4y^2 - x^2 = 4 + k^2$, a hyperbola. Thus the surface is a hyperboloid of two sheets with axis the y-axis.

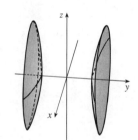

17. Traces: $x = k$, $k^2 + 4z^2 - y = 0$ or $y - k^2 = 4z^2$, a parabola; $y = k$, $x^2 + 4z^2 = k$, an ellipse for $k > 0$; $z = k$, $x^2 + 4k^2 - y = 0$ or $y - 4k^2 = x^2$, a parabola. Thus the surface is an elliptic paraboloid with axis the y-axis and vertex the origin.

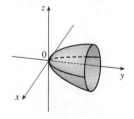

19. $y = z^2 - x^2$. The traces in $x = k$ are the parabolas $y = z^2 - k^2$; the traces in $y = k$ are $k = z^2 - x^2$, which are hyperbolas (note the hyperbolas are oriented differently for $k > 0$ than for $k < 0$); and the traces in $z = k$ are the parabolas $y = k^2 - x^2$. Thus, $\dfrac{y}{1} = \dfrac{z^2}{1^2} - \dfrac{x^2}{1^2}$ is a hyperbolic paraboloid.

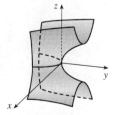

21. This is the equation of an ellipsoid: $x^2 + 4y^2 + 9z^2 = x^2 + \dfrac{y^2}{(1/2)^2} + \dfrac{z^2}{(1/3)^2} = 1$, with x-intercepts ± 1,

y-intercepts $\pm\frac{1}{2}$ and z-intercepts $\pm\frac{1}{3}$. So the major axis is the x-axis and the only possible graph is VII.

23. This is the equation of a hyperboloid of one sheet, with $a = b = c = 1$. Since the coefficient of y^2 is negative, the axis of the hyperboloid is the y-axis, hence the correct graph is II.

25. There are no real values of x and z that satisfy this equation for $y < 0$, so this surface does not extend to the left of the xz-plane. The surface intersects the plane $y = k > 0$ in an ellipse. Notice that y occurs to the first power whereas x and z occur to the second power. So the surface is an elliptic paraboloid with axis the y-axis. Its graph is VI.

27. This surface is a cylinder because the variable y is missing from the equation. The intersection of the surface and the xz-plane is an ellipse. So the graph is VIII.

29. $z^2 = 4x^2 + 9y^2 + 36$ or $-4x^2 - 9y^2 + z^2 = 36$

or $-\dfrac{x^2}{9} - \dfrac{y^2}{4} + \dfrac{z^2}{36} = 1$ represents a hyperboloid

of two sheets with axis the z-axis.

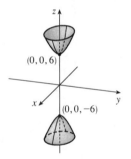

31. $x = 2y^2 + 3z^2$ or $x = \dfrac{y^2}{1/2} + \dfrac{z^2}{1/3}$ or

$\dfrac{x}{6} = \dfrac{y^2}{3} + \dfrac{z^2}{2}$ represents an elliptic paraboloid

with vertex $(0, 0, 0)$ and axis the x-axis.

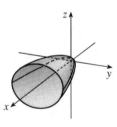

33. Completing squares in y and z gives

$4x^2 + (y - 2)^2 + 4(z - 3)^2 = 4$ or

$x^2 + \dfrac{(y - 2)^2}{4} + (z - 3)^2 = 1$, an ellipsoid with

center $(0, 2, 3)$.

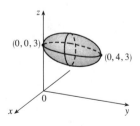

35. Completing squares in all three variables gives

$(x - 2)^2 - (y + 1)^2 + (z - 1)^2 = 0$ or

$(y + 1)^2 = (x - 2)^2 + (z - 1)^2$, a circular cone

with center $(2, -1, 1)$ and axis the horizontal line

$x = 2$, $z = 1$.

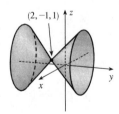

37.

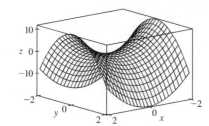

39.

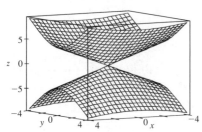

To restrict the z-range as in the second graph, we can use the option `view = -2..2` in Maple's `plot3d` command, or `PlotRange -> {-2,2}` in Mathematica's `Plot3D` command.

41.

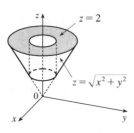

43. The surface is a paraboloid of revolution (circular paraboloid) with vertex at the origin, axis the y-axis and opens to the right. Thus the trace in the yz-plane is also a parabola: $y = z^2$, $x = 0$. The equation is $y = x^2 + z^2$.

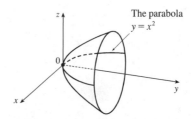

The parabola $y = x^2$

45. Let $P = (x, y, z)$ be an arbitrary point equidistant from $(-1, 0, 0)$ and the plane $x = 1$. Then the distance from P to $(-1, 0, 0)$ is $\sqrt{(x+1)^2 + y^2 + z^2}$ and the distance from P to the plane $x = 1$ is $|x - 1| / \sqrt{1^2} = |x - 1|$ (by Equation 13.5.9 [ET 12.5.9]). So $|x - 1| = \sqrt{(x+1)^2 + y^2 + z^2}$ $\Leftrightarrow$ $(x-1)^2 = (x+1)^2 + y^2 + z^2$ $\Leftrightarrow$ $x^2 - 2x + 1 = x^2 + 2x + 1 + y^2 + z^2$ $\Leftrightarrow$ $-4x = y^2 + z^2$. Thus the collection of all such points P is a circular paraboloid with vertex at the origin, axis the x-axis, which opens in the negative direction.

47. If (a, b, c) satisfies $z = y^2 - x^2$, then $c = b^2 - a^2$. L_1: $x = a + t$, $y = b + t$, $z = c + 2(b - a)t$, L_2: $x = a + t$, $y = b - t$, $z = c - 2(b + a)t$. Substitute the parametric equations of L_1 into the equation of the hyperbolic paraboloid in order to find the points of intersection: $z = y^2 - x^2$ $\Rightarrow$ $c + 2(b - a)t = (b + t)^2 - (a + t)^2 = b^2 - a^2 + 2(b - a)t$ $\Rightarrow$ $c = b^2 - a^2$. As this is true for all values of t, L_1 lies on $z = y^2 - x^2$. Performing similar operations with L_2 gives: $z = y^2 - x^2$ $\Rightarrow$ $c - 2(b + a)t = (b - t)^2 - (a + t)^2 = b^2 - a^2 - 2(b + a)t$ $\Rightarrow$ $c = b^2 - a^2$. This tells us that all of L_2 also lies on $z = y^2 - x^2$.

49.

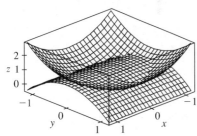

The curve of intersection looks like a bent ellipse. The projection of this curve onto the xy-plane is the set of points $(x, y, 0)$ which satisfy $x^2 + y^2 = 1 - y^2$ ⇔

$x^2 + 2y^2 = 1$ ⇔ $x^2 + \dfrac{y^2}{\left(1/\sqrt{2}\right)^2} = 1$. This is an

equation of an ellipse.

13.7 Cylindrical and Spherical Coordinates ET 12.7

1. See Figure 1 and the accompanying discussion; see the paragraph accompanying Figure 3.

3.

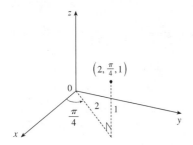

$x = 2\cos\dfrac{\pi}{4} = \sqrt{2},\, y = 2\sin\dfrac{\pi}{4} = \sqrt{2}$,

$z = 1$, so the point is $\left(\sqrt{2}, \sqrt{2}, 1\right)$ in rectangular coordinates.

5.

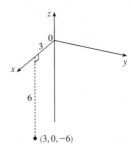

$x = 3\cos 0 = 3,\, y = 3\sin 0 = 0$, and

$z = -6$, so the point is $(3, 0, -6)$ in rectangular coordinates.

7.

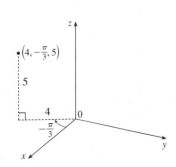

$x = 4\cos\left(-\dfrac{\pi}{3}\right) = 2,\, y = 4\sin\left(-\dfrac{\pi}{3}\right) = -2\sqrt{3}$, and

$z = 5$, so the point is $\left(2, -2\sqrt{3}, 5\right)$ in rectangular coordinates.

9. $r^2 = x^2 + y^2 = 1^2 + (-1)^2 = 2$ so $r = \sqrt{2}$; $\tan\theta = \dfrac{y}{x} = \dfrac{-1}{1} = -1$ and the point $(1, -1)$ is in the fourth quadrant of the xy-plane, so $\theta = \dfrac{7\pi}{4} + 2n\pi$; $z = 4$. Thus, one set of cylindrical coordinates is $\left(\sqrt{2}, \dfrac{7\pi}{4}, 4\right)$.

11. $r^2 = (-1)^2 + \left(-\sqrt{3}\right)^2 = 4$ so $r = 2$; $\tan\theta = \dfrac{-\sqrt{3}}{-1} = \sqrt{3}$ and the point $\left(-1, -\sqrt{3}\right)$ is in the third quadrant of the xy-plane, so $\theta = \dfrac{4\pi}{3} + 2n\pi$; $z = 2$. Thus, one set of cylindrical coordinates is $\left(2, \dfrac{4\pi}{3}, 2\right)$.

13.

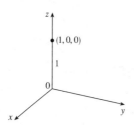

$x = \rho \sin \phi \cos \theta = (1) \sin 0 \cos 0 = 0$,

$y = \rho \sin \phi \sin \theta = (1) \sin 0 \sin 0 = 0$, and

$z = \rho \cos \phi = (1) \cos 0 = 1$ so the point is

$(0, 0, 1)$ in rectangular coordinates.

15.

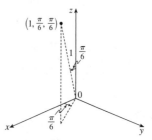

$x = \sin \frac{\pi}{6} \cos \frac{\pi}{6} = \frac{\sqrt{3}}{4}$, $y = \sin \frac{\pi}{6} \sin \frac{\pi}{6} = \frac{1}{4}$, and

$z = \cos \frac{\pi}{6} = \frac{\sqrt{3}}{2}$, so the point is $\left(\frac{\sqrt{3}}{4}, \frac{1}{4}, \frac{\sqrt{3}}{2} \right)$ in

rectangular coordinates.

17. $x = 2 \sin \frac{\pi}{4} \cos \frac{\pi}{3} = \frac{\sqrt{2}}{2}$, $y = 2 \sin \frac{\pi}{4} \sin \frac{\pi}{3} = \frac{\sqrt{6}}{2}$,

$z = 2 \cos \frac{\pi}{4} = \sqrt{2}$ so the point is $\left(\frac{\sqrt{2}}{2}, \frac{\sqrt{6}}{2}, \sqrt{2} \right)$ in rectangular

coordinates.

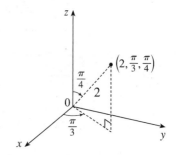

19. $\rho = \sqrt{x^2 + y^2 + z^2} = \sqrt{1 + 3 + 12} = 4$, $\cos \phi = \frac{z}{\rho} = \frac{2\sqrt{3}}{4} = \frac{\sqrt{3}}{2}$ $\Rightarrow$ $\phi = \frac{\pi}{6}$, and

$\cos \theta = \frac{x}{\rho \sin \phi} = \frac{1}{4 \sin(\pi/6)} = \frac{1}{2}$ $\Rightarrow$ $\theta = \frac{\pi}{3}$ (since $y > 0$). Thus spherical coordinates are $\left(4, \frac{\pi}{3}, \frac{\pi}{6} \right)$.

21. $\rho = \sqrt{0 + 1 + 1} = \sqrt{2}$, $\cos \phi = \frac{-1}{\sqrt{2}}$ $\Rightarrow$ $\phi = \frac{3\pi}{4}$, and $\cos \theta = \frac{0}{\sqrt{2} \sin(3\pi/4)} = 0$ $\Rightarrow$ $\theta = \frac{3\pi}{2}$

(since $y < 0$). Thus spherical coordinates are $\left(\sqrt{2}, \frac{3\pi}{2}, \frac{3\pi}{4} \right)$.

23. $\rho = \sqrt{x^2 + y^2 + z^2} = \sqrt{r^2 + z^2} = \sqrt{1 + 3} = 2$; $\theta = \frac{\pi}{6}$; $\cos \phi = \frac{z}{\rho} = \frac{\sqrt{3}}{2}$ $\Rightarrow$ $\phi = \frac{\pi}{6}$, thus in spherical

coordinates the point is $\left(2, \frac{\pi}{6}, \frac{\pi}{6} \right)$.

25. $\rho = \sqrt{r^2 + z^2} = \sqrt{3 + 1} = 2$; $\theta = \frac{\pi}{2}$; $\cos \phi = \frac{z}{\rho} = \frac{-1}{2}$ $\Rightarrow$ $\phi = \frac{2\pi}{3}$, so in spherical coordinates the point

is $\left(2, \frac{\pi}{2}, \frac{2\pi}{3} \right)$.

27. $z = \rho \cos \phi = 2 \cos 0 = 2$, $\rho^2 = x^2 + y^2 + z^2 = r^2 + z^2$ $\Rightarrow$ $r = \sqrt{\rho^2 - z^2} = \sqrt{2^2 - 2^2} = 0$,

(or $r = 2 \sin 0 = 0$), $\theta = 0$ and the point is $(0, 0, 2)$.

29. $z = 8 \cos \frac{\pi}{2} = 0$, $r = 8 \sin \frac{\pi}{2} = 8$, $\theta = \frac{\pi}{6}$ and the point is $\left(8, \frac{\pi}{6}, 0 \right)$.

31. Since $r = 3$, $x^2 + y^2 = 9$ and the surface is a circular cylinder with radius 3 and axis the z-axis.

33. Since $\phi = 0$, $x = 0$ and $y = 0$ while $z = \rho \geq 0$. Thus the "surface" is the positive z-axis including the origin.

35. Since $\phi = \frac{\pi}{3}$, the surface is the top half of the right circular cone with vertex at the origin and axis the positive z-axis.

37. $z = r^2 = x^2 + y^2$, so the surface is a circular paraboloid with vertex at the origin and axis the positive z-axis.

39. $2 = \rho \cos \phi = z$ is a plane through the point $(0, 0, 2)$ and parallel to the xy-plane.

41. $r = 2 \cos \theta \;\Rightarrow\; r^2 = x^2 + y^2 = 2r \cos \theta = 2x \;\Leftrightarrow\; (x - 1)^2 + y^2 = 1$, which is the equation of a circular cylinder with radius 1, whose axis is the vertical line $x = 1$, $y = 0$, $z = z$.

43. Since $r^2 + z^2 = 25$ and $r^2 = x^2 + y^2$, we have $x^2 + y^2 + z^2 = 25$, a sphere with radius 5 and center at the origin.

45. Since $x^2 = \rho^2 \sin^2 \phi \cos^2 \theta$ and $z^2 = \rho^2 \cos^2 \phi$, the equation of the surface in rectangular coordinates is $x^2 + z^2 = 4$. Thus the surface is a circular cylinder of radius 2 about the y-axis.

47. Since $r^2 - r = 0$, $r = 0$ or $r = 1$. But $x^2 + y^2 = r^2$. Thus the surface consists of the right circular cylinder of radius 1 and axis the z-axis along with the surface given by $x^2 + y^2 = 0$, that is, the z-axis.

49. (a) $x^2 + y^2 = r^2$, so the equation becomes $z = r^2$.

(b) $x = \rho \sin \phi \cos \theta$, $y = \rho \sin \phi \sin \theta$, and $z = \rho \cos \phi$, so the equation becomes
$\rho \cos \phi = (\rho \sin \phi \cos \theta)^2 + (\rho \sin \phi \sin \theta)^2$ or $\rho \cos \phi = \rho^2 \sin^2 \phi$ or $\rho \sin^2 \phi = \cos \phi$.

51. (a) $x = r \cos \theta$, so the equation becomes $r \cos \theta = 3$ or $r = 3 \sec \theta$ (since $\cos \theta \neq 0$ here).

(b) $x = \rho \sin \phi \cos \theta$, so the equation becomes $\rho \sin \phi \cos \theta = 3$.

53. (a) $r^2(\cos^2 \theta - \sin^2 \theta) - 2z^2 = 4$ or $2z^2 = r^2 \cos 2\theta - 4$.

(b) $\rho^2 (\sin^2 \phi \cos^2 \theta - \sin^2 \phi \sin^2 \theta - 2 \cos^2 \phi) = 4$ or $\rho^2 (\sin^2 \phi \cos 2\theta - 2 \cos^2 \phi) = 4$.

55. (a) $r^2 = 2r \sin \theta$ or $r = 2 \sin \theta$.

(b) $\rho^2 \sin^2 \phi (\cos^2 \theta + \sin^2 \theta) = 2\rho \sin \phi \sin \theta$ or $\rho \sin^2 \phi = 2 \sin \phi \sin \theta$ or $\rho \sin \phi = 2 \sin \theta$.

57.

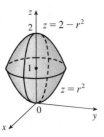

$z = r^2 = x^2 + y^2$ is a circular paraboloid with

vertex $(0, 0, 0)$, opening upward. $z = 2 - r^2$

$\Rightarrow\quad z - 2 = -(x^2 + y^2)$ is a circular paraboloid

with vertex $(0, 0, 2)$ opening downward. Thus

$r^2 \leq z \leq 2 - r^2$ is the solid region enclosed by

these two surfaces.

59.

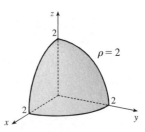

$\rho = 2$ represents a sphere of radius 2, centered at

the origin, so $\rho \leq 2$ is this sphere and its interior.

$0 \leq \phi \leq \frac{\pi}{2}$ restricts the solid to that portion of

the region that lies on or above the xy-plane, and

$0 \leq \theta \leq \frac{\pi}{2}$ further restricts the solid to the first

octant. Thus the solid is the portion in the first

octant of the solid ball centered at the origin with

radius 2.

61. $-\frac{\pi}{2} \le \theta \le \frac{\pi}{2}$ restricts the solid to the 4 octants in which x is

positive. $\rho = \sec \phi \;\Rightarrow\; \rho \cos \phi = z = 1$, which is the equation of

a horizontal plane. $0 \le \phi \le \frac{\pi}{6}$ describes a cone, opening upward. So

the solid lies above the cone $\phi = \frac{\pi}{6}$ and below the plane $z = 1$.

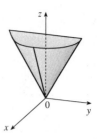

63. We can position the cylindrical shell vertically so that its axis coincides with the z-axis and its base lies in the
xy-plane. If we use centimeters as the unit of measurement, then cylindrical coordinates conveniently describe the
shell as $6 \le r \le 7, 0 \le \theta \le 2\pi, 0 \le z \le 20$.

65. $z \ge \sqrt{x^2 + y^2}$ because the solid lies above the cone. Squaring both sides of this inequality gives $z^2 \ge x^2 + y^2$
$\Rightarrow\; 2z^2 \ge x^2 + y^2 + z^2 = \rho^2 \;\Rightarrow\; z^2 = \rho^2 \cos^2 \phi \ge \frac{1}{2}\rho^2 \;\Rightarrow\; \cos^2 \phi \ge \frac{1}{2}$. The cone opens upward so that
the inequality is $\cos \phi \ge \frac{1}{\sqrt{2}}$, or equivalently $0 \le \phi \le \frac{\pi}{4}$. In spherical coordinates the sphere $z = x^2 + y^2 + z^2$ is
$\rho \cos \phi = \rho^2 \;\Rightarrow\; \rho = \cos \phi$. $0 \le \rho \le \cos \phi$ because the solid lies below the sphere. The solid can therefore be
described as the region in spherical coordinates satisfying $0 \le \rho \le \cos \phi, 0 \le \phi \le \frac{\pi}{4}$.

67. In cylindrical coordinates, the equation of the cylinder is $r = 3$,
$0 \le z \le 10$. The hemisphere is the upper part of the sphere radius 3,
center $(0, 0, 10)$, equation $r^2 + (z - 10)^2 = 3^2, z \ge 10$. In Maple, we
can use either the `coords=cylindrical` option in a regular `plot`
command, or the `plots[cylinderplot]` command. In
Mathematica, we can use `ParametricPlot3d`.

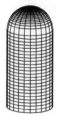

13 Review

ET 12

──────────────── CONCEPT CHECK ────────────────

1. A scalar is a real number, while a vector is a quantity that has both a real-valued magnitude and a direction.

2. To add two vectors geometrically, we can use either the Triangle Law or the Parallelogram Law, as illustrated in
Figures 3 and 4 in Section 13.2 [ET 12.2]. Algebraically, we add the corresponding components of the vectors.

3. For $c > 0$, $c\,\mathbf{a}$ is a vector with the same direction as $\mathbf{a}$ and length c times the length of $\mathbf{a}$. If $c < 0$, $c\mathbf{a}$ points in the
opposite direction as $\mathbf{a}$ and has length $|c|$ times the length of $\mathbf{a}$. (See Figures 7 and 15 in Section 13.2 [ET 12.2].)
Algebraically, to find $c\,\mathbf{a}$ we multiply each component of $\mathbf{a}$ by c.

4. See (1) in Section 13.2 [ET 12.2].

5. See Theorem 13.3.3 [ET 12.3.3] and Definition 13.3.1 [ET 12.3.1].

6. The dot product can be used to find the angle between two vectors and the scalar projection of one vector onto
another. In particular, the dot product can determine if two vectors are orthogonal. Also, the dot product can be used
to determine the work done moving an object given the force and displacement vectors.

7. See the boxed equations on page 847 [ET 811] as well as Figures 4 and 5 and the accompanying discussion on
pages 846–47 [ET 810–11].

8. See Theorem 13.4.6 [ET 12.4.6] and the preceding discussion; use either (1) or (4) in Section 13.4 [ET 12.4].

9. The cross product can be used to create a vector orthogonal to two given vectors as well as to determine if two vectors are parallel. The cross product can also be used to find the area of a parallelogram determined by two vectors. In addition, the cross product can be used to determine torque if the force and position vectors are known.

10. (a) The area of the parallelogram determined by **a** and **b** is the length of the cross product: $|\mathbf{a} \times \mathbf{b}|$.

(b) The volume of the parallelepiped determined by **a**, **b**, and **c** is the magnitude of their scalar triple product: $|\mathbf{a} \cdot (\mathbf{b} \times \mathbf{c})|$.

11. If an equation of the plane is known, it can be written as $ax + by + cz + d = 0$. A normal vector, which is perpendicular to the plane, is $\langle a, b, c \rangle$ (or any scalar multiple of $\langle a, b, c \rangle$). If an equation is not known, we can use points on the plane to find two non-parallel vectors which lie in the plane. The cross product of these vectors is a vector perpendicular to the plane.

12. The angle between two intersecting planes is defined as the acute angle between their normal vectors. We can find this angle using Corollary 13.3.6 [ET 12.3.6].

13. See (1), (2), and (3) in Section 13.5 [ET 12.5].

14. See (5), (6), and (7) in Section 13.5 [ET 12.5].

15. (a) Two (nonzero) vectors are parallel if and only if one is a scalar multiple of the other. In addition, two nonzero vectors are parallel if and only if their cross product is **0**.

(b) Two vectors are perpendicular if and only if their dot product is 0.

(c) Two planes are parallel if and only if their normal vectors are parallel.

16. (a) Determine the vectors $\overrightarrow{PQ} = \langle a_1, a_2, a_3 \rangle$ and $\overrightarrow{PR} = \langle b_1, b_2, b_3 \rangle$. If there is a scalar t such that $\langle a_1, a_2, a_3 \rangle = t \langle b_1, b_2, b_3 \rangle$, then the vectors are parallel and the points must all lie on the same line. Alternatively, if $\overrightarrow{PQ} \times \overrightarrow{PR} = \mathbf{0}$, then $\overrightarrow{PQ}$ and $\overrightarrow{PR}$ are parallel, so P, Q, and R are collinear. Thirdly, an algebraic method is to determine an equation of the line joining two of the points, and then check whether or not the third point satisfies this equation.

(b) Find the vectors $\overrightarrow{PQ} = \mathbf{a}$, $\overrightarrow{PR} = \mathbf{b}$, $\overrightarrow{PS} = \mathbf{c}$. $\mathbf{a} \times \mathbf{b}$ is normal to the plane formed by P, Q and R, and so S lies on this plane if $\mathbf{a} \times \mathbf{b}$ and $\mathbf{c}$ are orthogonal, that is, if $(\mathbf{a} \times \mathbf{b}) \cdot \mathbf{c} = 0$. (Or use the reasoning in Example 5 in Section 13.4 [ET 12.4].)

Alternatively, find an equation for the plane determined by three of the points and check whether or not the fourth point satisfies this equation.

17. (a) See Exercise 13.4.39 [ET 12.4.39].

(b) See Example 8 in Section 13.5 [ET 12.5].

(c) See Example 10 in Section 13.5 [ET 12.5].

18. The traces of a surface are the curves of intersection of the surface with planes parallel to the coordinate planes. We can find the trace in the plane $x = k$ (parallel to the yz-plane) by setting $x = k$ and determining the curve represented by the resulting equation. Traces in the planes $y = k$ (parallel to the xz-plane) and $z = k$ (parallel to the xy-plane) are found similarly.

19. See Table 1 in Section 13.6 [ET 12.6].

20. (a) See (1) and the discussion accompanying Figure 3 in Section 13.7 [ET 12.7].

(b) See (3) and Figures 6–8, and the accompanying discussion, in Section 13.7 [ET 12.7].

———————————————————————— TRUE-FALSE QUIZ ————————————————————————

1. True, by Theorem 13.3.2 [ET 12.3.2] #2.

3. True. If θ is the angle between $\mathbf{u}$ and $\mathbf{v}$, then by Theorem 13.4.6 [ET 12.4.6],
$$|\mathbf{u} \times \mathbf{v}| = |\mathbf{u}|\,|\mathbf{v}| \sin\theta = |\mathbf{v}|\,|\mathbf{u}| \sin\theta = |\mathbf{v} \times \mathbf{u}|.$$
(Or, by Theorem 13.4.8 [ET 12.4.8], $|\mathbf{u} \times \mathbf{v}| = |-\mathbf{v} \times \mathbf{u}| = |-1|\,|\mathbf{v} \times \mathbf{u}| = |\mathbf{v} \times \mathbf{u}|$.)

5. Theorem 13.4.8 [ET 12.4.8] #2 tells us that this is true.

7. This is true by Theorem 13.4.8 [ET 12.4.8] #5.

9. This is true because $\mathbf{u} \times \mathbf{v}$ is orthogonal to $\mathbf{u}$ (see Theorem 13.4.5 [ET 12.4.5]), and the dot product of two orthogonal vectors is 0.

11. If $|\mathbf{u}| = 1$, $|\mathbf{v}| = 1$ and θ is the angle between these two vectors (so $0 \le \theta \le \pi$), then by Theorem 13.4.6 [ET 12.4.6], $|\mathbf{u} \times \mathbf{v}| = |\mathbf{u}|\,|\mathbf{v}| \sin\theta = \sin\theta$, which is equal to 1 if and only if $\theta = \frac{\pi}{2}$ (that is, if and only if the two vectors are orthogonal). Therefore, the assertion that the cross product of two unit vectors is a unit vector is false.

13. This is false. In $\mathbb{R}^2$, $x^2 + y^2 = 1$ represents a circle, but $\left\{(x, y, z) \mid x^2 + y^2 = 1\right\}$ represents a *three-dimensional surface*, namely, a circular cylinder with axis the z-axis.

———————————————————————— EXERCISES ————————————————————————

1. (a) By the formula for an equation of a sphere (see Section 13.1 [ET 12.1]), an equation of the sphere with center $(1, -1, 2)$ and radius 3 is $(x - 1)^2 + (y + 1)^2 + (z - 2)^2 = 9$.

(b) Completing squares gives $(x + 2)^2 + (y + 3)^2 + (z - 5)^2 = -2 + 4 + 9 + 25 = 36$. Thus, the sphere is centered at $(-2, -3, 5)$ and has radius 6.

3. $\mathbf{u} \cdot \mathbf{v} = |\mathbf{u}|\,|\mathbf{v}| \cos 45° = (2)(3)\frac{\sqrt{2}}{2} = 3\sqrt{2}$. $|\mathbf{u} \times \mathbf{v}| = |\mathbf{u}|\,|\mathbf{v}| \sin 45° = (2)(3)\frac{\sqrt{2}}{2} = 3\sqrt{2}$. By the right-hand rule, $\mathbf{u} \times \mathbf{v}$ is directed out of the page.

5. For the two vectors to be orthogonal, we need $\langle 3, 2, x \rangle \cdot \langle 2x, 4, x \rangle = 0$ $\Leftrightarrow$
$(3)(2x) + (2)(4) + (x)(x) = 0$ $\Leftrightarrow$ $x^2 + 6x + 8 = 0$ $\Leftrightarrow$ $(x + 2)(x + 4) = 0$ $\Leftrightarrow$
$x = -2$ or $x = -4$.

7. (a) $(\mathbf{u} \times \mathbf{v}) \cdot \mathbf{w} = \mathbf{u} \cdot (\mathbf{v} \times \mathbf{w}) = 2$

(b) $\mathbf{u} \cdot (\mathbf{w} \times \mathbf{v}) = \mathbf{u} \cdot [-(\mathbf{v} \times \mathbf{w})] = -\mathbf{u} \cdot (\mathbf{v} \times \mathbf{w}) = -2$

(c) $\mathbf{v} \cdot (\mathbf{u} \times \mathbf{w}) = (\mathbf{v} \times \mathbf{u}) \cdot \mathbf{w} = -(\mathbf{u} \times \mathbf{v}) \cdot \mathbf{w} = -2$

(d) $(\mathbf{u} \times \mathbf{v}) \cdot \mathbf{v} = \mathbf{u} \cdot (\mathbf{v} \times \mathbf{v}) = \mathbf{u} \cdot \mathbf{0} = 0$

9. For simplicity, consider a unit cube positioned with its back left corner at the origin. Vector representations of the diagonals joining the points $(0, 0, 0)$ to $(1, 1, 1)$ and $(1, 0, 0)$ to $(0, 1, 1)$ are $\langle 1, 1, 1 \rangle$ and $\langle -1, 1, 1 \rangle$. Let θ be the angle between these two vectors. $\langle 1, 1, 1 \rangle \cdot \langle -1, 1, 1 \rangle = -1 + 1 + 1 = 1 = |\langle 1, 1, 1 \rangle|\,|\langle -1, 1, 1 \rangle| \cos\theta = 3 \cos\theta$
$\Rightarrow$ $\cos\theta = \frac{1}{3}$ $\Rightarrow$ $\theta = \cos^{-1}\left(\frac{1}{3}\right) \approx 71°$.

11. $\overrightarrow{AB} = \langle 1, 0, -1 \rangle$, $\overrightarrow{AC} = \langle 0, 4, 3 \rangle$, so

(a) a vector perpendicular to the plane is $\overrightarrow{AB} \times \overrightarrow{AC} = \langle 0 + 4, -(3 + 0), 4 - 0 \rangle = \langle 4, -3, 4 \rangle$.

(b) $\frac{1}{2}\left|\overrightarrow{AB} \times \overrightarrow{AC}\right| = \frac{1}{2}\sqrt{16 + 9 + 16} = \frac{\sqrt{41}}{2}$.

13. Let F_1 be the magnitude of the force directed $20°$ away from the direction of shore, and let F_2 be the magnitude of the other force. Separating these forces into components parallel to the direction of the resultant force and perpendicular to it gives $F_1 \cos 20° + F_2 \cos 30° = 255$ (1), and $F_1 \sin 20° - F_2 \sin 30° = 0$ $\Rightarrow$

$F_1 = F_2 \dfrac{\sin 30°}{\sin 20°}$ (2). Substituting (2) into (1) gives $F_2(\sin 30° \cot 20° + \cos 30°) = 255$ $\Rightarrow$ $F_2 \approx 114$ N.

Substituting this into (2) gives $F_1 \approx 166$ N.

15. The line has direction $\mathbf{v} = \langle -3, 2, 3 \rangle$. Letting $P_0 = (4, -1, 2)$, parametric equations are $x = 4 - 3t$, $y = -1 + 2t$, $z = 2 + 3t$.

17. A direction vector for the line is a normal vector for the plane, $\mathbf{n} = \langle 2, -1, 5 \rangle$, and parametric equations for the line are $x = -2 + 2t$, $y = 2 - t$, $z = 4 + 5t$.

19. Here the vectors $\mathbf{a} = \langle 4 - 3, 0 - (-1), 2 - 1 \rangle = \langle 1, 1, 1 \rangle$ and $\mathbf{b} = \langle 6 - 3, 3 - (-1), 1 - 1 \rangle = \langle 3, 4, 0 \rangle$ lie in the plane, so $\mathbf{n} = \mathbf{a} \times \mathbf{b} = \langle -4, 3, 1 \rangle$ is a normal vector to the plane and an equation of the plane is $-4(x - 3) + 3(y - (-1)) + 1(z - 1) = 0$ or $-4x + 3y + z = -14$.

21. Substitution of the parametric equations into the equation of the plane gives $2x - y + z = 2(2 - t) - (1 + 3t) + 4t = 2$ $\Rightarrow$ $-t + 3 = 2$ $\Rightarrow$ $t = 1$. When $t = 1$, the parametric equations give $x = 2 - 1 = 1$, $y = 1 + 3 = 4$ and $z = 4$. Therefore, the point of intersection is $(1, 4, 4)$.

23. Since the direction vectors $\langle 2, 3, 4 \rangle$ and $\langle 6, -1, 2 \rangle$ aren't parallel, neither are the lines. For the lines to intersect, the three equations $1 + 2t = -1 + 6s$, $2 + 3t = 3 - s$, $3 + 4t = -5 + 2s$ must be satisfied simultaneously. Solving the first two equations gives $t = \frac{1}{5}$, $s = \frac{2}{5}$ and checking we see these values don't satisfy the third equation. Thus the lines aren't parallel and they don't intersect, so they must be skew.

25. By Exercise 13.5.69 [ET 12.5.69], $D = \dfrac{|2 - 24|}{\sqrt{26}} = \dfrac{22}{\sqrt{26}}$.

27. The equation $x = z$ represents a plane perpendicular to the xz-plane and intersecting the xz-plane in the line $x = z$, $y = 0$.

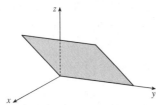

29. A (right elliptical) cone with vertex at the origin and axis the x-axis.

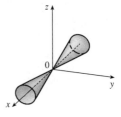

31. An equivalent equation is $-x^2 + \dfrac{y^2}{4} - z^2 = 1$, a hyperboloid of two sheets with axis the y-axis. For $|y| > 2$, traces parallel to the xz-plane are circles.

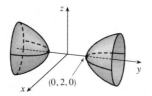

33. Completing the square in y gives
$$4x^2 + 4(y-1)^2 + z^2 = 4 \text{ or}$$
$$x^2 + (y-1)^2 + \frac{z^2}{4} = 1, \text{ an ellipsoid centered}$$
at $(0, 1, 0)$.

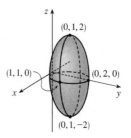

35. $4x^2 + y^2 = 16 \iff \dfrac{x^2}{4} + \dfrac{y^2}{16} = 1$. The equation of the ellipsoid is $\dfrac{x^2}{4} + \dfrac{y^2}{16} + \dfrac{z^2}{c^2} = 1$, since the horizontal trace in the plane $z = 0$ must be the original ellipse. The traces of the ellipsoid in the yz-plane must be circles since the surface is obtained by rotation about the x-axis. Therefore, $c^2 = 16$ and the equation of the ellipsoid is

$$\frac{x^2}{4} + \frac{y^2}{16} + \frac{z^2}{16} = 1 \iff 4x^2 + y^2 + z^2 = 16.$$

37. $x = r\cos\theta = 2\sqrt{3}\cos\frac{\pi}{3} = 2\sqrt{3} \cdot \frac{1}{2} = \sqrt{3}$, $y = r\sin\theta = 2\sqrt{3}\sin\frac{\pi}{3} = 2\sqrt{3} \cdot \frac{\sqrt{3}}{2} = 3$, $z = 2$, so in rectangular coordinates the point is $(\sqrt{3}, 3, 2)$. $\rho = \sqrt{r^2 + z^2} = \sqrt{12 + 4} = 4$, $\theta = \frac{\pi}{3}$, and $\cos\phi = \frac{z}{\rho} = \frac{1}{2}$, so $\phi = \frac{\pi}{3}$ and spherical coordinates are $\left(4, \frac{\pi}{3}, \frac{\pi}{3}\right)$.

39. $x = \rho\sin\phi\cos\theta = 8\sin\frac{\pi}{6}\cos\frac{\pi}{4} = 8 \cdot \frac{1}{2} \cdot \frac{\sqrt{2}}{2} = 2\sqrt{2}$, $y = \rho\sin\phi\sin\theta = 8\sin\frac{\pi}{6}\sin\frac{\pi}{4} = 2\sqrt{2}$, and $z = \rho\cos\phi = 8\cos\frac{\pi}{6} = 8 \cdot \frac{\sqrt{3}}{2} = 4\sqrt{3}$. Thus rectangular coordinates for the point are $\left(2\sqrt{2}, 2\sqrt{2}, 4\sqrt{3}\right)$. $r^2 = x^2 + y^2 = 8 + 8 = 16 \Rightarrow r = 4$, $\theta = \frac{\pi}{4}$, and $z = 4\sqrt{3}$, so cylindrical coordinates are $\left(4, \frac{\pi}{4}, 4\sqrt{3}\right)$.

41. $\theta = \frac{\pi}{4}$. In spherical coordinates, this is a half-plane including the z-axis and intersecting the xy-plane in the half-line $x = y$, $x > 0$.

43. Since $\rho = 3\sec\phi$, $\rho\cos\phi = 3$ or $z = 3$. Thus the surface is a plane parallel to the xy-plane and through the point $(0, 0, 3)$.

45. $x^2 + y^2 + z^2 = 4$. In cylindrical coordinates, this becomes $r^2 + z^2 = 4$. In spherical coordinates, it becomes $\rho^2 = 4$ or $\rho = 2$.

47. The resulting surface is a circular paraboloid with equation $z = 4x^2 + 4y^2$. Changing to cylindrical coordinates we have $z = 4(x^2 + y^2) = 4r^2$.

☐ PROBLEMS PLUS

1. Since three-dimensional situations are often difficult to visualize and work with, let us first try to find an analogous problem in two dimensions. The analogue of a cube is a square and the analogue of a sphere is a circle. Thus a similar problem in two dimensions is the following: if five circles with the same radius r are contained in a square of side 1 m so that the circles touch each other and four of the circles touch two sides of the square, find r.

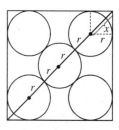

The diagonal of the square is $\sqrt{2}$. The diagonal is also $4r + 2x$. But x is the diagonal of a smaller square of side r.

Therefore $x = \sqrt{2}\,r \;\Rightarrow\; \sqrt{2} = 4r + 2x = 4r + 2\sqrt{2}\,r = (4 + 2\sqrt{2}\,)r \;\Rightarrow\; r = \frac{\sqrt{2}}{4 + 2\sqrt{2}}$.

Let us use these ideas to solve the original three-dimensional problem. The diagonal of the cube is $\sqrt{1^2 + 1^2 + 1^2} = \sqrt{3}$. The diagonal of the cube is also $4r + 2x$ where x is the diagonal of a smaller cube with edge r. Therefore $x = \sqrt{r^2 + r^2 + r^2} = \sqrt{3}\,r \;\Rightarrow\; \sqrt{3} = 4r + 2x = 4r + 2\sqrt{3}\,r = (4 + 2\sqrt{3}\,)r$. Thus $r = \dfrac{\sqrt{3}}{4 + 2\sqrt{3}} = \dfrac{2\sqrt{3} - 3}{2}$. The radius of each ball is $\left(\sqrt{3} - \frac{3}{2}\right)$ m.

3. (a) We find the line of intersection L as in Example 13.5.7(b) [ET 12.5.7(b)]. Observe that the point $(-1, c, c)$ lies on both planes. Now since L lies in both planes, it is perpendicular to both of the normal vectors $\mathbf{n}_1$ and $\mathbf{n}_2$, and thus parallel to their cross product $\mathbf{n}_1 \times \mathbf{n}_2 = \begin{vmatrix} \mathbf{i} & \mathbf{j} & \mathbf{k} \\ c & 1 & 1 \\ 1 & -c & c \end{vmatrix} = \langle 2c, -c^2 + 1, -c^2 - 1 \rangle$. So symmetric equations of L can be written as $\dfrac{x + 1}{-2c} = \dfrac{y - c}{c^2 - 1} = \dfrac{z - c}{c^2 + 1}$, provided that $c \neq 0, \pm 1$.

If $c = 0$, then the two planes are given by $y + z = 0$ and $x = -1$, so symmetric equations of L are $x = -1$, $y = -z$. If $c = -1$, then the two planes are given by $-x + y + z = -1$ and $x + y + z = -1$, and they intersect in the line $x = 0$, $y = -z - 1$. If $c = 1$, then the two planes are given by $x + y + z = 1$ and $x - y + z = 1$, and they intersect in the line $y = 0$, $x = 1 - z$.

(b) If we set $z = t$ in the symmetric equations and solve for x and y separately, we get $x + 1 = \dfrac{(t - c)(-2c)}{c^2 + 1}$, $y - c = \dfrac{(t - c)(c^2 - 1)}{c^2 + 1} \;\Rightarrow\; x = \dfrac{-2ct + (c^2 - 1)}{c^2 + 1}$, $y = \dfrac{(c^2 - 1)t + 2c}{c^2 + 1}$. Eliminating c from these equations, we have $x^2 + y^2 = t^2 + 1$. So the curve traced out by L in the plane $z = t$ is a circle with center at $(0, 0, t)$ and radius $\sqrt{t^2 + 1}$.

(c) The area of a horizontal cross-section of the solid is $A(z) = \pi(z^2 + 1)$, so $V = \int_0^1 A(z)\,dz = \pi\left[\frac{1}{3}z^3 + z\right]_0^1 = \frac{4\pi}{3}$.

5. (a) When $\theta = \theta_s$, the block is not moving, so the sum of the forces on the

block must be **0**, thus $\mathbf{N} + \mathbf{F} + \mathbf{W} = \mathbf{0}$. This relationship is

illustrated geometrically in the figure. Since the vectors form a right

triangle, we have $\tan(\theta_s) = \dfrac{|\mathbf{F}|}{|\mathbf{N}|} = \dfrac{\mu_s n}{n} = \mu_s$.

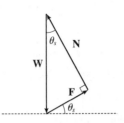

(b) We place the block at the origin and sketch the force vectors acting on the block, including the additional

horizontal force **H**, with initial points at the origin. We then rotate this system so that **F** lies along the positive

x-axis and the inclined plane is parallel to the x-axis.

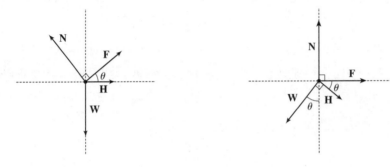

$|\mathbf{F}|$ is maximal, so $|\mathbf{F}| = \mu_s n$ for $\theta > \theta_s$. Then the vectors, in terms of components parallel and perpendicular

to the inclined plane, are

$$\mathbf{N} = n\,\mathbf{j} \qquad\qquad \mathbf{F} = (\mu_s n)\,\mathbf{i}$$

$$\mathbf{W} = (-mg\sin\theta)\,\mathbf{i} + (-mg\cos\theta)\,\mathbf{j}$$

$$\mathbf{H} = (h_{\min}\cos\theta)\,\mathbf{i} + (-h_{\min}\sin\theta)\,\mathbf{j}$$

Equating components, we have

$$\mu_s n - mg\sin\theta + h_{\min}\cos\theta = 0 \quad\Rightarrow\quad h_{\min}\cos\theta + \mu_s n = mg\sin\theta \tag{1}$$

$$n - mg\cos\theta - h_{\min}\sin\theta = 0 \quad\Rightarrow\quad h_{\min}\sin\theta + mg\cos\theta = n \tag{2}$$

(c) Since (2) is solved for n, we substitute into (1):

$$h_{\min}\cos\theta + \mu_s(h_{\min}\sin\theta + mg\cos\theta) = mg\sin\theta \quad\Rightarrow$$

$$h_{\min}\cos\theta + h_{\min}\mu_s\sin\theta = mg\sin\theta - mg\mu_s\cos\theta \quad\Rightarrow$$

$$h_{\min} = mg\left(\frac{\sin\theta - \mu_s\cos\theta}{\cos\theta + \mu_s\sin\theta}\right) = mg\left(\frac{\tan\theta - \mu_s}{1 + \mu_s\tan\theta}\right)$$

From part (a) we know $\mu_s = \tan\theta_s$, so this becomes $h_{\min} = mg\left(\dfrac{\tan\theta - \tan\theta_s}{1 + \tan\theta_s\tan\theta}\right)$ and using a

trigonometric identity, this is $mg\tan(\theta - \theta_s)$ as desired.

Note for $\theta = \theta_s$, $h_{min} = mg \tan 0 = 0$, which makes sense since the block is at rest for θ_s, thus no additional force $\mathbf{H}$ is necessary to prevent it from moving. As θ increases, the factor $\tan(\theta - \theta_s)$, and hence the value of h_{min}, increases slowly for small values of $\theta - \theta_s$ but much more rapidly as $\theta - \theta_s$ becomes significant. This seems reasonable, as the steeper the inclined plane, the less the horizontal components of the various forces affect the movement of the block, so we would need a much larger magnitude of horizontal force to keep the block motionless. If we allow $\theta \to 90\,^\circ$, corresponding to the inclined plane being placed vertically, the value of h_{min} is quite large; this is to be expected, as it takes a great amount of horizontal force to keep an object from moving vertically. In fact, without friction (so $\theta_s = 0$), we would have $\theta \to 90\,^\circ \;\Rightarrow\; h_{min} \to \infty$, and it would be impossible to keep the block from slipping.

(d) Since h_{max} is the largest value of h that keeps the block from slipping, the force of friction is keeping the block from moving *up* the inclined plane; thus, $\mathbf{F}$ is directed *down* the plane. Our system of forces is similar to that in part (b), then, except that we have $\mathbf{F} = -(\mu_s n)\,\mathbf{i}$. (Note that $|\mathbf{F}|$ is again maximal.) Following our procedure in parts (b) and (c), we equate components:

$$-\mu_s n - mg \sin \theta + h_{max} \cos \theta = 0 \quad\Rightarrow\quad h_{max} \cos \theta - \mu_s n = mg \sin \theta$$

$$n - mg \cos \theta - h_{max} \sin \theta = 0 \quad\Rightarrow\quad h_{max} \sin \theta + mg \cos \theta = n$$

Then substituting,

$$h_{max} \cos \theta - \mu_s(h_{max} \sin \theta + mg \cos \theta) = mg \sin \theta \quad\Rightarrow$$

$$h_{max} \cos \theta - h_{max}\mu_s \sin \theta = mg \sin \theta + mg\mu_s \cos \theta \quad\Rightarrow$$

$$h_{max} = mg\left(\frac{\sin \theta + \mu_s \cos \theta}{\cos \theta - \mu_s \sin \theta}\right) = mg\left(\frac{\tan \theta + \mu_s}{1 - \mu_s \tan \theta}\right)$$

$$= mg\left(\frac{\tan \theta + \tan \theta_s}{1 - \tan \theta_s \tan \theta}\right) = mg \tan(\theta + \theta_s)$$

We would expect h_{max} to increase as θ increases, with similar behavior as we established for h_{min}, but with h_{max} values always larger than h_{min}. We can see that this is the case if we graph h_{max} as a function of θ, as the curve is the graph of h_{min} translated $2\theta_s$ to the left, so the equation does seem reasonable. Notice that the equation predicts $h_{max} \to \infty$ as $\theta \to (90\,^\circ - \theta_s)$. In fact, as h_{max} increases, the normal force increases as well. When $(90\,^\circ - \theta_s) \le \theta \le 90\,^\circ$, the horizontal force is completely counteracted by the sum of the normal and frictional forces, so no part of the horizontal force contributes to moving the block up the plane no matter how large its magnitude.

14.1 Vector Functions and Space Curves

1. The component functions t^2, $\sqrt{t-1}$, and $\sqrt{5-t}$ are all defined when $t - 1 \geq 0 \;\Rightarrow\; t \geq 1$ and $5 - t \geq 0 \;\Rightarrow\; t \leq 5$, so the domain of $\mathbf{r}(t)$ is $[1, 5]$.

3. $\displaystyle\lim_{t\to0^+} \cos t = \cos 0 = 1$, $\displaystyle\lim_{t\to0^+} \sin t = \sin 0 = 0$, $\displaystyle\lim_{t\to0^+} t \ln t = \lim_{t\to0^+} \frac{\ln t}{1/t} = \lim_{t\to0^+} \frac{1/t}{-1/t^2} = \lim_{t\to0^+} -t = 0$

[by l'Hospital's Rule]. Thus $\displaystyle\lim_{t\to0^+} \langle \cos t, \sin t, t \ln t \rangle = \left\langle \lim_{t\to0^+} \cos t, \lim_{t\to0^+} \sin t, \lim_{t\to0^+} t \ln t \right\rangle = \langle 1, 0, 0 \rangle$.

5. $\displaystyle\lim_{t\to1} \sqrt{t+3} = 2$, $\displaystyle\lim_{t\to1} \frac{t-1}{t^2-1} = \lim_{t\to1} \frac{1}{t+1} = \frac{1}{2}$, $\displaystyle\lim_{t\to1} \left(\frac{\tan t}{t}\right) = \tan 1$.

Thus the given limit equals $\left\langle 2, \frac{1}{2}, \tan 1 \right\rangle$.

7. The corresponding parametric equations for this curve are

$x = t^4 + 1$, $y = t$. We can make a table of values, or we can

eliminate the parameter: $t = y \;\Rightarrow\; x = y^4 + 1$, with $y \in \mathbb{R}$.

By comparing different values of t, we find the direction in which

t increases as indicated in the graph.

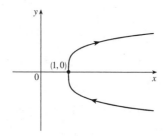

9. The corresponding parametric equations are $x = t$, $y = \cos 2t$,

$z = \sin 2t$. Note that $y^2 + z^2 = \cos^2 2t + \sin^2 2t = 1$, so the

curve lies on the circular cylinder $y^2 + z^2 = 1$. Since $x = t$, the

curve is a helix.

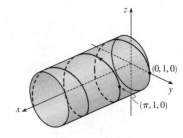

11. The parametric equations give $x^2 + z^2 = \sin^2 t + \cos^2 t = 1$, $y = 3$,

which is a circle of radius 1, center $(0, 3, 0)$ in the plane $y = 3$.

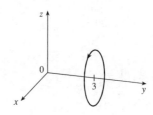

13. The parametric equations are $x = t^2$, $y = t^4$, $z = t^6$. These are
positive for $t \neq 0$ and 0 when $t = 0$. So the curve lies entirely in the
first quadrant. The projection of the graph onto the xy-plane is
$y = x^2$, $y > 0$, a half parabola. On the xz-plane $z = x^3$, $z > 0$, a
half cubic, and the yz-plane, $y^3 = z^2$.

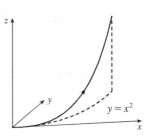

15. Taking $\mathbf{r}_0 = \langle 0, 0, 0 \rangle$ and $\mathbf{r}_1 = \langle 1, 2, 3 \rangle$, we have from Equation 13.5.4 [ET 12.5.4]
$\mathbf{r}(t) = (1 - t)\,\mathbf{r}_0 + t\,\mathbf{r}_1 = (1 - t)\,\langle 0, 0, 0 \rangle + t\,\langle 1, 2, 3 \rangle$, $0 \leq t \leq 1$ or $\mathbf{r}(t) = \langle t, 2t, 3t \rangle$, $0 \leq t \leq 1$.
Parametric equations are $x = t$, $y = 2t$, $z = 3t$, $0 \leq t \leq 1$.

17. Taking $\mathbf{r}_0 = \langle 1, -1, 2 \rangle$ and $\mathbf{r}_1 = \langle 4, 1, 7 \rangle$, we have $\mathbf{r}(t) = (1 - t)\,\mathbf{r}_0 + t\,\mathbf{r}_1 = (1 - t)\,\langle 1, -1, 2 \rangle + t\,\langle 4, 1, 7 \rangle$,
$0 \leq t \leq 1$ or $\mathbf{r}(t) = \langle 1 + 3t, -1 + 2t, 2 + 5t \rangle$, $0 \leq t \leq 1$. Parametric equations are $x = 1 + 3t$, $y = -1 + 2t$,
$z = 2 + 5t$, $0 \leq t \leq 1$.

19. $x = \cos 4t$, $y = t$, $z = \sin 4t$. At any point (x, y, z) on the curve, $x^2 + z^2 = \cos^2 4t + \sin^2 4t = 1$. So the curve
lies on a circular cylinder with axis the y-axis. Since $y = t$, this is a helix. So the graph is VI.

21. $x = t$, $y = 1/(1 + t^2)$, $z = t^2$. Note that y and z are positive for all t. The curve passes through $(0, 1, 0)$ when
$t = 0$. As $t \to \infty$, $(x, y, z) \to (\infty, 0, \infty)$, and as $t \to -\infty$, $(x, y, z) \to (-\infty, 0, \infty)$. So the graph is IV.

23. $x = \cos t$, $y = \sin t$, $z = \sin 5t$. $x^2 + y^2 = \cos^2 t + \sin^2 t = 1$, so the curve lies on a circular cylinder with axis
the z-axis. Each of x, y and z is periodic, and at $t = 0$ and $t = 2\pi$ the curve passes through the same point, so the
curve repeats itself and the graph is V.

25. If $x = t \cos t$, $y = t \sin t$, and $z = t$, then
$x^2 + y^2 = t^2 \cos^2 t + t^2 \sin^2 t = t^2 = z^2$, so the curve lies on the
cone $z^2 = x^2 + y^2$. Since $z = t$, the curve is a spiral on this cone.

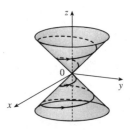

27. $\mathbf{r}(t) = \langle \sin t, \cos t, t^2 \rangle$

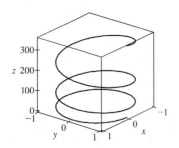

29. $\mathbf{r}(t) = \langle t^2, \sqrt{t - 1}, \sqrt{5 - t} \,\rangle$

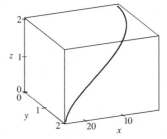

31.

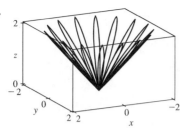

$x = (1 + \cos 16t) \cos t$, $y = (1 + \cos 16t) \sin t$, $z = 1 + \cos 16t$.

At any point on the graph,

$$x^2 + y^2 = (1 + \cos 16t)^2 \cos^2 t + (1 + \cos 16t)^2 \sin^2 t$$
$$= (1 + \cos 16t)^2 = z^2, \text{ so the graph lies on the cone}$$

$x^2 + y^2 = z^2$. From the graph at left, we see that this curve looks

like the projection of a leaved two-dimensional curve onto a cone.

33. If $t = -1$, then $x = 1, y = 4, z = 0$, so the curve passes through the point $(1, 4, 0)$. If $t = 3$, then

$x = 9, y = -8, z = 28$, so the curve passes through the point $(9, -8, 28)$. For the point $(4, 7, -6)$ to be on the

curve, we require $y = 1 - 3t = 7 \implies t = -2$. But then $z = 1 + (-2)^3 = -7 \neq -6$, so $(4, 7, -6)$ is not on the

curve.

35. Both equations are solved for z, so we can substitute to eliminate z: $\sqrt{x^2 + y^2} = 1 + y \implies$

$x^2 + y^2 = 1 + 2y + y^2 \implies x^2 = 1 + 2y \implies y = \frac{1}{2}(x^2 - 1)$. We can form parametric equations for the curve

C of intersection by choosing a parameter $x = t$, then $y = \frac{1}{2}(t^2 - 1)$ and $z = 1 + y = 1 + \frac{1}{2}(t^2 - 1) = \frac{1}{2}(t^2 + 1)$.

Thus a vector function representing C is $\mathbf{r}(t) = t \mathbf{i} + \frac{1}{2}(t^2 - 1) \mathbf{j} + \frac{1}{2}(t^2 + 1) \mathbf{k}$.

37.

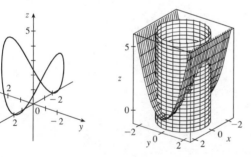

The projection of the curve C of intersection onto

the xy-plane is the circle $x^2 + y^2 = 4, z = 0$.

Then we can write $x = 2 \cos t, y = 2 \sin t$,

$0 \leq t \leq 2\pi$. Since C also lies on the surface

$z = x^2$, we have $z = x^2 = (2 \cos t)^2 = 4 \cos^2 t$.

Then parametric equations for C are $x = 2 \cos t$,

$y = 2 \sin t, z = 4 \cos^2 t, 0 \leq t \leq 2\pi$.

39. For the particles to collide, we require $\mathbf{r}_1(t) = \mathbf{r}_2(t) \iff \langle t^2, 7t - 12, t^2 \rangle = \langle 4t - 3, t^2, 5t - 6 \rangle$. Equating

components gives $t^2 = 4t - 3$, $7t - 12 = t^2$, and $t^2 = 5t - 6$. From the first equation, $t^2 - 4t + 3 = 0 \iff$

$(t - 3)(t - 1) = 0$ so $t = 1$ or $t = 3$. $t = 1$ does not satisfy the other two equations, but $t = 3$ does. The particles

collide when $t = 3$, at the point $(9, 9, 9)$.

41. Let $\mathbf{u}(t) = \langle u_1(t), u_2(t), u_3(t) \rangle$ and $\mathbf{v}(t) = \langle v_1(t), v_2(t), v_3(t) \rangle$. In each part of this problem the basic procedure

is to use Equation 1 and then analyze the individual component functions using the limit properties we have already

developed for real-valued functions.

(a) $\lim\limits_{t \to a} \mathbf{u}(t) + \lim\limits_{t \to a} \mathbf{v}(t) = \left\langle \lim\limits_{t \to a} u_1(t), \lim\limits_{t \to a} u_2(t), \lim\limits_{t \to a} u_3(t) \right\rangle + \left\langle \lim\limits_{t \to a} v_1(t), \lim\limits_{t \to a} v_2(t), \lim\limits_{t \to a} v_3(t) \right\rangle$ and the limits

of these component functions must each exist since the vector functions both possess limits as $t \to a$. Then adding the two vectors and using the addition property of limits for real-valued functions, we have that

$$\lim_{t \to a} \mathbf{u}(t) + \lim_{t \to a} \mathbf{v}(t) = \left\langle \lim_{t \to a} u_1(t) + \lim_{t \to a} v_1(t), \lim_{t \to a} u_2(t) + \lim_{t \to a} v_2(t), \lim_{t \to a} u_3(t) + \lim_{t \to a} v_3(t) \right\rangle$$

$$= \left\langle \lim_{t \to a} [u_1(t) + v_1(t)], \lim_{t \to a} [u_2(t) + v_2(t)], \lim_{t \to a} [u_3(t) + v_3(t)] \right\rangle$$

$$= \lim_{t \to a} \langle u_1(t) + v_1(t), u_2(t) + v_2(t), u_3(t) + v_3(t) \rangle \quad \text{[using (1) backward]}$$

$$= \lim_{t \to a} [\mathbf{u}(t) + \mathbf{v}(t)]$$

(b) $\displaystyle \lim_{t \to a} c\mathbf{u}(t) = \lim_{t \to a} \langle cu_1(t), cu_2(t), cu_3(t) \rangle = \left\langle \lim_{t \to a} cu_1(t), \lim_{t \to a} cu_2(t), \lim_{t \to a} cu_3(t) \right\rangle$

$$= \left\langle c \lim_{t \to a} u_1(t), c \lim_{t \to a} u_2(t), c \lim_{t \to a} u_3(t) \right\rangle = c \left\langle \lim_{t \to a} u_1(t), \lim_{t \to a} u_2(t), \lim_{t \to a} u_3(t) \right\rangle$$

$$= c \lim_{t \to a} \langle u_1(t), u_2(t), u_3(t) \rangle = c \lim_{t \to a} \mathbf{u}(t)$$

(c) $\displaystyle \lim_{t \to a} \mathbf{u}(t) \cdot \lim_{t \to a} \mathbf{v}(t) = \left\langle \lim_{t \to a} u_1(t), \lim_{t \to a} u_2(t), \lim_{t \to a} u_3(t) \right\rangle \cdot \left\langle \lim_{t \to a} v_1(t), \lim_{t \to a} v_2(t), \lim_{t \to a} v_3(t) \right\rangle$

$$= \left[\lim_{t \to a} u_1(t) \right] \left[\lim_{t \to a} v_1(t) \right] + \left[\lim_{t \to a} u_2(t) \right] \left[\lim_{t \to a} v_2(t) \right] + \left[\lim_{t \to a} u_3(t) \right] \left[\lim_{t \to a} v_3(t) \right]$$

$$= \lim_{t \to a} u_1(t)v_1(t) + \lim_{t \to a} u_2(t)v_2(t) + \lim_{t \to a} u_3(t)v_3(t)$$

$$= \lim_{t \to a} [u_1(t)v_1(t) + u_2(t)v_2(t) + u_3(t)v_3(t)] = \lim_{t \to a} [\mathbf{u}(t) \cdot \mathbf{v}(t)]$$

(d) $\displaystyle \lim_{t \to a} \mathbf{u}(t) \times \lim_{t \to a} \mathbf{v}(t) = \left\langle \lim_{t \to a} u_1(t), \lim_{t \to a} u_2(t), \lim_{t \to a} u_3(t) \right\rangle \times \left\langle \lim_{t \to a} v_1(t), \lim_{t \to a} v_2(t), \lim_{t \to a} v_3(t) \right\rangle$

$$= \left\langle \left[\lim_{t \to a} u_2(t) \right] \left[\lim_{t \to a} v_3(t) \right] - \left[\lim_{t \to a} u_3(t) \right] \left[\lim_{t \to a} v_2(t) \right], \right.$$

$$\left[\lim_{t \to a} u_3(t) \right] \left[\lim_{t \to a} v_1(t) \right] - \left[\lim_{t \to a} u_1(t) \right] \left[\lim_{t \to a} v_3(t) \right],$$

$$\left. \left[\lim_{t \to a} u_1(t) \right] \left[\lim_{t \to a} v_2(t) \right] - \left[\lim_{t \to a} u_2(t) \right] \left[\lim_{t \to a} v_1(t) \right] \right\rangle$$

$$= \left\langle \lim_{t \to a} [u_2(t)v_3(t) - u_3(t)v_2(t)], \lim_{t \to a} [u_3(t)v_1(t) - u_1(t)v_3(t)], \right.$$

$$\left. \lim_{t \to a} [u_1(t)v_2(t) - u_2(t)v_1(t)] \right\rangle$$

$$= \lim_{t \to a} \langle u_2(t)v_3(t) - u_3(t)v_2(t), u_3(t)\,v_1(t) - u_1(t)v_3(t),$$

$$u_1(t)v_2(t) - u_2(t)v_1(t) \rangle$$

$$= \lim_{t \to a} [\mathbf{u}(t) \times \mathbf{v}(t)]$$

43. Let $\mathbf{r}(t) = \langle f(t), g(t), h(t) \rangle$ and $\mathbf{b} = \langle b_1, b_2, b_3 \rangle$. If $\displaystyle\lim_{t \to a} \mathbf{r}(t) = \mathbf{b}$, then $\displaystyle\lim_{t \to a} \mathbf{r}(t)$ exists, so by (1),

$\mathbf{b} = \displaystyle\lim_{t \to a} \mathbf{r}(t) = \left\langle \lim_{t \to a} f(t), \lim_{t \to a} g(t), \lim_{t \to a} h(t) \right\rangle$. By the definition of equal vectors we have $\displaystyle\lim_{t \to a} f(t) = b_1$,

$\displaystyle\lim_{t \to a} g(t) = b_2$ and $\displaystyle\lim_{t \to a} h(t) = b_3$. But these are limits of real-valued functions, so by the definition of limits, for every $\varepsilon > 0$ there exists $\delta_1 > 0$, $\delta_2 > 0$, $\delta_3 > 0$ so $|f(t) - b_1| < \varepsilon/3$ whenever $0 < |t - a| < \delta_1$, $|g(t) - b_2| < \varepsilon/3$ whenever $0 < |t - a| < \delta_2$, and $|h(t) - b_3| < \varepsilon/3$ whenever $0 < |t - a| < \delta_3$. Letting $\delta = $ minimum of $\{\delta_1, \delta_2, \delta_3\}$, we have $|f(t) - b_1| + |g(t) - b_2| + |h(t) - b_3| < \varepsilon/3 + \varepsilon/3 + \varepsilon/3 = \varepsilon$ whenever $0 < |t - a| < \delta$. But $|\mathbf{r}(t) - \mathbf{b}| = |\langle f(t) - b_1, g(t) - b_2, h(t) - b_3\rangle|$

$= \sqrt{(f(t) - b_1)^2 + (g(t) - b_2)^2 + (h(t) - b_3)^2} \le \sqrt{[f(t) - b_1]^2} + \sqrt{[g(t) - b_2]^2} + \sqrt{[h(t) - b_3]^2}$

$= |f(t) - b_1| + |g(t) - b_2| + |h(t) - b_3|$. Thus for every $\varepsilon > 0$ there exists $\delta > 0$ such that

$|\mathbf{r}(t) - \mathbf{b}| \le |f(t) - b_1| + |g(t) - b_2| + |h(t) - b_3| < \varepsilon$ whenever $0 < |t - a| < \delta$. Conversely, if for every

$\varepsilon > 0$, there exists $\delta > 0$ such that $|\mathbf{r}(t) - \mathbf{b}| < \varepsilon$ whenever $0 < |t - a| < \delta$, then

$|\langle f(t) - b_1, g(t) - b_2, h(t) - b_3\rangle| < \varepsilon$ $\Leftrightarrow$ $\sqrt{[f(t) - b_1]^2 + [g(t) - b_2]^2 + [h(t) - b_3]^2} < \varepsilon$ $\Leftrightarrow$

$[f(t) - b_1]^2 + [g(t) - b_2]^2 + [h(t) - b_3]^2 < \varepsilon^2$ whenever $0 < |t - a| < \delta$. But each term on the left side of this

inequality is positive so $[f(t) - b_1]^2 < \varepsilon^2$, $[g(t) - b_2]^2 < \varepsilon^2$ and $[h(t) - b_3]^2 < \varepsilon^2$ whenever $0 < |t - a| < \delta$, or

taking the square root of both sides in each of the above we have $|f(t) - b_1| < \varepsilon$, $|g(t) - b_2| < \varepsilon$ and

$|h(t) - b_3| < \varepsilon$ whenever $0 < |t - a| < \delta$. And by definition of limits of real-valued functions we have

$\lim\limits_{t \to a} f(t) = b_1$, $\lim\limits_{t \to a} g(t) = b_2$ and $\lim\limits_{t \to a} h(t) = b_3$. But by (1), $\lim\limits_{t \to a} \mathbf{r}(t) = \left\langle \lim\limits_{t \to a} f(t), \lim\limits_{t \to a} g(t), \lim\limits_{t \to a} h(t)\right\rangle$,

so $\lim\limits_{t \to a} \mathbf{r}(t) = \langle b_1, b_2, b_3\rangle = \mathbf{b}$.

14.2 Derivatives and Integrals of Vector Functions   ET 13.2

1. (a)

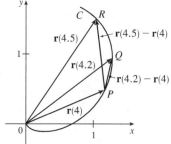

(b) $\dfrac{\mathbf{r}(4.5) - \mathbf{r}(4)}{0.5} = 2[\mathbf{r}(4.5) - \mathbf{r}(4)]$, so we draw a vector in the

same direction but with twice the length of the vector

$\mathbf{r}(4.5) - \mathbf{r}(4)$. $\dfrac{\mathbf{r}(4.2) - \mathbf{r}(4)}{0.2} = 5[\mathbf{r}(4.2) - \mathbf{r}(4)]$, so we draw a

vector in the same direction but with 5 times the length of the

vector $\mathbf{r}(4.2) - \mathbf{r}(4)$.

(c) By Definition 1, $\mathbf{r}'(4) = \lim\limits_{h \to 0} \dfrac{\mathbf{r}(4 + h) - \mathbf{r}(4)}{h}$.

$\mathbf{T}(4) = \dfrac{\mathbf{r}'(4)}{|\mathbf{r}'(4)|}$.

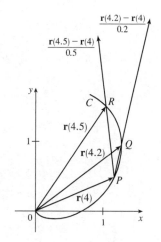

(d) $\mathbf{T}(4)$ is a unit vector in the same direction as $\mathbf{r}'(4)$, that is, parallel to the tangent line to the curve at $\mathbf{r}(4)$ with length 1.

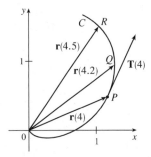

3. (a), (c)

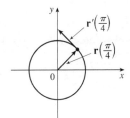

(b) $\mathbf{r}'(t) = \langle -\sin t, \cos t \rangle$

5. Since $(x-1)^2 = t^2 = y$, the curve is a parabola.

(a), (c)

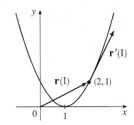

(b) $\mathbf{r}'(t) = \mathbf{i} + 2t\,\mathbf{j}$

7. (a), (c)

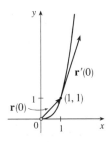

(b) $\mathbf{r}'(t) = e^t\,\mathbf{i} + 3e^{3t}\,\mathbf{j}$

9. $\mathbf{r}'(t) = \left\langle \dfrac{d}{dt}\,[t^2],\ \dfrac{d}{dt}\,[1-t],\ \dfrac{d}{dt}\,[\sqrt{t}] \right\rangle = \left\langle 2t, -1, \dfrac{1}{2\sqrt{t}} \right\rangle$

11. $\mathbf{r}(t) = \mathbf{i} - \mathbf{j} + e^{4t}\,\mathbf{k} \;\Rightarrow\; \mathbf{r}'(t) = 0\,\mathbf{i} + 0\,\mathbf{j} + 4e^{4t}\,\mathbf{k} = 4e^{4t}\,\mathbf{k}$

13. $\mathbf{r}(t) = e^{t^2}\,\mathbf{i} - \mathbf{j} + \ln(1+3t)\,\mathbf{k} \;\Rightarrow\; \mathbf{r}'(t) = 2te^{t^2}\,\mathbf{i} + \dfrac{3}{1+3t}\,\mathbf{k}$

15. $\mathbf{r}'(t) = 0 + \mathbf{b} + 2t\,\mathbf{c} = \mathbf{b} + 2t\,\mathbf{c}$ by Formulas 1 and 3 of Theorem 3.

17. $\mathbf{r}'(t) = \langle 30t^4, 12t^2, 2 \rangle \;\Rightarrow\; \mathbf{r}'(1) = \langle 30, 12, 2 \rangle$. So $|\mathbf{r}'(1)| = \sqrt{30^2 + 12^2 + 2^2} = \sqrt{1048} = 2\sqrt{262}$ and

$\mathbf{T}(1) = \dfrac{\mathbf{r}'(1)}{|\mathbf{r}'(1)|} = \dfrac{1}{2\sqrt{262}}\,\langle 30, 12, 2 \rangle = \left\langle \dfrac{15}{\sqrt{262}}, \dfrac{6}{\sqrt{262}}, \dfrac{1}{\sqrt{262}} \right\rangle.$

19. $\mathbf{r}'(t) = -\sin t\,\mathbf{i} + 3\,\mathbf{j} + 4\cos 2t\,\mathbf{k}$ $\Rightarrow$ $\mathbf{r}'(0) = 3\,\mathbf{j} + 4\,\mathbf{k}$. Thus

$$\mathbf{T}(0) = \frac{\mathbf{r}'(0)}{|\mathbf{r}'(0)|} = \frac{1}{\sqrt{0^2 + 3^2 + 4^2}}\,(3\,\mathbf{j} + 4\,\mathbf{k}) = \tfrac{1}{5}(3\,\mathbf{j} + 4\,\mathbf{k}) = \tfrac{3}{5}\,\mathbf{j} + \tfrac{4}{5}\,\mathbf{k}.$$

21. $\mathbf{r}(t) = \langle t, t^2, t^3 \rangle$ $\Rightarrow$ $\mathbf{r}'(t) = \langle 1, 2t, 3t^2 \rangle$. Then $\mathbf{r}'(1) = \langle 1, 2, 3 \rangle$ and $|\mathbf{r}'(1)| = \sqrt{1^2 + 2^2 + 3^2} = \sqrt{14}$, so

$$\mathbf{T}(1) = \frac{\mathbf{r}'(1)}{|\mathbf{r}'(1)|} = \tfrac{1}{\sqrt{14}}\langle 1, 2, 3 \rangle = \left\langle \tfrac{1}{\sqrt{14}}, \tfrac{2}{\sqrt{14}}, \tfrac{3}{\sqrt{14}} \right\rangle.\quad \mathbf{r}''(t) = \langle 0, 2, 6t \rangle, \text{ so}$$

$$\mathbf{r}'(t) \times \mathbf{r}''(t) = \begin{vmatrix} \mathbf{i} & \mathbf{j} & \mathbf{k} \\ 1 & 2t & 3t^2 \\ 0 & 2 & 6t \end{vmatrix} = \begin{vmatrix} 2t & 3t^2 \\ 2 & 6t \end{vmatrix}\mathbf{i} - \begin{vmatrix} 1 & 3t^2 \\ 0 & 6t \end{vmatrix}\mathbf{j} + \begin{vmatrix} 1 & 2t \\ 0 & 2 \end{vmatrix}\mathbf{k}$$

$$= (12t^2 - 6t^2)\,\mathbf{i} - (6t - 0)\,\mathbf{j} + (2 - 0)\,\mathbf{k} = \langle 6t^2, -6t, 2 \rangle.$$

23. The vector equation for the curve is $\mathbf{r}(t) = \langle t^5, t^4, t^3 \rangle$, so $\mathbf{r}'(t) = \langle 5t^4, 4t^3, 3t^2 \rangle$. The point $(1, 1, 1)$ corresponds to $t = 1$, so the tangent vector there is $\mathbf{r}'(1) = \langle 5, 4, 3 \rangle$. Thus, the tangent line goes through the point $(1, 1, 1)$ and is parallel to the vector $\langle 5, 4, 3 \rangle$. Parametric equations are $x = 1 + 5t,\ y = 1 + 4t,\ z = 1 + 3t$.

25. The vector equation for the curve is $\mathbf{r}(t) = \langle e^{-t}\cos t, e^{-t}\sin t, e^{-t} \rangle$, so

$$\mathbf{r}'(t) = \langle e^{-t}(-\sin t) + (\cos t)(-e^{-t}),\ e^{-t}\cos t + (\sin t)(-e^{-t}),\ (-e^{-t}) \rangle$$

$$= \langle -e^{-t}(\cos t + \sin t),\ e^{-t}(\cos t - \sin t),\ -e^{-t} \rangle.$$

The point $(1, 0, 1)$ corresponds to $t = 0$, so the tangent vector there is

$\mathbf{r}'(0) = \langle -e^0(\cos 0 + \sin 0),\ e^0(\cos 0 - \sin 0),\ -e^0 \rangle = \langle -1, 1, -1 \rangle$. Thus, the tangent line is parallel to the vector $\langle -1, 1, -1 \rangle$ and parametric equations are $x = 1 + (-1)t = 1 - t,\ y = 0 + 1 \cdot t = t$, $z = 1 + (-1)t = 1 - t$.

27. $\mathbf{r}(t) = \langle t, \sqrt{2}\cos t, \sqrt{2}\sin t \rangle$ $\Rightarrow$

$\mathbf{r}'(t) = \langle 1, -\sqrt{2}\sin t, \sqrt{2}\cos t \rangle$. At $\left(\tfrac{\pi}{4}, 1, 1\right)$, $t = \tfrac{\pi}{4}$ and

$\mathbf{r}'\!\left(\tfrac{\pi}{4}\right) = \langle 1, -1, 1 \rangle$. Thus, parametric equations of the tangent

line are $x = \tfrac{\pi}{4} + t,\ y = 1 - t,\ z = 1 + t$.

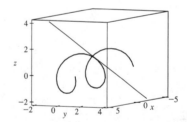

29. (a) $\mathbf{r}(t) = \langle t^3, t^4, t^5 \rangle$ $\Rightarrow$ $\mathbf{r}'(t) = \langle 3t^2, 4t^3, 5t^4 \rangle$, and since $\mathbf{r}'(0) = \langle 0, 0, 0 \rangle = \mathbf{0}$, the curve is not smooth.

(b) $\mathbf{r}(t) = \langle t^3 + t, t^4, t^5 \rangle$ $\Rightarrow$ $\mathbf{r}'(t) = \langle 3t^2 + 1, 4t^3, 5t^4 \rangle$. $\mathbf{r}'(t)$ is continuous since its component functions are continuous. Also, $\mathbf{r}'(t) \neq \mathbf{0}$, as the y- and z-components are 0 only for $t = 0$, but $\mathbf{r}'(0) = \langle 1, 0, 0 \rangle \neq \mathbf{0}$. Thus, the curve is smooth.

(c) $\mathbf{r}(t) = \langle \cos^3 t, \sin^3 t \rangle$ $\Rightarrow$ $\mathbf{r}'(t) = \langle -3\cos^2 t\sin t, 3\sin^2 t\cos t \rangle$. Since $\mathbf{r}'(0) = \langle -3\cos^2 0\sin 0, 3\sin^2 0\cos 0 \rangle = \langle 0, 0 \rangle = \mathbf{0}$, the curve is not smooth.

31. The angle of intersection of the two curves is the angle between the two tangent vectors to the curves at the point of intersection. Since $\mathbf{r}_1'(t) = \langle 1, 2t, 3t^2 \rangle$ and $t = 0$ at $(0, 0, 0)$, $\mathbf{r}_1'(0) = \langle 1, 0, 0 \rangle$ is a tangent vector to $\mathbf{r}_1$ at $(0, 0, 0)$. Similarly, $\mathbf{r}_2'(t) = \langle \cos t, 2\cos 2t, 1 \rangle$ and since $\mathbf{r}_2(0) = \langle 0, 0, 0 \rangle$, $\mathbf{r}_2'(0) = \langle 1, 2, 1 \rangle$ is a tangent vector to $\mathbf{r}_2$ at $(0, 0, 0)$. If θ is the angle between these two tangent vectors, then $\cos\theta = \tfrac{1}{\sqrt{1}\sqrt{6}}\langle 1, 0, 0 \rangle \cdot \langle 1, 2, 1 \rangle = \tfrac{1}{\sqrt{6}}$ and $\theta = \cos^{-1}\left(\tfrac{1}{\sqrt{6}}\right) \approx 66\,°$.

33. $\int_0^1 (16t^3\,\mathbf{i} - 9t^2\,\mathbf{j} + 25t^4\,\mathbf{k})\,dt = \left(\int_0^1 16t^3\,dt\right)\mathbf{i} - \left(\int_0^1 9t^2\,dt\right)\mathbf{j} + \left(\int_0^1 25t^4\,dt\right)\mathbf{k}$

$$= \left[4t^4\right]_0^1 \mathbf{i} - \left[3t^3\right]_0^1 \mathbf{j} + \left[5t^5\right]_0^1 \mathbf{k} = 4\,\mathbf{i} - 3\,\mathbf{j} + 5\,\mathbf{k}$$

35. $\int_0^{\pi/2}(3\sin^2 t\cos t\,\mathbf{i} + 3\sin t\cos^2 t\,\mathbf{j} + 2\sin t\cos t\,\mathbf{k})\,dt$

$$= \left(\int_0^{\pi/2} 3\sin^2 t\cos t\,dt\right)\mathbf{i} + \left(\int_0^{\pi/2} 3\sin t\cos^2 t\,dt\right)\mathbf{j} + \left(\int_0^{\pi/2} 2\sin t\cos t\,dt\right)\mathbf{k}$$

$$= \left[\sin^3 t\right]_0^{\pi/2}\mathbf{i} + \left[-\cos^3 t\right]_0^{\pi/2}\mathbf{j} + \left[\sin^2 t\right]_0^{\pi/2}\mathbf{k}$$

$$= (1-0)\,\mathbf{i} + (0+1)\,\mathbf{j} + (1-0)\,\mathbf{k} = \mathbf{i} + \mathbf{j} + \mathbf{k}$$

37. $\int (e^t\,\mathbf{i} + 2t\,\mathbf{j} + \ln t\,\mathbf{k})\,dt = \left(\int e^t\,dt\right)\mathbf{i} + \left(\int 2t\,dt\right)\mathbf{j} + \left(\int \ln t\,dt\right)\mathbf{k}$

$$= e^t\,\mathbf{i} + t^2\,\mathbf{j} + (t\ln t - t)\,\mathbf{k} + \mathbf{C}, \text{ where } \mathbf{C} \text{ is a vector constant of integration.}$$

39. $\mathbf{r}'(t) = t^2\,\mathbf{i} + 4t^3\,\mathbf{j} - t^2\,\mathbf{k} \;\Rightarrow\; \mathbf{r}(t) = \frac{1}{3}t^3\,\mathbf{i} + t^4\,\mathbf{j} - \frac{1}{3}t^3\,\mathbf{k} + \mathbf{C}$, where $\mathbf{C}$ is a constant vector.

But $\mathbf{j} = \mathbf{r}(0) = (0)\,\mathbf{i} + (0)\,\mathbf{j} - (0)\,\mathbf{k} + \mathbf{C}$. Thus $\mathbf{C} = \mathbf{j}$ and

$\mathbf{r}(t) = \frac{1}{3}t^3\,\mathbf{i} + t^4\,\mathbf{j} - \frac{1}{3}t^3\,\mathbf{k} + \mathbf{j} = \frac{1}{3}t^3\,\mathbf{i} + (t^4+1)\,\mathbf{j} - \frac{1}{3}t^3\,\mathbf{k}$.

For Exercises 41–44, let $\mathbf{u}(t) = \langle u_1(t), u_2(t), u_3(t)\rangle$ and $\mathbf{v}(t) = \langle v_1(t), v_2(t), v_3(t)\rangle$. In each of these exercises, the procedure is to apply Theorem 2 so that the corresponding properties of derivatives of real-valued functions can be used.

41. $\dfrac{d}{dt}\,[\mathbf{u}(t) + \mathbf{v}(t)] = \dfrac{d}{dt}\,\langle u_1(t) + v_1(t), u_2(t) + v_2(t), u_3(t) + v_3(t)\rangle$

$$= \left\langle \frac{d}{dt}\,[u_1(t) + v_1(t)], \frac{d}{dt}\,[u_2(t) + v_2(t)], \frac{d}{dt}\,[u_3(t) + v_3(t)] \right\rangle$$

$$= \langle u_1'(t) + v_1'(t), u_2'(t) + v_2'(t), u_3'(t) + v_3'(t)\rangle$$

$$= \langle u_1'(t), u_2'(t), u_3'(t)\rangle + \langle v_1'(t), v_2'(t), v_3'(t)\rangle = \mathbf{u}'(t) + \mathbf{v}'(t).$$

43. $\dfrac{d}{dt}\,[\mathbf{u}(t) \times \mathbf{v}(t)]$

$$= \frac{d}{dt}\,\langle u_2(t)v_3(t) - u_3(t)v_2(t), u_3(t)v_1(t) - u_1(t)v_3(t), u_1(t)v_2(t) - u_2(t)v_1(t)\rangle$$

$$= \langle u_2'v_3(t) + u_2(t)v_3'(t) - u_3'(t)v_2(t) - u_3(t)v_2'(t),$$

$$u_3'(t)v_1(t) + u_3(t)v_1'(t) - u_1'(t)v_3(t) - u_1(t)v_3'(t),$$

$$u_1'(t)v_2(t) + u_1(t)v_2'(t) - u_2'(t)v_1(t) - u_2(t)v_1'(t)\rangle$$

$$= \langle u_2'(t)v_3(t) - u_3'(t)v_2(t), u_3'(t)v_1(t) - u_1'(t)v_3(t), u_1'(t)v_2(t) - u_2'(t)v_1(t)\rangle$$

$$+ \langle u_2(t)v_3'(t) - u_3(t)v_2'(t), u_3(t)v_1'(t) - u_1(t)v_3'(t), u_1(t)v_2'(t) - u_2(t)v_1'(t)\rangle$$

$$= \mathbf{u}'(t) \times \mathbf{v}(t) + \mathbf{u}(t) \times \mathbf{v}'(t)$$

Alternate solution: Let $\mathbf{r}(t) = \mathbf{u}(t) \times \mathbf{v}(t)$. Then

$$\mathbf{r}(t+h) - \mathbf{r}(t) = [\mathbf{u}(t+h) \times \mathbf{v}(t+h)] - [\mathbf{u}(t) \times \mathbf{v}(t)]$$

$$= [\mathbf{u}(t+h) \times \mathbf{v}(t+h)] - [\mathbf{u}(t) \times \mathbf{v}(t)] + [\mathbf{u}(t+h) \times \mathbf{v}(t)] - [\mathbf{u}(t+h) \times \mathbf{v}(t)]$$

$$= \mathbf{u}(t+h) \times [\mathbf{v}(t+h) - \mathbf{v}(t)] + [\mathbf{u}(t+h) - \mathbf{u}(t)] \times \mathbf{v}(t)$$

(Be careful of the order of the cross product.)

Dividing through by h and taking the limit as $h \to 0$ we have

$$\mathbf{r}'(t) = \lim_{h \to 0} \frac{\mathbf{u}(t+h) \times [\mathbf{v}(t+h) - \mathbf{v}(t)]}{h} + \lim_{h \to 0} \frac{[\mathbf{u}(t+h) - \mathbf{u}(t)] \times \mathbf{v}(t)}{h}$$

$$= \mathbf{u}(t) \times \mathbf{v}'(t) + \mathbf{u}'(t) \times \mathbf{v}(t)$$

by Exercise 14.1.41(a) [ET 13.1.41(a)] and Definition 1.

45. $\dfrac{d}{dt} [\mathbf{u}(t) \cdot \mathbf{v}(t)] = \mathbf{u}'(t) \cdot \mathbf{v}(t) + \mathbf{u}(t) \cdot \mathbf{v}'(t)$ [by Formula 4 of Theorem 3]

$$= (-4t\,\mathbf{j} + 9t^2\,\mathbf{k}) \cdot (t\,\mathbf{i} + \cos t\,\mathbf{j} + \sin t\,\mathbf{k}) + (\mathbf{i} - 2t^2\,\mathbf{j} + 3t^3\,\mathbf{k}) \cdot (\mathbf{i} - \sin t\,\mathbf{j} + \cos t\,\mathbf{k})$$

$$= -4t \cos t + 9t^2 \sin t + 1 + 2t^2 \sin t + 3t^3 \cos t$$

$$= 1 - 4t \cos t + 11t^2 \sin t + 3t^3 \cos t$$

47. $\dfrac{d}{dt} [\mathbf{r}(t) \times \mathbf{r}'(t)] = \mathbf{r}'(t) \times \mathbf{r}'(t) + \mathbf{r}(t) \times \mathbf{r}''(t)$ by Formula 5 of Theorem 3. But $\mathbf{r}'(t) \times \mathbf{r}'(t) = \mathbf{0}$

(see Example 13.4.2 [ET 12.4.2]). Thus, $\dfrac{d}{dt} [\mathbf{r}(t) \times \mathbf{r}'(t)] = \mathbf{r}(t) \times \mathbf{r}''(t)$.

49. $\dfrac{d}{dt} |\mathbf{r}(t)| = \dfrac{d}{dt} [\mathbf{r}(t) \cdot \mathbf{r}(t)]^{1/2} = \tfrac{1}{2} [\mathbf{r}(t) \cdot \mathbf{r}(t)]^{-1/2} [2\mathbf{r}(t) \cdot \mathbf{r}'(t)] = \dfrac{1}{|\mathbf{r}(t)|} \mathbf{r}(t) \cdot \mathbf{r}'(t)$

51. Since $\mathbf{u}(t) = \mathbf{r}(t) \cdot [\mathbf{r}'(t) \times \mathbf{r}''(t)]$,

$$\mathbf{u}'(t) = \mathbf{r}'(t) \cdot [\mathbf{r}'(t) \times \mathbf{r}''(t)] + \mathbf{r}(t) \cdot \frac{d}{dt} [\mathbf{r}'(t) \times \mathbf{r}''(t)]$$

$$= 0 + \mathbf{r}(t) \cdot [\mathbf{r}''(t) \times \mathbf{r}''(t) + \mathbf{r}'(t) \times \mathbf{r}'''(t)] \qquad [\text{since } \mathbf{r}'(t) \perp \mathbf{r}'(t) \times \mathbf{r}''(t)]$$

$$= \mathbf{r}(t) \cdot [\mathbf{r}'(t) \times \mathbf{r}'''(t)] \qquad [\text{since } \mathbf{r}''(t) \times \mathbf{r}''(t) = \mathbf{0}]$$

14.3 Arc Length and Curvature   ET 13.3

1. $\mathbf{r}'(t) = \langle 2\cos t, 5, -2\sin t \rangle \;\Rightarrow\; |\mathbf{r}'(t)| = \sqrt{(2\cos t)^2 + 5^2 + (-2\sin t)^2} = \sqrt{29}$. Then using Formula 3, we have $L = \int_{-10}^{10} |\mathbf{r}'(t)|\, dt = \int_{-10}^{10} \sqrt{29}\, dt = \sqrt{29}\, t \big]_{-10}^{10} = 20\sqrt{29}$.

3. $\mathbf{r}'(t) = \sqrt{2}\,\mathbf{i} + e^t\,\mathbf{j} - e^{-t}\,\mathbf{k} \;\Rightarrow$

$|\mathbf{r}'(t)| = \sqrt{\left(\sqrt{2}\right)^2 + (e^t)^2 + (-e^{-t})^2} = \sqrt{2 + e^{2t} + e^{-2t}} = \sqrt{(e^t + e^{-t})^2} = e^t + e^{-t}$ (since $e^t + e^{-t} > 0$).

Then $L = \int_0^1 |\mathbf{r}'(t)|\, dt = \int_0^1 (e^t + e^{-t})\, dt = [e^t - e^{-t}]_0^1 = e - e^{-1}$.

5. $\mathbf{r}'(t) = 2t\,\mathbf{j} + 3t^2\,\mathbf{k} \quad \Rightarrow \quad |\mathbf{r}'(t)| = \sqrt{4t^2 + 9t^4} = t\sqrt{4 + 9t^2}$ (since $t \geq 0$). Then

$L = \int_0^1 |\mathbf{r}'(t)|\,dt = \int_0^1 t\sqrt{4 + 9t^2}\,dt = \frac{1}{18} \cdot \frac{2}{3}(4 + 9t^2)^{3/2}\Big]_0^1 = \frac{1}{27}(13^{3/2} - 4^{3/2}) = \frac{1}{27}(13^{3/2} - 8).$

7. The point $(2, 4, 8)$ corresponds to $t = 2$, so by Equation 2, $L = \int_0^2 \sqrt{(1)^2 + (2t)^2 + (3t^2)^2}\,dt$.

If $f(t) = \sqrt{1 + 4t^2 + 9t^4}$, then Simpson's Rule gives

$L \approx \dfrac{2 - 0}{10 \cdot 3}\,[f(0) + 4f(0.2) + 2f(0.4) + \cdots + 4f(1.8) + f(2)] \approx 9.5706.$

9. $\mathbf{r}'(t) = 2\,\mathbf{i} - 3\,\mathbf{j} + 4\,\mathbf{k}$ and $\frac{ds}{dt} = |\mathbf{r}'(t)| = \sqrt{4 + 9 + 16} = \sqrt{29}$. Then

$s = s(t) = \int_0^t |\mathbf{r}'(u)|\,du = \int_0^t \sqrt{29}\,du = \sqrt{29}\,t$. Therefore, $t = \frac{1}{\sqrt{29}}s$, and substituting for t in the original

equation, we have $\mathbf{r}(t(s)) = \frac{2}{\sqrt{29}}s\,\mathbf{i} + \left(1 - \frac{3}{\sqrt{29}}s\right)\mathbf{j} + \left(5 + \frac{4}{\sqrt{29}}s\right)\mathbf{k}$.

11. $|\mathbf{r}'(t)| = \sqrt{(3\cos t)^2 + 16 + (-3\sin t)^2} = \sqrt{9 + 16} = 5$ and $s(t) = \int_0^t |\mathbf{r}'(u)|\,du = \int_0^t 5\,du = 5t \quad \Rightarrow$

$t(s) = \frac{1}{5}s$. Therefore, $\mathbf{r}(t(s)) = 3\sin\left(\frac{1}{5}s\right)\mathbf{i} + \frac{4}{5}s\,\mathbf{j} + 3\cos\left(\frac{1}{5}s\right)\mathbf{k}$.

13. (a) $\mathbf{r}'(t) = \langle 2\cos t, 5, -2\sin t\rangle \quad \Rightarrow \quad |\mathbf{r}'(t)| = \sqrt{4\cos^2 t + 25 + 4\sin^2 t} = \sqrt{29}$. Then

$\mathbf{T}(t) = \dfrac{\mathbf{r}'(t)}{|\mathbf{r}'(t)|} = \frac{1}{\sqrt{29}}\langle 2\cos t, 5, -2\sin t\rangle$ or $\left\langle \frac{2}{\sqrt{29}}\cos t, \frac{5}{\sqrt{29}}, -\frac{2}{\sqrt{29}}\sin t\right\rangle$.

$\mathbf{T}'(t) = \frac{1}{\sqrt{29}}\langle -2\sin t, 0, -2\cos t\rangle \quad \Rightarrow \quad |\mathbf{T}'(t)| = \frac{1}{\sqrt{29}}\sqrt{4\sin^2 t + 0 + 4\cos^2 t} = \frac{2}{\sqrt{29}}$. Thus

$\mathbf{N}(t) = \dfrac{\mathbf{T}'(t)}{|\mathbf{T}'(t)|} = \dfrac{1/\sqrt{29}}{2/\sqrt{29}}\langle -2\sin t, 0, -2\cos t\rangle = \langle -\sin t, 0, -\cos t\rangle.$

(b) $\kappa(t) = \dfrac{|\mathbf{T}'(t)|}{|\mathbf{r}'(t)|} = \dfrac{2/\sqrt{29}}{\sqrt{29}} = \dfrac{2}{29}.$

15. (a) $\mathbf{r}'(t) = \langle \sqrt{2}, e^t, -e^{-t}\rangle \quad \Rightarrow \quad |\mathbf{r}'(t)| = \sqrt{2 + e^{2t} + e^{-2t}} = \sqrt{(e^t + e^{-t})^2} = e^t + e^{-t}$. Then

$\mathbf{T}(t) = \dfrac{\mathbf{r}'(t)}{|\mathbf{r}'(t)|} = \dfrac{1}{e^t + e^{-t}}\langle \sqrt{2}, e^t, -e^{-t}\rangle = \dfrac{1}{e^{2t} + 1}\langle \sqrt{2}e^t, e^{2t}, -1\rangle \quad \left(\text{after multiplying by } \dfrac{e^t}{e^t}\right)$ and

$$\mathbf{T}'(t) = \frac{1}{e^{2t} + 1}\langle \sqrt{2}e^t, 2e^{2t}, 0\rangle - \frac{2e^{2t}}{(e^{2t} + 1)^2}\langle \sqrt{2}e^t, e^{2t}, -1\rangle$$

$$= \frac{1}{(e^{2t} + 1)^2}\left[(e^{2t} + 1)\langle \sqrt{2}e^t, 2e^{2t}, 0\rangle - 2e^{2t}\langle \sqrt{2}e^t, e^{2t}, -1\rangle\right]$$

$$= \frac{1}{(e^{2t} + 1)^2}\langle \sqrt{2}e^t\,(1 - e^{2t}), 2e^{2t}, 2e^{2t}\rangle$$

Then

$$|\mathbf{T}'(t)| = \frac{1}{(e^{2t} + 1)^2}\sqrt{2e^{2t}(1 - 2e^{2t} + e^{4t}) + 4e^{4t} + 4e^{4t}} = \frac{1}{(e^{2t} + 1)^2}\sqrt{2e^{2t}(1 + 2e^{2t} + e^{4t})}$$

$$= \frac{1}{(e^{2t} + 1)^2}\sqrt{2e^{2t}\,(1 + e^{2t})^2} = \frac{\sqrt{2}e^t(1 + e^{2t})}{(e^{2t} + 1)^2} = \frac{\sqrt{2}\,e^t}{e^{2t} + 1}$$

Therefore

$$\mathbf{N}(t) = \frac{\mathbf{T}'(t)}{|\mathbf{T}'(t)|} = \frac{e^{2t}+1}{\sqrt{2}\,e^t} \frac{1}{(e^{2t}+1)^2} \left\langle \sqrt{2}\,e^t(1-e^{2t}), 2e^{2t}, 2e^{2t} \right\rangle$$

$$= \frac{1}{\sqrt{2}\,e^t(e^{2t}+1)} \left\langle \sqrt{2}\,e^t(1-e^{2t}), 2e^{2t}, 2e^{2t} \right\rangle$$

$$= \frac{1}{e^{2t}+1} \left\langle 1-e^{2t}, \sqrt{2}\,e^t, \sqrt{2}\,e^t \right\rangle$$

(b) $\kappa(t) = \dfrac{|\mathbf{T}'(t)|}{|\mathbf{r}'(t)|} = \dfrac{\sqrt{2}\,e^t}{e^{2t}+1} \cdot \dfrac{1}{e^t+e^{-t}} = \dfrac{\sqrt{2}\,e^t}{e^{3t}+2e^t+e^{-t}} = \dfrac{\sqrt{2}\,e^{2t}}{e^{4t}+2e^{2t}+1} = \dfrac{\sqrt{2}\,e^{2t}}{(e^{2t}+1)^2}.$

17. $\mathbf{r}'(t) = 2t\,\mathbf{i} + \mathbf{k}$, $\mathbf{r}''(t) = 2\,\mathbf{i}$, $|\mathbf{r}'(t)| = \sqrt{(2t)^2 + 0^2 + 1^2} = \sqrt{4t^2+1}$, $\mathbf{r}'(t) \times \mathbf{r}''(t) = 2\,\mathbf{j}$, $|\mathbf{r}'(t) \times \mathbf{r}''(t)| = 2$.

Then $\kappa(t) = \dfrac{|\mathbf{r}'(t) \times \mathbf{r}''(t)|}{|\mathbf{r}'(t)|^3} = \dfrac{2}{\left(\sqrt{4t^2+1}\right)^3} = \dfrac{2}{(4t^2+1)^{3/2}}$.

19. $\mathbf{r}'(t) = 3\,\mathbf{i} + 4\cos t\,\mathbf{j} - 4\sin t\,\mathbf{k}$, $\mathbf{r}''(t) = -4\sin t\,\mathbf{j} - 4\cos t\,\mathbf{k}$,

$|\mathbf{r}'(t)| = \sqrt{9 + 16\cos^2 t + 16\sin^2 t} = \sqrt{9+16} = 5$, $\mathbf{r}'(t) \times \mathbf{r}''(t) = -16\,\mathbf{i} + 12\cos t\,\mathbf{j} - 12\sin t\,\mathbf{k}$,

$|\mathbf{r}'(t) \times \mathbf{r}''(t)| = \sqrt{256 + 144\cos^2 t + 144\sin^2 t} = \sqrt{400} = 20$. Then $\kappa(t) = \dfrac{|\mathbf{r}'(t) \times \mathbf{r}''(t)|}{|\mathbf{r}'(t)|^3} = \dfrac{20}{5^3} = \dfrac{4}{25}$.

21. $\mathbf{r}'(t) = \langle 1, 2t, 3t^2 \rangle$. The point $(1, 1, 1)$ corresponds to $t = 1$, and

$\mathbf{r}'(1) = \langle 1, 2, 3 \rangle \;\Rightarrow\; |\mathbf{r}'(1)| = \sqrt{1+4+9} = \sqrt{14}$. $\mathbf{r}''(t) = \langle 0, 2, 6t \rangle \;\Rightarrow\; \mathbf{r}''(1) = \langle 0, 2, 6 \rangle$.

$\mathbf{r}'(1) \times \mathbf{r}''(1) = \langle 6, -6, 2 \rangle$, so $|\mathbf{r}'(1) \times \mathbf{r}''(1)| = \sqrt{36 + 36 + 4} = \sqrt{76}$. Then

$\kappa(1) = \dfrac{|\mathbf{r}'(1) \times \mathbf{r}''(1)|}{|\mathbf{r}'(1)|^3} = \dfrac{\sqrt{76}}{\sqrt{14}^3} = \dfrac{1}{7}\sqrt{\dfrac{19}{14}}$.

23. $f(x) = x^3$, $f'(x) = 3x^2$, $f''(x) = 6x$, $\kappa(x) = \dfrac{|f''(x)|}{\left[1 + (f'(x))^2\right]^{3/2}} = \dfrac{6\,|x|}{(1+9x^4)^{3/2}}$

25. $f(x) = 4x^{5/2}$, $f'(x) = 10x^{3/2}$, $f''(x) = 15x^{1/2}$,

$\kappa(x) = \dfrac{|f''(x)|}{[1 + (f'(x))^2]^{3/2}} = \dfrac{\left|15x^{1/2}\right|}{[1 + (10x^{3/2})^2]^{3/2}} = \dfrac{15\sqrt{x}}{(1+100x^3)^{3/2}}$

27. Since $y' = y'' = e^x$, the curvature is $\kappa(x) = \dfrac{|y''(x)|}{[1 + (y'(x))^2]^{3/2}} = \dfrac{e^x}{(1+e^{2x})^{3/2}} = e^x(1+e^{2x})^{-3/2}$.

To find the maximum curvature, we first find the critical numbers of $\kappa(x)$:

$\kappa'(x) = e^x(1+e^{2x})^{-3/2} + e^x\left(-\frac{3}{2}\right)(1+e^{2x})^{-5/2}(2e^{2x}) = e^x\dfrac{1 + e^{2x} - 3e^{2x}}{(1+e^{2x})^{5/2}} = e^x\dfrac{1 - 2e^{2x}}{(1+e^{2x})^{5/2}}$.

$\kappa'(x) = 0$ when $1 - 2e^{2x} = 0$, so $e^{2x} = \frac{1}{2}$ or $x = -\frac{1}{2}\ln 2$. And since $1 - 2e^{2x} > 0$ for $x < -\frac{1}{2}\ln 2$ and

$1 - 2e^{2x} < 0$ for $x > -\frac{1}{2}\ln 2$, the maximum curvature is attained at the point

$\left(-\frac{1}{2}\ln 2, e^{(-\ln 2)/2}\right) = \left(-\frac{1}{2}\ln 2, \frac{1}{\sqrt{2}}\right)$. Since $\lim\limits_{x \to \infty} e^x(1+e^{2x})^{-3/2} = 0$, $\kappa(x)$ approaches 0 as $x \to \infty$.

29. (a) C appears to be changing direction more quickly at P than Q, so we would expect the curvature to be greater at P.

(b) First we sketch approximate osculating circles at P and

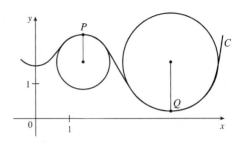

Q. Using the axes scale as a guide, we measure the radius of the osculating circle at P to be approximately 0.8 units,

thus $\rho = \dfrac{1}{\kappa} \ \Rightarrow \ \kappa = \dfrac{1}{\rho} \approx \dfrac{1}{0.8} \approx 1.3$. Similarly, we

estimate the radius of the osculating circle at Q to be

1.4 units, so $\kappa = \dfrac{1}{\rho} \approx \dfrac{1}{1.4} \approx 0.7$.

31. $y = x^4 \ \Rightarrow \ y' = 4x^3,\ y'' = 12x^2$, and

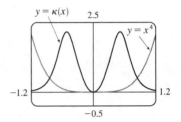

$\kappa(x) = \dfrac{|y''|}{[1 + (y')^2]^{3/2}} = \dfrac{12x^2}{(1 + 16x^6)^{3/2}}$. The appearance of

the two humps in this graph is perhaps a little surprising, but it
is explained by the fact that $y = x^4$ is very flat around the
origin, and so here the curvature is zero.

33. Notice that the curve b has two inflection points at which the graph appears almost straight. We would expect the curvature to be 0 or nearly 0 at these values, but the curve a isn't near 0 there. Thus, a must be the graph of $y = f(x)$ rather than the graph of curvature, and b is the graph of $y = \kappa(x)$.

35. Using a CAS, we find (after simplifying)

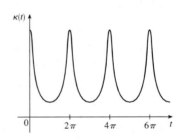

$\kappa(t) = \dfrac{6\sqrt{4\cos^2 t - 12\cos t + 13}}{(17 - 12\cos t)^{3/2}}$. (To compute cross

products in Maple, use the `Linalg` package and the

`crossprod(a,b)` command; in Mathematica, use

`Cross[a,b]`.) Curvature is largest at integer multiples of 2π.

37. $x = e^t \cos t \ \Rightarrow \ \dot{x} = e^t(\cos t - \sin t) \ \Rightarrow \ \ddot{x} = e^t(-\sin t - \cos t) + e^t(\cos t - \sin t) = -2e^t \sin t$,

$y = e^t \sin t \ \Rightarrow \ \dot{y} = e^t(\cos t + \sin t) \ \Rightarrow \ \ddot{y} = e^t(-\sin t + \cos t) + e^t(\cos t + \sin t) = 2e^t \cos t$. Then

$$\kappa(t) = \frac{|\dot{x}\ddot{y} - \dot{y}\ddot{x}|}{(\dot{x}^2 + \dot{y}^2)^{3/2}} = \frac{\left|e^t(\cos t - \sin t)(2e^t \cos t) - e^t(\cos t + \sin t)(-2e^t \sin t)\right|}{\left([e^t(\cos t - \sin t)]^2 + [e^t(\cos t + \sin t)]^2\right)^{3/2}}$$

$$= \frac{\left|2e^{2t}(\cos^2 t - \sin t \cos t + \sin t \cos t + \sin^2 t)\right|}{\left[e^{2t}(\cos^2 t - 2\cos t \sin t + \sin^2 t + \cos^2 t + 2\cos t \sin t + \sin^2 t)\right]^{3/2}}$$

$$= \frac{\left|2e^{2t}(1)\right|}{[e^{2t}(1+1)]^{3/2}} = \frac{2e^{2t}}{e^{3t}(2)^{3/2}} = \frac{1}{\sqrt{2}\,e^t}$$

39. $\left(1, \frac{2}{3}, 1\right)$ corresponds to $t = 1$. $\mathbf{T}(t) = \dfrac{\mathbf{r}'(t)}{|\mathbf{r}'(t)|} = \dfrac{\langle 2t, 2t^2, 1\rangle}{\sqrt{4t^2 + 4t^4 + 1}} = \dfrac{\langle 2t, 2t^2, 1\rangle}{2t^2 + 1}$, so $\mathbf{T}(1) = \left\langle \frac{2}{3}, \frac{2}{3}, \frac{1}{3}\right\rangle$.

$\mathbf{T}'(t) = -4t(2t^2 + 1)^{-2}\langle 2t, 2t^2, 1\rangle + (2t^2 + 1)^{-1}\langle 2, 4t, 0\rangle$ [by Theorem 14.2.3 [ET 13.2.3] #3]

$\quad = (2t^2 + 1)^{-2}\langle -8t^2 + 4t^2 + 2, -8t^3 + 8t^3 + 4t, -4t\rangle = 2(2t^2 + 1)^{-2}\langle 1 - 2t^2, 2t, -2t\rangle$

$$\mathbf{N}(t) = \frac{\mathbf{T}'(t)}{|\mathbf{T}'(t)|} = \frac{2(2t^2+1)^{-2}\langle 1-2t^2, 2t, -2t\rangle}{2(2t^2+1)^{-2}\sqrt{(1-2t^2)^2+(2t)^2+(-2t)^2}} = \frac{\langle 1-2t^2, 2t, -2t\rangle}{\sqrt{1-4t^2+4t^4+8t^2}}$$

$$= \frac{\langle 1-2t^2, 2t, -2t\rangle}{1+2t^2}$$

$\mathbf{N}(1) = \langle -\frac{1}{3}, \frac{2}{3}, -\frac{2}{3}\rangle$ and $\mathbf{B}(1) = \mathbf{T}(1) \times \mathbf{N}(1) = \langle -\frac{4}{9} - \frac{2}{9}, -\left(-\frac{4}{9}+\frac{1}{9}\right), \frac{4}{9}+\frac{2}{9}\rangle = \langle -\frac{2}{3}, \frac{1}{3}, \frac{2}{3}\rangle.$

41. $(0, \pi, -2)$ corresponds to $t = \pi$. $\mathbf{r}(t) = \langle 2\sin 3t, t, 2\cos 3t\rangle \quad \Rightarrow$

$$\mathbf{T}(t) = \frac{\mathbf{r}'(t)}{|\mathbf{r}'(t)|} = \frac{\langle 6\cos 3t, 1, -6\sin 3t\rangle}{\sqrt{36\cos^2 3t + 1 + 36\sin^2 3t}} = \frac{1}{\sqrt{37}}\langle 6\cos 3t, 1, -6\sin 3t\rangle.$$

$\mathbf{T}(\pi) = \frac{1}{\sqrt{37}}\langle -6, 1, 0\rangle$ is a normal vector for the normal plane, and so $\langle -6, 1, 0\rangle$ is also normal. Thus an equation

for the plane is $-6(x-0) + 1(y-\pi) + 0(z+2) = 0$ or $y - 6x = \pi$.

$$\mathbf{T}'(t) = \frac{1}{\sqrt{37}}\langle -18\sin 3t, 0, -18\cos 3t\rangle \quad \Rightarrow \quad |\mathbf{T}'(t)| = \frac{\sqrt{18^2\sin^2 3t + 18^2\cos^2 3t}}{\sqrt{37}} = \frac{18}{\sqrt{37}} \quad \Rightarrow$$

$$\mathbf{N}(t) = \frac{\mathbf{T}'(t)}{|\mathbf{T}'(t)|} = \langle -\sin 3t, 0, -\cos 3t\rangle. \text{ So } \mathbf{N}(\pi) = \langle 0, 0, 1\rangle \text{ and}$$

$\mathbf{B}(\pi) = \frac{1}{\sqrt{37}}\langle -6, 1, 0\rangle \times \langle 0, 0, 1\rangle = \frac{1}{\sqrt{37}}\langle 1, 6, 0\rangle.$ Since $\mathbf{B}(\pi)$ is a normal to the osculating plane, so is $\langle 1, 6, 0\rangle$

and an equation for the plane is $1(x-0) + 6(y-\pi) + 0(z+2) = 0$ or $x + 6y = 6\pi$.

43. The ellipse is given by the parametric equations $x = 2\cos t$, $y = 3\sin t$, so using the result from Exercise 36,

$$\kappa(t) = \frac{|\dot{x}\ddot{y} - \ddot{x}\dot{y}|}{(\dot{x}^2 + \dot{y}^2)^{3/2}} = \frac{|(-2\sin t)(-3\sin t) - (3\cos t)(-2\cos t)|}{(4\sin^2 t + 9\cos^2 t)^{3/2}} = \frac{6}{(4\sin^2 t + 9\cos^2 t)^{3/2}}$$

At $(2, 0)$, $t = 0$. Now $\kappa(0) = \frac{6}{27} = \frac{2}{9}$, so the radius of the

osculating circle is $1/\kappa(0) = \frac{9}{2}$ and its center is $\left(-\frac{5}{2}, 0\right)$. Its

equation is therefore $\left(x + \frac{5}{2}\right)^2 + y^2 = \frac{81}{4}$. At $(0, 3)$, $t = \frac{\pi}{2}$, and

$\kappa\left(\frac{\pi}{2}\right) = \frac{6}{8} = \frac{3}{4}$. So the radius of the osculating circle is $\frac{4}{3}$ and its

center is $\left(0, \frac{5}{3}\right)$. Hence its equation is $x^2 + \left(y - \frac{5}{3}\right)^2 = \frac{16}{9}$.

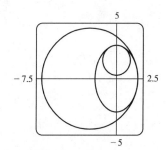

45. The tangent vector is normal to the normal plane, and the vector $\langle 6, 6, -8\rangle$ is normal to the given plane. But

$\mathbf{T}(t) \parallel \mathbf{r}'(t)$ and $\langle 6, 6, -8\rangle \parallel \langle 3, 3, -4\rangle$, so we need to find t such that $\mathbf{r}'(t) \parallel \langle 3, 3, -4\rangle$. $\mathbf{r}(t) = \langle t^3, 3t, t^4\rangle \quad \Rightarrow$

$\mathbf{r}'(t) = \langle 3t^2, 3, 4t^3\rangle \parallel \langle 3, 3, -4\rangle$ when $t = -1$. So the planes are parallel at the point $\mathbf{r}(-1) = (-1, -3, 1)$.

47. $\kappa = \left|\dfrac{d\mathbf{T}}{ds}\right| = \left|\dfrac{d\mathbf{T}/dt}{ds/dt}\right| = \dfrac{|d\mathbf{T}/dt|}{ds/dt}$ and $\mathbf{N} = \dfrac{d\mathbf{T}/dt}{|d\mathbf{T}/dt|}$, so $\kappa\mathbf{N} = \dfrac{\left|\dfrac{d\mathbf{T}}{dt}\right|\dfrac{d\mathbf{T}}{dt}}{\left|\dfrac{d\mathbf{T}}{dt}\right|\dfrac{ds}{dt}} = \dfrac{d\mathbf{T}/dt}{ds/dt} = \dfrac{d\mathbf{T}}{ds}$ by the Chain Rule.

49. (a) $|\mathbf{B}| = 1 \Rightarrow \mathbf{B} \cdot \mathbf{B} = 1 \Rightarrow \dfrac{d}{ds}(\mathbf{B} \cdot \mathbf{B}) = 0 \Rightarrow 2\dfrac{d\mathbf{B}}{ds} \cdot \mathbf{B} = 0 \Rightarrow \dfrac{d\mathbf{B}}{ds} \perp \mathbf{B}$

(b) $\mathbf{B} = \mathbf{T} \times \mathbf{N} \Rightarrow$

$$\frac{d\mathbf{B}}{ds} = \frac{d}{ds}(\mathbf{T} \times \mathbf{N}) = \frac{d}{dt}(\mathbf{T} \times \mathbf{N})\frac{1}{ds/dt} = \frac{d}{dt}(\mathbf{T} \times \mathbf{N})\frac{1}{|\mathbf{r}'(t)|}$$

$$= [(\mathbf{T}' \times \mathbf{N}) + (\mathbf{T} \times \mathbf{N}')]\frac{1}{|\mathbf{r}'(t)|} = \left[\left(\mathbf{T}' \times \frac{\mathbf{T}'}{|\mathbf{T}'|}\right) + (\mathbf{T} \times \mathbf{N}')\right]\frac{1}{|\mathbf{r}'(t)|} = \frac{\mathbf{T} \times \mathbf{N}'}{|\mathbf{r}'(t)|}$$

$$\Rightarrow \frac{d\mathbf{B}}{ds} \perp \mathbf{T}$$

(c) $\mathbf{B} = \mathbf{T} \times \mathbf{N} \Rightarrow \mathbf{T} \perp \mathbf{N}, \mathbf{B} \perp \mathbf{T}$ and $\mathbf{B} \perp \mathbf{N}$. So $\mathbf{B}, \mathbf{T}$ and $\mathbf{N}$ form an orthogonal set of vectors in the three-dimensional space $\mathbb{R}^3$. From parts (a) and (b), $d\mathbf{B}/ds$ is perpendicular to both $\mathbf{B}$ and $\mathbf{T}$, so $d\mathbf{B}/ds$ is parallel to $\mathbf{N}$. Therefore, $d\mathbf{B}/ds = -\tau(s)\mathbf{N}$, where $\tau(s)$ is a scalar.

(d) Since $\mathbf{B} = \mathbf{T} \times \mathbf{N}$, $\mathbf{T} \perp \mathbf{N}$ and both $\mathbf{T}$ and $\mathbf{N}$ are unit vectors, $\mathbf{B}$ is a unit vector mutually perpendicular to both $\mathbf{T}$ and $\mathbf{N}$. For a plane curve, $\mathbf{T}$ and $\mathbf{N}$ always lie in the plane of the curve, so that $\mathbf{B}$ is a constant unit vector always perpendicular to the plane. Thus $d\mathbf{B}/ds = \mathbf{0}$, but $d\mathbf{B}/ds = -\tau(s)\mathbf{N}$ and $\mathbf{N} \neq \mathbf{0}$, so $\tau(s) = 0$.

51. (a) $\mathbf{r}' = s'\,\mathbf{T} \Rightarrow \mathbf{r}'' = s''\,\mathbf{T} + s'\,\mathbf{T}' = s''\,\mathbf{T} + s'\dfrac{d\mathbf{T}}{ds}s' = s''\,\mathbf{T} + \kappa(s')^2\,\mathbf{N}$ by the first Serret-Frenet formula.

(b) Using part (a), we have

$$\mathbf{r}' \times \mathbf{r}'' = (s'\,\mathbf{T}) \times [s''\,\mathbf{T} + \kappa(s')^2\,\mathbf{N}]$$
$$= [(s'\,\mathbf{T}) \times (s''\,\mathbf{T})] + [(s'\mathbf{T}) \times (\kappa(s')^2\,\mathbf{N})] \qquad \text{[by Theorem 13.4.8 [ET 12.4.8] #3]}$$
$$= (s's'')(\mathbf{T} \times \mathbf{T}) + \kappa(s')^3(\mathbf{T} \times \mathbf{N}) = \mathbf{0} + \kappa(s')^3\,\mathbf{B} = \kappa(s')^3\,\mathbf{B}$$

(c) Using part (a), we have

$$\mathbf{r}''' = [s''\,\mathbf{T} + \kappa(s')^2\,\mathbf{N}]' = s'''\,\mathbf{T} + s''\,\mathbf{T}' + \kappa'(s')^2\,\mathbf{N} + 2\kappa s's''\,\mathbf{N} + \kappa(s')^2\,\mathbf{N}'$$
$$= s'''\,\mathbf{T} + s''\frac{d\mathbf{T}}{ds}s' + \kappa'(s')^2\,\mathbf{N} + 2\kappa s's''\,\mathbf{N} + \kappa(s')^2\frac{d\mathbf{N}}{ds}s'$$
$$= s'''\,\mathbf{T} + s''s'\kappa\mathbf{N} + \kappa'(s')^2\,\mathbf{N} + 2\kappa s's''\,\mathbf{N} + \kappa(s')^3(-\kappa\,\mathbf{T} + \tau\,\mathbf{B}) \qquad \text{[by the second formula]}$$
$$= [s''' - \kappa^2(s')^3]\,\mathbf{T} + [3\kappa s's'' + \kappa'(s')^2]\,\mathbf{N} + \kappa\tau(s')^3\,\mathbf{B}$$

(d) Using parts (b) and (c) and the facts that $\mathbf{B} \cdot \mathbf{T} = 0$, $\mathbf{B} \cdot \mathbf{N} = 0$, and $\mathbf{B} \cdot \mathbf{B} = 1$, we get

$$\frac{(\mathbf{r}' \times \mathbf{r}'') \cdot \mathbf{r}'''}{|\mathbf{r}' \times \mathbf{r}''|^2} = \frac{\kappa(s')^3\,\mathbf{B} \cdot \left\{[s''' - \kappa^2(s')^3]\,\mathbf{T} + [3\kappa s's'' + \kappa'(s')^2]\,\mathbf{N} + \kappa\tau(s')^3\,\mathbf{B}\right\}}{|\kappa(s')^3\,\mathbf{B}|^2}$$

$$= \frac{\kappa(s')^3\kappa\tau(s')^3}{[\kappa(s')^3]^2} = \tau$$

53. $\mathbf{r} = \left\langle t, \tfrac{1}{2}t^2, \tfrac{1}{3}t^3\right\rangle \Rightarrow \mathbf{r}' = \left\langle 1, t, t^2\right\rangle, \mathbf{r}'' = \left\langle 0, 1, 2t\right\rangle, \mathbf{r}''' = \left\langle 0, 0, 2\right\rangle \Rightarrow \mathbf{r}' \times \mathbf{r}'' = \left\langle t^2, -2t, 1\right\rangle \Rightarrow$

$$\tau = \frac{(\mathbf{r}' \times \mathbf{r}'') \cdot \mathbf{r}'''}{|\mathbf{r}' \times \mathbf{r}''|^2} = \frac{\left\langle t^2, -2t, 1\right\rangle \cdot \left\langle 0, 0, 2\right\rangle}{t^4 + 4t^2 + 1} = \frac{2}{t^4 + 4t^2 + 1}$$

55. For one helix, the vector equation is $\mathbf{r}(t) = \langle 10\cos t, 10\sin t, 34t/(2\pi) \rangle$ (measuring in angstroms), because the radius of each helix is 10 angstroms, and z increases by 34 angstroms for each increase of 2π in t. Using the arc length formula, letting t go from 0 to $2.9 \times 10^8 \times 2\pi$, we find the approximate length of each helix to be

$$L = \int_0^{2.9 \times 10^8 \times 2\pi} \left| \mathbf{r}'(t) \right| dt = \int_0^{2.9 \times 10^8 \times 2\pi} \sqrt{(-10\sin t)^2 + (10\cos t)^2 + \left(\tfrac{34}{2\pi}\right)^2}\, dt$$

$$= \left. \sqrt{100 + \left(\tfrac{34}{2\pi}\right)^2}\, t \right]_0^{2.9 \times 10^8 \times 2\pi} = 2.9 \times 10^8 \times 2\pi \sqrt{100 + \left(\tfrac{34}{2\pi}\right)^2}$$

$$\approx 2.07 \times 10^{10}\ \text{Å} \text{ — more than two meters!}$$

14.4 Motion in Space: Velocity and Acceleration   ET 13.4

1. (a) If $\mathbf{r}(t) = x(t)\,\mathbf{i} + y(t)\,\mathbf{j} + z(t)\,\mathbf{k}$ is the position vector of the particle at time t, then the average velocity over the time interval $[0, 1]$ is

$$\mathbf{v}_{\text{ave}} = \frac{\mathbf{r}(1) - \mathbf{r}(0)}{1 - 0} = \frac{(4.5\,\mathbf{i} + 6.0\,\mathbf{j} + 3.0\,\mathbf{k}) - (2.7\,\mathbf{i} + 9.8\,\mathbf{j} + 3.7\,\mathbf{k})}{1} = 1.8\,\mathbf{i} - 3.8\,\mathbf{j} - 0.7\,\mathbf{k}. \text{ Similarly,}$$

over the other intervals we have

$$[0.5, 1]: \quad \mathbf{v}_{\text{ave}} = \frac{\mathbf{r}(1) - \mathbf{r}(0.5)}{1 - 0.5} = \frac{(4.5\,\mathbf{i} + 6.0\,\mathbf{j} + 3.0\,\mathbf{k}) - (3.5\,\mathbf{i} + 7.2\,\mathbf{j} + 3.3\,\mathbf{k})}{0.5}$$

$$= 2.0\,\mathbf{i} - 2.4\,\mathbf{j} - 0.6\,\mathbf{k}$$

$$[1, 2]: \quad \mathbf{v}_{\text{ave}} = \frac{\mathbf{r}(2) - \mathbf{r}(1)}{2 - 1} = \frac{(7.3\,\mathbf{i} + 7.8\,\mathbf{j} + 2.7\,\mathbf{k}) - (4.5\,\mathbf{i} + 6.0\,\mathbf{j} + 3.0\,\mathbf{k})}{1}$$

$$= 2.8\,\mathbf{i} + 1.8\,\mathbf{j} - 0.3\,\mathbf{k}$$

$$[1, 1.5]: \quad \mathbf{v}_{\text{ave}} = \frac{\mathbf{r}(1.5) - \mathbf{r}(1)}{1.5 - 1} = \frac{(5.9\,\mathbf{i} + 6.4\,\mathbf{j} + 2.8\,\mathbf{k}) - (4.5\,\mathbf{i} + 6.0\,\mathbf{j} + 3.0\,\mathbf{k})}{0.5}$$

$$= 2.8\,\mathbf{i} + 0.8\,\mathbf{j} - 0.4\,\mathbf{k}$$

(b) We can estimate the velocity at $t = 1$ by averaging the average velocities over the time intervals $[0.5, 1]$ and $[1, 1.5]$: $\mathbf{v}(1) \approx \frac{1}{2}[(2\,\mathbf{i} - 2.4\,\mathbf{j} - 0.6\,\mathbf{k}) + (2.8\,\mathbf{i} + 0.8\,\mathbf{j} - 0.4\,\mathbf{k})] = 2.4\,\mathbf{i} - 0.8\,\mathbf{j} - 0.5\,\mathbf{k}$. Then the speed is $|\mathbf{v}(1)| \approx \sqrt{(2.4)^2 + (-0.8)^2 + (-0.5)^2} \approx 2.58$.

3. $\mathbf{r}(t) = \langle t^2 - 1, t \rangle \quad \Rightarrow$ At $t = 1$:

$\mathbf{v}(t) = \mathbf{r}'(t) = \langle 2t, 1 \rangle,$ $\mathbf{v}(1) = \langle 2, 1 \rangle$

$\mathbf{a}(t) = \mathbf{r}''(t) = \langle 2, 0 \rangle,$ $\mathbf{a}(1) = \langle 2, 0 \rangle$

$|\mathbf{v}(t)| = \sqrt{4t^2 + 1}$

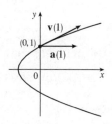

5. $\mathbf{r}(t) = e^t\,\mathbf{i} + e^{-t}\,\mathbf{j} \;\Rightarrow$ At $t = 0$:

$\mathbf{v}(t) = e^t\,\mathbf{i} - e^{-t}\,\mathbf{j},$ $\mathbf{v}(0) = \mathbf{i} - \mathbf{j},$

$\mathbf{a}(t) = e^t\,\mathbf{i} + e^{-t}\,\mathbf{j}$ $\mathbf{a}(0) = \mathbf{i} + \mathbf{j}$

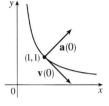

$|\mathbf{v}(t)| = \sqrt{e^{2t} + e^{-2t}} = e^{-t}\sqrt{e^{4t} + 1}$

Since $x = e^t$, $t = \ln x$ and $y = e^{-t} = e^{-\ln x} = 1/x$, and $x > 0$, $y > 0$.

7. $\mathbf{r}(t) = \sin t\,\mathbf{i} + t\,\mathbf{j} + \cos t\,\mathbf{k} \;\Rightarrow$

$\mathbf{v}(t) = \cos t\,\mathbf{i} + \mathbf{j} - \sin t\,\mathbf{k},\, \mathbf{v}(0) = \mathbf{i} + \mathbf{j}$

$\mathbf{a}(t) = -\sin t\,\mathbf{i} - \cos t\,\mathbf{k},\, \mathbf{a}(0) = -\mathbf{k}$

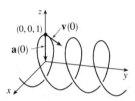

$|\mathbf{v}(t)| = \sqrt{\cos^2 t + 1 + \sin^2 t} = \sqrt{2}$

Since $x^2 + z^2 = 1$, $y = t$, the path of the particle is a helix about the y-axis.

9. $\mathbf{r}(t) = \langle t^2 + 1, t^3, t^2 - 1 \rangle \;\Rightarrow\; \mathbf{v}(t) = \mathbf{r}'(t) = \langle 2t, 3t^2, 2t \rangle,\, \mathbf{a}(t) = \mathbf{v}'(t) = \langle 2, 6t, 2 \rangle,$

$|\mathbf{v}(t)| = \sqrt{(2t)^2 + (3t^2)^2 + (2t)^2} = \sqrt{9t^4 + 8t^2} = |t|\sqrt{9t^2 + 8}.$

11. $\mathbf{r}(t) = \sqrt{2}\,t\,\mathbf{i} + e^t\,\mathbf{j} + e^{-t}\,\mathbf{k} \;\Rightarrow\; \mathbf{v}(t) = \mathbf{r}'(t) = \sqrt{2}\,\mathbf{i} + e^t\,\mathbf{j} - e^{-t}\,\mathbf{k},\, \mathbf{a}(t) = \mathbf{v}'(t) = e^t\,\mathbf{j} + e^{-t}\,\mathbf{k},$

$|\mathbf{v}(t)| = \sqrt{2 + e^{2t} + e^{-2t}} = \sqrt{(e^t + e^{-t})^2} = e^t + e^{-t}.$

13. $\mathbf{r}(t) = e^t\langle \cos t, \sin t, t \rangle \;\Rightarrow$

$\mathbf{v}(t) = \mathbf{r}'(t) = e^t\langle \cos t, \sin t, t \rangle + e^t\langle -\sin t, \cos t, 1 \rangle = e^t\langle \cos t - \sin t, \sin t + \cos t, t + 1 \rangle$

$\mathbf{a}(t) = \mathbf{v}'(t) = e^t\langle \cos t - \sin t - \sin t - \cos t, \sin t + \cos t + \cos t - \sin t, t + 1 + 1 \rangle$

$\qquad = e^t\langle -2\sin t, 2\cos t, t + 2 \rangle$

$|\mathbf{v}(t)| = e^t\sqrt{\cos^2 t + \sin^2 t - 2\cos t\sin t + \sin^2 t + \cos^2 t + 2\sin t\cos t + t^2 + 2t + 1}$

$\qquad = e^t\sqrt{t^2 + 2t + 3}$

15. $\mathbf{a}(t) = \mathbf{k} \;\Rightarrow\; \mathbf{v}(t) = \int \mathbf{a}(t)\,dt = \int \mathbf{k}\,dt = t\,\mathbf{k} + \mathbf{c}_1$ and $\mathbf{i} - \mathbf{j} = \mathbf{v}(0) = 0\,\mathbf{k} + \mathbf{c}_1$, so $\mathbf{c}_1 = \mathbf{i} - \mathbf{j}$ and

$\mathbf{v}(t) = \mathbf{i} - \mathbf{j} + t\,\mathbf{k}.$ $\mathbf{r}(t) = \int \mathbf{v}(t)\,dt = \int (\mathbf{i} - \mathbf{j} + t\,\mathbf{k})\,dt = t\,\mathbf{i} - t\,\mathbf{j} + \frac{1}{2}t^2\,\mathbf{k} + \mathbf{c}_2$. But $\mathbf{0} = \mathbf{r}(0) = \mathbf{0} + \mathbf{c}_2$,

so $\mathbf{c}_2 = \mathbf{0}$ and $\mathbf{r}(t) = t\,\mathbf{i} - t\,\mathbf{j} + \frac{1}{2}t^2\,\mathbf{k}.$

17. (a) $\mathbf{a}(t) = \mathbf{i} + 2\,\mathbf{j} + 2t\,\mathbf{k} \;\Rightarrow$ (b)

$\mathbf{v}(t) = \int (\mathbf{i} + 2\,\mathbf{j} + 2t\,\mathbf{k})\,dt = t\,\mathbf{i} + 2t\,\mathbf{j} + t^2\,\mathbf{k} + \mathbf{c}_1$, and

$\mathbf{0} = \mathbf{v}(0) = \mathbf{0} + \mathbf{c}_1$, so $\mathbf{c}_1 = \mathbf{0}$ and $\mathbf{v}(t) = \mathbf{i} + 2t\,\mathbf{j} + t^2\,\mathbf{k}.$

$\mathbf{r}(t) = \int (t\,\mathbf{i} + 2t\,\mathbf{j} + t^2\,\mathbf{k})\,dt = \frac{1}{2}t^2\,\mathbf{i} + t^2\,\mathbf{j} + \frac{1}{3}t^3\,\mathbf{k} + \mathbf{c}_2.$

But $\mathbf{i} + \mathbf{k} = \mathbf{r}(0) = \mathbf{0} + \mathbf{c}_2$, so $\mathbf{c}_2 = \mathbf{i} + \mathbf{k}$ and

$\mathbf{r}(t) = \left(1 + \frac{1}{2}t^2\right)\mathbf{i} + t^2\,\mathbf{j} + \left(1 + \frac{1}{3}t^3\right)\mathbf{k}.$

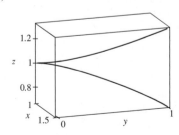

19. $\mathbf{r}(t) = \langle t^2, 5t, t^2 - 16t \rangle \quad \Rightarrow \quad \mathbf{v}(t) = \langle 2t, 5, 2t - 16 \rangle$,

$|\mathbf{v}(t)| = \sqrt{4t^2 + 25 + 4t^2 - 64t + 256} = \sqrt{8t^2 - 64t + 281}$ and

$\dfrac{d}{dt}|\mathbf{v}(t)| = \frac{1}{2}(8t^2 - 64t + 281)^{-1/2}(16t - 64)$. This is zero if and only if the numerator is zero, that is,

$16t - 64 = 0$ or $t = 4$. Since $\dfrac{d}{dt}|\mathbf{v}(t)| < 0$ for $t < 4$ and $\dfrac{d}{dt}|\mathbf{v}(t)| > 0$ for $t > 4$, the minimum speed of $\sqrt{153}$ is

attained at $t = 4$ units of time.

21. $|\mathbf{F}(t)| = 20$ N in the direction of the positive z-axis, so $\mathbf{F}(t) = 20\,\mathbf{k}$. Also $m = 4$ kg, $\mathbf{r}(0) = \mathbf{0}$ and $\mathbf{v}(0) = \mathbf{i} - \mathbf{j}$.
Since $20\mathbf{k} = \mathbf{F}(t) = 4\,\mathbf{a}(t)$, $\mathbf{a}(t) = 5\,\mathbf{k}$. Then $\mathbf{v}(t) = 5t\,\mathbf{k} + \mathbf{c}_1$ where $\mathbf{c}_1 = \mathbf{i} - \mathbf{j}$ so $\mathbf{v}(t) = \mathbf{i} - \mathbf{j} + 5t\,\mathbf{k}$ and the
speed is $|\mathbf{v}(t)| = \sqrt{1 + 1 + 25t^2} = \sqrt{25t^2 + 2}$. Also $\mathbf{r}(t) = t\,\mathbf{i} - t\,\mathbf{j} + \frac{5}{2}t^2\,\mathbf{k} + \mathbf{c}_2$ and $\mathbf{0} = \mathbf{r}(0)$, so $\mathbf{c}_2 = \mathbf{0}$ and
$\mathbf{r}(t) = t\,\mathbf{i} - t\,\mathbf{j} + \frac{5}{2}t^2\,\mathbf{k}$.

23. $|\mathbf{v}(0)| = 500$ m/s and since the angle of elevation is $30°$, the direction of the velocity is $\frac{1}{2}\left(\sqrt{3}\,\mathbf{i} + \mathbf{j}\right)$. Thus
$\mathbf{v}(0) = 250\left(\sqrt{3}\,\mathbf{i} + \mathbf{j}\right)$ and if we set up the axes so the projectile starts at the origin, then $\mathbf{r}(0) = \mathbf{0}$. Ignoring air
resistance, the only force is that due to gravity, so $\mathbf{F}(t) = -mg\,\mathbf{j}$ where $g \approx 9.8$ m/s². Thus $\mathbf{a}(t) = -g\,\mathbf{j}$ and
$\mathbf{v}(t) = -gt\,\mathbf{j} + \mathbf{c}_1$. But $250\left(\sqrt{3}\,\mathbf{i} + \mathbf{j}\right) = \mathbf{v}(0) = \mathbf{c}_1$, so $\mathbf{v}(t) = 250\sqrt{3}\,\mathbf{i} + (250 - gt)\,\mathbf{j}$ and
$\mathbf{r}(t) = 250\sqrt{3}\,t\,\mathbf{i} + \left(250t - \frac{1}{2}gt^2\right)\mathbf{j} + \mathbf{c}_2$ where $\mathbf{0} = \mathbf{r}(0) = \mathbf{c}_2$. Thus $\mathbf{r}(t) = 250\sqrt{3}\,t\,\mathbf{i} + \left(250t - \frac{1}{2}gt^2\right)\mathbf{j}$.

(a) Setting $250t - \frac{1}{2}gt^2 = 0$ gives $t = 0$ or $t = \frac{500}{g} \approx 51.0$ s. So the range is $250\sqrt{3} \cdot \frac{500}{g} \approx 22$ km.

(b) $0 = \dfrac{d}{dt}\left(250t - \frac{1}{2}gt^2\right) = 250 - gt$ implies that the maximum height is attained when $t = 250/g \approx 25.5$ s.

Thus, the maximum height is $(250)(250/g) - g(250/g)^2\frac{1}{2} = (250)^2/(2g) \approx 3.2$ km.

(c) From part (a), impact occurs at $t = 500/g \approx 51.0$. Thus, the velocity at impact is
$\mathbf{v}(500/g) = 250\sqrt{3}\,\mathbf{i} + [250 - g(500/g)]\,\mathbf{j} = 250\sqrt{3}\,\mathbf{i} - 250\,\mathbf{j}$ and the speed is
$|\mathbf{v}(500/g)| = 250\sqrt{3 + 1} = 500$ m/s.

25. As in Example 5, $\mathbf{r}(t) = (v_0 \cos 45°)t\,\mathbf{i} + \left[(v_0 \sin 45°)t - \frac{1}{2}gt^2\right]\mathbf{j} = \frac{1}{2}\left[v_0\sqrt{2}\,t\,\mathbf{i} + \left(v_0\sqrt{2}\,t - gt^2\right)\mathbf{j}\right]$. Then the

ball lands at $t = \dfrac{v_0\sqrt{2}}{g}$ s. Now since it lands 90 m away, $90 = \frac{1}{2}v_0\sqrt{2}\,\dfrac{v_0\sqrt{2}}{g}$ or $v_0^2 = 90g$ and the initial velocity

is $v_0 = \sqrt{90g} \approx 30$ m/s.

27. Let α be the angle of elevation. Then $v_0 = 150$ m/s and from Example 5, the horizontal distance traveled by the

projectile is $d = \dfrac{v_0^2 \sin 2\alpha}{g}$. Thus $\dfrac{150^2 \sin 2\alpha}{g} = 800 \quad \Rightarrow \quad \sin 2\alpha = \dfrac{800g}{150^2} \approx 0.3484 \quad \Rightarrow \quad 2\alpha \approx 20.4°$ or

$180 - 20.4 = 159.6°$. Two angles of elevation then are $\alpha \approx 10.2°$ and $\alpha \approx 79.8°$.

29. (a) After t seconds, the boat will be $5t$ meters west of

point A. The velocity of the water at that location is

$\frac{3}{400}(5t)(40 - 5t)\,\mathbf{j}$. The velocity of the boat in still

water is $5\,\mathbf{i}$, so the resultant velocity of the boat is

$\mathbf{v}(t) = 5\,\mathbf{i} + \frac{3}{400}(5t)(40 - 5t)\,\mathbf{j} = 5\mathbf{i} + \left(\frac{3}{2}t - \frac{3}{16}t^2\right)\mathbf{j}$.

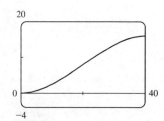

Integrating, we obtain $\mathbf{r}(t) = 5t\,\mathbf{i} + \left(\frac{3}{4}t^2 - \frac{1}{16}t^3\right)\mathbf{j} + \mathbf{C}$.

If we place the origin at A (and consider $\mathbf{j}$ to coincide with the northern direction) then $\mathbf{r}(0) = \mathbf{0} \quad \Rightarrow \quad \mathbf{C} = \mathbf{0}$

and we have $\mathbf{r}(t) = 5t\,\mathbf{i} + \left(\frac{3}{4}t^2 - \frac{1}{16}t^3\right)\mathbf{j}$. The boat reaches the east bank after 8 s, and it is located at

$\mathbf{r}(8) = 5(8)\mathbf{i} + \left(\frac{3}{4}(8)^2 - \frac{1}{16}(8)^3\right)\mathbf{j} = 40\,\mathbf{i} + 16\,\mathbf{j}$. Thus the boat is 16 m downstream.

(b) Let α be the angle north of east that the boat heads. Then the velocity of the boat in still water is given by

$5(\cos\alpha)\,\mathbf{i} + 5(\sin\alpha)\,\mathbf{j}$. At t seconds, the boat is $5(\cos\alpha)t$ meters from the west bank, at which point the

velocity of the water is $\frac{3}{400}[5(\cos\alpha)t][40 - 5(\cos\alpha)t]\,\mathbf{j}$. The resultant velocity of the boat is given by

$$\mathbf{v}(t) = 5(\cos\alpha)\,\mathbf{i} + \left[5\sin\alpha + \frac{3}{400}(5t\cos\alpha)(40 - 5t\cos\alpha)\right]\mathbf{j}$$

$$= (5\cos\alpha)\,\mathbf{i} + \left(5\sin\alpha + \frac{3}{2}t\cos\alpha - \frac{3}{16}t^2\cos^2\alpha\right)\mathbf{j}$$

Integrating, $\mathbf{r}(t) = (5t\cos\alpha)\,\mathbf{i} + \left(5t\sin\alpha + \frac{3}{4}t^2\cos\alpha - \frac{1}{16}t^3\cos^2\alpha\right)\mathbf{j}$ (where we have again placed

the origin at A). The boat will reach the east bank when $5t\cos\alpha = 40 \;\Rightarrow\; t = \dfrac{40}{5\cos\alpha} = \dfrac{8}{\cos\alpha}$.

In order to land at point $B\,(40, 0)$ we need $5t\sin\alpha + \frac{3}{4}t^2\cos\alpha - \frac{1}{16}t^3\cos^2\alpha = 0 \;\Rightarrow$

$5\left(\dfrac{8}{\cos\alpha}\right)\sin\alpha + \dfrac{3}{4}\left(\dfrac{8}{\cos\alpha}\right)^2\cos\alpha - \dfrac{1}{16}\left(\dfrac{8}{\cos\alpha}\right)^3\cos^2\alpha = 0 \;\Rightarrow\; \dfrac{1}{\cos\alpha}(40\sin\alpha + 48 - 32) = 0 \;\Rightarrow$

$40\sin\alpha + 16 = 0 \;\Rightarrow\; \sin\alpha = -\frac{2}{5}$. Thus $\alpha = \sin^{-1}\left(-\frac{2}{5}\right) \approx -23.6°$, so the boat should head 23.6° south

of east (upstream).

The path does seem realistic. The boat initially heads upstream to

counteract the effect of the current. Near the center of the river, the

current is stronger and the boat is pushed downstream. When the

boat nears the eastern bank, the current is slower and the boat is

able to progress upstream to arrive at point B.

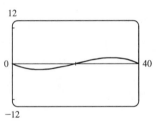

31. $\mathbf{r}(t) = (3t - t^3)\,\mathbf{i} + 3t^2\,\mathbf{j} \;\Rightarrow\; \mathbf{r}'(t) = (3 - 3t^2)\,\mathbf{i} + 6t\,\mathbf{j}$,

$|\mathbf{r}'(t)| = \sqrt{(3 - 3t^2)^2 + (6t)^2} = \sqrt{9 + 18t^2 + 9t^4} = \sqrt{(3 - 3t^2)^2} = 3 + 3t^2$,

$\mathbf{r}''(t) = -6t\,\mathbf{i} + 6\,\mathbf{j},\; \mathbf{r}'(t) \times \mathbf{r}''(t) = (18 + 18t^2)\,\mathbf{k}$. Then Equation 9 gives

$a_T = \dfrac{\mathbf{r}'(t) \cdot \mathbf{r}''(t)}{|\mathbf{r}'(t)|} = \dfrac{(3 - 3t^2)(-6t) + (6t)(6)}{3 + 3t^2} = \dfrac{18t + 18t^3}{3 + 3t^2} = \dfrac{18t(1 + t^2)}{3(1 + t^2)} = 6t$ $\Big[$or by Equation 8,

$a_T = v' = \dfrac{d}{dt}\left[3 + 3t^2\right] = 6t\Big]$ and Equation 10 gives $a_N = \dfrac{|\mathbf{r}'(t) \times \mathbf{r}''(t)|}{|\mathbf{r}'(t)|} = \dfrac{18 + 18t^2}{3 + 3t^2} = \dfrac{18(1 + t^2)}{3(1 + t^2)} = 6$.

33. $\mathbf{r}(t) = \cos t\,\mathbf{i} + \sin t\,\mathbf{j} + t\,\mathbf{k} \;\Rightarrow\; \mathbf{r}'(t) = -\sin t\,\mathbf{i} + \cos t\,\mathbf{j} + \mathbf{k},\; |\mathbf{r}'(t)| = \sqrt{\sin^2 t + \cos^2 t + 1} = \sqrt{2}$,

$\mathbf{r}''(t) = -\cos t\,\mathbf{i} - \sin t\,\mathbf{j},\; \mathbf{r}'(t) \times \mathbf{r}''(t) = \sin t\,\mathbf{i} - \cos t\,\mathbf{j} + \mathbf{k}$.

Then $a_T = \dfrac{\mathbf{r}'(t) \cdot \mathbf{r}''(t)}{|\mathbf{r}'(t)|} = \dfrac{\sin t\cos t - \sin t\cos t}{\sqrt{2}} = 0$ and

$a_N = \dfrac{|\mathbf{r}'(t) \times \mathbf{r}''(t)|}{|\mathbf{r}'(t)|} = \dfrac{\sqrt{\sin^2 t + \cos^2 t + 1}}{\sqrt{2}} = \dfrac{\sqrt{2}}{\sqrt{2}} = 1$.

35. $\mathbf{r}(t) = e^t\,\mathbf{i} + \sqrt{2}\,t\,\mathbf{j} + e^{-t}\,\mathbf{k} \quad \Rightarrow \quad \mathbf{r}'(t) = e^t\,\mathbf{i} + \sqrt{2}\,\mathbf{j} - e^{-t}\,\mathbf{k},$

$|\mathbf{r}(t)| = \sqrt{e^{2t} + 2 + e^{-2t}} = \sqrt{(e^t + e^{-t})^2} = e^t + e^{-t}, \ \mathbf{r}''(t) = e^t\,\mathbf{i} + e^{-t}\,\mathbf{k}.$

Then $a_T = \dfrac{e^{2t} - e^{-2t}}{e^t + e^{-t}} = \dfrac{(e^t + e^{-t})(e^t - e^{-t})}{e^t + e^{-t}} = e^t - e^{-t} = 2\sinh t$ and

$a_N = \dfrac{\left|\sqrt{2}\,e^{-t}\,\mathbf{i} - 2\,\mathbf{j} - \sqrt{2}\,e^t\,\mathbf{k}\right|}{e^t + e^{-t}} = \dfrac{\sqrt{2(e^{-2t} + 2 + e^{2t})}}{e^t + e^{-t}} = \sqrt{2}\,\dfrac{e^t + e^{-t}}{e^t + e^{-t}} = \sqrt{2}.$

37. The tangential component of $\mathbf{a}$ is the length of the projection of $\mathbf{a}$ onto $\mathbf{T}$,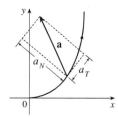
so we sketch the scalar projection of $\mathbf{a}$ in the tangential direction to the
curve and estimate its length to be 4.5 (using the fact that $\mathbf{a}$ has length 10
as a guide). Similarly, the normal component of $\mathbf{a}$ is the length of the
projection of $\mathbf{a}$ onto $\mathbf{N}$, so we sketch the scalar projection of $\mathbf{a}$ in the
normal direction to the curve and estimate its length to be 9.0. Thus

$a_T \approx 4.5\ \text{cm/s}^2$ and $a_N \approx 9.0\ \text{cm/s}^2.$

39. If the engines are turned off at time t, then the spacecraft will continue to travel in the direction of $\mathbf{v}(t)$, so we need

a t such that for some scalar $s > 0$, $\mathbf{r}(t) + s\,\mathbf{v}(t) = \langle 6, 4, 9\rangle.$ $\quad \mathbf{v}(t) = \mathbf{r}'(t) = \mathbf{i} + \dfrac{1}{t}\,\mathbf{j} + \dfrac{8t}{(t^2+1)^2}\,\mathbf{k} \quad \Rightarrow$

$\mathbf{r}(t) + s\,\mathbf{v}(t) = \left\langle 3 + t + s,\ 2 + \ln t + \dfrac{s}{t},\ 7 - \dfrac{4}{t^2+1} + \dfrac{8st}{(t^2+1)^2} \right\rangle \quad \Rightarrow \quad 3 + t + s = 6 \quad \Rightarrow \quad s = 3 - t,$

so $7 - \dfrac{4}{t^2+1} + \dfrac{8(3-t)t}{(t^2+1)^2} = 9 \quad \Leftrightarrow \quad \dfrac{24t - 12t^2 - 4}{(t^2+1)^2} = 2 \quad \Leftrightarrow \quad t^4 + 8t^2 - 12t + 3 = 0.$ It is easily seen that

$t = 1$ is a root of this polynomial. Also $2 + \ln 1 + \dfrac{3-1}{1} = 4$, so $t = 1$ is the desired solution.

14 Review ET 13

CONCEPT CHECK

1. A vector function is a function whose domain is a set of real numbers and whose range is a set of vectors. To find
the derivative or integral, we can differentiate or integrate each component of the vector function.

2. The tip of the moving vector $\mathbf{r}(t)$ of a continuous vector function traces out a space curve.

3. (a) A curve represented by the vector function $\mathbf{r}(t)$ is smooth if $\mathbf{r}'(t)$ is continuous and $\mathbf{r}'(t) \ne \mathbf{0}$ on its parametric
domain (except possibly at the endpoints).

(b) The tangent vector to a smooth curve at a point P with position vector $\mathbf{r}(t)$ is the vector $\mathbf{r}'(t)$. The tangent line
at P is the line through P parallel to the tangent vector $\mathbf{r}'(t)$. The unit tangent vector is $\mathbf{T}(t) = \dfrac{\mathbf{r}'(t)}{|\mathbf{r}'(t)|}.$

4. (a)–(f) See Theorem 14.2.3 [ET 13.2.3].

5. Use Formula 14.3.2 [ET 13.3.2], or equivalently 14.3.3 [ET 13.3.3].

6. (a) The curvature of a curve is $\kappa = \left|\dfrac{d\mathbf{T}}{ds}\right|$ where $\mathbf{T}$ is the unit tangent vector.

(b) $\kappa(t) = \left|\dfrac{\mathbf{T}'(t)}{\mathbf{r}'(t)}\right|$ (c) $\kappa(t) = \dfrac{|\mathbf{r}'(t) \times \mathbf{r}''(t)|}{|\mathbf{r}'(t)|^3}$ (d) $\kappa(x) = \dfrac{|f''(x)|}{[1 + (f'(x))^2]^{3/2}}$

7. (a) The unit normal vector: $\mathbf{N}(t) = \dfrac{\mathbf{T}'(t)}{|\mathbf{T}'(t)|}$. The binormal vector: $\mathbf{B}(t) = \mathbf{T}(t) \times \mathbf{N}(t)$.

(b) See the discussion preceding Example 7 in Section 14.3 [ET 13.3].

8. (a) If $\mathbf{r}(t)$ is the position vector of the particle on the space curve, the velocity $\mathbf{v}(t) = \mathbf{r}'(t)$, the speed is given by $|\mathbf{v}(t)|$, and the acceleration $\mathbf{a}(t) = \mathbf{v}'(t) = \mathbf{r}''(t)$.

(b) $\mathbf{a} = a_T \mathbf{T} + a_N \mathbf{N}$ where $a_T = v'$ and $a_N = \kappa v^2$.

9. See the statement of Kepler's Laws on page 912 [ET 876].

─────────────────────── TRUE-FALSE QUIZ ───────────────────────

1. True. If we reparametrize the curve by replacing $u = t^3$, we have $\mathbf{r}(u) = u\,\mathbf{i} + 2u\,\mathbf{j} + 3u\,\mathbf{k}$, which is a line through the origin with direction vector $\mathbf{i} + 2\,\mathbf{j} + 3\,\mathbf{k}$.

3. False. $\mathbf{r}'(t) = \langle -\sin t, 2t, 4t^3 \rangle$, and since $\mathbf{r}'(0) = \langle 0, 0, 0 \rangle = \mathbf{0}$, the curve is not smooth.

5. False. By Formula 5 of Theorem 14.2.3 [ET 13.2.3], $\dfrac{d}{dt}[\mathbf{u}(t) \times \mathbf{v}(t)] = \mathbf{u}'(t) \times \mathbf{v}(t) + \mathbf{u}(t) \times \mathbf{v}'(t)$.

7. False. κ is the magnitude of the rate of change of the unit tangent vector $\mathbf{T}$ with respect to arc length s, not with respect to t.

9. True. See the discussion preceding Example 7 in Section 14.3 [ET 13.3].

─────────────────────── EXERCISES ───────────────────────

1. (a) The corresponding parametric equations for the curve are $x = t$, $y = \cos \pi t$, $z = \sin \pi t$. Since $y^2 + z^2 = 1$, the curve is contained in a circular cylinder with axis the x-axis. Since $x = t$, the curve is a helix.

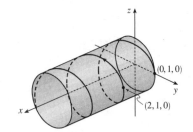

(b) $\mathbf{r}(t) = t\,\mathbf{i} + \cos \pi t\,\mathbf{j} + \sin \pi t\,\mathbf{k} \;\Rightarrow\; \mathbf{r}'(t) = \mathbf{i} - \pi \sin \pi t\,\mathbf{j} + \pi \cos \pi t\,\mathbf{k} \;\Rightarrow$
$\mathbf{r}''(t) = -\pi^2 \cos \pi t\,\mathbf{j} - \pi^2 \sin \pi t\,\mathbf{k}$

3. The projection of the curve C of intersection onto the xy-plane is the circle $x^2 + y^2 = 16$, $z = 0$. So we can write $x = 4 \cos t$, $y = 4 \sin t$, $0 \le t \le 2\pi$. From the equation of the plane, we have $z = 5 - x = 5 - 4 \cos t$, so parametric equations for C are $x = 4 \cos t$, $y = 4 \sin t$, $z = 5 - 4 \cos t$, $0 \le t \le 2\pi$, and the corresponding vector function is $\mathbf{r}(t) = 4 \cos t\,\mathbf{i} + 4 \sin t\,\mathbf{j} + (5 - 4 \cos t)\,\mathbf{k}$, $0 \le t \le 2\pi$.

5. $\displaystyle \int_0^1 (t^2\,\mathbf{i} + t \cos \pi t\,\mathbf{j} + \sin \pi t\,\mathbf{k})\,dt = \left(\int_0^1 t^2\,dt \right)\mathbf{i} + \left(\int_0^1 t \cos \pi t\,dt \right)\mathbf{j} + \left(\int_0^1 \sin \pi t\,dt \right)\mathbf{k}$

$\qquad = \left[\tfrac{1}{3}t^3 \right]_0^1 \mathbf{i} + \left(\tfrac{t}{\pi} \sin \pi t \right]_0^1 - \int_0^1 \tfrac{1}{\pi} \sin \pi t\,dt \right)\mathbf{j} + \left[-\tfrac{1}{\pi} \cos \pi t \right]_0^1 \mathbf{k}$

$\qquad = \tfrac{1}{3}\,\mathbf{i} + \left[\tfrac{1}{\pi^2} \cos \pi t \right]_0^1 \mathbf{j} + \tfrac{2}{\pi}\,\mathbf{k} = \tfrac{1}{3}\,\mathbf{i} - \tfrac{2}{\pi^2}\,\mathbf{j} + \tfrac{2}{\pi}\,\mathbf{k}$

where we integrated by parts in the y-component.

7. $t = 1$ at $(1, 4, 2)$ and $t = 4$ at $(2, 1, 17)$, so

$$L = \int_1^4 \sqrt{\frac{1}{4t} + \frac{16}{t^4} + 4t^2}\, dt$$

$$\approx \frac{4-1}{3 \cdot 4}\left[\sqrt{\tfrac{1}{4} + 16 + 4} \; + \; 4 \cdot \sqrt{\frac{1}{4 \cdot \frac{7}{4}} + \frac{16}{\left(\frac{7}{4}\right)^4} + 4\left(\tfrac{7}{4}\right)^2} \; + \; 2 \cdot \sqrt{\frac{1}{4 \cdot \frac{10}{4}} + \frac{16}{\left(\frac{10}{4}\right)^4} + 4\left(\tfrac{10}{4}\right)^2}\right.$$

$$\left. + \; 4 \cdot \sqrt{\frac{1}{4 \cdot \frac{13}{4}} + \frac{16}{\left(\frac{13}{4}\right)^4} + 4\left(\tfrac{13}{4}\right)^2} \; + \; \sqrt{\frac{1}{4 \cdot 4} + \frac{16}{4^4} + 4 \cdot 4^2}\right]$$

$$\approx 15.9241$$

9. The angle of intersection of the two curves, θ, is the angle between their respective tangents at the point of intersection. For both curves the point $(1, 0, 0)$ occurs when $t = 0$. $\quad \mathbf{r}_1'(t) = -\sin t\,\mathbf{i} + \cos t\,\mathbf{j} + \mathbf{k} \quad \Rightarrow$ $\mathbf{r}_1'(0) = \mathbf{j} + \mathbf{k}$ and $\mathbf{r}_2'(t) = \mathbf{i} + 2t\,\mathbf{j} + 3t^2\,\mathbf{k} \quad \Rightarrow \quad \mathbf{r}_2'(0) = \mathbf{i}$. $\quad \mathbf{r}_1'(0) \cdot \mathbf{r}_2'(0) = (\mathbf{j} + \mathbf{k}) \cdot \mathbf{i} = 0$. Therefore, the curves intersect in a right angle, that is, $\theta = \frac{\pi}{2}$.

11. (a) $\mathbf{T}(t) = \dfrac{\mathbf{r}'(t)}{|\mathbf{r}'(t)|} = \dfrac{\langle t^2, t, 1\rangle}{|\langle t^2, t, 1\rangle|} = \dfrac{\langle t^2, t, 1\rangle}{\sqrt{t^4 + t^2 + 1}}$

(b) $\mathbf{T}'(t) = -\frac{1}{2}(t^4 + t^2 + 1)^{-3/2}(4t^3 + 2t)\langle t^2, t, 1\rangle + (t^4 + t^2 + 1)^{-1/2}\langle 2t, 1, 0\rangle$

$$= \frac{-2t^3 - t}{(t^4 + t^2 + 1)^{3/2}}\langle t^2, t, 1\rangle + \frac{1}{(t^4 + t^2 + 1)^{1/2}}\langle 2t, 1, 0\rangle$$

$$= \frac{\langle -2t^5 - t^3, -2t^4 - t^2, -2t^3 - t\rangle + \langle 2t^5 + 2t^3 + 2t, t^4 + t^2 + 1, 0\rangle}{(t^4 + t^2 + 1)^{3/2}}$$

$$= \frac{\langle 2t, -t^4 + 1, -2t^3 - t\rangle}{(t^4 + t^2 + 1)^{3/2}}$$

$$|\mathbf{T}'(t)| = \frac{\sqrt{4t^2 + t^8 - 2t^4 + 1 + 4t^6 + 4t^4 + t^2}}{(t^4 + t^2 + 1)^{3/2}} = \frac{\sqrt{t^8 + 4t^6 + 2t^4 + 5t^2}}{(t^4 + t^2 + 1)^{3/2}}, \text{ and}$$

$$\mathbf{N}(t) = \frac{\langle 2t, 1 - t^4, -2t^3 - t\rangle}{\sqrt{t^8 + 4t^6 + 2t^4 + 5t^2}}.$$

(c) $\kappa(t) = \dfrac{|\mathbf{T}'(t)|}{|\mathbf{r}'(t)|} = \dfrac{\sqrt{t^8 + 4t^6 + 2t^4 + 5t^2}}{(t^4 + t^2 + 1)^2}$

13. $y' = 4x^3$, $y'' = 12x^2$ and $\kappa(x) = \dfrac{|y''|}{[1 + (y')^2]^{3/2}} = \dfrac{|12x^2|}{(1 + 16x^6)^{3/2}}$, so $\kappa(1) = \dfrac{12}{17^{3/2}}$.

15. $\mathbf{r}(t) = \langle \sin 2t, t, \cos 2t\rangle \quad \Rightarrow \quad \mathbf{r}'(t) = \langle 2\cos 2t, 1, -2\sin 2t\rangle \quad \Rightarrow \quad \mathbf{T}(t) = \frac{1}{\sqrt{5}}\langle 2\cos 2t, 1, -2\sin 2t\rangle \quad \Rightarrow$ $\mathbf{T}'(t) = \frac{1}{\sqrt{5}}\langle -4\sin 2t, 0, -4\cos 2t\rangle \quad \Rightarrow \quad \mathbf{N}(t) = \langle -\sin 2t, 0, -\cos 2t\rangle$. So $\mathbf{N} = \mathbf{N}(\pi) = \langle 0, 0, -1\rangle$ and $\mathbf{B} = \mathbf{T} \times \mathbf{N} = \frac{1}{\sqrt{5}}\langle -1, 2, 0\rangle$. So a normal to the osculating plane is $\langle -1, 2, 0\rangle$ and an equation is $-1(x - 0) + 2(y - \pi) + 0(z - 1) = 0$ or $x - 2y + 2\pi = 0$.

17. $r(t) = t \ln t \, i + t \, j + e^{-t} \, k$, $v(t) = r'(t) = (1 + \ln t) \, i + j - e^{-t} \, k$,

$|v(t)| = \sqrt{(1 + \ln t)^2 + 1^2 + (-e^{-t})^2} = \sqrt{2 + 2 \ln t + (\ln t)^2 + e^{-2t}}$, $a(t) = v'(t) = \frac{1}{t} \, i + e^{-t} \, k$.

19. We set up the axes so that the shot leaves the athlete's hand 7 ft above the origin. Then we are given $r(0) = 7j$,

$|v(0)| = 43$ ft/s, and $v(0)$ has direction given by a $45°$ angle of elevation. Then a unit vector in the direction of

$v(0)$ is $\frac{1}{\sqrt{2}}(i + j)$ $\Rightarrow$ $v(0) = \frac{43}{\sqrt{2}}(i + j)$. Assuming air resistance is negligible, the only external force is due

to gravity, so as in Example 14.4.5 [ET 13.4.5] we have $a = -gj$ where here $g \approx 32$ ft/s². Since $v'(t) = a(t)$, we

integrate, giving $v(t) = -gt \, j + C$ where $C = v(0) = \frac{43}{\sqrt{2}}(i + j)$ $\Rightarrow$ $v(t) = \frac{43}{\sqrt{2}} \, i + \left(\frac{43}{\sqrt{2}} - gt \right) j$. Since

$r'(t) = v(t)$ we integrate again, so $r(t) = \frac{43}{\sqrt{2}} t \, i + \left(\frac{43}{\sqrt{2}} t - \frac{1}{2} gt^2 \right) j + D$. But $D = r(0) = 7j$ $\Rightarrow$

$r(t) = \frac{43}{\sqrt{2}} t \, i + \left(\frac{43}{\sqrt{2}} t - \frac{1}{2} gt^2 + 7 \right) j$.

(a) At 2 seconds, the shot is at $r(2) = \frac{43}{\sqrt{2}}(2) \, i + \left(\frac{43}{\sqrt{2}}(2) - \frac{1}{2} g(2)^2 + 7 \right) j \approx 60.8 \, i + 3.8 \, j$, so the shot is about

3.8 ft above the ground, at a horizontal distance of 60.8 ft from the athlete.

(b) The shot reaches its maximum height when the vertical component of velocity is 0: $\frac{43}{\sqrt{2}} - gt = 0$ $\Rightarrow$

$t = \dfrac{43}{\sqrt{2} \, g} \approx 0.95$ s. Then $r(0.95) \approx 28.9 \, i + 21.4 \, j$, so the maximum height is approximately 21.4 ft.

(c) The shot hits the ground when the vertical component of $r(t)$ is 0, so $\frac{43}{\sqrt{2}} t - \frac{1}{2} gt^2 + 7 = 0$ $\Rightarrow$

$-16t^2 + \frac{43}{\sqrt{2}} t + 7 = 0$ $\Rightarrow$ $t \approx 2.11$ s. $r(2.11) \approx 64.2 \, i - 0.08 \, j$, thus the shot lands approximately 64.2 ft

from the athlete.

21. (a) Instead of proceeding directly, we use Formula 3 of Theorem 14.2.3 [ET 13.2.3]:

$r(t) = t \, R(t)$ $\Rightarrow$ $v = r'(t) = R(t) + t \, R'(t) = \cos \omega t \, i + \sin \omega t \, j + t \, v_d$.

(b) Using the same method as in part (a) and starting with $v = R(t) + t \, R'(t)$, we have

$a = v' = R'(t) + R'(t) + t \, R''(t) = 2 \, R'(t) + t \, R''(t) = 2 \, v_d + t \, a_d$.

(c) Here we have $r(t) = e^{-t} \cos \omega t \, i + e^{-t} \sin \omega t \, j = e^{-t} \, R(t)$. So, as in parts (a) and (b),

$v = r'(t) = e^{-t} \, R'(t) - e^{-t} \, R(t) = e^{-t}[R'(t) - R(t)]$ $\Rightarrow$

$a = v' = e^{-t}[R''(t) - R'(t)] - e^{-t}[R'(t) - R(t)] = e^{-t}[R''(t) - 2 \, R'(t) + R(t)]$

$= e^{-t} \, a_d - 2e^{-t} \, v_d + e^{-t} \, R$

Thus, the Coriolis acceleration (the sum of the "extra" terms not involving a_d) is $-2e^{-t} \, v_d + e^{-t} \, R$.

□ PROBLEMS PLUS

1. (a) $\mathbf{r}(t) = R\cos\omega t\,\mathbf{i} + R\sin\omega t\,\mathbf{j}$ $\Rightarrow$ $\mathbf{v} = \mathbf{r}'(t) = -\omega R\sin\omega t\,\mathbf{i} + \omega R\cos\omega t\,\mathbf{j}$, so $\mathbf{r} = R(\cos\omega t\,\mathbf{i} + \sin\omega t\,\mathbf{j})$
and $\mathbf{v} = \omega R(-\sin\omega t\,\mathbf{i} + \cos\omega t\,\mathbf{j})$. $\mathbf{v}\cdot\mathbf{r} = \omega R^2(-\cos\omega t\sin\omega t + \sin\omega t\cos\omega t) = 0$, so $\mathbf{v}\perp\mathbf{r}$. Since $\mathbf{r}$
points along a radius of the circle, and $\mathbf{v}\perp\mathbf{r}$, $\mathbf{v}$ is tangent to the circle. Because it is a velocity vector, $\mathbf{v}$ points
in the direction of motion.

(b) In (a), we wrote $\mathbf{v}$ in the form $\omega R\,\mathbf{u}$, where $\mathbf{u}$ is the unit vector $-\sin\omega t\,\mathbf{i} + \cos\omega t\,\mathbf{j}$. Clearly
$|\mathbf{v}| = \omega R\,|\mathbf{u}| = \omega R$. At speed ωR, the particle completes one revolution, a distance $2\pi R$, in time
$$T = \frac{2\pi R}{\omega R} = \frac{2\pi}{\omega}.$$

(c) $\mathbf{a} = \dfrac{d\mathbf{v}}{dt} = -\omega^2 R\cos\omega t\,\mathbf{i} - \omega^2 R\sin\omega t\,\mathbf{j} = -\omega^2 R(\cos\omega t\,\mathbf{i} + \sin\omega t\,\mathbf{j})$, so $\mathbf{a} = -\omega^2\mathbf{r}$. This shows that $\mathbf{a}$ is
proportional to $\mathbf{r}$ and points in the opposite direction (toward the origin). Also, $|\mathbf{a}| = \omega^2\,|\mathbf{r}| = \omega^2 R$.

(d) By Newton's Second Law (see Section 14.4 [ET 13.4]), $\mathbf{F} = m\mathbf{a}$, so
$$|\mathbf{F}| = m\,|\mathbf{a}| = mR\omega^2 = \frac{m\,(\omega R)^2}{R} = \frac{m\,|\mathbf{v}|^2}{R}.$$

3. (a) The projectile reaches maximum height when $0 = \dfrac{dy}{dt} = \dfrac{d}{dt}\left[(v_0\sin\alpha)t - \tfrac{1}{2}gt^2\right] = v_0\sin\alpha - gt$; that is, when
$t = \dfrac{v_0\sin\alpha}{g}$ and $y = (v_0\sin\alpha)\left(\dfrac{v_0\sin\alpha}{g}\right) - \dfrac{1}{2}g\left(\dfrac{v_0\sin\alpha}{g}\right)^2 = \dfrac{v_0^2\sin^2\alpha}{2g}$. This is the maximum height
attained when the projectile is fired with an angle of elevation α. This maximum height is largest when $\alpha = \frac{\pi}{2}$.
In that case, $\sin\alpha = 1$ and the maximum height is $\dfrac{v_0^2}{2g}$.

(b) Let $R = v_0^2/g$. We are asked to consider the parabola $x^2 + 2Ry - R^2 = 0$ which can be rewritten as
$y = -\dfrac{1}{2R}x^2 + \dfrac{R}{2}$. The points on or inside this parabola are those for which $-R \le x \le R$ and
$0 \le y \le \dfrac{-1}{2R}x^2 + \dfrac{R}{2}$. When the projectile is fired at angle of elevation α, the points (x, y) along its path
satisfy the relations $x = (v_0\cos\alpha)\,t$ and $y = (v_0\sin\alpha)t - \tfrac{1}{2}gt^2$, where $0 \le t \le (2v_0\sin\alpha)/g$ (as in
Example 14.4.5 [ET 13.4.5]). Thus $|x| \le \left|v_0\cos\alpha\left(\dfrac{2v_0\sin\alpha}{g}\right)\right| = \left|\dfrac{v_0^2}{g}\sin 2\alpha\right| \le \left|\dfrac{v_0^2}{g}\right| = |R|$. This shows
that $-R \le x \le R$.

For t in the specified range, we also have $y = t\left(v_0\sin\alpha - \tfrac{1}{2}gt\right) = \tfrac{1}{2}gt\left(\dfrac{2v_0\sin\alpha}{g} - t\right) \ge 0$ and
$$y = (v_0\sin\alpha)\frac{x}{v_0\cos\alpha} - \frac{g}{2}\left(\frac{x}{v_0\cos\alpha}\right)^2 = (\tan\alpha)\,x - \frac{g}{2v_0^2\cos^2\alpha}x^2 = -\frac{1}{2R\cos^2\alpha}x^2 + (\tan\alpha)\,x.$$
Thus
$$y - \left(\frac{-1}{2R}x^2 + \frac{R}{2}\right) = \frac{-1}{2R\cos^2\alpha}x^2 + \frac{1}{2R}x^2 + (\tan\alpha)\,x - \frac{R}{2}$$
$$= \frac{x^2}{2R}\left(1 - \frac{1}{\cos^2\alpha}\right) + (\tan\alpha)\,x - \frac{R}{2} = \frac{x^2(1 - \sec^2\alpha) + 2R(\tan\alpha)\,x - R^2}{2R}$$
$$= \frac{-(\tan^2\alpha)\,x^2 + 2R(\tan\alpha)\,x - R^2}{2R} = \frac{-[(\tan\alpha)\,x - R]^2}{2R} \le 0$$

We have shown that every target that can be hit by the projectile lies on or inside the parabola

$y = -\dfrac{1}{2R}x^2 + \dfrac{R}{2}$. Now let (a, b) be any point on or inside the parabola $y = -\dfrac{1}{2R}x^2 + \dfrac{R}{2}$. Then

$-R \le a \le R$ and $0 \le b \le -\dfrac{1}{2R}a^2 + \dfrac{R}{2}$. We seek an angle α such that (a, b) lies in the path of the projectile;

that is, we wish to find an angle α such that $b = -\dfrac{1}{2R\cos^2\alpha}a^2 + (\tan\alpha)\,a$ or

equivalently $b = \dfrac{-1}{2R}(\tan^2\alpha + 1)a^2 + (\tan\alpha)\,a$. Rearranging this equation we get

$\dfrac{a^2}{2R}\tan^2\alpha - a\tan\alpha + \left(\dfrac{a^2}{2R} + b\right) = 0$ or $a^2(\tan\alpha)^2 - 2aR(\tan\alpha) + (a^2 + 2bR) = 0$ $(*)$. This quadratic

equation for $\tan\alpha$ has real solutions exactly when the discriminant is nonnegative. Now $B^2 - 4AC \ge 0$ $\Leftrightarrow$

$(-2aR)^2 - 4a^2(a^2 + 2bR) \ge 0$ $\Leftrightarrow$ $4a^2(R^2 - a^2 - 2bR) \ge 0$ $\Leftrightarrow$ $-a^2 - 2bR + R^2 \ge 0$ $\Leftrightarrow$

$b \le \dfrac{1}{2R}(R^2 - a^2)$ $\Leftrightarrow$ $b \le \dfrac{-1}{2R}a^2 + \dfrac{R}{2}$. This condition is satisfied since (a, b) is on or inside the parabola

$y = -\dfrac{1}{2R}x^2 + \dfrac{R}{2}$. It follows that (a, b) lies in the path of the projectile when $\tan\alpha$ satisfies $(*)$, that is, when

$\tan\alpha = \dfrac{2aR \pm \sqrt{4a^2(R^2 - a^2 - 2bR)}}{2a^2} = \dfrac{R \pm \sqrt{R^2 - 2bR - a^2}}{a}.$

(c)

If the gun is pointed at a target with height h at a distance D downrange,

then $\tan\alpha = h/D$. When the projectile reaches a distance D downrange

(remember we are assuming that it doesn't hit the ground first), we have

$$D = x = (v_0\cos\alpha)t, \text{ so } t = \dfrac{D}{v_0\cos\alpha} \text{ and}$$

$y = (v_0\sin\alpha)t - \tfrac{1}{2}gt^2 = D\tan\alpha - \dfrac{gD^2}{2v_0^2\cos^2\alpha}$. Meanwhile, the target, whose x-coordinate is also D, has

fallen from height h to height $h - \tfrac{1}{2}gt^2 = D\tan\alpha - \dfrac{gD^2}{2v_0^2\cos^2\alpha}$. Thus the projectile hits the target.

5. (a) $m\dfrac{d^2\mathbf{R}}{dt^2} = -mg\mathbf{j} - k\dfrac{d\mathbf{R}}{dt}$ $\Rightarrow$ $\dfrac{d}{dt}\left(m\dfrac{d\mathbf{R}}{dt} + k\mathbf{R} + mgt\mathbf{j}\right) = 0$ $\Rightarrow$ $m\dfrac{d\mathbf{R}}{dt} + k\mathbf{R} + mgt\mathbf{j} = \mathbf{c}$

(**c** is a constant vector in the xy-plane). At $t = 0$, this says that $m\mathbf{v}(0) + k\mathbf{R}(0) = \mathbf{c}$. Since $\mathbf{v}(0) = \mathbf{v}_0$ and

$\mathbf{R}(0) = \mathbf{0}$, we have $\mathbf{c} = m\mathbf{v}_0$. Therefore $\dfrac{d\mathbf{R}}{dt} + \dfrac{k}{m}\mathbf{R} + gt\mathbf{j} = \mathbf{v}_0$, or $\dfrac{d\mathbf{R}}{dt} + \dfrac{k}{m}\mathbf{R} = \mathbf{v}_0 - gt\mathbf{j}$.

(**b**) Multiplying by $e^{(k/m)t}$ gives $e^{(k/m)t}\dfrac{d\mathbf{R}}{dt} + \dfrac{k}{m}e^{(k/m)t}\mathbf{R} = e^{(k/m)t}\mathbf{v}_0 - gte^{(k/m)t}\mathbf{j}$ or

$\dfrac{d}{dt}\left(e^{(k/m)t}\mathbf{R}\right) = e^{(k/m)t}\mathbf{v}_0 - gte^{(k/m)t}\mathbf{j}$. Integrating gives

$e^{(k/m)t}\mathbf{R} = \dfrac{m}{k}e^{(k/m)t}\mathbf{v}_0 - \left[\dfrac{mg}{k}te^{(k/m)t} - \dfrac{m^2g}{k^2}e^{(k/m)t}\right]\mathbf{j} + \mathbf{b}$ for some constant vector $\mathbf{b}$.

Setting $t = 0$ yields the relation $\mathbf{R}(0) = \dfrac{m}{k}\mathbf{v}_0 + \dfrac{m^2 g}{k^2}\mathbf{j} + \mathbf{b}$, so $\mathbf{b} = -\dfrac{m}{k}\mathbf{v}_0 - \dfrac{m^2 g}{k^2}\mathbf{j}$. Thus

$$e^{(k/m)t}\,\mathbf{R} = \frac{m}{k}\left[e^{(k/m)t} - 1\right]\mathbf{v}_0 - \left[\frac{mg}{k}te^{(k/m)t} - \frac{m^2 g}{k^2}\left(e^{(k/m)t} - 1\right)\right]\mathbf{j} \text{ and}$$

$$\mathbf{R}(t) = \frac{m}{k}\left[1 - e^{-kt/m}\right]\mathbf{v}_0 + \frac{mg}{k}\left[\frac{m}{k}(1 - e^{-kt/m}) - t\right]\mathbf{j}.$$

7. (a) $\mathbf{a} = -g\,\mathbf{j} \;\Rightarrow\; \mathbf{v} = \mathbf{v}_0 - gt\,\mathbf{j} = 2\,\mathbf{i} - gt\,\mathbf{j} \;\Rightarrow\; \mathbf{s} = \mathbf{s}_0 + 2t\,\mathbf{i} - \tfrac{1}{2}gt^2\,\mathbf{j} = 3.5\,\mathbf{j} + 2t\,\mathbf{i} - \tfrac{1}{2}gt^2\,\mathbf{j} \;\Rightarrow\;$

$\mathbf{s} = 2t\,\mathbf{i} + \left(3.5 - \tfrac{1}{2}gt^2\right)\mathbf{j}$. Therefore $y = 0$ when $t = \sqrt{7/g}$ seconds. At that instant, the ball is

$2\sqrt{7/g} \approx 0.94$ ft to the right of the table top. Its coordinates (relative to an origin on the floor directly under

the table's edge) are $(0.94, 0)$. At impact, the velocity is $\mathbf{v} = 2\,\mathbf{i} - \sqrt{7g}\,\mathbf{j}$, so the speed is

$|\mathbf{v}| = \sqrt{4 + 7g} \approx 15$ ft/s.

(b) The slope of the curve when $t = \sqrt{\dfrac{7}{g}}$ is $\dfrac{dy}{dx} = \dfrac{dy/dt}{dx/dt} = \dfrac{-gt}{2} = \dfrac{-g\sqrt{7/g}}{2} = \dfrac{-\sqrt{7g}}{2}$. Thus $\cot\theta = \dfrac{\sqrt{7g}}{2}$

and $\theta \approx 7.6°$.

(c) From (a), $|\mathbf{v}| = \sqrt{4 + 7g}$. So the ball rebounds with speed $0.8\sqrt{4 + 7g} \approx 12.08$ ft/s at angle of inclination

$90° - \theta \approx 82.3886°$. By Example 14.4.5 [ET 13.4.5], the horizontal distance traveled between bounces is

$d = \dfrac{v_0^2 \sin 2\alpha}{g}$, where $v_0 \approx 12.08$ ft/s and $\alpha \approx 82.3886°$. Therefore, $d \approx 1.197$ ft. So the ball strikes the floor

at about $2\sqrt{7/g} + 1.197 \approx 2.13$ ft to the right of the table's edge.

15 □ PARTIAL DERIVATIVES

15.1 Functions of Several Variables

1. (a) From Table 1, $f(-15, 40) = -27$, which means that if the temperature is $-15°C$ and the wind speed is 40 km/h, then the air would feel equivalent to approximately $-27°C$ without wind.

   (b) The question is asking: when the temperature is $-20°C$, what wind speed gives a wind-chill index of $-30°C$? From Table 1, the speed is 20 km/h.

   (c) The question is asking: when the wind speed is 20 km/h, what temperature gives a wind-chill index of $-49°C$? From Table 1, the temperature is $-35°C$.

   (d) The function $W = f(-5, v)$ means that we fix T at -5 and allow v to vary, resulting in a function of one variable. In other words, the function gives wind-chill index values for different wind speeds when the temperature is $-5°C$. From Table 1 (look at the row corresponding to $T = -5$), the function decreases and appears to approach a constant value as v increases.

   (e) The function $W = f(T, 50)$ means that we fix v at 50 and allow T to vary, again giving a function of one variable. In other words, the function gives wind-chill index values for different temperatures when the wind speed is 50 km/h . From Table 1 (look at the column corresponding to $v = 50$), the function increases almost linearly as T increases.

3. If the amounts of labor and capital are both doubled, we replace L, K in the function with $2L, 2K$, giving

$$P(2L, 2K) = 1.01(2L)^{0.75}(2K)^{0.25} = 1.01(2^{0.75})(2^{0.25})L^{0.75}K^{0.25} = (2^1)1.01L^{0.75}K^{0.25}$$
$$= 2P(L, K)$$

Thus, the production is doubled. It is also true for the general case $P(L, K) = bL^{\alpha}K^{1-\alpha}$:

$$P(2L, 2K) = b(2L)^{\alpha}(2K)^{1-\alpha} = b(2^{\alpha})(2^{1-\alpha})L^{\alpha}K^{1-\alpha} = (2^{\alpha+1-\alpha})bL^{\alpha}K^{1-\alpha} = 2P(L, K).$$

5. (a) According to the table, $f(40, 15) = 25$, which means that if a 40-knot wind has been blowing in the open sea for 15 hours, it will create waves with estimated heights of 25 feet.

   (b) $h = f(30, t)$ means we fix v at 30 and allow t to vary, resulting in a function of one variable. Thus here, $h = f(30, t)$ gives the wave heights produced by 30-knot winds blowing for t hours. From the table (look at the row corresponding to $v = 30$), the function increases but at a declining rate as t increases. In fact, the function values appear to be approaching a limiting value of approximately 19, which suggests that 30-knot winds cannot produce waves higher than about 19 feet.

   (c) $h = f(v, 30)$ means we fix t at 30, again giving a function of one variable. So, $h = f(v, 30)$ gives the wave heights produced by winds of speed v blowing for 30 hours. From the table (look at the column corresponding to $t = 30$), the function appears to increase at an increasing rate, with no apparent limiting value. This suggests that faster winds (lasting 30 hours) always create higher waves.

7. (a) $f(2,0) = 2^2 e^{3(2)(0)} = 4(1) = 4$

(b) Since both x^2 and the exponential function are defined everywhere, $x^2 e^{3xy}$ is defined for all choices of values for x and y. Thus, the domain of f is $\mathbb{R}^2$.

(c) Because the range of $g(x,y) = 3xy$ is $\mathbb{R}$, and the range of e^x is $(0, \infty)$, the range of $e^{g(x,y)} = e^{3xy}$ is $(0, \infty)$. The range of x^2 is $[0, \infty)$, so the range of the product $x^2 e^{3xy}$ is $[0, \infty)$.

9. (a) $f(2,-1,6) = e^{\sqrt{6-2^2-(-1)^2}} = e^{\sqrt{1}} = e$.

(b) $e^{\sqrt{z-x^2-y^2}}$ is defined when $z - x^2 - y^2 \geq 0 \Rightarrow z \geq x^2 + y^2$. Thus the domain of f is
$$\{(x,y,z) \mid z \geq x^2 + y^2\}.$$

(c) Since $\sqrt{z - x^2 - y^2} \geq 0$, we have $e^{\sqrt{z-x^2-y^2}} \geq 1$. Thus the range of f is $[1, \infty)$.

11. $\sqrt{x+y}$ is defined only when $x + y \geq 0$, or $y \geq -x$. So the domain of f is $\{(x,y) \mid y \geq -x\}$.

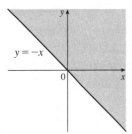

13. $\ln(9 - x^2 - 9y^2)$ is defined only when $9 - x^2 - 9y^2 > 0$, or $\frac{1}{9}x^2 + y^2 < 1$. So the domain of f is $\{(x,y) \mid \frac{1}{9}x^2 + y^2 < 1\}$, the interior of an ellipse.

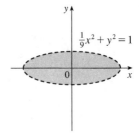

15. $\dfrac{3x+5y}{x^2+y^2-4}$ is defined only when $x^2 + y^2 - 4 \neq 0$, or $x^2 + y^2 \neq 4$. So the domain of f is $\{(x,y) \mid x^2 + y^2 \neq 4\}$.

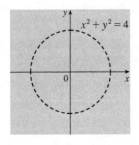

17. $\sqrt{y - x^2}$ is defined only when $y - x^2 \geq 0$, or $y \geq x^2$. In addition, f is not defined if $1 - x^2 = 0 \Rightarrow x = \pm 1$. Thus the domain of f is
$$\{(x,y) \mid y \geq x^2, x \neq \pm 1\}.$$

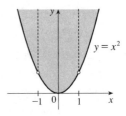

19. We need $1 - x^2 - y^2 - z^2 \geq 0$ or

$x^2 + y^2 + z^2 \leq 1$, so

$D = \{(x, y, z) \mid x^2 + y^2 + z^2 \leq 1\}$

(the points inside or on the sphere of radius 1, center the origin).

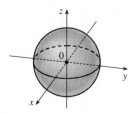

21. $z = 3$, a horizontal plane through the point $(0, 0, 3)$.

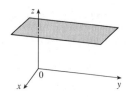

23. $z = 1 - x - y$ or $x + y + z = 1$, a plane with intercepts 1, 1, and 1.

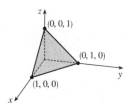

25. $z = 1 - x^2$, a parabolic cylinder.

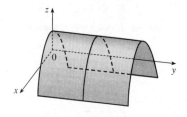

27. $z = 4x^2 + y^2 + 1$, an elliptic paraboloid with vertex at $(0, 0, 1)$.

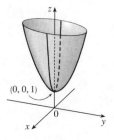

29. $z = \sqrt{x^2 + y^2}$ so $x^2 + y^2 = z^2$ and $z \geq 0$, the top half of a right circular cone.

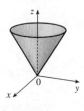

31. The point $(-3, 3)$ lies between the level curves with z-values 50 and 60. Since the point is a little closer to the level curve with $z = 60$, we estimate that $f(-3, 3) \approx 56$. The point $(3, -2)$ appears to be just about halfway between the level curves with z-values 30 and 40, so we estimate $f(3, -2) \approx 35$. The graph rises as we approach the origin, gradually from above, steeply from below.

33. Near A, the level curves are very close together, indicating that the terrain is quite steep. At B, the level curves are much farther apart, so we would expect the terrain to be much less steep than near A, perhaps almost flat.

35.

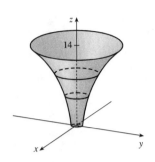

37. The level curves are $xy = k$. For $k = 0$ the curves are the coordinate axis; if $k > 0$, they are hyperbolas in the first and third quadrants; if $k < 0$, they are hyperbolas in the second and fourth quadrants.

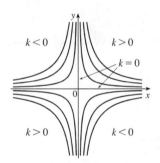

39. The level curves are $y - \ln x = k$ or $y = \ln x + k$.

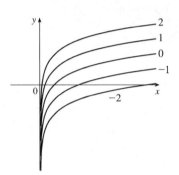

41. $k = \sqrt{x + y}$ or for $x + y \geq 0$, $k^2 = x + y$, or $y = -x + k^2$.

Note: $k \geq 0$ since $k = \sqrt{x + y}$.

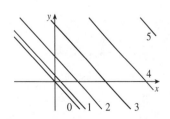

43. $k = x - y^2$, or $x - k = y^2$, a family of parabolas with vertex $(k, 0)$.

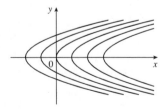

45. The contour map consists of the level curves $k = x^2 + 9y^2$, a family
of ellipses with major axis the x-axis. (Or, if $k = 0$, the origin.)
The graph of $f(x, y)$ is the surface $z = x^2 + 9y^2$, an elliptic
paraboloid.

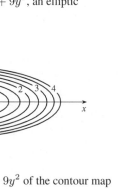

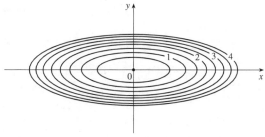

If we visualize lifting each ellipse $k = x^2 + 9y^2$ of the contour map
to the plane $z = k$, we have horizontal traces that indicate the shape
of the graph of f.

47. The isothermals are given by $k = 100/(1 + x^2 + 2y^2)$ or
$x^2 + 2y^2 = (100 - k)/k$ $(0 < k \le 100)$, a family of ellipses.

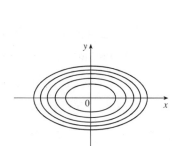

49. $f(x, y) = x^3 + y^3$

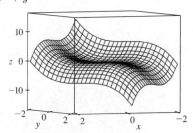

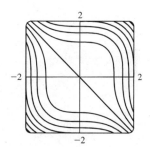

Note that the function is 0 along the line $y = -x$.

51. $f(x, y) = xy^2 - x^3$

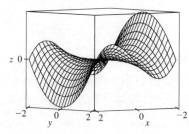

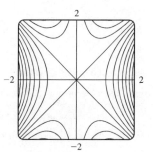

The traces parallel to the yz-plane (such as the left-front trace in the graph above) are parabolas; those parallel to the xz-plane (such as the right-front trace) are cubic curves. The surface is called a monkey saddle because a monkey sitting on the surface near the origin has places for both legs and tail to rest.

53. (a) B   *Reasons:* This function is constant on any circle centered at the origin, a description which matches
(b) III   only B and III.

55. (a) F   *Reasons:* z increases without bound as we use points closer to the origin, a condition satisfied only
(b) V   by F and V.

57. (a) D   *Reasons:* This function is periodic in both x and y, with period 2π in each variable.
(b) IV

59. $k = x + 3y + 5z$ is a family of parallel planes with normal vector $\langle 1, 3, 5 \rangle$.

61. $k = x^2 - y^2 + z^2$ are the equations of the level surfaces. For $k = 0$, the surface is a right circular cone with vertex the origin and axis the y-axis. For $k > 0$, we have a family of hyperboloids of one sheet with axis the y-axis. For $k < 0$, we have a family of hyperboloids of two sheets with axis the y-axis.

63. (a) The graph of g is the graph of f shifted upward 2 units.

(b) The graph of g is the graph of f stretched vertically by a factor of 2.

(c) The graph of g is the graph of f reflected about the xy-plane.

(d) The graph of $g(x, y) = -f(x, y) + 2$ is the graph of f reflected about the xy-plane and then shifted upward 2 units.

65. $f(x, y) = 3x - x^4 - 4y^2 - 10xy$

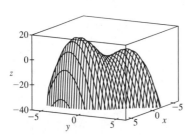

Three-dimensional view

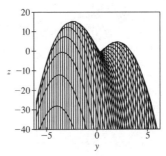

Front view

It does appear that the function has a maximum value, at the higher of the two "hilltops." From the front view graph, the maximum value appears to be approximately 15. Both hilltops could be considered local maximum points, as the values of f there are larger than at the neighboring points. There does not appear to be any local minimum point; although the valley shape between the two peaks looks like a minimum of some kind, some neighboring points have lower function values.

67.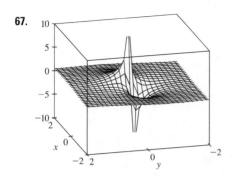

$f(x, y) = \dfrac{x + y}{x^2 + y^2}$. As both x and y become large, the function values appear to approach 0, regardless of which direction is considered. As (x, y) approaches the origin, the graph exhibits asymptotic behavior. From some directions, $f(x, y) \to \infty$, while in others $f(x, y) \to -\infty$. (These are the vertical spikes visible in the graph.) If the graph is examined carefully, however, one can see that $f(x, y)$ approaches 0 along the line $y = -x$.

69. $f(x, y) = e^{cx^2 + y^2}$. First, if $c = 0$, the graph is the cylindrical surface $z = e^{y^2}$ (whose level curves are parallel lines). When $c > 0$, the vertical trace above the y-axis remains fixed while the sides of the surface in the x-direction "curl" upward, giving the graph a shape resembling an elliptic paraboloid. The level curves of the surface are ellipses centered at the origin.

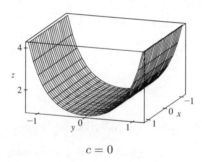

$c = 0$

For $0 < c < 1$, the ellipses have major axis the x-axis and the eccentricity increases as $c \to 0$.

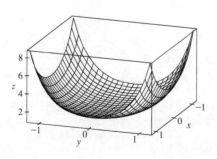

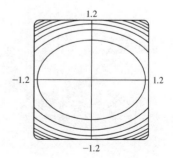

$c = 0.5$ (level curves in increments of 1)

For $c = 1$ the level curves are circles centered at the origin.

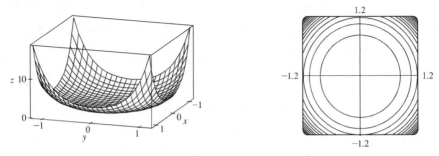

$c = 1$ (level curves in increments of 1)

When $c > 1$, the level curves are ellipses with major axis the y-axis, and the eccentricity increases as c increases.

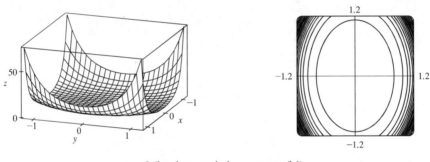

$c = 2$ (level curves in increments of 4)

For values of $c < 0$, the sides of the surface in the x-direction curl downward and approach the xy-plane (while the vertical trace $x = 0$ remains fixed), giving a saddle-shaped appearance to the graph near the point $(0, 0, 1)$. The level curves consist of a family of hyperbolas. As c decreases, the surface becomes flatter in the x-direction and the surface's approach to the curve in the trace $x = 0$ becomes steeper, as the graphs demonstrate.

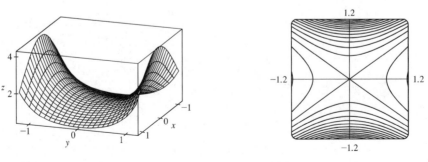

$c = -0.5$ (level curves in increments of 0.25)

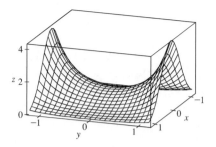 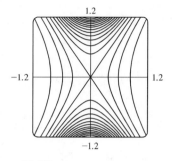

$c = -2$ (level curves in increments of 0.25)

71. (a) $P = bL^\alpha K^{1-\alpha}$ $\Rightarrow$ $\dfrac{P}{K} = bL^\alpha K^{-\alpha}$ $\Rightarrow$ $\dfrac{P}{K} = b\left(\dfrac{L}{K}\right)^\alpha$ $\Rightarrow$ $\ln\dfrac{P}{K} = \ln\left(b\left(\dfrac{L}{K}\right)^\alpha\right)$ $\Rightarrow$

$\ln\dfrac{P}{K} = \ln b + \alpha\ln\left(\dfrac{L}{K}\right)$

(b) We list the values for $\ln(L/K)$ and $\ln(P/K)$ for the years 1899–1922. (Historically, these values were rounded to 2 decimal places.)

Year	$x = \ln(L/K)$	$y = \ln(P/K)$		Year	$x = \ln(L/K)$	$y = \ln(P/K)$
1899	0	0		1911	−0.38	−0.34
1900	−0.02	−0.06		1912	−0.38	−0.24
1901	−0.04	−0.02		1913	−0.41	−0.25
1902	−0.04	0		1914	−0.47	−0.37
1903	−0.07	−0.05		1915	−0.53	−0.34
1904	−0.13	−0.12		1916	−0.49	−0.28
1905	−0.18	−0.04		1917	−0.53	−0.39
1906	−0.20	−0.07		1918	−0.60	−0.50
1907	−0.23	−0.15		1919	−0.68	−0.57
1908	−0.41	−0.38		1920	−0.74	−0.57
1909	−0.33	−0.24		1921	−1.05	−0.85
1910	−0.35	−0.27		1922	−0.98	−0.59

After entering the (x, y) pairs into a calculator or CAS, the resulting least squares regression line through the points is approximately $y = 0.75136x + 0.01053$, which we round to $y = 0.75x + 0.01$.

(c) Comparing the regression line from part (b) to the equation $y = \ln b + \alpha x$ with $x = \ln(L/K)$ and $y = \ln(P/K)$, we have $\alpha = 0.75$ and $\ln b = 0.01$ $\Rightarrow$ $b = e^{0.01} \approx 1.01$. Thus, the Cobb-Douglas production function is $P = bL^\alpha K^{1-\alpha} = 1.01L^{0.75}K^{0.25}$.

15.2 Limits and Continuity ET 14.2

1. In general, we can't say anything about $f(3, 1)$! $\displaystyle\lim_{(x,y)\to(3,1)} f(x, y) = 6$ means that the values of $f(x, y)$ approach

6 as (x, y) approaches, but is not equal to, $(3, 1)$. If f is continuous, we know that $\displaystyle\lim_{(x,y)\to(a,b)} f(x, y) = f(a, b)$, so

$\displaystyle\lim_{(x,y)\to(3,1)} f(x, y) = f(3, 1) = 6$.

3. We make a table of values of $f(x, y) = \dfrac{x^2 y^3 + x^3 y^2 - 5}{2 - xy}$ for a set of (x, y) points near the origin.

y / x	−0.2	−0.1	−0.05	0	0.05	0.1	0.2
−0.2	−2.551	−2.525	−2.513	−2.500	−2.488	−2.475	−2.451
−0.1	−2.525	−2.513	−2.506	−2.500	−2.494	−2.488	−2.475
−0.05	−2.513	−2.506	−2.503	−2.500	−2.497	−2.494	−2.488
0	−2.500	−2.500	−2.500		−2.500	−2.500	−2.500
0.05	−2.488	−2.494	−2.497	−2.500	−2.503	−2.506	−2.513
0.1	−2.475	−2.488	−2.494	−2.500	−2.506	−2.513	−2.525
0.2	−2.451	−2.475	−2.488	−2.500	−2.513	−2.525	−2.551

As the table shows, the values of $f(x, y)$ seem to approach -2.5 as (x, y) approaches the origin from a variety of different directions. This suggests that $\displaystyle\lim_{(x,y)\to(0,0)} f(x, y) = -2.5$.

Since f is a rational function, it is continuous on its domain. f is defined at $(0, 0)$, so we can use direct substitution

to establish that $\displaystyle\lim_{(x,y)\to(0,0)} f(x, y) = \dfrac{0^2 0^3 + 0^3 0^2 - 5}{2 - 0 \cdot 0} = -\dfrac{5}{2}$, verifying our guess.

5. $f(x, y) = x^5 + 4x^3 y - 5xy^2$ is a polynomial, and hence continuous, so

$\displaystyle\lim_{(x,y)\to(5,-2)} f(x, y) = f(5, -2) = 5^5 + 4(5)^3(-2) - 5(5)(-2)^2 = 2025$.

7. $f(x, y) = x^2/(x^2 + y^2)$. First approach $(0, 0)$ along the x-axis. Then $f(x, 0) = x^2/x^2 = 1$ for $x \neq 0$, so $f(x, y) \to 1$. Now approach $(0, 0)$ along the y-axis. Then for $y \neq 0$, $f(0, y) = 0$, so $f(x, y) \to 0$. Since f has two different limits along two different lines, the limit does not exist.

9. $f(x, y) = (xy \cos y)/(3x^2 + y^2)$. On the x-axis, $f(x, 0) = 0$ for $x \neq 0$, so $f(x, y) \to 0$ as $(x, y) \to (0, 0)$ along the x-axis. Approaching $(0, 0)$ along the line $y = x$, $f(x, x) = (x^2 \cos x)/4x^2 = \frac{1}{4} \cos x$ for $x \neq 0$, so $f(x, y) \to \frac{1}{4}$ along this line. Thus the limit does not exist.

11. $f(x, y) = \dfrac{xy}{\sqrt{x^2 + y^2}}$. We can see that the limit along any line through $(0, 0)$ is 0, as well as along other paths

through $(0, 0)$ such as $x = y^2$ and $y = x^2$. So we suspect that the limit exists and equals 0; we use the Squeeze

Theorem to prove our assertion. $0 \leq \left| \dfrac{xy}{\sqrt{x^2 + y^2}} \right| \leq |x|$ since $|y| \leq \sqrt{x^2 + y^2}$, and $|x| \to 0$ as $(x, y) \to (0, 0)$.

So $\displaystyle\lim_{(x,y)\to(0,0)} f(x, y) = 0$.

13. Let $f(x, y) = \dfrac{2x^2 y}{x^4 + y^2}$. Then $f(x, 0) = 0$ for $x \neq 0$, so $f(x, y) \to 0$ as $(x, y) \to (0, 0)$ along the x-axis. But

$f(x, x^2) = \dfrac{2x^4}{2x^4} = 1$ for $x \neq 0$, so $f(x, y) \to 1$ as $(x, y) \to (0, 0)$ along the parabola $y = x^2$. Thus the limit

doesn't exist.

15. $\displaystyle \lim_{(x,y) \to (0,0)} \frac{x^2 + y^2}{\sqrt{x^2 + y^2 + 1} - 1} = \lim_{(x,y) \to (0,0)} \frac{x^2 + y^2}{\sqrt{x^2 + y^2 + 1} - 1} \cdot \frac{\sqrt{x^2 + y^2 + 1} + 1}{\sqrt{x^2 + y^2 + 1} + 1}$

$$= \lim_{(x,y) \to (0,0)} \frac{\left(x^2 + y^2\right)\left(\sqrt{x^2 + y^2 + 1} + 1\right)}{x^2 + y^2}$$

$$= \lim_{(x,y) \to (0,0)} \left(\sqrt{x^2 + y^2 + 1} + 1\right) = 2$$

17. e^{-xy} and $\sin(\pi z / 2)$ are each compositions of continuous functions, and hence continuous, so their product

$f(x, y, z) = e^{-xy} \sin(\pi z / 2)$ is a continuous function. Then

$$\lim_{(x,y,z) \to (3,0,1)} f(x, y, z) = f(3, 0, 1) = e^{-(3)(0)} \sin(\pi \cdot 1/2) = 1.$$

19. $f(x, y, z) = \dfrac{xy + yz^2 + xz^2}{x^2 + y^2 + z^4}$. Then $f(x, 0, 0) = 0/x^2 = 0$ for $x \neq 0$, so as $(x, y, z) \to (0, 0, 0)$ along the x-axis,

$f(x, y, z) \to 0$. But $f(x, x, 0) = x^2/(2x^2) = \frac{1}{2}$ for $x \neq 0$, so as $(x, y, z) \to (0, 0, 0)$ along the line $y = x, z = 0$,

$f(x, y, z) \to \frac{1}{2}$. Thus the limit doesn't exist.

21.

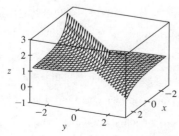

From the ridges on the graph, we see that as $(x, y) \to (0, 0)$ along

the lines under the two ridges, $f(x, y)$ approaches different values.

So the limit does not exist.

23. $h(x, y) = g(f(x, y)) = (2x + 3y - 6)^2 + \sqrt{2x + 3y - 6}$. Since f is a polynomial, it is continuous on $\mathbb{R}^2$

and g is continuous on its domain $\{t \mid t \geq 0\}$. Thus h is continuous on its domain

$D = \{(x, y) \mid 2x + 3y - 6 \geq 0\} = \{(x, y) \mid y \geq -\frac{2}{3}x + 2\}$, which consists of all points on or above the

line $y = -\frac{2}{3}x + 2$.

25.

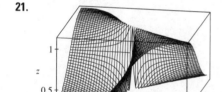

From the graph, it appears that f is discontinuous along the line

$y = x$. If we consider $f(x, y) = e^{1/(x-y)}$ as a composition of

functions, $g(x, y) = 1/(x - y)$ is a rational function and therefore

continuous except where $x - y = 0 \Rightarrow y = x$. Since the

function $h(t) = e^t$ is continuous everywhere, the composition

$h(g(x, y)) = e^{1/(x-y)} = f(x, y)$ is continuous except along the

line $y = x$, as we suspected.

27. The functions $\sin(xy)$ and $e^x - y^2$ are continuous everywhere, so $F(x, y) = \dfrac{\sin(xy)}{e^x - y^2}$ is continuous except where

$e^x - y^2 = 0 \ \Rightarrow \ y^2 = e^x \ \Rightarrow \ y = \pm\sqrt{e^x} = \pm e^{\frac{1}{2}x}$. Thus F is continuous on its domain
$\{(x, y) \mid y \neq \pm e^{x/2}\}$.

29. $F(x, y) = \arctan\left(x + \sqrt{y}\right) = g(f(x, y))$ where $f(x, y) = x + \sqrt{y}$, continuous on its domain $\{(x, y) \mid y \geq 0\}$,
and $g(t) = \arctan t$ is continuous everywhere. Thus F is continuous on its domain $\{(x, y) \mid y \geq 0\}$.

31. $G(x, y) = \ln\left(x^2 + y^2 - 4\right) = g(f(x, y))$ where $f(x, y) = x^2 + y^2 - 4$, continuous on $\mathbb{R}^2$, and
$g(t) = \ln t$, continuous on its domain $\{t \mid t > 0\}$. Thus G is continuous on its domain
$\{(x, y) \mid x^2 + y^2 - 4 > 0\} = \{(x, y) \mid x^2 + y^2 > 4\}$, the exterior of the circle $x^2 + y^2 = 4$.

33. $\sqrt{y}$ is continuous on its domain $\{y \mid y \geq 0\}$ and $x^2 - y^2 + z^2$ is continuous everywhere, so

$f(x, y, z) = \dfrac{\sqrt{y}}{x^2 - y^2 + z^2}$ is continuous for $y \geq 0$ and $x^2 - y^2 + z^2 \neq 0 \ \Rightarrow \ y^2 \neq x^2 + z^2$, that is,
$\left\{(x, y, z) \mid y \geq 0, y \neq \sqrt{x^2 + z^2}\,\right\}$.

35. $f(x, y) = \begin{cases} \dfrac{x^2 y^3}{2x^2 + y^2} & \text{if } (x, y) \neq (0, 0) \\ 1 & \text{if } (x, y) = (0, 0) \end{cases}$ The first piece of f is a rational function defined everywhere except

at the origin, so f is continuous on $\mathbb{R}^2$ except possibly at the origin. Since $x^2 \leq 2x^2 + y^2$, we have
$\left|x^2 y^3/(2x^2 + y^2)\right| \leq \left|y^3\right|$. We know that $\left|y^3\right| \to 0$ as $(x, y) \to (0, 0)$. So, by the Squeeze Theorem,

$\lim\limits_{(x,y) \to (0,0)} f(x, y) = \lim\limits_{(x,y) \to (0,0)} \dfrac{x^2 y^3}{2x^2 + y^2} = 0$. But $f(0, 0) = 1$, so f is discontinuous at $(0, 0)$. Therefore, f is

continuous on the set $\{(x, y) \mid (x, y) \neq (0, 0)\}$.

37. $\lim\limits_{(x,y) \to (0,0)} \dfrac{x^3 + y^3}{x^2 + y^2} = \lim\limits_{r \to 0^+} \dfrac{(r\cos\theta)^3 + (r\sin\theta)^3}{r^2} = \lim\limits_{r \to 0^+} \left(r\cos^3\theta + r\sin^3\theta\right) = 0$

39. $\lim\limits_{(x,y,z) \to (0,0,0)} \dfrac{xyz}{x^2 + y^2 + z^2} = \lim\limits_{\rho \to 0^+} \dfrac{(\rho\sin\phi\cos\theta)(\rho\sin\phi\sin\theta)(\rho\cos\phi)}{\rho^2}$

$= \lim\limits_{\rho \to 0^+} \left(\rho\sin^2\phi\cos\phi\sin\theta\cos\theta\right) = 0$

41. Since $|\mathbf{x} - \mathbf{a}|^2 = |\mathbf{x}|^2 + |\mathbf{a}|^2 - 2\,|\mathbf{x}|\,|\mathbf{a}|\cos\theta \geq |\mathbf{x}|^2 + |\mathbf{a}|^2 - 2\,|\mathbf{x}|\,|\mathbf{a}| = (|\mathbf{x}| - |\mathbf{a}|)^2$, we have
$\big||\mathbf{x}| - |\mathbf{a}|\big| \leq |\mathbf{x} - \mathbf{a}|$. Let $\epsilon > 0$ be given and set $\delta = \epsilon$. Then whenever $0 < |\mathbf{x} - \mathbf{a}| < \delta$,
$\big||\mathbf{x}| - |\mathbf{a}|\big| \leq |\mathbf{x} - \mathbf{a}| < \delta = \epsilon$. Hence $\lim\limits_{\mathbf{x} \to \mathbf{a}} |\mathbf{x}| = |\mathbf{a}|$ and $f(\mathbf{x}) = |\mathbf{x}|$ is continuous on $\mathbb{R}^n$.

15.3 Partial Derivatives ET 14.3

1. (a) $\partial T/\partial x$ represents the rate of change of T when we fix y and t and consider T as a function of the single

variable x, which describes how quickly the temperature changes when longitude changes but latitude and time

are constant. $\partial T/\partial y$ represents the rate of change of T when we fix x and t and consider T as a function of y,

which describes how quickly the temperature changes when latitude changes but longitude and time are

constant. $\partial T/\partial t$ represents the rate of change of T when we fix x and y and consider T as a function of t,

which describes how quickly the temperature changes over time for a constant longitude and latitude.

(b) $f_x(158, 21, 9)$ represents the rate of change of temperature at longitude $158°$W, latitude $21°$N at 9:00 A.M. when only longitude varies. Since the air is warmer to the west than to the east, increasing longitude results in an increased air temperature, so we would expect $f_x(158, 21, 9)$ to be positive. $f_y(158, 21, 9)$ represents the rate of change of temperature at the same time and location when only latitude varies. Since the air is warmer to the south and cooler to the north, increasing latitude results in a decreased air temperature, so we would expect $f_y(158, 21, 9)$ to be negative. $f_t(158, 21, 9)$ represents the rate of change of temperature at the same time and location when only time varies. Since typically air temperature increases from the morning to the afternoon as the sun warms it, we would expect $f_t(158, 21, 9)$ to be positive.

3. (a) By Definition 4, $f_T(-15, 30) = \lim\limits_{h \to 0} \dfrac{f(-15 + h, 30) - f(-15, 30)}{h}$, which we can approximate by

considering $h = 5$ and $h = -5$ and using the values given in the table:

$$f_T(-15, 30) \approx \frac{f(-10, 30) - f(-15, 30)}{5} = \frac{-20 - (-26)}{5} = \frac{6}{5} = 1.2,$$

$$f_T(-15, 30) \approx \frac{f(-20, 30) - f(-15, 30)}{-5} = \frac{-33 - (-26)}{-5} = \frac{-7}{-5} = 1.4.$$ Averaging these values, we

estimate $f_T(-15, 30)$ to be approximately 1.3. Thus, when the actual temperature is $-15°$C and the wind speed is 30 km/h, the apparent temperature rises by about $1.3°$C for every degree that the actual temperature rises.

Similarly, $f_v(-15, 30) = \lim\limits_{h \to 0} \dfrac{f(-15, 30 + h) - f(-15, 30)}{h}$ which we can approximate by considering

$h = 10$ and $h = -10$: $f_v(-15, 30) \approx \dfrac{f(-15, 40) - f(-15, 30)}{10} = \dfrac{-27 - (-26)}{10} = \dfrac{-1}{10} = -0.1,$

$f_v(-15, 30) \approx \dfrac{f(-15, 20) - f(-15, 30)}{-10} = \dfrac{-24 - (-26)}{-10} = \dfrac{2}{-10} = -0.2.$ Averaging these values, we

estimate $f_v(-15, 30)$ to be approximately -0.15. Thus, when the actual temperature is $-15°$C and the wind speed is 30 km/h, the apparent temperature decreases by about $0.15°$C for every km/h that the wind speed increases.

(b) For a fixed wind speed v, the values of the wind-chill index W increase as temperature T increases (look at a

column of the table), so $\dfrac{\partial W}{\partial T}$ is positive. For a fixed temperature T, the values of W decrease (or remain

constant) as v increases (look at a row of the table), so $\dfrac{\partial W}{\partial v}$ is negative (or perhaps 0).

(c) For fixed values of T, the function values $f(T, v)$ appear to become constant (or nearly constant) as v increases, so the corresponding rate of change is 0 or near 0 as v increases. This suggests that $\lim\limits_{v \to \infty} (\partial W / \partial v) = 0$.

5. (a) If we start at $(1, 2)$ and move in the positive x-direction, the graph of f increases. Thus $f_x(1, 2)$ is positive.

(b) If we start at $(1, 2)$ and move in the positive y-direction, the graph of f decreases. Thus $f_y(1, 2)$ is negative.

7. First of all, if we start at the point $(3, -3)$ and move in the positive y-direction, we see that both b and c decrease, while a increases. Both b and c have a low point at about $(3, -1.5)$, while a is 0 at this point. So a is definitely the graph of f_y, and one of b and c is the graph of f. To see which is which, we start at the point $(-3, -1.5)$ and move in the positive x-direction. b traces out a line with negative slope, while c traces out a parabola opening downward. This tells us that b is the x-derivative of c. So c is the graph of f, b is the graph of f_x, and a is the graph of f_y.

9. $f(x, y) = 16 - 4x^2 - y^2 \Rightarrow f_x(x, y) = -8x$ and $f_y(x, y) = -2y \Rightarrow f_x(1, 2) = -8$ and $f_y(1, 2) = -4$.

The graph of f is the paraboloid $z = 16 - 4x^2 - y^2$ and the vertical plane $y = 2$ intersects it in the parabola

$z = 12 - 4x^2$, $y = 2$ (the curve C_1 in the first figure).

The slope of the tangent line to this parabola at $(1, 2, 8)$ is $f_x(1, 2) = -8$. Similarly the plane $x = 1$ intersects the

paraboloid in the parabola $z = 12 - y^2$, $x = 1$ (the curve C_2 in the second figure) and the slope of the tangent line

at $(1, 2, 8)$ is $f_y(1, 2) = -4$.

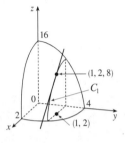 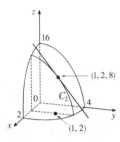

11. $f(x, y) = x^2 + y^2 + x^2 y \Rightarrow f_x = 2x + 2xy$, $f_y = 2y + x^2$

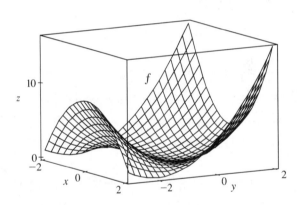 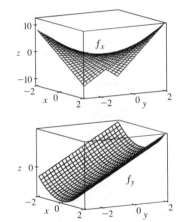

Note that the traces of f in planes parallel to the xz-plane are parabolas which open downward for $y < -1$ and
upward for $y > -1$, and the traces of f_x in these planes are straight lines, which have negative slopes for $y < -1$
and positive slopes for $y > -1$. The traces of f in planes parallel to the yz-plane are parabolas which always open
upward, and the traces of f_y in these planes are straight lines with positive slopes.

13. $f(x, y) = 3x - 2y^4 \Rightarrow f_x(x, y) = 3 - 0 = 3$, $f_y(x, y) = 0 - 8y^3 = -8y^3$

15. $z = xe^{3y} \Rightarrow \dfrac{\partial z}{\partial x} = e^{3y}$, $\dfrac{\partial z}{\partial y} = 3xe^{3y}$

17. $f(x, y) = \dfrac{x - y}{x + y} \Rightarrow f_x(x, y) = \dfrac{(1)(x + y) - (x - y)(1)}{(x + y)^2} = \dfrac{2y}{(x + y)^2}$,

$f_y(x, y) = \dfrac{(-1)(x + y) - (x - y)(1)}{(x + y)^2} = -\dfrac{2x}{(x + y)^2}$

19. $w = \sin \alpha \cos \beta \Rightarrow \dfrac{\partial w}{\partial \alpha} = \cos \alpha \cos \beta, \dfrac{\partial w}{\partial \beta} = -\sin \alpha \sin \beta$

21. $f(r, s) = r \ln(r^2 + s^2) \Rightarrow f_r(r, s) = r \cdot \dfrac{2r}{r^2 + s^2} + \ln(r^2 + s^2) \cdot 1 = \dfrac{2r^2}{r^2 + s^2} + \ln(r^2 + s^2)$,

$f_s(r, s) = r \cdot \dfrac{2s}{r^2 + s^2} + 0 = \dfrac{2rs}{r^2 + s^2}$

23. $u = te^{w/t} \Rightarrow \dfrac{\partial u}{\partial t} = t \cdot e^{w/t}(-wt^{-2}) + e^{w/t} \cdot 1 = e^{w/t} - \dfrac{w}{t}e^{w/t} = e^{w/t}\left(1 - \dfrac{w}{t}\right), \dfrac{\partial u}{\partial w} = te^{w/t} \cdot \dfrac{1}{t} = e^{w/t}$

25. $f(x, y, z) = xy^2 z^3 + 3yz \Rightarrow f_x(x, y, z) = y^2 z^3, f_y(x, y, z) = 2xyz^3 + 3z, f_z(x, y, z) = 3xy^2 z^2 + 3y$

27. $w = \ln(x + 2y + 3z) \Rightarrow \dfrac{\partial w}{\partial x} = \dfrac{1}{x + 2y + 3z}, \dfrac{\partial w}{\partial y} = \dfrac{2}{x + 2y + 3z}, \dfrac{\partial w}{\partial z} = \dfrac{3}{x + 2y + 3z}$

29. $u = xe^{-t} \sin \theta \Rightarrow \dfrac{\partial u}{\partial x} = e^{-t} \sin \theta, \dfrac{\partial u}{\partial t} = -xe^{-t} \sin \theta, \dfrac{\partial u}{\partial \theta} = xe^{-t} \cos \theta$

31. $f(x, y, z, t) = xyz^2 \tan(yt) \Rightarrow f_x(x, y, z, t) = yz^2 \tan(yt)$,

$f_y(x, y, z, t) = xyz^2 \cdot \sec^2(yt) \cdot t + xz^2 \tan(yt) = xyz^2 t \sec^2(yt) + xz^2 \tan(yt)$,

$f_z(x, y, z, t) = 2xyz \tan(yt), f_t(x, y, z, t) = xyz^2 \sec^2(yt) \cdot y = xy^2 z^2 \sec^2(yt)$.

33. $u = \sqrt{x_1^2 + x_2^2 + \cdots + x_n^2}$. For each $i = 1, \ldots, n$,

$u_{x_i} = \frac{1}{2}\left(x_1^2 + x_2^2 + \cdots + x_n^2\right)^{-1/2}(2x_i) = \dfrac{x_i}{\sqrt{x_1^2 + x_2^2 + \cdots + x_n^2}}$.

35. $f(x, y) = \sqrt{x^2 + y^2} \Rightarrow f_x(x, y) = \frac{1}{2}(x^2 + y^2)^{-1/2}(2x) = \dfrac{x}{\sqrt{x^2 + y^2}}$, so $f_x(3, 4) = \dfrac{3}{\sqrt{3^2 + 4^2}} = \dfrac{3}{5}$.

37. $f(x, y, z) = \dfrac{x}{y + z} = x(y + z)^{-1} \Rightarrow f_z(x, y, z) = x(-1)(y + z)^{-2} = -\dfrac{x}{(y + z)^2}$, so

$f_z(3, 2, 1) = -\dfrac{3}{(2 + 1)^2} = -\dfrac{1}{3}$.

39. $f(x, y) = x^2 - xy + 2y^2 \Rightarrow$

$$f_x(x, y) = \lim_{h \to 0} \dfrac{f(x + h, y) - f(x, y)}{h} = \lim_{h \to 0} \dfrac{(x + h)^2 - (x + h)y + 2y^2 - (x^2 - xy + 2y^2)}{h}$$

$$= \lim_{h \to 0} \dfrac{h(2x - y + h)}{h} = \lim_{h \to 0}(2x - y + h) = 2x - y$$

$$f_y(x, y) = \lim_{h \to 0} \dfrac{f(x, y + h) - f(x, y)}{h} = \lim_{h \to 0} \dfrac{x^2 - x(y + h) + 2(y + h)^2 - (x^2 - xy + 2y^2)}{h}$$

$$= \lim_{h \to 0} \dfrac{h(4y - x + 2h)}{h} = \lim_{h \to 0}(4y - x + 2h) = 4y - x$$

41. $x^2 + y^2 + z^2 = 3xyz \Rightarrow \dfrac{\partial}{\partial x}(x^2 + y^2 + z^2) = \dfrac{\partial}{\partial x}(3xyz) \Rightarrow 2x + 0 + 2z\dfrac{\partial z}{\partial x} = 3y\left(x\dfrac{\partial z}{\partial x} + z\cdot 1\right)$

$\Leftrightarrow 2z\dfrac{\partial z}{\partial x} - 3xy\dfrac{\partial z}{\partial x} = 3yz - 2x \Leftrightarrow (2z - 3xy)\dfrac{\partial z}{\partial x} = 3yz - 2x$, so $\dfrac{\partial z}{\partial x} = \dfrac{3yz - 2x}{2z - 3xy}$.

$\dfrac{\partial}{\partial y}(x^2 + y^2 + z^2) = \dfrac{\partial}{\partial y}(3xyz) \Rightarrow 0 + 2y + 2z\dfrac{\partial z}{\partial y} = 3x\left(y\dfrac{\partial z}{\partial y} + z\cdot 1\right) \Leftrightarrow$

$2z\dfrac{\partial z}{\partial y} - 3xy\dfrac{\partial z}{\partial y} = 3xz - 2y \Leftrightarrow (2z - 3xy)\dfrac{\partial z}{\partial y} = 3xz - 2y$, so $\dfrac{\partial z}{\partial y} = \dfrac{3xz - 2y}{2z - 3xy}$.

43. $x - z = \arctan(yz) \Rightarrow \dfrac{\partial}{\partial x}(x - z) = \dfrac{\partial}{\partial x}(\arctan(yz)) \Rightarrow 1 - \dfrac{\partial z}{\partial x} = \dfrac{1}{1 + (yz)^2}\cdot y\dfrac{\partial z}{\partial x} \Leftrightarrow$

$1 = \left(\dfrac{y}{1 + y^2z^2} + 1\right)\dfrac{\partial z}{\partial x} \Leftrightarrow 1 = \left(\dfrac{y + 1 + y^2z^2}{1 + y^2z^2}\right)\dfrac{\partial z}{\partial x}$, so $\dfrac{\partial z}{\partial x} = \dfrac{1 + y^2z^2}{1 + y + y^2z^2}$.

$\dfrac{\partial}{\partial y}(x - z) = \dfrac{\partial}{\partial y}(\arctan(yz)) \Rightarrow 0 - \dfrac{\partial z}{\partial y} = \dfrac{1}{1 + (yz)^2}\cdot\left(y\dfrac{\partial z}{\partial y} + z\cdot 1\right) \Leftrightarrow$

$-\dfrac{z}{1 + y^2z^2} = \left(\dfrac{y}{1 + y^2z^2} + 1\right)\dfrac{\partial z}{\partial y} \Leftrightarrow -\dfrac{z}{1 + y^2z^2} = \left(\dfrac{y + 1 + y^2z^2}{1 + y^2z^2}\right)\dfrac{\partial z}{\partial y} \Leftrightarrow \dfrac{\partial z}{\partial y} = -\dfrac{z}{1 + y + y^2z^2}$.

45. (a) $z = f(x) + g(y) \Rightarrow \dfrac{\partial z}{\partial x} = f'(x), \dfrac{\partial z}{\partial y} = g'(y)$

(b) $z = f(x + y)$. Let $u = x + y$. Then $\dfrac{\partial z}{\partial x} = \dfrac{df}{du}\dfrac{\partial u}{\partial x} = \dfrac{df}{du}(1) = f'(u) = f'(x + y)$,

$\dfrac{\partial z}{\partial y} = \dfrac{df}{du}\dfrac{\partial u}{\partial y} = \dfrac{df}{du}(1) = f'(u) = f'(x + y)$.

47. $f(x, y) = x^4 - 3x^2y^3 \Rightarrow f_x(x, y) = 4x^3 - 6xy^3, f_y(x, y) = -9x^2y^2$. Then $f_{xx}(x, y) = 12x^2 - 6y^3$,

$f_{xy}(x, y) = -18xy^2, f_{yx}(x, y) = -18xy^2$, and $f_{yy}(x, y) = -18x^2y$.

49. $z = \dfrac{x}{x + y} = x(x + y)^{-1} \Rightarrow z_x = \dfrac{1(x + y) - 1(x)}{(x + y)^2} = \dfrac{y}{(x + y)^2}, z_y = x(-1)(x + y)^{-2} = -\dfrac{x}{(x + y)^2}$.

Then $z_{xx} = y(-2)(x + y)^{-3} = -\dfrac{2y}{(x + y)^3}, z_{xy} = \dfrac{1(x + y)^2 - y(2)(x + y)}{[(x + y)^2]^2} = \dfrac{x + y - 2y}{(x + y)^3} = \dfrac{x - y}{(x + y)^3}$,

$z_{yx} = -\dfrac{1(x + y)^2 - x(2)(x + y)}{[(x + y)^2]^2} = -\dfrac{-x^2 + xy + y^2}{(x + y)^2} = \dfrac{(x + y)(x - y)}{(x + y)^2} = \dfrac{x - y}{(x + y)^3}$, and

$z_{yy} = -x(-2)(x + y)^{-3} = \dfrac{2x}{(x + y)^3}$.

51. $u = e^{-s}\sin t \Rightarrow u_s = -e^{-s}\sin t, u_t = e^{-s}\cos t$. Then $u_{ss} = e^{-s}\sin t, u_{st} = -e^{-s}\cos t$,

$u_{ts} = -e^{-s}\cos t$, and $u_{tt} = -e^{-s}\sin t$.

53. $u = x\sin(x + 2y) \Rightarrow u_x = x\cdot\cos(x + 2y)(1) + \sin(x + 2y)\cdot 1 = x\cos(x + 2y) + \sin(x + 2y)$,

$u_{xy} = x(-\sin(x + 2y)(2)) + \cos(x + 2y)(2) = 2\cos(x + 2y) - 2x\sin(x + 2y)$ and

$u_y = x\cos(x + 2y)(2) = 2x\cos(x + 2y)$,

$u_{yx} = 2x\cdot(-\sin(x + 2y)(1)) + \cos(x + 2y)\cdot 2 = 2\cos(x + 2y) - 2x\sin(x + 2y)$. Thus $u_{xy} = u_{yx}$.

55. $u = \ln \sqrt{x^2 + y^2} = \ln(x^2 + y^2)^{1/2} = \frac{1}{2}\ln(x^2 + y^2)$ $\Rightarrow$ $u_x = \frac{1}{2}\frac{1}{x^2+y^2}\cdot 2x = \frac{x}{x^2+y^2}$,

$u_{xy} = x(-1)(x^2+y^2)^{-2}(2y) = -\frac{2xy}{(x^2+y^2)^2}$ and $u_y = \frac{1}{2}\frac{1}{x^2+y^2}\cdot 2y = \frac{y}{x^2+y^2}$,

$u_{yx} = y(-1)(x^2+y^2)^{-2}(2x) = -\frac{2xy}{(x^2+y^2)^2}$. Thus $u_{xy} = u_{yx}$.

57. $f(x,y) = 3xy^4 + x^3y^2$ $\Rightarrow$ $f_x = 3y^4 + 3x^2y^2$, $f_{xx} = 6xy^2$, $f_{xxy} = 12xy$ and $f_y = 12xy^3 + 2x^3y$,

$f_{yy} = 36xy^2 + 2x^3$, $f_{yyy} = 72xy$.

59. $f(x,y,z) = \cos(4x + 3y + 2z)$ $\Rightarrow$

$f_x = -\sin(4x + 3y + 2z)(4) = -4\sin(4x + 3y + 2z)$,

$f_{xy} = -4\cos(4x + 3y + 2z)(3) = -12\cos(4x + 3y + 2z)$,

$f_{xyz} = -12(-\sin(4x + 3y + 2z))(2) = 24\sin(4x + 3y + 2z)$ and

$f_y = -\sin(4x + 3y + 2z)(3) = -3\sin(4x + 3y + 2z)$,

$f_{yz} = -3\cos(4x + 3y + 2z)(2) = -6\cos(4x + 3y + 2z)$,

$f_{yzz} = -6(-\sin(4x + 3y + 2z))(2) = 12\sin(4x + 3y + 2z)$.

61. $u = e^{r\theta}\sin\theta$ $\Rightarrow$ $\dfrac{\partial u}{\partial \theta} = e^{r\theta}\cos\theta + \sin\theta\cdot e^{r\theta}(r) = e^{r\theta}(\cos\theta + r\sin\theta)$,

$\dfrac{\partial^2 u}{\partial r\,\partial\theta} = e^{r\theta}(\sin\theta) + (\cos\theta + r\sin\theta)e^{r\theta}(\theta) = e^{r\theta}(\sin\theta + \theta\cos\theta + r\theta\sin\theta)$,

$\dfrac{\partial^3 u}{\partial r^2\,\partial\theta} = e^{r\theta}(\theta\sin\theta) + (\sin\theta + \theta\cos\theta + r\theta\sin\theta)\cdot e^{r\theta}(\theta) = \theta e^{r\theta}(2\sin\theta + \theta\cos\theta + r\theta\sin\theta)$.

63. $w = \dfrac{x}{y+2z} = x(y+2z)^{-1}$ $\Rightarrow$ $\dfrac{\partial w}{\partial x} = (y+2z)^{-1}$, $\dfrac{\partial^2 w}{\partial y\,\partial x} = -(y+2z)^{-2}(1) = -(y+2z)^{-2}$,

$\dfrac{\partial^3 w}{\partial z\,\partial y\,\partial x} = -(-2)(y+2z)^{-3}(2) = 4(y+2z)^{-3} = \dfrac{4}{(y+2z)^3}$ and

$\dfrac{\partial w}{\partial y} = x(-1)(y+2z)^{-2}(1) = -x(y+2z)^{-2}$, $\dfrac{\partial^2 w}{\partial x\,\partial y} = -(y+2z)^{-2}$, $\dfrac{\partial^3 w}{\partial x^2\,\partial y} = 0$.

65. By Definition 4, $f_x(3,2) = \lim\limits_{h\to 0}\dfrac{f(3+h,2) - f(3,2)}{h}$ which we can approximate by considering $h = 0.5$

and $h = -0.5$: $f_x(3,2) \approx \dfrac{f(3.5,2) - f(3,2)}{0.5} = \dfrac{22.4 - 17.5}{0.5} = 9.8$,

$f_x(3,2) \approx \dfrac{f(2.5,2) - f(3,2)}{-0.5} = \dfrac{10.2 - 17.5}{-0.5} = 14.6$. Averaging these values, we estimate $f_x(3,2)$ to be

approximately 12.2. Similarly, $f_x(3,2.2) = \lim\limits_{h\to 0}\dfrac{f(3+h,2.2) - f(3,2.2)}{h}$ which we can approximate by considering $h = 0.5$ and $h = -0.5$: $f_x(3, 2.2) \approx \dfrac{f(3.5, 2.2) - f(3, 2.2)}{0.5} = \dfrac{26.1 - 15.9}{0.5} = 20.4$,

$f_x(3, 2.2) \approx \dfrac{f(2.5, 2.2) - f(3, 2.2)}{-0.5} = \dfrac{9.3 - 15.9}{-0.5} = 13.2$. Averaging these values, we have $f_x(3, 2.2) \approx 16.8$.

To estimate $f_{xy}(3, 2)$, we first need an estimate for $f_x(3, 1.8)$:

$f_x(3, 1.8) \approx \dfrac{f(3.5, 1.8) - f(3, 1.8)}{0.5} = \dfrac{20.0 - 18.1}{0.5} = 3.8$,

$f_x(3, 1.8) \approx \dfrac{f(2.5, 1.8) - f(3, 1.8)}{-0.5} = \dfrac{12.5 - 18.1}{-0.5} = 11.2$. Averaging these values, we get $f_x(3, 1.8) \approx 7.5$.

Now $f_{xy}(x, y) = \dfrac{\partial}{\partial y}[f_x(x, y)]$ and $f_x(x, y)$ is itself a function of 2 variables, so Definition 4 says that

$f_{xy}(x, y) = \dfrac{\partial}{\partial y}[f_x(x, y)] = \lim\limits_{h \to 0} \dfrac{f_x(x, y + h) - f_x(x, y)}{h} \quad \Rightarrow \quad f_{xy}(3, 2) = \lim\limits_{h \to 0} \dfrac{f_x(3, 2 + h) - f_x(3, 2)}{h}$.

We can estimate this value using our previous work with $h = 0.2$ and $h = -0.2$:

$f_{xy}(3, 2) \approx \dfrac{f_x(3, 2.2) - f_x(3, 2)}{0.2} = \dfrac{16.8 - 12.2}{0.2} = 23$,

$f_{xy}(3, 2) \approx \dfrac{f_x(3, 1.8) - f_x(3, 2)}{-0.2} = \dfrac{7.5 - 12.2}{-0.2} = 23.5$. Averaging these values, we estimate $f_{xy}(3, 2)$ to be

approximately 23.25.

67. $u = e^{-\alpha^2 k^2 t} \sin kx \quad \Rightarrow \quad u_x = k e^{-\alpha^2 k^2 t} \cos kx$, $u_{xx} = -k^2 e^{-\alpha^2 k^2 t} \sin kx$, and $u_t = -\alpha^2 k^2 e^{-\alpha^2 k^2 t} \sin kx$.

Thus $\alpha^2 u_{xx} = u_t$.

69. $u = \dfrac{1}{\sqrt{x^2 + y^2 + z^2}} \quad \Rightarrow \quad u_x = \left(-\frac{1}{2}\right)(x^2 + y^2 + z^2)^{-3/2}(2x) = -x(x^2 + y^2 + z^2)^{-3/2}$ and

$u_{xx} = -(x^2 + y^2 + z^2)^{-3/2} - x\left(-\frac{3}{2}\right)(x^2 + y^2 + z^2)^{-5/2}(2x) = \dfrac{2x^2 - y^2 - z^2}{(x^2 + y^2 + z^2)^{5/2}}$.

By symmetry, $u_{yy} = \dfrac{2y^2 - x^2 - z^2}{(x^2 + y^2 + z^2)^{5/2}}$ and $u_{zz} = \dfrac{2z^2 - x^2 - y^2}{(x^2 + y^2 + z^2)^{5/2}}$.

Thus $u_{xx} + u_{yy} + u_{zz} = \dfrac{2x^2 - y^2 - z^2 + 2y^2 - x^2 - z^2 + 2z^2 - x^2 - y^2}{(x^2 + y^2 + z^2)^{5/2}} = 0$.

71. Let $v = x + at$, $w = x - at$. Then $u_t = \dfrac{\partial[f(v) + g(w)]}{\partial t} = \dfrac{df(v)}{dv}\dfrac{\partial v}{\partial t} + \dfrac{dg(w)}{dw}\dfrac{\partial w}{\partial t} = af'(v) - ag'(w)$ and

$u_{tt} = \dfrac{\partial[af'(v) - ag'(w)]}{\partial t} = a[af''(v) + ag''(w)] = a^2[f''(v) + g''(w)]$. Similarly, by using the Chain Rule we

have $u_x = f'(v) + g'(w)$ and $u_{xx} = f''(v) + g''(w)$. Thus $u_{tt} = a^2 u_{xx}$.

73. $z_x = e^y + ye^x$, $z_{xx} = ye^x$, $\partial^3 z/\partial x^3 = ye^x$. By symmetry $z_y = xe^y + e^x$, $z_{yy} = xe^y$ and $\partial^3 z/\partial y^3 = xe^y$.

Then $\partial^3 z/\partial x \partial y^2 = e^y$ and $\partial^3 z/\partial x^2 \partial y = e^x$. Thus $z = xe^y + ye^x$ satisfies the given partial differential

equation.

75. If we fix $K = K_0$, $P(L, K_0)$ is a function of a single variable L, and $\dfrac{dP}{dL} = \alpha \dfrac{P}{L}$ is a separable differential

equation. Then $\dfrac{dP}{P} = \alpha \dfrac{dL}{L} \;\Rightarrow\; \displaystyle\int \dfrac{dP}{P} = \int \alpha \dfrac{dL}{L} \;\Rightarrow\; \ln |P| = \alpha \ln |L| + C\,(K_0)$, where $C(K_0)$

can depend on K_0. Then $|P| = e^{\alpha \ln |L| + C(K_0)}$, and since $P > 0$ and $L > 0$, we have

$P = e^{\alpha \ln L} e^{C(K_0)} = e^{C(K_0)} e^{\ln L^\alpha} = C_1(K_0) L^\alpha$ where $C_1(K_0) = e^{C(K_0)}$.

77. By the Chain Rule, taking the partial derivative of both sides with respect to R_1 gives

$$\frac{\partial R^{-1}}{\partial R}\frac{\partial R}{\partial R_1} = \frac{\partial \left[(1/R_1) + (1/R_2) + (1/R_3) \right]}{\partial R_1} \quad \text{or} \quad -R^{-2}\frac{\partial R}{\partial R_1} = -R_1^{-2}. \text{ Thus } \frac{\partial R}{\partial R_1} = \frac{R^2}{R_1^2}.$$

79. By Exercise 78, $PV = mRT \;\Rightarrow\; P = \dfrac{mRT}{V}$, so $\dfrac{\partial P}{\partial T} = \dfrac{mR}{V}$. Also, $PV = mRT \;\Rightarrow\; V = \dfrac{mRT}{P}$

and $\dfrac{\partial V}{\partial T} = \dfrac{mR}{P}$. Since $T = \dfrac{PV}{mR}$, we have $T \dfrac{\partial P}{\partial T}\dfrac{\partial V}{\partial T} = \dfrac{PV}{mR}\cdot\dfrac{mR}{V}\cdot\dfrac{mR}{P} = mR$.

81. $\dfrac{\partial K}{\partial m} = \tfrac{1}{2}V^2$, $\dfrac{\partial K}{\partial V} = mV$, $\dfrac{\partial^2 K}{\partial V^2} = m$. Thus $\dfrac{\partial K}{\partial m}\cdot\dfrac{\partial^2 K}{\partial V^2} = \tfrac{1}{2}V^2 m = K$.

83. $f_x(x, y) = x + 4y \;\Rightarrow\; f_{xy}(x, y) = 4$ and $f_y(x, y) = 3x - y \;\Rightarrow\; f_{yx}(x, y) = 3$. Since f_{xy} and f_{yx} are

continuous everywhere but $f_{xy}(x, y) \neq f_{yx}(x, y)$, Clairaut's Theorem implies that such a function $f(x, y)$ does not

exist.

85. By the geometry of partial derivatives, the slope of the tangent line is $f_x(1, 2)$. By implicit differentiation of

$4x^2 + 2y^2 + z^2 = 16$, we get $8x + 2z\,(\partial z/\partial x) = 0 \;\Rightarrow\; \partial z/\partial x = -4x/z$, so when $x = 1$ and $z = 2$ we have

$\partial z/\partial x = -2$. So the slope is $f_x(1, 2) = -2$. Thus the tangent line is given by $z - 2 = -2(x - 1)$, $y = 2$. Taking

the parameter to be $t = x - 1$, we can write parametric equations for this line: $x = 1 + t$, $y = 2$, $z = 2 - 2t$.

87. By Clairaut's Theorem, $f_{xyy} = (f_{xy})_y = (f_{yx})_y = f_{yxy} = (f_y)_{xy} = (f_y)_{yx} = f_{yyx}$.

89. Let $g(x) = f(x, 0) = x(x^2)^{-3/2}e^0 = x\,|x|^{-3}$. But we are using the point $(1, 0)$, so near $(1, 0)$, $g(x) = x^{-2}$. Then

$g'(x) = -2x^{-3}$ and $g'(1) = -2$, so using (1) we have $f_x(1, 0) = g'(1) = -2$.

91. (a)

(b) For $(x, y) \neq (0, 0)$, $f_x(x, y) = \dfrac{(3x^2 y - y^3)(x^2 + y^2) - (x^3 y - xy^3)(2x)}{(x^2 + y^2)^2} = \dfrac{x^4 y + 4x^2 y^3 - y^5}{(x^2 + y^2)^2}$, and by

symmetry $f_y(x, y) = \dfrac{x^5 - 4x^3 y^2 - xy^4}{(x^2 + y^2)^2}$.

(c) $f_x(0,0) = \lim\limits_{h \to 0} \dfrac{f(h,0) - f(0,0)}{h} = \lim\limits_{h \to 0} \dfrac{(0/h^2) - 0}{h} = 0$ and $f_y(0,0) = \lim\limits_{h \to 0} \dfrac{f(0,h) - f(0,0)}{h} = 0$.

(d) By (3), $f_{xy}(0,0) = \dfrac{\partial f_x}{\partial y} = \lim\limits_{h \to 0} \dfrac{f_x(0,h) - f_x(0,0)}{h} = \lim\limits_{h \to 0} \dfrac{(-h^5 - 0)/h^4}{h} = -1$ while by (2),

$f_{yx}(0,0) = \dfrac{\partial f_y}{\partial x} = \lim\limits_{h \to 0} \dfrac{f_y(h,0) - f_y(0,0)}{h} = \lim\limits_{h \to 0} \dfrac{h^5/h^4}{h} = 1$.

(e) For $(x,y) \neq (0,0)$, we use a CAS to compute

$$f_{xy}(x,y) = \frac{x^6 + 9x^4y^2 - 4x^2y^4 + 4y^6}{(x^2 + y^2)^3}.$$

Now as $(x,y) \to (0,0)$ along the x-axis, $f_{xy}(x,y) \to 1$ while as $(x,y) \to (0,0)$ along the y-axis, $f_{xy}(x,y) \to 4$. Thus f_{xy} isn't continuous at $(0,0)$ and Clairaut's Theorem doesn't apply, so there is no contradiction. The graphs of f_{xy} and f_{yx} are identical except at the origin, where we observe the discontinuity.

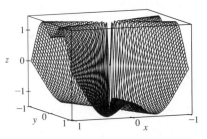

15.4 Tangent Planes and Linear Approximations   ET 14.4

1. $z = f(x,y) = 4x^2 - y^2 + 2y \ \Rightarrow\ f_x(x,y) = 8x,\ f_y(x,y) = -2y + 2$, so $f_x(-1,2) = -8,\ f_y(-1,2) = -2$. By Equation 2, an equation of the tangent plane is $z - 4 = f_x(-1,2)[x - (-1)] + f_y(-1,2)(y - 2) \ \Rightarrow$ $z - 4 = -8(x + 1) - 2(y - 2)$ or $z = -8x - 2y$.

3. $z = f(x,y) = \sqrt{4 - x^2 - 2y^2} \ \Rightarrow\ f_x(x,y) = \tfrac{1}{2}(4 - x^2 - 2y^2)^{-1/2}(-2x) = -\dfrac{x}{\sqrt{4 - x^2 - 2y^2}}$,

$f_y(x,y) = \tfrac{1}{2}(4 - x^2 - 2y^2)^{-1/2}(-4y) = -\dfrac{2y}{\sqrt{4 - x^2 - 2y^2}}$, so $f_x(1,-1) = -1$ and $f_y(1,-1) = 2$. Thus, an

equation of the tangent plane is $z - 1 = f_x(1,-1)(x - 1) + f_y(1,-1)[y - (-1)] \ \Rightarrow$ $z - 1 = -1(x - 1) + 2(y + 1)$ or $x - 2y + z = 4$.

5. $z = f(x,y) = y\cos(x - y) \ \Rightarrow\ f_x = y(-\sin(x - y)(1)) = -y\sin(x - y)$, $f_y = y(-\sin(x - y)(-1)) + \cos(x - y) = y\sin(x - y) + \cos(x - y)$, so $f_x(2,2) = -2\sin(0) = 0$, $f_y(2,2) = 2\sin(0) + \cos(0) = 1$ and an equation of the tangent plane is $z - 2 = 0(x - 2) + 1(y - 2)$ or $z = y$.

7. $z = f(x,y) = x^2 + xy + 3y^2$, so $f_x(x,y) = 2x + y \ \Rightarrow\ f_x(1,1) = 3,\ f_y(x,y) = x + 6y \ \Rightarrow\ f_y(1,1) = 7$ and an equation of the tangent plane is $z - 5 = 3(x - 1) + 7(y - 1)$ or $z = 3x + 7y - 5$. After zooming in, the surface and the tangent plane become almost indistinguishable. (Here, the tangent plane is below the surface.) If we zoom in farther, the surface and the tangent plane will appear to coincide.

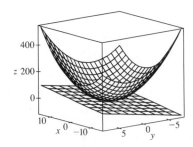
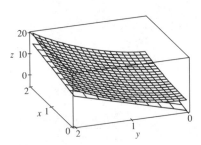

9. $f(x, y) = e^{-(x^2+y^2)/15}(\sin^2 x + \cos^2 y)$. A CAS gives

$f_x = -\frac{2}{15}e^{-(x^2+y^2)/15}(x\sin^2 x + x\cos^2 y - 15\sin x \cos x)$ and

$f_y = -\frac{2}{15}e^{-(x^2+y^2)/15}(y\sin^2 x + y\cos^2 y + 15\sin y \cos y)$. We use the CAS to evaluate these at $(2,3)$, and then substitute the results into Equation 2 in order to plot the tangent plane. After zooming in, the surface and the tangent plane become almost indistinguishable. (Here, the tangent plane is above the surface.) If we zoom in farther, the surface and the tangent plane will appear to coincide.

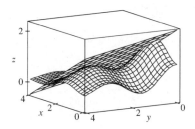
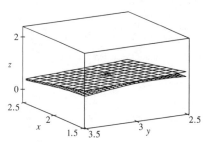

11. $f(x, y) = x\sqrt{y}$. The partial derivatives are $f_x(x, y) = \sqrt{y}$ and $f_y(x, y) = \dfrac{x}{2\sqrt{y}}$, so $f_x(1, 4) = 2$ and

$f_y(1, 4) = \frac{1}{4}$. Both f_x and f_y are continuous functions for $y > 0$, so by Theorem 8, f is differentiable at $(1, 4)$. By Equation 3, the linearization of f at $(1, 4)$ is given by

$L(x, y) = f(1, 4) + f_x(1, 4)(x - 1) + f_y(1, 4)(y - 4) = 2 + 2(x - 1) + \frac{1}{4}(y - 4) = 2x + \frac{1}{4}y - 1$.

13. $f(x, y) = e^x \cos xy$. The partial derivatives are $f_x(x, y) = e^x(\cos xy - y\sin xy)$ and $f_y(x, y) = -xe^x \sin xy$, so $f_x(0, 0) = 1$ and $f_y(0, 0) = 0$. Both f_x and f_y are continuous functions, so f is differentiable at $(0, 0)$ by Theorem 8. The linearization of f at $(0, 0)$ is given by

$L(x, y) = f(0, 0) + f_x(0, 0)(x - 0) + f_y(0, 0)(y - 0) = 1 + 1(x - 0) + 0(y - 0) = x + 1$.

15. $f(x, y) = \tan^{-1}(x + 2y)$. The partial derivatives are $f_x(x, y) = \dfrac{1}{1 + (x + 2y)^2}$ and

$f_y(x, y) = \dfrac{2}{1 + (x + 2y)^2}$, so $f_x(1, 0) = \frac{1}{2}$ and $f_y(1, 0) = 1$. Both f_x and f_y are continuous

functions, so f is differentiable at $(1, 0)$, and the linearization of f at $(1, 0)$ is

$L(x, y) = f(1, 0) + f_x(1, 0)(x - 1) + f_y(1, 0)(y - 0) = \frac{\pi}{4} + \frac{1}{2}(x - 1) + 1(y) = \frac{1}{2}x + y + \frac{\pi}{4} - \frac{1}{2}$.

17. $f(x, y) = \sqrt{20 - x^2 - 7y^2}$ $\Rightarrow$ $f_x(x, y) = -\dfrac{x}{\sqrt{20 - x^2 - 7y^2}}$ and $f_y(x, y) = -\dfrac{7y}{\sqrt{20 - x^2 - 7y^2}}$,

so $f_x(2, 1) = -\frac{2}{3}$ and $f_y(2, 1) = -\frac{7}{3}$. Then the linear approximation of f at $(2, 1)$ is given by

$$f(x, y) \approx f(2, 1) + f_x(2, 1)(x - 2) + f_y(2, 1)(y - 1) = 3 - \frac{2}{3}(x - 2) - \frac{7}{3}(y - 1)$$

$$= -\frac{2}{3}x - \frac{7}{3}y + \frac{20}{3}$$

Thus $f(1.95, 1.08) \approx -\frac{2}{3}(1.95) - \frac{7}{3}(1.08) + \frac{20}{3} = 2.84\overline{6}$.

19. $f(x, y, z) = \sqrt{x^2 + y^2 + z^2}$ $\Rightarrow$ $f_x(x, y, z) = \dfrac{x}{\sqrt{x^2 + y^2 + z^2}}$, $f_y(x, y, z) = \dfrac{y}{\sqrt{x^2 + y^2 + z^2}}$, and

$f_z(x, y, z) = \dfrac{z}{\sqrt{x^2 + y^2 + z^2}}$, so $f_x(3, 2, 6) = \frac{3}{7}$, $f_y(3, 2, 6) = \frac{2}{7}$, and $f_z(3, 2, 6) = \frac{6}{7}$. Then the linear approximation of f at $(3, 2, 6)$ is given by

$$f(x, y, z) \approx f(3, 2, 6) + f_x(3, 2, 6)(x - 3) + f_y(3, 2, 6)(y - 2) + f_z(3, 2, 6)(z - 6)$$

$$= 7 + \tfrac{3}{7}(x - 3) + \tfrac{2}{7}(y - 2) + \tfrac{6}{7}(z - 6) = \tfrac{3}{7}x + \tfrac{2}{7}y + \tfrac{6}{7}z$$

Thus $\sqrt{(3.02)^2 + (1.97)^2 + (5.99)^2} = f(3.02, 1.97, 5.99) \approx \tfrac{3}{7}(3.02) + \tfrac{2}{7}(1.97) + \tfrac{6}{7}(5.99) \approx 6.9914$.

21. From the table, $f(94, 80) = 127$. To estimate $f_T(94, 80)$ and $f_H(94, 80)$ we follow the procedure used in

Section 15.3 [ET 14.3]. Since $f_T(94, 80) = \lim\limits_{h \to 0} \dfrac{f(94 + h, 80) - f(94, 80)}{h}$, we approximate this quantity

with $h = \pm 2$ and use the values given in the table: $f_T(94, 80) \approx \dfrac{f(96, 80) - f(94, 80)}{2} = \dfrac{135 - 127}{2} = 4$,

$f_T(94, 80) \approx \dfrac{f(92, 80) - f(94, 80)}{-2} = \dfrac{119 - 127}{-2} = 4$.

Averaging these values gives $f_T(94, 80) \approx 4$. Similarly, $f_H(94, 80) = \lim\limits_{h \to 0} \dfrac{f(94, 80 + h) - f(94, 80)}{h}$,

so we use $h = \pm 5$: $f_H(94, 80) \approx \dfrac{f(94, 85) - f(94, 80)}{5} = \dfrac{132 - 127}{5} = 1$,

$f_H(94, 80) \approx \dfrac{f(94, 75) - f(94, 80)}{-5} = \dfrac{122 - 127}{-5} = 1$. Averaging these values gives $f_H(94, 80) \approx 1$. The

linear approximation, then, is

$$f(T, H) \approx f(94, 80) + f_T(94, 80)(T - 94) + f_H(94, 80)(H - 80)$$

$$\approx 127 + 4(T - 94) + 1(H - 80)$$

Thus when $T = 95$ and $H = 78$, $f(95, 78) \approx 127 + 4(95 - 94) + 1(78 - 80) = 129$, so we estimate the heat

index to be approximately $129\,°\mathrm{F}$.

23. $z = x^3 \ln(y^2) \;\; \Rightarrow$

$$dz = \frac{\partial z}{\partial x}\,dx + \frac{\partial z}{\partial y}\,dy = 3x^2 \ln(y^2)\,dx + x^3 \cdot \frac{1}{y^2}(2y)\,dy = 3x^2 \ln(y^2)\,dx + \frac{2x^3}{y}\,dy.$$

25. $u = e^t \sin\theta \;\; \Rightarrow \;\; du = \dfrac{\partial u}{\partial t}\,dt + \dfrac{\partial u}{\partial \theta}\,d\theta = e^t \sin\theta\,dt + e^t \cos\theta\,d\theta$

27. $w = \ln\sqrt{x^2 + y^2 + z^2} \;\; \Rightarrow$

$$dw = \frac{\partial w}{\partial x}\,dx + \frac{\partial w}{\partial y}\,dy + \frac{\partial w}{\partial z}\,dz$$

$$= \left(\frac{1}{2}\right) \frac{2x(x^2 + y^2 + z^2)^{-1/2}\,dx + 2y(x^2 + y^2 + z^2)^{-1/2}\,dy + 2z(x^2 + y^2 + z^2)^{-1/2}\,dz}{(x^2 + y^2 + z^2)^{1/2}}$$

$$= \frac{x\,dx + y\,dy + z\,dz}{x^2 + y^2 + z^2}$$

29. $dx = \Delta x = 0.05$, $dy = \Delta y = 0.1$, $z = 5x^2 + y^2$, $z_x = 10x$, $z_y = 2y$. Thus when $x = 1$ and

$y = 2$, $dz = z_x(1, 2)\,dx + z_y(1, 2)\,dy = (10)(0.05) + (4)(0.1) = 0.9$ while

$\Delta z = f(1.05, 2.1) - f(1, 2) = 5(1.05)^2 + (2.1)^2 - 5 - 4 = 0.9225$.

31. $dA = \dfrac{\partial A}{\partial x}\,dx + \dfrac{\partial A}{\partial y}\,dy = y\,dx + x\,dy$ and $|\Delta x| \leq 0.1$, $|\Delta y| \leq 0.1$. We use $dx = 0.1$, $dy = 0.1$ with

$x = 30$, $y = 24$; then the maximum error in the area is about $dA = 24(0.1) + 30(0.1) = 5.4\ \mathrm{cm}^2$.

33. The volume of a can is $V = \pi r^2 h$ and $\Delta V \approx dV$ is an estimate of the amount of tin. Here
$dV = 2\pi rh\, dr + \pi r^2\, dh$, so put $dr = 0.04$, $dh = 0.08$ (0.04 on top, 0.04 on bottom) and then
$\Delta V \approx dV = 2\pi(48)(0.04) + \pi(16)(0.08) \approx 16.08$ cm^3. Thus the amount of tin is about 16 cm^3.

35. The area of the rectangle is $A = xy$, and $\Delta A \approx dA$ is an estimate of the area of paint in the stripe. Here
$dA = y\, dx + x\, dy$, so with $dx = dy = \frac{3+3}{12} = \frac{1}{2}$, $\Delta A \approx dA = (100)\left(\frac{1}{2}\right) + (200)\left(\frac{1}{2}\right) = 150$ ft^2. Thus there are
approximately 150 ft^2 of paint in the stripe.

37. First we find $\dfrac{\partial R}{\partial R_1}$ implicitly by taking partial derivatives of both sides with respect to R_1:

$$\frac{\partial}{\partial R_1}\left[\frac{1}{R}\right] = \frac{\partial\left[(1/R_1) + (1/R_2) + (1/R_3)\right]}{\partial R_1} \quad\Rightarrow\quad -R^{-2}\frac{\partial R}{\partial R_1} = -R_1^{-2} \quad\Rightarrow\quad \frac{\partial R}{\partial R_1} = \frac{R^2}{R_1^2}. \text{ Then by}$$

symmetry, $\dfrac{\partial R}{\partial R_2} = \dfrac{R^2}{R_2^2}$, $\dfrac{\partial R}{\partial R_3} = \dfrac{R^2}{R_3^2}$. When $R_1 = 25$, $R_2 = 40$ and $R_3 = 50$, $\dfrac{1}{R} = \dfrac{17}{200} \Leftrightarrow R = \frac{200}{17}$ ohms.

Since the possible error for each R_i is 0.5%, the maximum error of R is attained by setting $\Delta R_i = 0.005 R_i$. So

$$\Delta R \approx dR = \frac{\partial R}{\partial R_1}\Delta R_1 + \frac{\partial R}{\partial R_2}\Delta R_2 + \frac{\partial R}{\partial R_3}\Delta R_3 = (0.005)R^2\left[\frac{1}{R_1} + \frac{1}{R_2} + \frac{1}{R_3}\right]$$

$$= (0.005)R = \frac{1}{17} \approx 0.059 \text{ ohms}$$

39. $\Delta z = f(a + \Delta x, b + \Delta y) - f(a, b) = (a + \Delta x)^2 + (b + \Delta y)^2 - (a^2 + b^2)$

$\quad = a^2 + 2a\,\Delta x + (\Delta x)^2 + b^2 + 2b\,\Delta y + (\Delta y)^2 - a^2 - b^2 = 2a\,\Delta x + (\Delta x)^2 + 2b\,\Delta y + (\Delta y)^2$

But $f_x(a, b) = 2a$ and $f_y(a, b) = 2b$ and so $\Delta z = f_x(a, b)\,\Delta x + f_y(a, b)\,\Delta y + \Delta x\,\Delta x + \Delta y\,\Delta y$, which is
Definition 7 with $\varepsilon_1 = \Delta x$ and $\varepsilon_2 = \Delta y$. Hence f is differentiable.

41. To show that f is continuous at (a, b) we need to show that $\displaystyle\lim_{(x,y)\to(a,b)} f(x, y) = f(a, b)$ or equivalently

$\displaystyle\lim_{(\Delta x, \Delta y)\to(0,0)} f(a + \Delta x, b + \Delta y) = f(a, b)$. Since f is differentiable at (a, b),

$f(a + \Delta x, b + \Delta y) - f(a, b) = \Delta z = f_x(a, b)\,\Delta x + f_y(a, b)\,\Delta y + \varepsilon_1\,\Delta x + \varepsilon_2\,\Delta y$, where ε_1 and $\varepsilon_2 \to 0$ as
$(\Delta x, \Delta y) \to (0, 0)$. Thus $f(a + \Delta x, b + \Delta y) = f(a, b) + f_x(a, b)\,\Delta x + f_y(a, b)\,\Delta y + \varepsilon_1\,\Delta x + \varepsilon_2\,\Delta y$. Taking
the limit of both sides as $(\Delta x, \Delta y) \to (0, 0)$ gives $\displaystyle\lim_{(\Delta x, \Delta y)\to(0,0)} f(a + \Delta x, b + \Delta y) = f(a, b)$. Thus f is

continuous at (a, b).

15.5 The Chain Rule

ET 14.5

1. $z = x^2 y + xy^2$, $x = 2 + t^4$, $y = 1 - t^3$ $\Rightarrow$

$\dfrac{dz}{dt} = \dfrac{\partial z}{\partial x}\dfrac{dx}{dt} + \dfrac{\partial z}{\partial y}\dfrac{dy}{dt} = (2xy + y^2)(4t^3) + (x^2 + 2xy)(-3t^2) = 4(2xy + y^2)t^3 - 3(x^2 + 2xy)t^2$

3. $z = \sin x \cos y$, $x = \pi t$, $y = \sqrt{t}$ $\Rightarrow$

$\dfrac{dz}{dt} = \dfrac{\partial z}{\partial x}\dfrac{dx}{dt} + \dfrac{\partial z}{\partial y}\dfrac{dy}{dt} = \cos x \cos y \cdot \pi + \sin x\,(-\sin y) \cdot \frac{1}{2}t^{-1/2} = \pi\cos x \cos y - \dfrac{1}{2\sqrt{t}}\sin x \sin y$

5. $w = xe^{y/z}$, $x = t^2$, $y = 1 - t$, $z = 1 + 2t$ $\Rightarrow$

$\dfrac{dw}{dt} = \dfrac{\partial w}{\partial x}\dfrac{dx}{dt} + \dfrac{\partial w}{\partial y}\dfrac{dy}{dt} + \dfrac{\partial w}{\partial z}\dfrac{dz}{dt} = e^{y/z} \cdot 2t + xe^{y/z}\left(\dfrac{1}{z}\right) \cdot (-1) + xe^{y/z}\left(-\dfrac{y}{z^2}\right) \cdot 2 = e^{y/z}\left(2t - \dfrac{x}{z} - \dfrac{2xy}{z^2}\right)$

7. $z = x^2 + xy + y^2$, $x = s + t$, $y = st$ $\Rightarrow$

$$\frac{\partial z}{\partial s} = \frac{\partial z}{\partial x}\frac{\partial x}{\partial s} + \frac{\partial z}{\partial y}\frac{\partial y}{\partial s} = (2x + y)(1) + (x + 2y)(t) = 2x + y + xt + 2yt$$

$$\frac{\partial z}{\partial t} = \frac{\partial z}{\partial x}\frac{\partial x}{\partial t} + \frac{\partial z}{\partial y}\frac{\partial y}{\partial t} = (2x + y)(1) + (x + 2y)(s) = 2x + y + xs + 2ys$$

9. $z = \arctan(2x + y)$, $x = s^2 t$, $y = s \ln t$ $\Rightarrow$

$$\frac{\partial z}{\partial s} = \frac{\partial z}{\partial x}\frac{\partial x}{\partial s} + \frac{\partial z}{\partial y}\frac{\partial y}{\partial s} = \frac{2}{1 + (2x + y)^2} \cdot 2st + \frac{1}{1 + (2x + y)^2} \cdot \ln t = \frac{4st + \ln t}{1 + (2x + y)^2}$$

$$\frac{\partial z}{\partial t} = \frac{\partial z}{\partial x}\frac{\partial x}{\partial t} + \frac{\partial z}{\partial y}\frac{\partial y}{\partial t} = \frac{2}{1 + (2x + y)^2} \cdot s^2 + \frac{1}{1 + (2x + y)^2} \cdot \frac{s}{t} = \frac{2s^2 + s/t}{1 + (2x + y)^2}$$

11. $z = e^r \cos \theta$, $r = st$, $\theta = \sqrt{s^2 + t^2}$ $\Rightarrow$

$$\frac{\partial z}{\partial s} = \frac{\partial z}{\partial r}\frac{\partial r}{\partial s} + \frac{\partial z}{\partial \theta}\frac{\partial \theta}{\partial s} = e^r \cos \theta \cdot t + e^r(-\sin \theta) \cdot \tfrac{1}{2}(s^2 + t^2)^{-1/2}(2s)$$

$$= te^r \cos \theta - e^r \sin \theta \cdot \frac{s}{\sqrt{s^2 + t^2}} = e^r\left(t \cos \theta - \frac{s}{\sqrt{s^2 + t^2}} \sin \theta\right)$$

$$\frac{\partial z}{\partial t} = \frac{\partial z}{\partial r}\frac{\partial r}{\partial t} + \frac{\partial z}{\partial \theta}\frac{\partial \theta}{\partial t} = e^r \cos \theta \cdot s + e^r(-\sin \theta) \cdot \tfrac{1}{2}(s^2 + t^2)^{-1/2}(2t)$$

$$= se^r \cos \theta - e^r \sin \theta \cdot \frac{t}{\sqrt{s^2 + t^2}} = e^r\left(s \cos \theta - \frac{t}{\sqrt{s^2 + t^2}} \sin \theta\right)$$

13. When $t = 3$, $x = g(3) = 2$ and $y = h(3) = 7$. By the Chain Rule (2),

$$\frac{dz}{dt} = \frac{\partial f}{\partial x}\frac{dx}{dt} + \frac{\partial f}{\partial y}\frac{dy}{dt} = f_x(2, 7)g'(3) + f_y(2, 7)\,h'(3) = (6)(5) + (-8)(-4) = 62.$$

15. $g(u, v) = f(x(u, v), y(u, v))$ where $x = e^u + \sin v$, $y = e^u + \cos v$ $\Rightarrow$ $\dfrac{\partial x}{\partial u} = e^u$, $\dfrac{\partial x}{\partial v} = \cos v$, $\dfrac{\partial y}{\partial u} = e^u$,

$\dfrac{\partial y}{\partial v} = -\sin v$. By the Chain Rule (3), $\dfrac{\partial g}{\partial u} = \dfrac{\partial f}{\partial x}\dfrac{\partial x}{\partial u} + \dfrac{\partial f}{\partial y}\dfrac{\partial y}{\partial u}$. Then

$$g_u(0, 0) = f_x(x(0, 0), y(0, 0))\, x_u(0, 0) + f_y(x(0, 0), y(0, 0))\, y_u(0, 0)$$

$$= f_x(1, 2)(e^0) + f_y(1, 2)(e^0) = 2(1) + 5(1) = 7$$

Similarly $\dfrac{\partial g}{\partial v} = \dfrac{\partial f}{\partial x}\dfrac{\partial x}{\partial v} + \dfrac{\partial f}{\partial y}\dfrac{\partial y}{\partial v}$. Then

$$g_v(0, 0) = f_x(x(0, 0), y(0, 0))\, x_v(0, 0) + f_y(x(0, 0), y(0, 0))\, y_v(0, 0)$$

$$= f_x(1, 2)(\cos 0) + f_y(1, 2)(-\sin 0) = 2(1) + 5(0) = 2$$

17.

$u = f(x, y)$, $x = x(r, s, t)$, $y = y(r, s, t)$ $\Rightarrow$

$$\frac{\partial u}{\partial r} = \frac{\partial u}{\partial x}\frac{\partial x}{\partial r} + \frac{\partial u}{\partial y}\frac{\partial y}{\partial r}, \quad \frac{\partial u}{\partial s} = \frac{\partial u}{\partial x}\frac{\partial x}{\partial s} + \frac{\partial u}{\partial y}\frac{\partial y}{\partial s},$$

$$\frac{\partial u}{\partial t} = \frac{\partial u}{\partial x}\frac{\partial x}{\partial t} + \frac{\partial u}{\partial y}\frac{\partial y}{\partial t}$$

19.

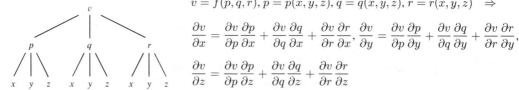

$$v = f(p, q, r), p = p(x, y, z), q = q(x, y, z), r = r(x, y, z) \implies$$

$$\frac{\partial v}{\partial x} = \frac{\partial v}{\partial p}\frac{\partial p}{\partial x} + \frac{\partial v}{\partial q}\frac{\partial q}{\partial x} + \frac{\partial v}{\partial r}\frac{\partial r}{\partial x}, \frac{\partial v}{\partial y} = \frac{\partial v}{\partial p}\frac{\partial p}{\partial y} + \frac{\partial v}{\partial q}\frac{\partial q}{\partial y} + \frac{\partial v}{\partial r}\frac{\partial r}{\partial y},$$

$$\frac{\partial v}{\partial z} = \frac{\partial v}{\partial p}\frac{\partial p}{\partial z} + \frac{\partial v}{\partial q}\frac{\partial q}{\partial z} + \frac{\partial v}{\partial r}\frac{\partial r}{\partial z}$$

21. $z = x^2 + xy^3, x = uv^2 + w^3, y = u + ve^w \implies \dfrac{\partial z}{\partial u} = \dfrac{\partial z}{\partial x}\dfrac{\partial x}{\partial u} + \dfrac{\partial z}{\partial y}\dfrac{\partial y}{\partial u} = (2x + y^3)(v^2) + (3xy^2)(1),$

$\dfrac{\partial z}{\partial v} = \dfrac{\partial z}{\partial x}\dfrac{\partial x}{\partial v} + \dfrac{\partial z}{\partial y}\dfrac{\partial y}{\partial v} = (2x + y^3)(2uv) + (3xy^2)(e^w),$

$\dfrac{\partial z}{\partial w} = \dfrac{\partial z}{\partial x}\dfrac{\partial x}{\partial w} + \dfrac{\partial z}{\partial y}\dfrac{\partial y}{\partial w} = (2x + y^3)(3w^2) + (3xy^2)(ve^w).$ When $u = 2$, $v = 1$, and $w = 0$, we have $x = 2$,

$y = 3$, so $\dfrac{\partial z}{\partial u} = (31)(1) + (54)(1) = 85,\ \dfrac{\partial z}{\partial v} = (31)(4) + (54)(1) = 178,\ \dfrac{\partial z}{\partial w} = (31)(0) + (54)(1) = 54.$

23. $R = \ln(u^2 + v^2 + w^2), u = x + 2y, v = 2x - y, w = 2xy \implies$

$$\frac{\partial R}{\partial x} = \frac{\partial R}{\partial u}\frac{\partial u}{\partial x} + \frac{\partial R}{\partial v}\frac{\partial v}{\partial x} + \frac{\partial R}{\partial w}\frac{\partial w}{\partial x} = \frac{2u}{u^2 + v^2 + w^2}(1) + \frac{2v}{u^2 + v^2 + w^2}(2) + \frac{2w}{u^2 + v^2 + w^2}(2y)$$

$$= \frac{2u + 4v + 4wy}{u^2 + v^2 + w^2},$$

$$\frac{\partial R}{\partial y} = \frac{\partial R}{\partial u}\frac{\partial u}{\partial y} + \frac{\partial R}{\partial v}\frac{\partial v}{\partial y} + \frac{\partial R}{\partial w}\frac{\partial w}{\partial y} = \frac{2u}{u^2 + v^2 + w^2}(2) + \frac{2v}{u^2 + v^2 + w^2}(-1) + \frac{2w}{u^2 + v^2 + w^2}(2x)$$

$$= \frac{4u - 2v + 4wx}{u^2 + v^2 + w^2}.$$

When $x = y = 1$ we have $u = 3$, $v = 1$, and $w = 2$, so $\dfrac{\partial R}{\partial x} = \dfrac{9}{7}$ and $\dfrac{\partial R}{\partial y} = \dfrac{9}{7}$.

25. $u = x^2 + yz, x = pr\cos\theta, y = pr\sin\theta, z = p + r \implies$

$$\frac{\partial u}{\partial p} = \frac{\partial u}{\partial x}\frac{\partial x}{\partial p} + \frac{\partial u}{\partial y}\frac{\partial y}{\partial p} + \frac{\partial u}{\partial z}\frac{\partial z}{\partial p} = (2x)(r\cos\theta) + (z)(r\sin\theta) + (y)(1) = 2xr\cos\theta + zr\sin\theta + y,$$

$$\frac{\partial u}{\partial r} = \frac{\partial u}{\partial x}\frac{\partial x}{\partial r} + \frac{\partial u}{\partial y}\frac{\partial y}{\partial r} + \frac{\partial u}{\partial z}\frac{\partial z}{\partial r} = (2x)(p\cos\theta) + (z)(p\sin\theta) + (y)(1) = 2xp\cos\theta + zp\sin\theta + y,$$

$$\frac{\partial u}{\partial \theta} = \frac{\partial u}{\partial x}\frac{\partial x}{\partial \theta} + \frac{\partial u}{\partial y}\frac{\partial y}{\partial \theta} + \frac{\partial u}{\partial z}\frac{\partial z}{\partial \theta} = (2x)(-pr\sin\theta) + (z)(pr\cos\theta) + (y)(0) = -2xpr\sin\theta + zpr\cos\theta.$$

When $p = 2$, $r = 3$, and $\theta = 0$ we have $x = 6$, $y = 0$, and $z = 5$, so $\dfrac{\partial u}{\partial p} = 36,\ \dfrac{\partial u}{\partial r} = 24,$ and $\dfrac{\partial u}{\partial \theta} = 30.$

27. $\sqrt{xy} = 1 + x^2y$, so let $F(x, y) = (xy)^{1/2} - 1 - x^2y = 0$. Then by Equation 6

$$\frac{dy}{dx} = -\frac{F_x}{F_y} = -\frac{\frac{1}{2}(xy)^{-1/2}(y) - 2xy}{\frac{1}{2}(xy)^{-1/2}(x) - x^2} = -\frac{y - 4xy\sqrt{xy}}{x - 2x^2\sqrt{xy}} = \frac{4(xy)^{3/2} - y}{x - 2x^2\sqrt{xy}}.$$

29. $\cos(x - y) = xe^y$, so let $F(x, y) = \cos(x - y) - xe^y = 0$.

Then $\dfrac{dy}{dx} = -\dfrac{F_x}{F_y} = -\dfrac{-\sin(x - y) - e^y}{-\sin(x - y)(-1) - xe^y} = \dfrac{\sin(x - y) + e^y}{\sin(x - y) - xe^y}.$

31. $x^2 + y^2 + z^2 = 3xyz$, so let $F(x, y, z) = x^2 + y^2 + z^2 - 3xyz = 0$. Then by Equations 7

$$\frac{\partial z}{\partial x} = -\frac{F_x}{F_z} = -\frac{2x - 3yz}{2z - 3xy} = \frac{3yz - 2x}{2z - 3xy} \quad \text{and} \quad \frac{\partial z}{\partial y} = -\frac{F_y}{F_z} = -\frac{2y - 3xz}{2z - 3xy} = \frac{3xz - 2y}{2z - 3xy}.$$

33. $x - z = \arctan(yz)$, so let $F(x, y, z) = x - z - \arctan(yz) = 0$. Then

$$\frac{\partial z}{\partial x} = -\frac{F_x}{F_z} = -\frac{1}{-1 - \dfrac{1}{1 + (yz)^2}(y)} = \frac{1 + y^2 z^2}{1 + y + y^2 z^2} \quad \text{and}$$

$$\frac{\partial z}{\partial y} = -\frac{F_y}{F_z} = -\frac{-\dfrac{1}{1 + (yz)^2}(z)}{-1 - \dfrac{1}{1 + (yz)^2}(y)} = -\frac{\dfrac{z}{1 + y^2 z^2}}{\dfrac{1 + y^2 z^2 + y}{1 + y^2 z^2}} = -\frac{z}{1 + y + y^2 z^2}.$$

35. Since x and y are each functions of t, $T(x, y)$ is a function of t, so by the Chain Rule, $\dfrac{dT}{dt} = \dfrac{\partial T}{\partial x}\dfrac{dx}{dt} + \dfrac{\partial T}{\partial y}\dfrac{dy}{dt}$.

After 3 seconds, $x = \sqrt{1 + t} = \sqrt{1 + 3} = 2$, $y = 2 + \frac{1}{3}t = 2 + \frac{1}{3}(3) = 3$, $\dfrac{dx}{dt} = \dfrac{1}{2\sqrt{1 + t}} = \dfrac{1}{2\sqrt{1 + 3}} = \dfrac{1}{4}$,

and $\dfrac{dy}{dt} = \dfrac{1}{3}$. Then $\dfrac{dT}{dt} = T_x(2, 3)\dfrac{dx}{dt} + T_y(2, 3)\dfrac{dy}{dt} = 4\left(\frac{1}{4}\right) + 3\left(\frac{1}{3}\right) = 2$. Thus the temperature is rising at a rate

of $2°\text{C/s}$.

37. $C = 1449.2 + 4.6T - 0.055T^2 + 0.00029T^3 + 0.016D$, so $\dfrac{\partial C}{\partial T} = 4.6 - 0.11T + 0.00087T^2$ and

$\dfrac{\partial C}{\partial D} = 0.016$. According to the graph, the diver is experiencing a temperature of approximately $12.5°\text{C}$ at

$t = 20$ minutes, so $\dfrac{\partial C}{\partial T} = 4.6 - 0.11(12.5) + 0.00087(12.5)^2 \approx 3.36$. By sketching tangent lines at $t = 20$ to the

graphs given, we estimate $\dfrac{dD}{dt} \approx \dfrac{1}{2}$ and $\dfrac{dT}{dt} \approx -\dfrac{1}{10}$. Then, by the Chain Rule,

$\dfrac{dC}{dt} = \dfrac{\partial C}{\partial T}\dfrac{dT}{dt} + \dfrac{\partial C}{\partial D}\dfrac{dD}{dt} \approx (3.36)\left(-\frac{1}{10}\right) + (0.016)\left(\frac{1}{2}\right) \approx -0.33$. Thus the speed of sound experienced by the

diver is decreasing at a rate of approximately 0.33 m/s per minute.

39. (a) $V = \ell w h$, so by the Chain Rule,

$$\frac{dV}{dt} = \frac{\partial V}{\partial \ell}\frac{d\ell}{dt} + \frac{\partial V}{\partial w}\frac{dw}{dt} + \frac{\partial V}{\partial h}\frac{dh}{dt} = wh\frac{d\ell}{dt} + \ell h\frac{dw}{dt} + \ell w\frac{dh}{dt}$$

$$= 2 \cdot 2 \cdot 2 + 1 \cdot 2 \cdot 2 + 1 \cdot 2 \cdot (-3) = 6 \text{ m}^3/\text{s}$$

(b) $S = 2(\ell w + \ell h + wh)$, so by the Chain Rule,

$$\frac{dS}{dt} = \frac{\partial S}{\partial \ell}\frac{d\ell}{dt} + \frac{\partial S}{\partial w}\frac{dw}{dt} + \frac{\partial S}{\partial h}\frac{dh}{dt} = 2(w + h)\frac{d\ell}{dt} + 2(\ell + h)\frac{dw}{dt} + 2(\ell + w)\frac{dh}{dt}$$

$$= 2(2 + 2)2 + 2(1 + 2)2 + 2(1 + 2)(-3) = 10 \text{ m}^2/\text{s}$$

(c) $L^2 = \ell^2 + w^2 + h^2 \;\Rightarrow\; 2L\dfrac{dL}{dt} = 2\ell\dfrac{d\ell}{dt} + 2w\dfrac{dw}{dt} + 2h\dfrac{dh}{dt} = 2(1)(2) + 2(2)(2) + 2(2)(-3) = 0 \;\Rightarrow$

$dL/dt = 0 \text{ m/s}$.

41. $\dfrac{dP}{dt} = 0.05$, $\dfrac{dT}{dt} = 0.15$, $V = 8.31\dfrac{T}{P}$ and $\dfrac{dV}{dt} = 8.31\dfrac{dT}{P\,dt} - 8.31\dfrac{T}{P^2}\dfrac{dP}{dt}$. Thus when $P = 20$ and $T = 320$,

$\dfrac{dV}{dt} = 8.31\left[\dfrac{0.15}{20} - \dfrac{(0.05)(320)}{400}\right] \approx -0.27$ L/s.

43. (a) By the Chain Rule, $\dfrac{\partial z}{\partial r} = \dfrac{\partial z}{\partial x}\cos\theta + \dfrac{\partial z}{\partial y}\sin\theta$, $\dfrac{\partial z}{\partial \theta} = \dfrac{\partial z}{\partial x}(-r\sin\theta) + \dfrac{\partial z}{\partial y}r\cos\theta$.

(b) $\left(\dfrac{\partial z}{\partial r}\right)^2 = \left(\dfrac{\partial z}{\partial x}\right)^2\cos^2\theta + 2\dfrac{\partial z}{\partial x}\dfrac{\partial z}{\partial y}\cos\theta\sin\theta + \left(\dfrac{\partial z}{\partial y}\right)^2\sin^2\theta$,

$\left(\dfrac{\partial z}{\partial \theta}\right)^2 = \left(\dfrac{\partial z}{\partial x}\right)^2 r^2\sin^2\theta - 2\dfrac{\partial z}{\partial x}\dfrac{\partial z}{\partial y}r^2\cos\theta\sin\theta + \left(\dfrac{\partial z}{\partial y}\right)^2 r^2\cos^2\theta$. Thus

$\left(\dfrac{\partial z}{\partial r}\right)^2 + \dfrac{1}{r^2}\left(\dfrac{\partial z}{\partial \theta}\right)^2 = \left[\left(\dfrac{\partial z}{\partial x}\right)^2 + \left(\dfrac{\partial z}{\partial y}\right)^2\right](\cos^2\theta + \sin^2\theta) = \left(\dfrac{\partial z}{\partial x}\right)^2 + \left(\dfrac{\partial z}{\partial y}\right)^2$.

45. Let $u = x - y$. Then $\dfrac{\partial z}{\partial x} = \dfrac{dz}{du}\dfrac{\partial u}{\partial x} = \dfrac{dz}{du}$ and $\dfrac{\partial z}{\partial y} = \dfrac{dz}{du}(-1)$. Thus $\dfrac{\partial z}{\partial x} + \dfrac{\partial z}{\partial y} = 0$.

47. Let $u = x + at$, $v = x - at$. Then $z = f(u) + g(v)$, so $\partial z/\partial u = f'(u)$ and $\partial z/\partial v = g'(v)$.

Thus $\dfrac{\partial z}{\partial t} = \dfrac{\partial z}{\partial u}\dfrac{\partial u}{\partial t} + \dfrac{\partial z}{\partial v}\dfrac{\partial v}{\partial t} = af'(u) - ag'(v)$ and

$\dfrac{\partial^2 z}{\partial t^2} = a\dfrac{\partial}{\partial t}[f'(u) - g'(v)] = a\left(\dfrac{df'(u)}{du}\dfrac{\partial u}{\partial t} - \dfrac{dg'(v)}{dv}\dfrac{\partial v}{\partial t}\right) = a^2 f''(u) + a^2 g''(v)$.

Similarly $\dfrac{\partial z}{\partial x} = f'(u) + g'(v)$ and $\dfrac{\partial^2 z}{\partial x^2} = f''(u) + g''(v)$. Thus $\dfrac{\partial^2 z}{\partial t^2} = a^2\dfrac{\partial^2 z}{\partial x^2}$.

49. $\dfrac{\partial z}{\partial s} = \dfrac{\partial z}{\partial x}2s + \dfrac{\partial z}{\partial y}2r$. Then

$\dfrac{\partial^2 z}{\partial r\,\partial s} = \dfrac{\partial}{\partial r}\left(\dfrac{\partial z}{\partial x}2s\right) + \dfrac{\partial}{\partial r}\left(\dfrac{\partial z}{\partial y}2r\right)$

$= \dfrac{\partial^2 z}{\partial x^2}\dfrac{\partial x}{\partial r}2s + \dfrac{\partial}{\partial y}\left(\dfrac{\partial z}{\partial x}\right)\dfrac{\partial y}{\partial r}2s + \dfrac{\partial z}{\partial x}\dfrac{\partial}{\partial r}2s + \dfrac{\partial^2 z}{\partial y^2}\dfrac{\partial y}{\partial r}2r + \dfrac{\partial}{\partial x}\left(\dfrac{\partial z}{\partial y}\right)\dfrac{\partial x}{\partial r}2r + \dfrac{\partial z}{\partial y}2$

$= 4rs\dfrac{\partial^2 z}{\partial x^2} + \dfrac{\partial^2 z}{\partial y\,\partial x}4s^2 + 0 + 4rs\dfrac{\partial^2 z}{\partial y^2} + \dfrac{\partial^2 z}{\partial x\,\partial y}4r^2 + 2\dfrac{\partial z}{\partial y}$

By the continuity of the partials, $\dfrac{\partial^2 z}{\partial r\partial s} = 4rs\dfrac{\partial^2 z}{\partial x^2} + 4rs\dfrac{\partial^2 z}{\partial y^2} + (4r^2 + 4s^2)\dfrac{\partial^2 z}{\partial x\,\partial y} + 2\dfrac{\partial z}{\partial y}$.

51. $\dfrac{\partial z}{\partial r} = \dfrac{\partial z}{\partial x}\cos\theta + \dfrac{\partial z}{\partial y}\sin\theta$ and $\dfrac{\partial z}{\partial \theta} = -\dfrac{\partial z}{\partial x}r\sin\theta + \dfrac{\partial z}{\partial y}r\cos\theta$. Then

$\dfrac{\partial^2 z}{\partial r^2} = \cos\theta\left(\dfrac{\partial^2 z}{\partial x^2}\cos\theta + \dfrac{\partial^2 z}{\partial y\,\partial x}\sin\theta\right) + \sin\theta\left(\dfrac{\partial^2 z}{\partial y^2}\sin\theta + \dfrac{\partial^2 z}{\partial x\,\partial y}\cos\theta\right)$

$= \cos^2\theta\dfrac{\partial^2 z}{\partial x^2} + 2\cos\theta\sin\theta\dfrac{\partial^2 z}{\partial x\,\partial y} + \sin^2\theta\dfrac{\partial^2 z}{\partial y^2}$

and

$$\frac{\partial^2 z}{\partial \theta^2} = -r \cos \theta \, \frac{\partial z}{\partial x} + (-r \sin \theta) \left(\frac{\partial^2 z}{\partial x^2} (-r \sin \theta) + \frac{\partial^2 z}{\partial y \, \partial x} r \cos \theta \right)$$

$$-r \sin \theta \, \frac{\partial z}{\partial y} + r \cos \theta \left(\frac{\partial^2 z}{\partial y^2} r \cos \theta + \frac{\partial^2 z}{\partial x \, \partial y} (-r \sin \theta) \right)$$

$$= -r \cos \theta \, \frac{\partial z}{\partial x} - r \sin \theta \, \frac{\partial z}{\partial y} + r^2 \sin^2 \theta \, \frac{\partial^2 z}{\partial x^2} - 2r^2 \cos \theta \sin \theta \, \frac{\partial^2 z}{\partial x \, \partial y} + r^2 \cos^2 \theta \, \frac{\partial^2 z}{\partial y^2}$$

Thus

$$\frac{\partial^2 z}{\partial r^2} + \frac{1}{r^2} \frac{\partial^2 z}{\partial \theta^2} + \frac{1}{r} \frac{\partial z}{\partial r} = (\cos^2 \theta + \sin^2 \theta) \frac{\partial^2 z}{\partial x^2} + (\sin^2 \theta + \cos^2 \theta) \frac{\partial^2 z}{\partial y^2} - \frac{1}{r} \cos \theta \, \frac{\partial z}{\partial x}$$

$$-\frac{1}{r} \sin \theta \, \frac{\partial z}{\partial y} + \frac{1}{r} \left(\cos \theta \, \frac{\partial z}{\partial x} + \sin \theta \, \frac{\partial z}{\partial y} \right)$$

$$= \frac{\partial^2 z}{\partial x^2} + \frac{\partial^2 z}{\partial y^2} \text{ as desired.}$$

53. (a) Since f is a polynomial, it has continuous second-order partial derivatives, and

$$f(tx, ty) = (tx)^2(ty) + 2(tx)(ty)^2 + 5(ty)^3 = t^3 x^2 y + 2t^3 xy^2 + 5t^3 y^3$$
$$= t^3(x^2 y + 2xy^2 + 5y^3) = t^3 f(x, y)$$

Thus, f is homogeneous of degree 3.

(b) Differentiating both sides of $f(tx, ty) = t^n f(x, y)$ with respect to t using the Chain Rule, we get

$$\frac{\partial}{\partial t} f(tx, ty) = \frac{\partial}{\partial t} [t^n f(x, y)] \quad \Leftrightarrow$$

$$\frac{\partial}{\partial(tx)} f(tx, ty) \cdot \frac{\partial(tx)}{\partial t} + \frac{\partial}{\partial(ty)} f(tx, ty) \cdot \frac{\partial(ty)}{\partial t} = x \frac{\partial}{\partial(tx)} f(tx, ty) + y \frac{\partial}{\partial(ty)} f(tx, ty) = nt^{n-1} f(x, y).$$

Setting $t = 1$: $x \frac{\partial}{\partial x} f(x, y) + y \frac{\partial}{\partial y} f(x, y) = nf(x, y)$.

55. Differentiating both sides of $f(tx, ty) = t^n f(x, y)$ with respect to x using the Chain Rule, we get

$$\frac{\partial}{\partial x} f(tx, ty) = \frac{\partial}{\partial x} [t^n f(x, y)] \quad \Leftrightarrow$$

$$\frac{\partial}{\partial(tx)} f(tx, ty) \cdot \frac{\partial(tx)}{\partial x} + \frac{\partial}{\partial(ty)} f(tx, ty) \cdot \frac{\partial(ty)}{\partial x} = t^n \frac{\partial}{\partial x} f(x, y) \quad \Leftrightarrow \quad tf_x(tx, ty) = t^n f_x(x, y).$$

Thus $f_x(tx, ty) = t^{n-1} f_x(x, y)$.

15.6 Directional Derivatives and the Gradient Vector ET 14.6

1. First we draw a line passing through Raleigh and the eye of the hurricane. We can approximate the directional
derivative at Raleigh in the direction of the eye of the hurricane by the average rate of change of pressure between
the points where this line intersects the contour lines closest to Raleigh. In the direction of the eye of the hurricane,
the pressure changes from 996 millibars to 992 millibars. We estimate the distance between these two points to be
approximately 40 miles, so the rate of change of pressure in the direction given is approximately

$\frac{992 - 996}{40} = -0.1$ millibar/mi.

3. $D_{\mathbf{u}} f(-20, 30) = \nabla f(-20, 30) \cdot \mathbf{u} = f_T(-20, 30)\left(\frac{1}{\sqrt{2}}\right) + f_v(-20, 30)\left(\frac{1}{\sqrt{2}}\right).$

$f_T(-20, 30) = \lim\limits_{h \to 0} \dfrac{f(-20 + h, 30) - f(-20, 30)}{h}$, so we can approximate $f_T(-20, 30)$ by considering $h = \pm 5$

and using the values given in the table: $f_T(-20, 30) \approx \dfrac{f(-15, 30) - f(-20, 30)}{5} = \dfrac{-26 - (-33)}{5} = 1.4$,

$f_T(-20, 30) \approx \dfrac{f(-25, 30) - f(-20, 30)}{-5} = \dfrac{-39 - (-33)}{-5} = 1.2$. Averaging these values gives

$f_T(-20, 30) \approx 1.3$. Similarly, $f_v(-20, 30) = \lim\limits_{h \to 0} \dfrac{f(-20, 30 + h) - f(-20, 30)}{h}$, so we can approximate

$f_v(-20, 30)$ with $h = \pm 10$:

$f_v(-20, 30) \approx \dfrac{f(-20, 40) - f(-20, 30)}{10} = \dfrac{-34 - (-33)}{10} = -0.1$,

$f_v(-20, 30) \approx \dfrac{f(-20, 20) - f(-20, 30)}{-10} = \dfrac{-30 - (-33)}{-10} = -0.3$. Averaging these values gives

$f_v(-20, 30) \approx -0.2$. Then $D_{\mathbf{u}} f(-20, 30) \approx 1.3\left(\frac{1}{\sqrt{2}}\right) + (-0.2)\left(\frac{1}{\sqrt{2}}\right) \approx 0.778$.

5. $f(x, y) = \sqrt{5x - 4y} \quad \Rightarrow \quad f_x(x, y) = \frac{1}{2}(5x - 4y)^{-1/2}(5) = \dfrac{5}{2\sqrt{5x - 4y}}$ and

$f_y(x, y) = \frac{1}{2}(5x - 4y)^{-1/2}(-4) = -\dfrac{2}{\sqrt{5x - 4y}}$. If $\mathbf{u}$ is a unit vector in the direction of $\theta = -\frac{\pi}{6}$, then from

Equation 6, $D_{\mathbf{u}} f(4, 1) = f_x(4, 1) \cos\left(-\frac{\pi}{6}\right) + f_y(4, 1) \sin\left(-\frac{\pi}{6}\right) = \frac{5}{8} \cdot \frac{\sqrt{3}}{2} + \left(-\frac{1}{2}\right)\left(-\frac{1}{2}\right) = \frac{5\sqrt{3}}{16} + \frac{1}{4}$.

7. $f(x, y) = 5xy^2 - 4x^3 y$

(a) $\nabla f(x, y) = \langle f_x(x, y), f_y(x, y) \rangle = \langle 5y^2 - 12x^2 y, 10xy - 4x^3 \rangle$

(b) $\nabla f(1, 2) = \langle 5(2)^2 - 12(1)^2(2), 10(1)(2) - 4(1)^3 \rangle = \langle -4, 16 \rangle$

(c) By Equation 9, $D_{\mathbf{u}} f(1, 2) = \nabla f(1, 2) \cdot \mathbf{u} = \langle -4, 16 \rangle \cdot \langle \frac{5}{13}, \frac{12}{13} \rangle = (-4)\left(\frac{5}{13}\right) + (16)\left(\frac{12}{13}\right) = \frac{172}{13}$.

9. $f(x, y, z) = xe^{2yz}$

(a) $\nabla f(x, y, z) = \langle f_x(x, y, z), f_y(x, y, z), f_z(x, y, z) \rangle = \langle e^{2yz}, 2xze^{2yz}, 2xye^{2yz} \rangle$

(b) $\nabla f(3, 0, 2) = \langle 1, 12, 0 \rangle$

(c) By Equation 14, $D_{\mathbf{u}} f(3, 0, 2) = \nabla f(3, 0, 2) \cdot \mathbf{u} = \langle 1, 12, 0 \rangle \cdot \langle \frac{2}{3}, -\frac{2}{3}, \frac{1}{3} \rangle = \frac{2}{3} - \frac{24}{3} + 0 = -\frac{22}{3}$.

11. $f(x, y) = 1 + 2x\sqrt{y} \quad \Rightarrow \quad \nabla f(x, y) = \langle 2\sqrt{y}, 2x \cdot \frac{1}{2}y^{-1/2} \rangle = \langle 2\sqrt{y}, x/\sqrt{y} \rangle$, $\nabla f(3, 4) = \langle 4, \frac{3}{2} \rangle$,

and a unit vector in the direction of $\mathbf{v}$ is $\mathbf{u} = \dfrac{1}{\sqrt{4^2 + (-3)^2}}\langle 4, -3 \rangle = \langle \frac{4}{5}, -\frac{3}{5} \rangle$, so

$D_{\mathbf{u}} f(3, 4) = \nabla f(3, 4) \cdot \mathbf{u} = \langle 4, \frac{3}{2} \rangle \cdot \langle \frac{4}{5}, -\frac{3}{5} \rangle = \frac{23}{10}$.

13. $g(s, t) = s^2 e^t \quad \Rightarrow \quad \nabla g(s, t) = 2se^t \mathbf{i} + s^2 e^t \mathbf{j}$, $\nabla g(2, 0) = 4\mathbf{i} + 4\mathbf{j}$, and a unit vector in the direction of $\mathbf{v}$ is

$\mathbf{u} = \frac{1}{\sqrt{2}}(\mathbf{i} + \mathbf{j})$, so $D_{\mathbf{u}} g(2, 0) = \nabla g(2, 0) \cdot \mathbf{u} = (4\mathbf{i} + 4\mathbf{j}) \cdot \frac{1}{\sqrt{2}}(\mathbf{i} + \mathbf{j}) = \frac{8}{\sqrt{2}} = 4\sqrt{2}$.

15. $f(x, y, z) = \sqrt{x^2 + y^2 + z^2} \;\Rightarrow\; \nabla f(x, y, z) = \left\langle \dfrac{x}{\sqrt{x^2 + y^2 + z^2}}, \dfrac{y}{\sqrt{x^2 + y^2 + z^2}}, \dfrac{z}{\sqrt{x^2 + y^2 + z^2}} \right\rangle,$

$\nabla f(1, 2, -2) = \left\langle \frac{1}{3}, \frac{2}{3}, -\frac{2}{3} \right\rangle,$ and a unit vector in the direction of $\mathbf{v}$ is $\mathbf{u} = \frac{1}{9}\langle -6, 6, -3 \rangle = \left\langle -\frac{2}{3}, \frac{2}{3}, -\frac{1}{3} \right\rangle,$ so

$D_{\mathbf{u}} f(1, 2, -2) = \nabla f(1, 2, -2) \cdot \mathbf{u} = \left\langle \frac{1}{3}, \frac{2}{3}, -\frac{2}{3} \right\rangle \cdot \left\langle -\frac{2}{3}, \frac{2}{3}, -\frac{1}{3} \right\rangle = \frac{4}{9}.$

17. $g(x, y, z) = (x + 2y + 3z)^{3/2} \;\Rightarrow$

$\nabla g(x, y, z) = \left\langle \frac{3}{2}(x + 2y + 3z)^{1/2}(1), \frac{3}{2}(x + 2y + 3z)^{1/2}(2), \frac{3}{2}(x + 2y + 3z)^{1/2}(3) \right\rangle$

$\qquad = \left\langle \frac{3}{2}\sqrt{x + 2y + 3z}, 3\sqrt{x + 2y + 3z}, \frac{9}{2}\sqrt{x + 2y + 3z} \right\rangle,\; \nabla g(1, 1, 2) = \left\langle \frac{9}{2}, 9, \frac{27}{2} \right\rangle,$

and a unit vector in the direction of $\mathbf{v} = 2\mathbf{j} - \mathbf{k}$ is $\mathbf{u} = \frac{2}{\sqrt{5}}\mathbf{j} - \frac{1}{\sqrt{5}}\mathbf{k},$ so

$D_{\mathbf{u}} g(1, 1, 2) = \left\langle \frac{9}{2}, 9, \frac{27}{2} \right\rangle \cdot \left\langle 0, \frac{2}{\sqrt{5}}, -\frac{1}{\sqrt{5}} \right\rangle = \frac{18}{\sqrt{5}} - \frac{27}{2\sqrt{5}} = \frac{9}{2\sqrt{5}}.$

19. $f(x, y) = \sqrt{xy} \;\Rightarrow\; \nabla f(x, y) = \left\langle \frac{1}{2}(xy)^{-1/2}(y), \frac{1}{2}(xy)^{-1/2}(x) \right\rangle = \left\langle \dfrac{y}{2\sqrt{xy}}, \dfrac{x}{2\sqrt{xy}} \right\rangle,$ so

$\nabla f(2, 8) = \left\langle 1, \frac{1}{4} \right\rangle.$ The unit vector in the direction of $\overrightarrow{PQ} = \langle 5 - 2, 4 - 8 \rangle = \langle 3, -4 \rangle$ is $\mathbf{u} = \left\langle \frac{3}{5}, -\frac{4}{5} \right\rangle,$ so

$D_{\mathbf{u}} f(2, 8) = \nabla f(2, 8) \cdot \mathbf{u} = \left\langle 1, \frac{1}{4} \right\rangle \cdot \left\langle \frac{3}{5}, -\frac{4}{5} \right\rangle = \frac{2}{5}.$

21. $f(x, y) = y^2/x = y^2 x^{-1} \;\Rightarrow\; \nabla f(x, y) = \left\langle -y^2 x^{-2}, 2yx^{-1} \right\rangle = \left\langle -y^2/x^2, 2y/x \right\rangle.$

$\nabla f(2, 4) = \langle -4, 4 \rangle,$ or equivalently $\langle -1, 1 \rangle,$ is the direction of maximum rate of change, and the maximum rate

is $|\nabla f(2, 4)| = \sqrt{16 + 16} = 4\sqrt{2}.$

23. $f(x, y) = \sin(xy) \;\Rightarrow\; \nabla f(x, y) = \langle y\cos(xy), x\cos(xy) \rangle,\; \nabla f(1, 0) = \langle 0, 1 \rangle.$ Thus the maximum rate of

change is $|\nabla f(1, 0)| = 1$ in the direction $\langle 0, 1 \rangle.$

25. $f(x, y, z) = \ln(xy^2 z^3) \;\Rightarrow\; \nabla f(x, y, z) = \left\langle \dfrac{y^2 z^3}{xy^2 z^3}, \dfrac{2xyz^3}{xy^2 z^3}, \dfrac{3xy^2 z^2}{xy^2 z^3} \right\rangle = \left\langle \dfrac{1}{x}, \dfrac{2}{y}, \dfrac{3}{z} \right\rangle.$

$\nabla f(1, -2, -3) = \langle 1, -1, -1 \rangle$ is the direction of maximum rate of change and the maximum rate

is $|\nabla f(1, -2, -3)| = \sqrt{3}.$

27. (a) As in the proof of Theorem 15, $D_{\mathbf{u}} f = |\nabla f| \cos\theta.$ Since the minimum value of $\cos\theta$ is -1 occurring when

$\theta = \pi,$ the minimum value of $D_{\mathbf{u}} f$ is $-|\nabla f|$ occurring when $\theta = \pi,$ that is when $\mathbf{u}$ is in the opposite direction

of ∇f (assuming $\nabla f \neq \mathbf{0}$).

(b) $f(x, y) = x^4 y - x^2 y^3 \;\Rightarrow\; \nabla f(x, y) = \left\langle 4x^3 y - 2xy^3, x^4 - 3x^2 y^2 \right\rangle,$ so f decreases fastest at the point

$(2, -3)$ in the direction $-\nabla f(2, -3) = -\langle 12, -92 \rangle = \langle -12, 92 \rangle.$

29. The direction of fastest change is $\nabla f(x, y) = (2x - 2)\mathbf{i} + (2y - 4)\mathbf{j},$ so we need to find all points (x, y) where

$\nabla f(x, y)$ is parallel to $\mathbf{i} + \mathbf{j} \;\Leftrightarrow\; (2x - 2)\mathbf{i} + (2y - 4)\mathbf{j} = k(\mathbf{i} + \mathbf{j}) \;\Leftrightarrow\; k = 2x - 2$ and $k = 2y - 4.$ Then

$2x - 2 = 2y - 4 \;\Rightarrow\; y = x + 1,$ so the direction of fastest change is $\mathbf{i} + \mathbf{j}$ at all points on the line $y = x + 1.$

31. $T = \dfrac{k}{\sqrt{x^2 + y^2 + z^2}}$ and $120 = T(1, 2, 2) = \dfrac{k}{3}$ so $k = 360$.

(a) $\mathbf{u} = \dfrac{\langle 1, -1, 1 \rangle}{\sqrt{3}}$,

$$D_{\mathbf{u}}T(1, 2, 2) = \nabla T(1, 2, 2) \cdot \mathbf{u} = \left[-360(x^2 + y^2 + z^2)^{-3/2} \langle x, y, z \rangle \right]_{(1,2,2)} \cdot \mathbf{u}$$

$$= -\tfrac{40}{3} \langle 1, 2, 2 \rangle \cdot \tfrac{1}{\sqrt{3}} \langle 1, -1, 1 \rangle = -\tfrac{40}{3\sqrt{3}}$$

(b) From (a), $\nabla T = -360(x^2 + y^2 + z^2)^{-3/2} \langle x, y, z \rangle$, and since $\langle x, y, z \rangle$ is the position vector of the point (x, y, z), the vector $-\langle x, y, z \rangle$, and thus ∇T, always points toward the origin.

33. $\nabla V(x, y, z) = \langle 10x - 3y + yz, xz - 3x, xy \rangle$, $\nabla V(3, 4, 5) = \langle 38, 6, 12 \rangle$

(a) $D_{\mathbf{u}} V(3, 4, 5) = \langle 38, 6, 12 \rangle \cdot \tfrac{1}{\sqrt{3}} \langle 1, 1, -1 \rangle = \tfrac{32}{\sqrt{3}}$

(b) $\nabla V(3, 4, 5) = \langle 38, 6, 12 \rangle$ or equivalently $\langle 19, 3, 6 \rangle$.

(c) $|\nabla V(3, 4, 5)| = \sqrt{38^2 + 6^2 + 12^2} = \sqrt{1624} = 2\sqrt{406}$

35. A unit vector in the direction of $\overrightarrow{AB}$ is $\mathbf{i}$ and a unit vector in the direction of $\overrightarrow{AC}$ is $\mathbf{j}$.

Thus $D_{\overrightarrow{AB}} f(1, 3) = f_x(1, 3) = 3$ and $D_{\overrightarrow{AC}} f(1, 3) = f_y(1, 3) = 26$. Therefore

$\nabla f(1, 3) = \langle f_x(1, 3), f_y(1, 3) \rangle = \langle 3, 26 \rangle$, and by definition, $D_{\overrightarrow{AD}} f(1, 3) = \nabla f \cdot \mathbf{u}$ where $\mathbf{u}$ is a unit vector in the

direction of $\overrightarrow{AD}$, which is $\left\langle \tfrac{5}{13}, \tfrac{12}{13} \right\rangle$. Therefore, $D_{\overrightarrow{AD}} f(1, 3) = \langle 3, 26 \rangle \cdot \left\langle \tfrac{5}{13}, \tfrac{12}{13} \right\rangle = 3 \cdot \tfrac{5}{13} + 26 \cdot \tfrac{12}{13} = \tfrac{327}{13}$.

37. (a) $\nabla(au + bv) = \left\langle \dfrac{\partial(au + bv)}{\partial x}, \dfrac{\partial(au + bv)}{\partial y} \right\rangle = \left\langle a\dfrac{\partial u}{\partial x} + b\dfrac{\partial v}{\partial x}, a\dfrac{\partial u}{\partial y} + b\dfrac{\partial v}{\partial y} \right\rangle$

$$= a \left\langle \dfrac{\partial u}{\partial x}, \dfrac{\partial u}{\partial y} \right\rangle + b \left\langle \dfrac{\partial v}{\partial x}, \dfrac{\partial v}{\partial y} \right\rangle = a\nabla u + b\nabla v$$

(b) $\nabla(uv) = \left\langle v\dfrac{\partial u}{\partial x} + u\dfrac{\partial v}{\partial x}, v\dfrac{\partial u}{\partial y} + u\dfrac{\partial v}{\partial y} \right\rangle = v \left\langle \dfrac{\partial u}{\partial x}, \dfrac{\partial u}{\partial y} \right\rangle + u \left\langle \dfrac{\partial v}{\partial x}, \dfrac{\partial v}{\partial y} \right\rangle = v\nabla u + u\nabla v$

(c) $\nabla\left(\dfrac{u}{v}\right) = \left\langle \dfrac{v\dfrac{\partial u}{\partial x} - u\dfrac{\partial v}{\partial x}}{v^2}, \dfrac{v\dfrac{\partial u}{\partial y} - u\dfrac{\partial v}{\partial y}}{v^2} \right\rangle = \dfrac{v \left\langle \dfrac{\partial u}{\partial x}, \dfrac{\partial u}{\partial y} \right\rangle - u \left\langle \dfrac{\partial v}{\partial x}, \dfrac{\partial v}{\partial y} \right\rangle}{v^2} = \dfrac{v\nabla u - u\nabla v}{v^2}$

(d) $\nabla u^n = \left\langle \dfrac{\partial(u^n)}{\partial x}, \dfrac{\partial(u^n)}{\partial y} \right\rangle = \left\langle nu^{n-1}\dfrac{\partial u}{\partial x}, nu^{n-1}\dfrac{\partial u}{\partial y} \right\rangle = nu^{n-1}\nabla u$.

39. Let $F(x, y, z) = x^2 + 2y^2 + 3z^2$. Then $x^2 + 2y^2 + 3z^2 = 21$ is a level surface of F. $F_x(x, y, z) = 2x \Rightarrow$

$F_x(4, -1, 1) = 8$, $F_y(x, y, z) = 4y \Rightarrow F_y(4, -1, 1) = -4$, and $F_z(x, y, z) = 6z \Rightarrow F_z(4, -1, 1) = 6$.

(a) Equation 19 gives an equation of the tangent plane at $(4, -1, 1)$ as $8(x - 4) - 4[y - (-1)] + 6(z - 1) = 0$ or $4x - 2y + 3z = 21$.

(b) By Equation 20, the normal line has symmetric equations

$$\dfrac{x - 4}{8} = \dfrac{y + 1}{-4} = \dfrac{z - 1}{6} \text{ or } \dfrac{x - 4}{4} = \dfrac{y + 1}{-2} = \dfrac{z - 1}{3}.$$

41. Let $F(x, y, z) = x^2 - 2y^2 + z^2 + yz$. Then $x^2 - 2y^2 + z^2 + yz = 2$ is a level surface of F

and $\nabla F(x, y, z) = \langle 2x, -4y + z, 2z + y \rangle$.

(a) $\nabla F(2, 1, -1) = \langle 4, -5, -1 \rangle$ is a normal vector for the tangent plane at $(2, 1, -1)$, so an equation of the tangent
plane is $4(x - 2) - 5(y - 1) - 1(z + 1) = 0$ or $4x - 5y - z = 4$.

(b) The normal line has direction $\langle 4, -5, -1 \rangle$, so parametric equations are $x = 2 + 4t$, $y = 1 - 5t$, $z = -1 - t$,

and symmetric equations are $\dfrac{x - 2}{4} = \dfrac{y - 1}{-5} = \dfrac{z + 1}{-1}$.

43. $F(x, y, z) = -z + xe^y \cos z \quad \Rightarrow \quad \nabla F(x, y, z) = \langle e^y \cos z, xe^y \cos z, -1 - xe^y \sin z \rangle$,

$\nabla F(1, 0, 0) = \langle 1, 1, -1 \rangle$

(a) $1(x - 1) + 1(y - 0) - 1(z - 0) = 0$ or $x + y - z = 1$

(b) $x - 1 = y = -z$

45. $F(x, y, z) = xy + yz + zx$,

$\nabla F(x, y, z) = \langle y + z, x + z, y + x \rangle$,

$\nabla F(1, 1, 1) = \langle 2, 2, 2 \rangle$, so an equation of the tangent plane

is $2x + 2y + 2z = 6$ or $x + y + z = 3$, and the normal line is

given by $x - 1 = y - 1 = z - 1$ or $x = y = z$.

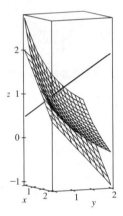

47. $\nabla f(x, y) = \langle 2x, 8y \rangle$, $\nabla f(2, 1) = \langle 4, 8 \rangle$. The tangent line has

equation $\nabla f(2, 1) \cdot \langle x - 2, y - 1 \rangle = 0 \quad \Rightarrow$

$4(x - 2) + 8(y - 1) = 0$, which simplifies to $x + 2y = 4$.

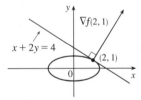

49. $\nabla F(x_0, y_0, z_0) = \left\langle \dfrac{2x_0}{a^2}, \dfrac{2y_0}{b^2}, \dfrac{2z_0}{c^2} \right\rangle$. Thus an equation of the tangent plane at (x_0, y_0, z_0) is

$\dfrac{2x_0}{a^2} x + \dfrac{2y_0}{b^2} y + \dfrac{2z_0}{c^2} z = 2\left(\dfrac{x_0^2}{a^2} + \dfrac{y_0^2}{b^2} + \dfrac{z_0^2}{c^2} \right) = 2(1) = 2$ since (x_0, y_0, z_0) is a point on the ellipsoid. Hence

$\dfrac{x_0}{a^2} x + \dfrac{y_0}{b^2} y + \dfrac{z_0}{c^2} z = 1$ is an equation of the tangent plane.

51. $\nabla F(x_0, y_0, z_0) = \left\langle \dfrac{2x_0}{a^2}, \dfrac{2y_0}{b^2}, \dfrac{-1}{c} \right\rangle$, so an equation of the tangent plane is

$\dfrac{2x_0}{a^2} x + \dfrac{2y_0}{b^2} y - \dfrac{1}{c} z = \dfrac{2x_0^2}{a^2} + \dfrac{2y_0^2}{b^2} - \dfrac{z_0}{c}$ or $\dfrac{2x_0}{a^2} x + \dfrac{2y_0}{b^2} y = \dfrac{z}{c} + 2\left(\dfrac{x_0^2}{a^2} + \dfrac{y_0^2}{b^2} \right) - \dfrac{z_0}{c}$. But $\dfrac{z_0}{c} = \dfrac{x_0^2}{a^2} + \dfrac{y_0^2}{b^2}$, so

the equation can be written as $\dfrac{2x_0}{a^2} x + \dfrac{2y_0}{b^2} y = \dfrac{z + z_0}{c}$.

53. $\nabla f(x_0, y_0, z_0) = \langle 2x_0, -2y_0, 4z_0 \rangle$ and the given line has direction numbers $2, 4, 6$, so

$\langle 2x_0, -2y_0, 4z_0 \rangle = k\langle 2, 4, 6 \rangle$ or $x_0 = k$, $y_0 = -2k$ and $z_0 = \frac{3}{2}k$. But $x_0^2 - y_0^2 + 2z_0^2 = 1$ or $\left(1 - 4 + \frac{9}{2}\right)k^2 = 1$,

so $k = \pm\sqrt{\frac{2}{3}} = \pm\frac{\sqrt{6}}{3}$ and there are two such points: $\left(\pm\frac{\sqrt{6}}{3}, \mp\frac{2\sqrt{6}}{3}, \pm\frac{\sqrt{6}}{2}\right)$.

55. Let (x_0, y_0, z_0) be a point on the cone [other than $(0,0,0)$]. Then an equation of the tangent plane to the cone at this

point is $2x_0 x + 2y_0 y - 2z_0 z = 2\left(x_0^2 + y_0^2 - z_0^2\right)$. But $x_0^2 + y_0^2 = z_0^2$ so the tangent plane is given by

$x_0 x + y_0 y - z_0 z = 0$, a plane which always contains the origin.

57. Let (x_0, y_0, z_0) be a point on the surface. Then an equation of the tangent plane at the point is

$$\frac{x}{2\sqrt{x_0}} + \frac{y}{2\sqrt{y_0}} + \frac{z}{2\sqrt{z_0}} = \frac{\sqrt{x_0} + \sqrt{y_0} + \sqrt{z_0}}{2}.$$ But $\sqrt{x_0} + \sqrt{y_0} + \sqrt{z_0} = \sqrt{c}$, so the equation is

$$\frac{x}{\sqrt{x_0}} + \frac{y}{\sqrt{y_0}} + \frac{z}{\sqrt{z_0}} = \sqrt{c}.$$ The x-, y-, and z-intercepts are $\sqrt{cx_0}$, $\sqrt{cy_0}$ and $\sqrt{cz_0}$ respectively. (The x-intercept

is found by setting $y = z = 0$ and solving the resulting equation for x, and the y- and z-intercepts are found

similarly.) So the sum of the intercepts is $\sqrt{c}(\sqrt{x_0} + \sqrt{y_0} + \sqrt{z_0}) = c$, a constant.

59. If $f(x, y, z) = z - x^2 - y^2$ and $g(x, y, z) = 4x^2 + y^2 + z^2$, then the tangent line is perpendicular to both ∇f and

∇g at $(-1, 1, 2)$. The vector $\mathbf{v} = \nabla f \times \nabla g$ will therefore be parallel to the tangent line. We have:

$\nabla f(x, y, z) = \langle -2x, -2y, 1 \rangle \quad \Rightarrow \quad \nabla f(-1, 1, 2) = \langle 2, -2, 1 \rangle$, and $\nabla g(x, y, z) = \langle 8x, 2y, 2z \rangle \quad \Rightarrow$

$\nabla g(-1, 1, 2) = \langle -8, 2, 4 \rangle$. Hence $\mathbf{v} = \nabla f \times \nabla g = \begin{vmatrix} \mathbf{i} & \mathbf{j} & \mathbf{k} \\ 2 & -2 & 1 \\ -8 & 2 & 4 \end{vmatrix} = -10\,\mathbf{i} - 16\,\mathbf{j} - 12\,\mathbf{k}$. Parametric equations

are: $x = -1 - 10t$, $y = 1 - 16t$, $z = 2 - 12t$.

61. (a) The direction of the normal line of F is given by ∇F, and that of G by ∇G. Assuming that

$\nabla F \neq 0 \neq \nabla G$, the two normal lines are perpendicular at P if $\nabla F \cdot \nabla G = 0$ at $P \quad \Leftrightarrow$

$\langle \partial F/\partial x, \partial F/\partial y, \partial F/\partial z \rangle \cdot \langle \partial G/\partial x, \partial G/\partial y, \partial G/\partial z \rangle = 0$ at $P \quad \Leftrightarrow \quad F_x G_x + F_y G_y + F_z G_z = 0$ at P.

(b) Here $F = x^2 + y^2 - z^2$ and $G = x^2 + y^2 + z^2 - r^2$, so

$\nabla F \cdot \nabla G = \langle 2x, 2y, -2z \rangle \cdot \langle 2x, 2y, 2z \rangle = 4x^2 + 4y^2 - 4z^2 = 4F = 0$, since the point $\langle x, y, z \rangle$ lies on the

graph of $F = 0$. To see that this is true without using calculus, note that $G = 0$ is the equation of a sphere

centered at the origin and $F = 0$ is the equation of a right circular cone with vertex at the origin (which is

generated by lines through the origin). At any point of intersection, the sphere's normal line (which passes

through the origin) lies on the cone, and thus is perpendicular to the cone's normal line. So the surfaces with

equations $F = 0$ and $G = 0$ are everywhere orthogonal.

63. Let $\mathbf{u} = \langle a, b \rangle$ and $\mathbf{v} = \langle c, d \rangle$. Then we know that at the given point, $D_{\mathbf{u}}\, f = \nabla f \cdot \mathbf{u} = af_x + bf_y$ and

$D_{\mathbf{v}}\, f = \nabla f \cdot \mathbf{v} = cf_x + df_y$. But these are just two linear equations in the two unknowns f_x and f_y, and since $\mathbf{u}$

and $\mathbf{v}$ are not parallel, we can solve the equations to find $\nabla f = \langle f_x, f_y \rangle$ at the given point. In fact,

$$\nabla f = \left\langle \frac{d\, D_{\mathbf{u}}\, f - b\, D_{\mathbf{v}}\, f}{ad - bc}, \frac{a\, D_{\mathbf{v}}\, f - c\, D_{\mathbf{u}}\, f}{ad - bc} \right\rangle.$$

15.7 Maximum and Minimum Values

1. (a) First we compute $D(1,1) = f_{xx}(1,1) f_{yy}(1,1) - [f_{xy}(1,1)]^2 = (4)(2) - (1)^2 = 7$. Since $D(1,1) > 0$ and $f_{xx}(1,1) > 0$, f has a local minimum at $(1,1)$ by the Second Derivatives Test.

(b) $D(1,1) = f_{xx}(1,1) f_{yy}(1,1) - [f_{xy}(1,1)]^2 = (4)(2) - (3)^2 = -1$. Since $D(1,1) < 0$, f has a saddle point at $(1,1)$ by the Second Derivatives Test.

3. In the figure, a point at approximately $(1,1)$ is enclosed by level curves which are oval in shape and indicate that as we move away from the point in any direction the values of f are increasing. Hence we would expect a local minimum at or near $(1,1)$. The level curves near $(0,0)$ resemble hyperbolas, and as we move away from the origin, the values of f increase in some directions and decrease in others, so we would expect to find a saddle point there.

To verify our predictions, we have $f(x,y) = 4 + x^3 + y^3 - 3xy \Rightarrow f_x(x,y) = 3x^2 - 3y$, $f_y(x,y) = 3y^2 - 3x$. We have critical points where these partial derivatives are equal to 0: $3x^2 - 3y = 0$, $3y^2 - 3x = 0$. Substituting $y = x^2$ from the first equation into the second equation gives $3(x^2)^2 - 3x = 0 \Rightarrow 3x(x^3 - 1) = 0 \Rightarrow x = 0$ or $x = 1$. Then we have two critical points, $(0,0)$ and $(1,1)$. The second partial derivatives are $f_{xx}(x,y) = 6x$, $f_{xy}(x,y) = -3$, and $f_{yy}(x,y) = 6y$, so $D(x,y) = f_{xx}(x,y) f_{yy}(x,y) - [f_{xy}(x,y)]^2 = (6x)(6y) - (-3)^2 = 36xy - 9$. Then $D(0,0) = 36(0)(0) - 9 = -9$, and $D(1,1) = 36(1)(1) - 9 = 27$. Since $D(0,0) < 0$, f has a saddle point at $(0,0)$ by the Second Derivatives Test. Since $D(1,1) > 0$ and $f_{xx}(1,1) > 0$, f has a local minimum at $(1,1)$.

5. $f(x,y) = 9 - 2x + 4y - x^2 - 4y^2 \Rightarrow f_x = -2 - 2x$, $f_y = 4 - 8y$, $f_{xx} = -2$, $f_{xy} = 0$, $f_{yy} = -8$. Then $f_x = 0$ and $f_y = 0$ imply $x = -1$ and $y = \frac{1}{2}$, and the only critical point is $\left(-1, \frac{1}{2}\right)$. $D(x,y) = f_{xx}f_{yy} - (f_{xy})^2 = (-2)(-8) - 0^2 = 16$, and since $D\left(-1, \frac{1}{2}\right) = 16 > 0$ and $f_{xx}\left(-1, \frac{1}{2}\right) = -2 < 0$, $f\left(-1, \frac{1}{2}\right) = 11$ is a local maximum by the Second Derivatives Test.

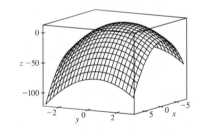

7. $f(x,y) = x^4 + y^4 - 4xy + 2 \Rightarrow f_x = 4x^3 - 4y$, $f_y = 4y^3 - 4x$, $f_{xx} = 12x^2$, $f_{xy} = -4$, $f_{yy} = 12y^2$. Then $f_x = 0$ implies $y = x^3$, and substitution into $f_y = 0 \Rightarrow x = y^3$ gives $x^9 - x = 0 \Rightarrow x(x^8 - 1) = 0 \Rightarrow x = 0$ or $x = \pm 1$. Thus the critical points are $(0,0)$, $(1,1)$, and $(-1,-1)$. Now $D(0,0) = 0 \cdot 0 - (-4)^2 = -16 < 0$, so $(0,0)$ is a saddle point. $D(1,1) = (12)(12) - (-4)^2 > 0$ and $f_{xx}(1,1) = 12 > 0$, so $f(1,1) = 0$ is a local minimum. $D(-1,-1) = (12)(12) - (-4)^2 > 0$ and $f_{xx} = (-1,-1) = 12 > 0$, so $f(-1,-1) = 0$ is also a local minimum.

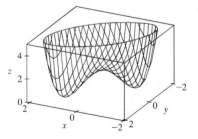

9. $f(x, y) = (1 + xy)(x + y) = x + y + x^2y + xy^2$ ⇒

$f_x = 1 + 2xy + y^2$, $f_y = 1 + x^2 + 2xy$, $f_{xx} = 2y$, $f_{xy} = 2x + 2y$,

$f_{yy} = 2x$. Then $f_x = 0$ implies $1 + 2xy + y^2 = 0$ and $f_y = 0$

implies $1 + x^2 + 2xy = 0$. Subtracting the second equation from the

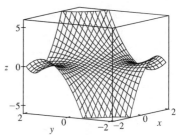

first gives $y^2 - x^2 = 0$ ⇒ $y = \pm x$, but if $y = x$ then

$1 + 2xy + y^2 = 0$ ⇒ $1 + 3x^2 = 0$ which has no real solution. If

$y = -x$ then $1 + 2xy + y^2 = 0$ ⇒ $1 - x^2 = 0$ ⇒ $x = \pm 1$,

so critical points are $(1, -1)$ and $(-1, 1)$. $D(1, -1) = (-2)(2) - 0 < 0$ and $D(-1, 1) = (2)(-2) - 0 < 0$,

so $(-1, 1)$ and $(1, -1)$ are saddle points.

11. $f(x, y) = 1 + 2xy - x^2 - y^2$ ⇒ $f_x = 2y - 2x$,

$f_y = 2x - 2y$, $f_{xx} = f_{yy} = -2$, $f_{xy} = 2$. Then $f_x = 0$ and

$f_y = 0$ implies $x = y$ so the critical points are all points of the form

(x_0, x_0). But $D(x_0, x_0) = 4 - 4 = 0$ so the Second Derivatives

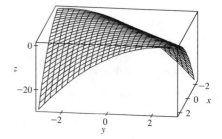

Test gives no information. However

$1 + 2xy - x^2 - y^2 = 1 - (x - y)^2$ and $1 - (x - y)^2 \le 1$ for all

(x, y), with equality if and only if $x = y$. Thus $f(x_0, x_0) = 1$ are

local maxima.

13. $f(x, y) = e^x \cos y$ ⇒ $f_x = e^x \cos y$, $f_y = -e^x \sin y$.

Now $f_x = 0$ implies $\cos y = 0$ or $y = \frac{\pi}{2} + n\pi$ for n an integer.

But $\sin\left(\frac{\pi}{2} + n\pi\right) \ne 0$, so there are no critical points.

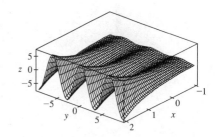

15. $f(x, y) = x \sin y$ ⇒ $f_x = \sin y$, $f_y = x \cos y$, $f_{xx} = 0$,

$f_{yy} = -x \sin y$ and $f_{xy} = \cos y$. Then $f_x = 0$ if and only if

$y = n\pi$, n an integer, and substituting into $f_y = 0$ requires $x = 0$

for each of these y-values. Thus the critical points are $(0, n\pi)$, n

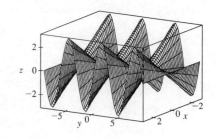

an integer. But $D(0, n\pi) = -\cos^2(n\pi) < 0$ so each critical point

is a saddle point.

17. $f(x, y) = (x^2 + y^2)e^{y^2 - x^2}$ $\Rightarrow$ $f_x = (x^2 + y^2)e^{y^2 - x^2}(-2x) + 2xe^{y^2 - x^2} = 2xe^{y^2 - x^2}(1 - x^2 - y^2)$,

$f_y = (x^2 + y^2)e^{y^2 - x^2}(2y) + 2ye^{y^2 - x^2} = 2ye^{y^2 - x^2}(1 + x^2 + y^2)$,

$f_{xx} = 2xe^{y^2 - x^2}(-2x) + (1 - x^2 - y^2)\left(2x\left(-2xe^{y^2 - x^2}\right) + 2e^{y^2 - x^2}\right)$

$\quad = 2e^{y^2 - x^2}((1 - x^2 - y^2)(1 - 2x^2) - 2x^2)$,

$f_{xy} = 2xe^{y^2 - x^2}(-2y) + 2x(2y)e^{y^2 - x^2}(1 - x^2 - y^2) = -4xye^{y^2 - x^2}(x^2 + y^2)$,

$f_{yy} = 2ye^{y^2 - x^2}(2y) + (1 + x^2 + y^2)\left(2y\left(2ye^{y^2 - x^2}\right) + 2e^{y^2 - x^2}\right)$

$\quad = 2e^{y^2 - x^2}((1 + x^2 + y^2)(1 + 2y^2) + 2y^2)$.

$f_y = 0$ implies $y = 0$, and substituting into $f_x = 0$ gives

$2xe^{-x^2}(1 - x^2) = 0$ $\Rightarrow$ $x = 0$ or $x = \pm 1$.

Thus the critical points are $(0, 0)$ and $(\pm 1, 0)$.

$D(0, 0) = (2)(2) - 0 > 0$ and $f_{xx}(0, 0) = 2 > 0$, so $f(0, 0) = 0$

is a local minimum. $D(\pm 1, 0) = (-4e^{-1})(4e^{-1}) - 0 < 0$ so

$(\pm 1, 0)$ are saddle points.

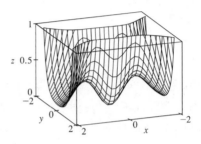

19. $f(x, y) = 3x^2 y + y^3 - 3x^2 - 3y^2 + 2$

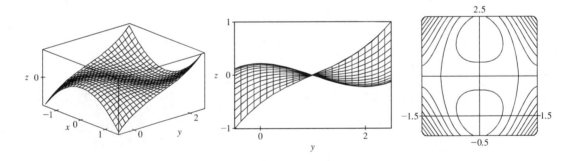

From the graphs, it appears that f has a local maximum $f(0, 0) \approx 2$ and a local minimum $f(0, 2) \approx -2$. There

appear to be saddle points near $(\pm 1, 1)$.

$f_x = 6xy - 6x$, $f_y = 3x^2 + 3y^2 - 6y$. Then $f_x = 0$ implies $x = 0$ or $y = 1$ and when $x = 0$, $f_y = 0$ implies

$y = 0$ or $y = 2$; when $y = 1$, $f_y = 0$ implies $x^2 = 1$ or $x = \pm 1$. Thus the critical points are $(0, 0)$, $(0, 2)$, $(\pm 1, 1)$.

Now $f_{xx} = 6y - 6$, $f_{yy} = 6y - 6$ and $f_{xy} = 6x$, so $D(0, 0) = D(0, 2) = 36 > 0$ while $D(\pm 1, 1) = -36 < 0$

and $f_{xx}(0, 0) = -6$, $f_{xx}(0, 2) = 6$. Hence $(\pm 1, 1)$ are saddle points while $f(0, 0) = 2$ is a local maximum and

$f(0, 2) = -2$ is a local minimum.

21. $f(x,y) = \sin x + \sin y + \sin(x+y), 0 \le x \le 2\pi, 0 \le y \le 2\pi$

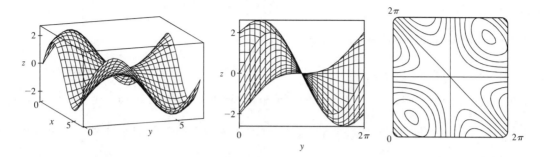

From the graphs it appears that f has a local maximum at about $(1,1)$ with value approximately 2.6, a local minimum at about $(5,5)$ with value approximately -2.6, and a saddle point at about $(3,3)$.

$f_x = \cos x + \cos(x+y)$, $f_y = \cos y + \cos(x+y)$, $f_{xx} = -\sin x - \sin(x+y)$, $f_{yy} = -\sin y - \sin(x+y)$, $f_{xy} = -\sin(x+y)$. Setting $f_x = 0$ and $f_y = 0$ and subtracting gives $\cos x - \cos y = 0$ or $\cos x = \cos y$. Thus $x = y$ or $x = 2\pi - y$. If $x = y$, $f_x = 0$ becomes $\cos x + \cos 2x = 0$ or $2\cos^2 x + \cos x - 1 = 0$, a quadratic in $\cos x$. Thus $\cos x = -1$ or $\frac{1}{2}$ and $x = \pi$, $\frac{\pi}{3}$, or $\frac{5\pi}{3}$, yielding the critical points (π, π), $\left(\frac{\pi}{3}, \frac{\pi}{3}\right)$ and $\left(\frac{5\pi}{3}, \frac{5\pi}{3}\right)$. Similarly if $x = 2\pi - y$, $f_x = 0$ becomes $(\cos x) + 1 = 0$ and the resulting critical point is (π, π). Now $D(x,y) = \sin x \sin y + \sin x \sin(x+y) + \sin y \sin(x+y)$. So $D(\pi, \pi) = 0$ and the Second Derivatives Test doesn't apply. $D\left(\frac{\pi}{3}, \frac{\pi}{3}\right) = \frac{9}{4} > 0$ and $f_{xx}\left(\frac{\pi}{3}, \frac{\pi}{3}\right) < 0$ so $f\left(\frac{\pi}{3}, \frac{\pi}{3}\right) = \frac{3\sqrt{3}}{2}$ is a local maximum while $D\left(\frac{5\pi}{3}, \frac{5\pi}{3}\right) = \frac{9}{4} > 0$ and $f_{xx}\left(\frac{5\pi}{3}, \frac{5\pi}{3}\right) > 0$, so $f\left(\frac{5\pi}{3}, \frac{5\pi}{3}\right) = -\frac{3\sqrt{3}}{2}$ is a local minimum.

23. $f(x,y) = x^4 - 5x^2 + y^2 + 3x + 2$ $\Rightarrow$ $f_x(x,y) = 4x^3 - 10x + 3$ and $f_y(x,y) = 2y$. $f_y = 0$ $\Rightarrow$ $y = 0$, and the graph of f_x shows that the roots of $f_x = 0$ are approximately $x = -1.714, 0.312$ and 1.402. (Alternatively, we could have used a calculator or a CAS to find these roots.) So to three decimal places, the critical points are $(-1.714, 0)$, $(1.402, 0)$, and $(0.312, 0)$. Now since $f_{xx} = 12x^2 - 10$, $f_{xy} = 0$, $f_{yy} = 2$, and $D = 24x^2 - 20$, we have $D(-1.714, 0) > 0$, $f_{xx}(-1.714, 0) > 0$, $D(1.402, 0) > 0$, $f_{xx}(1.402, 0) > 0$, and $D(0.312, 0) < 0$. Therefore $f(-1.714, 0) \approx -9.200$ and $f(1.402, 0) \approx 0.242$ are local minima, and $(0.312, 0)$ is a saddle point. The lowest point on the graph is approximately $(-1.714, 0, -9.200)$.

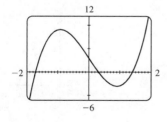

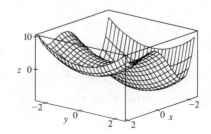

25. $f(x, y) = 2x + 4x^2 - y^2 + 2xy^2 - x^4 - y^4$ ⇒ $f_x(x, y) = 2 + 8x + 2y^2 - 4x^3$,

$f_y(x, y) = -2y + 4xy - 4y^3$. Now $f_y = 0$ ⇔ $2y(2y^2 - 2x + 1) = 0$ ⇔ $y = 0$ or $y^2 = x - \frac{1}{2}$.

The first of these implies that $f_x = -4x^3 + 8x + 2$, and the second implies that

$f_x = 2 + 8x + 2(x - \frac{1}{2}) - 4x^3 = -4x^3 + 10x + 1$. From the graphs, we see that the first possibility for f_x has

roots at approximately -1.267, -0.259, and 1.526, and the second has a root at approximately 1.629 (the negative

roots do not give critical points, since $y^2 = x - \frac{1}{2}$ must be positive). So to three decimal places, f has critical points

at $(-1.267, 0)$, $(-0.259, 0)$, $(1.526, 0)$, and $(1.629, \pm1.063)$. Now since $f_{xx} = 8 - 12x^2$, $f_{xy} = 4y$,

$f_{yy} = 4x - 12y^2$, and $D = (8 - 12x^2)(4x - 12y^2) - 16y^2$, we have $D(-1.267, 0) > 0$, $f_{xx}(-1.267, 0) > 0$,

$D(-0.259, 0) < 0$, $D(1.526, 0) < 0$, $D(1.629, \pm1.063) > 0$, and $f_{xx}(1.629, \pm1.063) < 0$. Therefore, to three

decimal places, $f(-1.267, 0) \approx 1.310$ and $f(1.629, \pm1.063) \approx 8.105$ are local maxima, and $(-0.259, 0)$ and

$(1.526, 0)$ are saddle points. The highest points on the graph are approximately $(1.629, \pm1.063, 8.105)$.

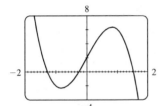
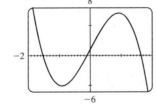
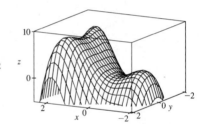

27. Since f is a polynomial it is continuous on D, so an absolute maximum and minimum exist. Here $f_x = 4$, $f_y = -5$

so there are no critical points inside D. Thus the absolute extrema must both occur on the boundary. Along L_1,

$x = 0$ and $f(0, y) = 1 - 5y$ for $0 \le y \le 3$, a decreasing function in y, so the maximum value is $f(0, 0) = 1$ and

the minimum value is $f(0, 3) = -14$. Along L_2, $y = 0$ and $f(x, 0) = 1 + 4x$ for $0 \le x \le 2$, an increasing

function in x, so the minimum value is $f(0, 0) = 1$ and the

maximum value is $f(2, 0) = 9$. Along L_3, $y = -\frac{3}{2}x + 3$ and

$f\left(x, -\frac{3}{2}x + 3\right) = \frac{23}{2}x - 14$ for $0 \le x \le 2$, an increasing function

in x, so the minimum value is $f(0, 3) = -14$ and the maximum

value is $f(2, 0) = 9$. Thus the absolue maximum of f on D is

$f(2, 0) = 9$ and the absolute minimum is $f(0, 3) = -14$.

29. $f_x(x, y) = 2x + 2xy$, $f_y(x, y) = 2y + x^2$, and setting $f_x = f_y = 0$

gives $(0, 0)$ as the only critical point in D, with $f(0, 0) = 4$.

On L_1: $y = -1$, $f(x, -1) = 5$, a constant.

On L_2: $x = 1$, $f(1, y) = y^2 + y + 5$, a quadratic in y which attains its

maximum at $(1, 1)$, $f(1, 1) = 7$ and its minimum at $\left(1, -\frac{1}{2}\right)$, $f\left(1, -\frac{1}{2}\right) = \frac{17}{4}$.

On L_3: $f(x, 1) = 2x^2 + 5$ which attains its maximum at $(-1, 1)$ and $(1, 1)$

with $f(\pm 1, 1) = 7$ and its minimum at $(0, 1)$, $f(0, 1) = 5$.

On L_4: $f(-1, y) = y^2 + y + 5$ with maximum at $(-1, 1)$, $f(-1, 1) = 7$ and

minimum at $\left(-1, -\frac{1}{2}\right)$, $f\left(-1, -\frac{1}{2}\right) = \frac{17}{4}$. Thus the absolute maximum is attained at both $(\pm 1, 1)$ with

$f(\pm 1, 1) = 7$ and the absolute minimum on D is attained at $(0, 0)$ with $f(0, 0) = 4$.

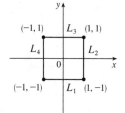

31. $f(x, y) = x^4 + y^4 - 4xy + 2$ is a polynomial and hence continuous

on D, so it has an absolute maximum and minimum on D. In

Exercise 7, we found the critical points of f; only $(1, 1)$ with

$f(1, 1) = 0$ is inside D. On L_1: $y = 0$, $f(x, 0) = x^4 + 2$,

$0 \le x \le 3$, a polynomial in x which attains its maximum at $x = 3$,

$f(3, 0) = 83$, and its minimum at $x = 0$, $f(0, 0) = 2$. On L_2:

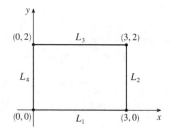

$x = 3$, $f(3, y) = y^4 - 12y + 83$, $0 \le y \le 2$, a polynomial in y

which attains its minimum at $y = \sqrt[3]{3}$, $f(3, \sqrt[3]{3}) = 83 - 9\sqrt[3]{3} \approx 70.0$, and its maximum at $y = 0$, $f(3, 0) = 83$.

On L_3: $y = 2$, $f(x, 2) = x^4 - 8x + 18$, $0 \le x \le 3$, a polynomial in x which attains its minimum at $x = \sqrt[3]{2}$,

$f(\sqrt[3]{2}, 2) = 18 - 6\sqrt[3]{2} \approx 10.4$, and its maximum at $x = 3$, $f(3, 2) = 75$. On L_4: $x = 0$, $f(0, y) = y^4 + 2$,

$0 \le y \le 2$, a polynomial in y which attains its maximum at $y = 2$, $f(0, 2) = 18$, and its minimum at $y = 0$,

$f(0, 0) = 2$. Thus the absolute maximum of f on D is $f(3, 0) = 83$ and the absolute minimum is $f(1, 1) = 0$.

33. $f_x(x, y) = 6x^2$ and $f_y(x, y) = 4y^3$. And so $f_x = 0$ and $f_y = 0$ only occur when $x = y = 0$. Hence, the only

critical point inside the disk is at $x = y = 0$ where $f(0, 0) = 0$. Now on the circle $x^2 + y^2 = 1$, $y^2 = 1 - x^2$ so let

$g(x) = f(x, y) = 2x^3 + (1 - x^2)^2 = x^4 + 2x^3 - 2x^2 + 1$, $-1 \le x \le 1$. Then $g'(x) = 4x^3 + 6x^2 - 4x = 0 \Rightarrow$

$x = 0, -2$, or $\frac{1}{2}$. $f(0, \pm 1) = g(0) = 1$, $f\left(\frac{1}{2}, \pm\frac{\sqrt{3}}{2}\right) = g\left(\frac{1}{2}\right) = \frac{13}{16}$, and $(-2, -3)$ is not in D. Checking the

endpoints, we get $f(-1, 0) = g(-1) = -2$ and $f(1, 0) = g(1) = 2$. Thus the absolute maximum and minimum

of f on D are $f(1, 0) = 2$ and $f(-1, 0) = -2$.

Another method: On the boundary $x^2 + y^2 = 1$ we can write $x = \cos\theta$, $y = \sin\theta$, so

$f(\cos\theta, \sin\theta) = 2\cos^3\theta + \sin^4\theta$, $0 \le \theta \le 2\pi$.

35. $f(x, y) = -(x^2 - 1)^2 - (x^2 y - x - 1)^2$ $\Rightarrow$ $f_x(x, y) = -2(x^2 - 1)(2x) - 2(x^2 y - x - 1)(2xy - 1)$ and

$f_y(x, y) = -2(x^2 y - x - 1)x^2$. Setting $f_y(x, y) = 0$ gives either $x = 0$ or $x^2 y - x - 1 = 0$. There are no

critical points for $x = 0$, since $f_x(0, y) = -2$, so we set $x^2 y - x - 1 = 0$ $\Leftrightarrow$ $y = \dfrac{x + 1}{x^2}$ $(x \neq 0)$, so

$f_x\left(x, \dfrac{x+1}{x^2}\right) = -2(x^2 - 1)(2x) - 2\left(x^2 \dfrac{x+1}{x^2} - x - 1\right)\left(2x \dfrac{x+1}{x^2} - 1\right) = -4x(x^2 - 1)$. Therefore

$f_x(x, y) = f_y(x, y) = 0$ at the points $(1, 2)$ and $(-1, 0)$. To classify these critical points, we calculate

$f_{xx}(x, y) = -12x^2 - 12x^2 y^2 + 12xy + 4y + 2$, $f_{yy}(x, y) = -2x^4$, and $f_{xy}(x, y) = -8x^3 y + 6x^2 + 4x$.

In order to use the Second Derivatives Test we calculate

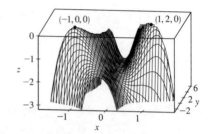

$D(-1, 0) = f_{xx}(-1, 0) f_{yy}(-1, 0) - [f_{xy}(-1, 0)]^2$

$\quad = 16 > 0,$

$f_{xx}(-1, 0) = -10 < 0$, $D(1, 2) = 16 > 0$, and

$f_{xx}(1, 2) = -26 < 0$, so both $(-1, 0)$ and $(1, 2)$ give local

maxima.

37. Let d be the distance from $(2, 1, -1)$ to any point (x, y, z) on the plane $x + y - z = 1$, so

$d = \sqrt{(x - 2)^2 + (y - 1)^2 + (z + 1)^2}$ where $z = x + y - 1$, and we minimize

$d^2 = f(x, y) = (x - 2)^2 + (y - 1)^2 + (x + y)^2$. Then $f_x(x, y) = 2(x - 2) + 2(x + y) = 4x + 2y - 4$,

$f_y(x, y) = 2(y - 1) + 2(x + y) = 2x + 4y - 2$. Solving $4x + 2y - 4 = 0$ and $2x + 4y - 2 = 0$ simultaneously

gives $x = 1$, $y = 0$. An absolute minimum exists (since there is a minimum distance from the point to the plane)

and it must occur at a critical point, so the shortest distance occurs for $x = 1$, $y = 0$ for which

$d = \sqrt{(1 - 2)^2 + (0 - 1)^2 + (1 + 0)^2} = \sqrt{3}$.

39. Minimize $d^2 = x^2 + y^2 + z^2 = x^2 + y^2 + xy + 1$. Then $f_x = 2x + y$, $f_y = 2y + x$ so the critical point is $(0, 0)$

and $D(0, 0) = 4 - 1 > 0$ with $f_{xx}(0, 0) = 2$ so this is a minimum. Thus $z^2 = 1$ or $z = \pm 1$ and the points on the

surface are $(0, 0, \pm 1)$.

41. $x + y + z = 100$, so maximize $f(x, y) = xy(100 - x - y)$. $f_x = 100y - 2xy - y^2$, $f_y = 100x - x^2 - 2xy$,

$f_{xx} = -2y$, $f_{yy} = -2x$, $f_{xy} = 100 - 2x - 2y$. Then $f_x = 0$ implies $y = 0$ or $y = 100 - 2x$. Substituting $y = 0$

into $f_y = 0$ gives $x = 0$ or $x = 100$ and substituting $y = 100 - 2x$ into $f_y = 0$ gives $3x^2 - 100x = 0$ so $x = 0$ or

$\frac{100}{3}$. Thus the critical points are $(0, 0)$, $(100, 0)$, $(0, 100)$ and $\left(\frac{100}{3}, \frac{100}{3}\right)$.

$D(0, 0) = D(100, 0) = D(0, 100) = -10{,}000$ while $D\left(\frac{100}{3}, \frac{100}{3}\right) = \frac{10{,}000}{3}$ and $f_{xx}\left(\frac{100}{3}, \frac{100}{3}\right) = -\frac{200}{3} < 0$.

Thus $(0, 0)$, $(100, 0)$ and $(0, 100)$ are saddle points whereas $f\left(\frac{100}{3}, \frac{100}{3}\right)$ is a local maximum. Thus the numbers are

$x = y = z = \frac{100}{3}$.

43. Maximize $f(x, y) = xy(36 - 9x^2 - 36y^2)^{1/2} / 2$ with (x, y, z) in first octant. Then

$f_x = \dfrac{y(36 - 9x^2 - 36y^2)^{1/2}}{2} + \dfrac{-9x^2 y(36 - 9x^2 - 36y^2)^{-1/2}}{2} = \dfrac{(36y - 18x^2 y - 36y^3)}{2(36 - 9x^2 - 36y^2)^{1/2}}$ and

$f_y = \dfrac{36x - 9x^3 - 72xy^2}{2(36 - 9x^2 - 36y^2)^{1/2}}$. Setting $f_x = 0$ gives $y = 0$ or $y^2 = \dfrac{2 - x^2}{2}$ but $y > 0$, so only the latter solution

applies. Substituting this y into $f_y = 0$ gives $x^2 = \frac{4}{3}$ or $x = \frac{2}{\sqrt{3}}$, $y = \frac{1}{\sqrt{3}}$ and then $z^2 = (36 - 12 - 12)/4 = 3$.

The fact that this gives a maximum volume follows from the geometry. This maximum volume is

$$V = (2x)(2y)(2z) = 8\left(\tfrac{2}{\sqrt{3}}\right)\left(\tfrac{1}{\sqrt{3}}\right)(\sqrt{3}) = \tfrac{16}{\sqrt{3}}.$$

45. Maximize $f(x, y) = \dfrac{xy}{3}(6 - x - 2y)$, then the maximum volume is $V = xyz$.

$f_x = \frac{1}{3}(6y - 2xy - y^2) = \frac{1}{3}y(6 - 2x - 2y)$ and $f_y = \frac{1}{3}x(6 - x - 4y)$. Setting $f_x = 0$ and $f_y = 0$ gives

critical point $(2, 1)$ which geometrically must yield a maximum. Thus the volume of the largest such box is

$V = (2)(1)\left(\frac{2}{3}\right) = \frac{4}{3}$.

47. Let the dimensions be x, y, and z; then $4x + 4y + 4z = c$ and the volume is

$V = xyz = xy\left(\frac{1}{4}c - x - y\right) = \frac{1}{4}cxy - x^2y - xy^2$, $x > 0$, $y > 0$. Then $V_x = \frac{1}{4}cy - 2xy - y^2$ and

$V_y = \frac{1}{4}cx - x^2 - 2xy$, so $V_x = 0 = V_y$ when $2x + y = \frac{1}{4}c$ and $x + 2y = \frac{1}{4}c$. Solving, we get $x = \frac{1}{12}c$, $y = \frac{1}{12}c$

and $z = \frac{1}{4}c - x - y = \frac{1}{12}c$. From the geometrical nature of the problem, this critical point must give an absolute

maximum. Thus the box is a cube with edge length $\frac{1}{12}c$.

49. Let the dimensions be x, y and z, then minimize $xy + 2(xz + yz)$ if $xyz = 32{,}000$ m^3. Then

$f(x, y) = xy + [64{,}000(x + y)/xy] = xy + 64{,}000(x^{-1} + y^{-1})$, $f_x = y - 64{,}000x^{-2}$, $f_y = x - 64{,}000y^{-2}$.

And $f_x = 0$ implies $y = 64{,}000/x^2$; substituting into $f_y = 0$ implies $x^3 = 64{,}000$ or $x = 40$ and then $y = 40$.

Now $D(x, y) = [(2)(64{,}000)]^2 x^{-3}y^{-3} - 1 > 0$ for $(40, 40)$ and $f_{xx}(40, 40) > 0$ so this is indeed a minimum.

Thus the dimensions of the box are $x = y = 40$ cm, $z = 20$ cm.

51. Let x, y, z be the dimensions of the rectangular box. Then the volume of the box is xyz and

$L = \sqrt{x^2 + y^2 + z^2}$ $\Rightarrow$ $L^2 = x^2 + y^2 + z^2$ $\Rightarrow$ $z = \sqrt{L^2 - x^2 - y^2}$. Substituting, we have volume

$V(x, y) = xy\sqrt{L^2 - x^2 - y^2}$, $x, y > 0$.

$V_x = xy \cdot \frac{1}{2}(L^2 - x^2 - y^2)^{-1/2}(-2x) + y\sqrt{L^2 - x^2 - y^2} = y\sqrt{L^2 - x^2 - y^2} - \dfrac{x^2y}{\sqrt{L^2 - x^2 - y^2}}$,

$V_y = x\sqrt{L^2 - x^2 - y^2} - \dfrac{xy^2}{\sqrt{L^2 - x^2 - y^2}}$,

$V_x = 0$ implies $y(L^2 - x^2 - y^2) = x^2y$ $\Rightarrow$ $y(L^2 - 2x^2 - y^2) = 0$ $\Rightarrow$ $2x^2 + y^2 = L^2$ (since $y > 0$), and

$V_y = 0$ implies $x(L^2 - x^2 - y^2) = xy^2$ $\Rightarrow$ $x(L^2 - x^2 - 2y^2) = 0$ $\Rightarrow$ $x^2 + 2y^2 = L^2$ (since $x > 0$).

Substituting $y^2 = L^2 - 2x^2$ into $x^2 + 2y^2 = L^2$ gives $x^2 + 2L^2 - 4x^2 = L^2$ $\Rightarrow$ $3x^2 = L^2$ $\Rightarrow$ $x = L/\sqrt{3}$

(since $x > 0$) and then $y = \sqrt{L^2 - 2\left(L/\sqrt{3}\right)^2} = L/\sqrt{3}$. So the only critical point is $\left(L/\sqrt{3}, L/\sqrt{3}\right)$ which,

from the geometrical nature of the problem, must give an absolute maximum. Thus the maximum volume is

$V\left(L/\sqrt{3}, L/\sqrt{3}\right) = \left(L/\sqrt{3}\right)^2 \sqrt{L^2 - \left(L/\sqrt{3}\right)^2 - \left(L/\sqrt{3}\right)^2} = L^3/(3\sqrt{3})$ cubic units.

53. Note that here the variables are m and b, and $f(m, b) = \sum\limits_{i=1}^{n} [y_i - (mx_i + b)]^2$. Then

$$f_m = \sum_{i=1}^{n} -2x_i[y_i - (mx_i + b)] = 0 \text{ implies } \sum_{i=1}^{n} (x_iy_i - mx_i^2 - bx_i) = 0 \text{ or } \sum_{i=1}^{n} x_iy_i = m\sum_{i=1}^{n} x_i^2 + b\sum_{i=1}^{n} x_i$$

and $f_b = \sum\limits_{i=1}^{n} -2[y_i - (mx_i + b)] = 0$ implies $\sum\limits_{i=1}^{n} y_i = m\sum\limits_{i=1}^{n} x_i + \sum\limits_{i=1}^{n} b = m\left(\sum\limits_{i=1}^{n} x_i\right) + nb$. Thus we have

the two desired equations. Now $f_{mm} = \sum\limits_{i=1}^{n} 2x_i^2$, $f_{bb} = \sum\limits_{i=1}^{n} 2 = 2n$ and $f_{mb} = \sum\limits_{i=1}^{n} 2x_i$. And $f_{mm}(m, b) > 0$

always and $D(m, b) = 4n\left(\sum\limits_{i=1}^{n} x_i^2\right) - 4\left(\sum\limits_{i=1}^{n} x_i\right)^2 = 4\left[n\left(\sum\limits_{i=1}^{n} x_i^2\right) - \left(\sum\limits_{i=1}^{n} x_i\right)^2\right] > 0$ always so the

solutions of these two equations do indeed minimize $\sum\limits_{i=1}^{n} d_i^2$.

15.8 Lagrange Multipliers   ET 14.8

1. At the extreme values of f, the level curves of f just touch the curve $g(x, y) = 8$ with a common tangent line.
(See Figure 1 and the accompanying discussion.) We can observe several such occurrences on the contour map, but
the level curve $f(x, y) = c$ with the largest value of c which still intersects the curve $g(x, y) = 8$ is approximately
$c = 59$, and the smallest value of c corresponding to a level curve which intersects $g(x, y) = 8$ appears to be
$c = 30$. Thus we estimate the maximum value of f subject to the constraint $g(x, y) = 8$ to be about 59 and the
minimum to be 30.

3. $f(x, y) = x^2 - y^2$, $g(x, y) = x^2 + y^2 = 1$ $\Rightarrow$ $\nabla f = \langle 2x, -2y \rangle$, $\lambda\nabla g = \langle 2\lambda x, 2\lambda y \rangle$. Then $2x = 2\lambda x$ implies
$x = 0$ or $\lambda = 1$. If $x = 0$, then $x^2 + y^2 = 1$ implies $y = \pm 1$ and if $\lambda = 1$, then $-2y = 2\lambda y$ implies $y = 0$ and thus
$x = \pm 1$. Thus the possible points for the extreme values of f are $(\pm 1, 0)$, $(0, \pm 1)$. But $f(\pm 1, 0) = 1$ while
$f(0, \pm 1) = -1$ so the maximum value of f on $x^2 + y^2 = 1$ is $f(\pm 1, 0) = 1$ and the minimum value
is $f(0, \pm 1) = -1$.

5. $f(x, y) = x^2 y$, $g(x, y) = x^2 + 2y^2 = 6$ $\Rightarrow$ $\nabla f = \langle 2xy, x^2 \rangle$, $\lambda\nabla g = \langle 2\lambda x, 4\lambda y \rangle$. Then $2xy = 2\lambda x$ implies
$x = 0$ or $\lambda = y$. If $x = 0$, then $x^2 = 4\lambda y$ implies $\lambda = 0$ or $y = 0$. However, if $y = 0$ then $g(x, y) = 0$, a
contradiction. So $\lambda = 0$ and then $g(x, y) = 6$ $\Rightarrow$ $y = \pm\sqrt{3}$. If $\lambda = y$, then $x^2 = 4\lambda y$ implies $x^2 = 4y^2$, and so
$g(x, y) = 6$ $\Rightarrow$ $4y^2 + 2y^2 = 6$ $\Rightarrow$ $y^2 = 1$ $\Rightarrow$ $y = \pm 1$. Thus f has possible extreme values at the points
$(0, \pm\sqrt{3})$, $(\pm 2, 1)$, and $(\pm 2, -1)$. After evaluating f at these points, we find the maximum value to be
$f(\pm 2, 1) = 4$ and the minimum to be $f(\pm 2, -1) = -4$.

7. $f(x, y, z) = 2x + 6y + 10z$, $g(x, y, z) = x^2 + y^2 + z^2 = 35$ $\Rightarrow$ $\nabla f = \langle 2, 6, 10 \rangle$,

$\lambda\nabla g = \langle 2\lambda x, 2\lambda y, 2\lambda z \rangle$. Then $2\lambda x = 2$, $2\lambda y = 6$, $2\lambda z = 10$ imply $x = \dfrac{1}{\lambda}$, $y = \dfrac{3}{\lambda}$, and $z = \dfrac{5}{\lambda}$. But

$35 = x^2 + y^2 + z^2 = \left(\dfrac{1}{\lambda}\right)^2 + \left(\dfrac{3}{\lambda}\right)^2 + \left(\dfrac{5}{\lambda}\right)^2$ $\Rightarrow$ $35 = \dfrac{35}{\lambda^2}$ $\Rightarrow$ $\lambda = \pm 1$, so f has possible extreme values at the points $(1, 3, 5)$, $(-1, -3, -5)$. The maximum value of f on $x^2 + y^2 + z^2 = 35$ is $f(1, 3, 5) = 70$, and the minimum is $f(-1, -3, -5) = -70$.

9. $f(x, y, z) = xyz$, $g(x, y, z) = x^2 + 2y^2 + 3z^2 = 6 \Rightarrow \nabla f = \langle yz, xz, xy \rangle$, $\lambda \nabla g = \langle 2\lambda x, 4\lambda y, 6\lambda z \rangle$. Then $\nabla f = \lambda \nabla g$ implies $\lambda = (yz)/(2x) = (xz)/(4y) = (xy)/(6z)$ or $x^2 = 2y^2$ and $z^2 = \frac{2}{3}y^2$. Thus $x^2 + 2y^2 + 3z^2 = 6$ implies $6y^2 = 6$ or $y = \pm 1$. Then the possible points are $\left(\sqrt{2}, \pm 1, \sqrt{\frac{2}{3}} \right)$, $\left(\sqrt{2}, \pm 1, -\sqrt{\frac{2}{3}} \right)$, $\left(-\sqrt{2}, \pm 1, \sqrt{\frac{2}{3}} \right)$, $\left(-\sqrt{2}, \pm 1, -\sqrt{\frac{2}{3}} \right)$. The maximum value of f on the ellipsoid is $\frac{2}{\sqrt{3}}$, occurring when all coordinates are positive or exactly two are negative and the minimum is $-\frac{2}{\sqrt{3}}$ occurring when 1 or 3 of the coordinates are negative.

11. $f(x, y, z) = x^2 + y^2 + z^2$, $g(x, y, z) = x^4 + y^4 + z^4 = 1 \Rightarrow$
$\nabla f = \langle 2x, 2y, 2z \rangle$, $\lambda \nabla g = \langle 4\lambda x^3, 4\lambda y^3, 4\lambda z^3 \rangle$.

Case 1: If $x \neq 0$, $y \neq 0$ and $z \neq 0$, then $\nabla f = \lambda \nabla g$ implies $\lambda = 1/(2x^2) = 1/(2y^2) = 1/(2z^2)$ or $x^2 = y^2 = z^2$ and $3x^4 = 1$ or $x = \pm \frac{1}{\sqrt[4]{3}}$ giving the points $\left(\pm \frac{1}{\sqrt[4]{3}}, \frac{1}{\sqrt[4]{3}}, \frac{1}{\sqrt[4]{3}} \right)$, $\left(\pm \frac{1}{\sqrt[4]{3}}, -\frac{1}{\sqrt[4]{3}}, \frac{1}{\sqrt[4]{3}} \right)$, $\left(\pm \frac{1}{\sqrt[4]{3}}, \frac{1}{\sqrt[4]{3}}, -\frac{1}{\sqrt[4]{3}} \right)$, $\left(\pm \frac{1}{\sqrt[4]{3}}, -\frac{1}{\sqrt[4]{3}}, -\frac{1}{\sqrt[4]{3}} \right)$ all with an f-value of $\sqrt{3}$.

Case 2: If one of the variables equals zero and the other two are not zero, then the squares of the two nonzero coordinates are equal with common value $\frac{1}{\sqrt{2}}$ and corresponding f value of $\sqrt{2}$.

Case 3: If exactly two of the variables are zero, then the third variable has value ± 1 with the corresponding f value of 1. Thus on $x^4 + y^4 + z^4 = 1$, the maximum value of f is $\sqrt{3}$ and the minimum value is 1.

13. $f(x, y, z, t) = x + y + z + t$, $g(x, y, z, t) = x^2 + y^2 + z^2 + t^2 = 1 \Rightarrow \langle 1, 1, 1, 1 \rangle = \langle 2\lambda x, 2\lambda y, 2\lambda z, 2\lambda t \rangle$, so $\lambda = 1/(2x) = 1/(2y) = 1/(2z) = 1/(2t)$ and $x = y = z = t$. But $x^2 + y^2 + z^2 + t^2 = 1$, so the possible points are $\left(\pm \frac{1}{2}, \pm \frac{1}{2}, \pm \frac{1}{2}, \pm \frac{1}{2} \right)$. Thus the maximum value of f is $f\left(\frac{1}{2}, \frac{1}{2}, \frac{1}{2}, \frac{1}{2} \right) = 2$ and the minimum value is $f\left(-\frac{1}{2}, -\frac{1}{2}, -\frac{1}{2}, -\frac{1}{2} \right) = -2$.

15. $f(x, y, z) = x + 2y$, $g(x, y, z) = x + y + z = 1$, $h(x, y, z) = y^2 + z^2 = 4 \Rightarrow \nabla f = \langle 1, 2, 0 \rangle$, $\lambda \nabla g = \langle \lambda, \lambda, \lambda \rangle$ and $\mu \nabla h = \langle 0, 2\mu y, 2\mu z \rangle$. Then $1 = \lambda$, $2 = \lambda + 2\mu y$ and $0 = \lambda + 2\mu z$ so $\mu y = \frac{1}{2} = -\mu z$ or $y = 1/(2\mu)$, $z = -1/(2\mu)$. Thus $x + y + z = 1$ implies $x = 1$ and $y^2 + z^2 = 4$ implies $\mu = \pm \frac{1}{2\sqrt{2}}$. Then the possible points are $\left(1, \pm\sqrt{2}, \mp\sqrt{2} \right)$ and the maximum value is $f\left(1, \sqrt{2}, -\sqrt{2} \right) = 1 + 2\sqrt{2}$ and the minimum value is $f\left(1, -\sqrt{2}, \sqrt{2} \right) = 1 - 2\sqrt{2}$.

17. $f(x, y, z) = yz + xy$, $g(x, y, z) = xy = 1$, $h(x, y, z) = y^2 + z^2 = 1 \Rightarrow \nabla f = \langle y, x + z, y \rangle$, $\lambda \nabla g = \langle \lambda y, \lambda x, 0 \rangle$, $\mu \nabla h = \langle 0, 2\mu y, 2\mu z \rangle$. Then $y = \lambda y$ implies $\lambda = 1$ [$y \neq 0$ since $g(x, y, z) = 1$],

$x + z = \lambda x + 2\mu y$ and $y = 2\mu z$. Thus $\mu = z/(2y) = y/(2y)$ or $y^2 = z^2$, and so $y^2 + z^2 = 1$ implies $y = \pm\frac{1}{\sqrt{2}}$,

$z = \pm\frac{1}{\sqrt{2}}$. Then $xy = 1$ implies $x = \pm\sqrt{2}$ and the possible points are $\left(\pm\sqrt{2}, \pm\frac{1}{\sqrt{2}}, \frac{1}{\sqrt{2}}\right), \left(\pm\sqrt{2}, \pm\frac{1}{\sqrt{2}}, -\frac{1}{\sqrt{2}}\right)$.

Hence the maximum of f subject to the constraints is $f\left(\pm\sqrt{2}, \pm\frac{1}{\sqrt{2}}, \pm\frac{1}{\sqrt{2}}\right) = \frac{3}{2}$ and the minimum is

$f\left(\pm\sqrt{2}, \pm\frac{1}{\sqrt{2}}, \mp\frac{1}{\sqrt{2}}\right) = \frac{1}{2}$.

Note: Since $xy = 1$ is one of the constraints we could have solved the problem by solving $f(y, z) = yz + 1$ subject to $y^2 + z^2 = 1$.

19. $f(x, y) = e^{-xy}$. For the interior of the region, we find the critical points: $f_x = -ye^{-xy}$, $f_y = -xe^{-xy}$,

so the only critical point is $(0, 0)$, and $f(0, 0) = 1$. For the boundary, we use Lagrange multipliers.

$g(x, y) = x^2 + 4y^2 = 1 \Rightarrow \lambda\nabla g = \langle 2\lambda x, 8\lambda y \rangle$, so setting $\nabla f = \lambda\nabla g$ we get $-ye^{-xy} = 2\lambda x$ and

$-xe^{-xy} = 8\lambda y$. The first of these gives $e^{-xy} = -2\lambda x/y$, and then the second gives $-x(-2\lambda x/y) = 8\lambda y \Rightarrow$

$x^2 = 4y^2$. Solving this last equation with the constraint $x^2 + 4y^2 = 1$ gives $x = \pm\frac{1}{\sqrt{2}}$ and $y = \pm\frac{1}{2\sqrt{2}}$. Now

$f\left(\pm\frac{1}{\sqrt{2}}, \mp\frac{1}{2\sqrt{2}}\right) = e^{1/4} \approx 1.284$ and $f\left(\pm\frac{1}{\sqrt{2}}, \pm\frac{1}{2\sqrt{2}}\right) = e^{-1/4} \approx 0.779$. The former are the maxima on the

region and the latter are the minima.

21. $P(L, K) = bL^\alpha K^{1-\alpha}$, $g(L, K) = mL + nK = p \Rightarrow \nabla P = \langle \alpha bL^{\alpha-1}K^{1-\alpha}, (1-\alpha)bL^\alpha K^{-\alpha} \rangle$,

$\lambda\nabla g = \langle \lambda m, \lambda n \rangle$. Then $\alpha b(K/L)^{1-\alpha} = \lambda m$ and $(1-\alpha)b(L/K)^\alpha = \lambda n$ and $mL + nK = p$, so

$\alpha b(K/L)^{1-\alpha}/m = (1-\alpha)b(L/K)^\alpha/n$ or $n\alpha/[m(1-\alpha)] = (L/K)^\alpha(L/K)^{1-\alpha}$ or $L = Kn\alpha/[m(1-\alpha)]$.

Substituting into $mL + nK = p$ gives $K = (1-\alpha)p/n$ and $L = \alpha p/m$ for the maximum production.

23. Let the sides of the rectangle be x and y. Then $f(x, y) = xy$, $g(x, y) = 2x + 2y = p \Rightarrow \nabla f(x, y) = \langle y, x \rangle$,

$\lambda\nabla g = \langle 2\lambda, 2\lambda \rangle$. Then $\lambda = \frac{1}{2}y = \frac{1}{2}x$ implies $x = y$ and the rectangle with maximum area is a square with side

length $\frac{1}{4}p$.

25. Let $f(x, y, z) = d^2 = (x-2)^2 + (y-1)^2 + (z+1)^2$, then we want to minimize f subject to the constraint

$g(x, y, z) = x + y - z = 1$. $\nabla f = \lambda\nabla g \Rightarrow \langle 2(x-2), 2(y-1), 2(z+1) \rangle = \lambda\langle 1, 1, -1 \rangle$, so $x = (\lambda+4)/2$,

$y = (\lambda+2)/2$, $z = -(\lambda+2)/2$. Substituting into the constraint equation gives $\dfrac{\lambda+4}{2} + \dfrac{\lambda+2}{2} + \dfrac{\lambda+2}{2} = 1$

$\Rightarrow 3\lambda + 8 = 2 \Rightarrow \lambda = -2$, so $x = 1$, $y = 0$, and $z = 0$. This must correspond to a minimum, so the shortest

distance is $d = \sqrt{(1-2)^2 + (0-1)^2 + (0+1)^2} = \sqrt{3}$.

27. $f(x, y, z) = x^2 + y^2 + z^2$, $g(x, y, z) = z^2 - xy - 1 = 0 \Rightarrow \nabla f = \langle 2x, 2y, 2z \rangle = \lambda\nabla g = \langle -\lambda y, -\lambda x, 2\lambda z \rangle$.

Then $2z = 2\lambda z$ implies $z = 0$ or $\lambda = 1$. If $z = 0$ then $g(x, y, z) = 1$ implies $xy = -1$ or $x = -1/y$. Thus

$2x = -\lambda y$ and $2y = -\lambda x$ imply $\lambda = 2/y^2 = 2y^2$ or $y = \pm 1$, $x = \pm 1$. If $\lambda = 1$, then $2x = -y$ and $2y = -x$

imply $x = y = 0$, so $z = \pm 1$. Hence the possible points are $(\pm 1, \mp 1, 0)$, $(0, 0, \pm 1)$ and the minimum value of f is

$f(0, 0, \pm 1) = 1$, so the points closest to the origin are $(0, 0, \pm 1)$.

29. $f(x, y, z) = xyz$, $g(x, y, z) = x + y + z = 100$ $\Rightarrow$ $\nabla f = \langle yz, xz, xy \rangle = \lambda \nabla g = \langle \lambda, \lambda, \lambda \rangle$. Then

$\lambda = yz = xz = xy$ implies $x = y = z = \frac{100}{3}$.

31. If the dimensions are $2x$, $2y$ and $2z$, then $f(x, y, z) = 8xyz$ and $g(x, y, z) = 9x^2 + 36y^2 + 4z^2 = 36$ $\Rightarrow$

$\nabla f = \langle 8yz, 8xz, 8xy \rangle = \lambda \nabla g = \langle 18\lambda x, 72\lambda y, 8\lambda z \rangle$. Thus $18\lambda x = 8yz$, $72\lambda y = 8xz$, $8\lambda z = 8xy$ so $x^2 = 4y^2$,

$z^2 = 9y^2$ and $36y^2 + 36y^2 + 36y^2 = 36$ or $y = \frac{1}{\sqrt{3}}$ ($y > 0$). Thus the volume of the largest such rectangle is

$8\left(\frac{1}{\sqrt{3}}\right)\left(\frac{2}{\sqrt{3}}\right)\left(\frac{3}{\sqrt{3}}\right) = 16\sqrt{3}$.

33. $f(x, y, z) = xyz$, $g(x, y, z) = x + 2y + 3z = 6$ $\Rightarrow$ $\nabla f = \langle yz, xz, xy \rangle = \lambda \nabla g = \langle \lambda, 2\lambda, 3\lambda \rangle$.

Then $\lambda = yz = \frac{1}{2}xz = \frac{1}{3}xy$ implies $x = 2y$, $z = \frac{2}{3}y$. But $2y + 2y + 2y = 6$ so $y = 1$, $x = 2$, $z = \frac{2}{3}$ and the

volume is $V = \frac{4}{3}$.

35. $f(x, y, z) = xyz$, $g(x, y, z) = 4(x + y + z) = c$ $\Rightarrow$ $\nabla f = \langle yz, xz, xy \rangle$, $\lambda \nabla g = \langle 4\lambda, 4\lambda, 4\lambda \rangle$. Thus

$4\lambda = yz = xz = xy$ or $x = y = z = \frac{1}{12}c$ are the dimensions giving the maximum volume.

37. If the dimensions of the box are given by x, y, and z, then we need to find the maximum value of $f(x, y, z) = xyz$

$(x, y, z > 0)$ subject to the constraint $L = \sqrt{x^2 + y^2 + z^2}$ or $g(x, y, z) = x^2 + y^2 + z^2 = L^2$. $\nabla f = \lambda \nabla g$ $\Rightarrow$

$\langle yz, xz, xy \rangle = \lambda \langle 2x, 2y, 2z \rangle$, so $yz = 2\lambda x$ $\Rightarrow$ $\lambda = \dfrac{yz}{2x}$, $xz = 2\lambda y$ $\Rightarrow$ $\lambda = \dfrac{xz}{2y}$, and $xy = 2\lambda z$ $\Rightarrow$

$\lambda = \dfrac{xy}{2z}$. Thus $\lambda = \dfrac{yz}{2x} = \dfrac{xz}{2y}$ $\Rightarrow$ $x^2 = y^2$ [since $z \neq 0$] $\Rightarrow$ $x = y$ and $\lambda = \dfrac{yz}{2x} = \dfrac{xy}{2z}$ $\Rightarrow$ $x = z$

[since $y \neq 0$]. Substituting into the constraint equation gives $x^2 + x^2 + x^2 = L^2$ $\Rightarrow$ $x^2 = L^2/3$ $\Rightarrow$

$x = L/\sqrt{3} = y = z$ and the maximum volume is $\left(L/\sqrt{3}\right)^3 = L^3/\left(3\sqrt{3}\right)$.

39. We need to find the extreme values of $f(x, y, z) = x^2 + y^2 + z^2$ subject to the two constraints

$g(x, y, z) = x + y + 2z = 2$ and $h(x, y, z) = x^2 + y^2 - z = 0$. $\nabla f = \langle 2x, 2y, 2z \rangle$, $\lambda \nabla g = \langle \lambda, \lambda, 2\lambda \rangle$ and

$\mu \nabla h = \langle 2\mu x, 2\mu y, -\mu \rangle$. Thus we need (1) $2x = \lambda + 2\mu x$, (2) $2y = \lambda + 2\mu y$, (3) $2z = 2\lambda - \mu$,

(4) $x + y + 2z = 2$, and (5) $x^2 + y^2 - z = 0$. From (1) and (2), $2(x - y) = 2\mu(x - y)$, so if $x \neq y$, $\mu = 1$.

Putting this in (3) gives $2z = 2\lambda - 1$ or $\lambda = z + \frac{1}{2}$, but putting $\mu = 1$ into (1) says $\lambda = 0$. Hence $z + \frac{1}{2} = 0$ or

$z = -\frac{1}{2}$. Then (4) and (5) become $x + y - 3 = 0$ and $x^2 + y^2 + \frac{1}{2} = 0$. The last equation cannot be true, so this

case gives no solution. So we must have $x = y$. Then (4) and (5) become $2x + 2z = 2$ and $2x^2 - z = 0$ which

imply $z = 1 - x$ and $z = 2x^2$. Thus $2x^2 = 1 - x$ or $2x^2 + x - 1 = (2x - 1)(x + 1) = 0$ so $x = \frac{1}{2}$ or $x = -1$.

The two points to check are $\left(\frac{1}{2}, \frac{1}{2}, \frac{1}{2}\right)$ and $(-1, -1, 2)$: $f\left(\frac{1}{2}, \frac{1}{2}, \frac{1}{2}\right) = \frac{3}{4}$ and $f(-1, -1, 2) = 6$. Thus $\left(\frac{1}{2}, \frac{1}{2}, \frac{1}{2}\right)$ is

the point on the ellipse nearest the origin and $(-1, -1, 2)$ is the one farthest from the origin.

41. $f(x, y, z) = ye^{x-z}$, $g(x, y, z) = 9x^2 + 4y^2 + 36z^2 = 36$, $h(x, y, z) = xy + yz = 1$.

$\nabla f = \lambda \nabla g + \mu \nabla h$ $\Rightarrow$ $\langle ye^{x-z}, e^{x-z}, -ye^{x-z} \rangle = \lambda \langle 18x, 8y, 72z \rangle + \mu \langle y, x + z, y \rangle$, so $ye^{x-z} = 18\lambda x + \mu y$,

$e^{x-z} = 8\lambda y + \mu(x + z)$, $-ye^{x-z} = 72\lambda z + \mu y$, $9x^2 + 4y^2 + 36z^2 = 36$, $xy + yz = 1$. Using a CAS to solve these 5 equations simultaneously for x, y, z, λ, and μ (in Maple, use the `allvalues` command), we get 4 real-valued solutions:

$$x \approx 0.222444, \quad y \approx -2.157012, \quad z \approx -0.686049, \quad \lambda \approx -0.200401, \quad \mu \approx 2.108584$$

$$x \approx -1.951921, \quad y \approx -0.545867, \quad z \approx 0.119973, \quad \lambda \approx 0.003141, \quad \mu \approx -0.076238$$

$$x \approx 0.155142, \quad y \approx 0.904622, \quad z \approx 0.950293, \quad \lambda \approx -0.012447, \quad \mu \approx 0.489938$$

$$x \approx 1.138731, \quad y \approx 1.768057, \quad z \approx -0.573138, \quad \lambda \approx 0.317141, \quad \mu \approx 1.862675$$

Substituting these values into f gives $f(0.222444, -2.157012, -0.686049) \approx -5.3506$,

$f(-1.951921, -0.545867, 0.119973) \approx -0.0688$, $f(0.155142, 0.904622, 0.950293) \approx 0.4084$,

$f(1.138731, 1.768057, -0.573138) \approx 9.7938$. Thus the maximum is approximately 9.7938, and the mininum is approximately -5.3506.

43. (a) We wish to maximize $f(x_1, x_2, \ldots, x_n) = \sqrt[n]{x_1 x_2 \cdots x_n}$ subject to

$g(x_1, x_2, \ldots, x_n) = x_1 + x_2 + \cdots + x_n = c$ and $x_i > 0$.

$$\nabla f = \left\langle \frac{1}{n}(x_1 x_2 \cdots x_n)^{\frac{1}{n}-1}(x_2 \cdots x_n), \frac{1}{n}(x_1 x_2 \cdots x_n)^{\frac{1}{n}-1}(x_1 x_3 \cdots x_n), \ldots, \right.$$
$$\left. \frac{1}{n}(x_1 x_2 \cdots x_n)^{\frac{1}{n}-1}(x_1 \cdots x_{n-1}) \right\rangle$$

and $\lambda \nabla g = \langle \lambda, \lambda, \ldots, \lambda \rangle$, so we need to solve the system of equations

$$\frac{1}{n}(x_1 x_2 \cdots x_n)^{\frac{1}{n}-1}(x_2 \cdots x_n) = \lambda \quad \Rightarrow \quad x_1^{1/n} x_2^{1/n} \cdots x_n^{1/n} = n\lambda x_1$$

$$\frac{1}{n}(x_1 x_2 \cdots x_n)^{\frac{1}{n}-1}(x_1 x_3 \cdots x_n) = \lambda \quad \Rightarrow \quad x_1^{1/n} x_2^{1/n} \cdots x_n^{1/n} = n\lambda x_2$$

$$\vdots$$

$$\frac{1}{n}(x_1 x_2 \cdots x_n)^{\frac{1}{n}-1}(x_1 \cdots x_{n-1}) = \lambda \quad \Rightarrow \quad x_1^{1/n} x_2^{1/n} \cdots x_n^{1/n} = n\lambda x_n$$

This implies $n\lambda x_1 = n\lambda x_2 = \cdots = n\lambda x_n$. Note $\lambda \neq 0$, otherwise we can't have all $x_i > 0$. Thus $x_1 = x_2 = \cdots = x_n$. But $x_1 + x_2 + \cdots + x_n = c \Rightarrow nx_1 = c \Rightarrow x_1 = \dfrac{c}{n} = x_2 = x_3 = \cdots = x_n$.

Then the only point where f can have an extreme value is $\left(\dfrac{c}{n}, \dfrac{c}{n}, \ldots, \dfrac{c}{n}\right)$. Since we can choose values for $(x_1, x_2, \ldots, x_n)$ that make f as close to zero (but not equal) as we like, f has no minimum value. Thus the maximum value is $f\left(\dfrac{c}{n}, \dfrac{c}{n}, \ldots, \dfrac{c}{n}\right) = \sqrt[n]{\dfrac{c}{n} \cdot \dfrac{c}{n} \cdot \ldots \cdot \dfrac{c}{n}} = \dfrac{c}{n}$.

(b) From part (a), $\dfrac{c}{n}$ is the maximum value of f. Thus $f(x_1, x_2, \ldots, x_n) = \sqrt[n]{x_1 x_2 \cdots x_n} \leq \dfrac{c}{n}$. But

$x_1 + x_2 + \cdots + x_n = c$, so $\sqrt[n]{x_1 x_2 \cdots x_n} \leq \dfrac{x_1 + x_2 + \cdots + x_n}{n}$. These two means are equal when f attains

its maximum value $\dfrac{c}{n}$, but this can occur only at the point $\left(\dfrac{c}{n}, \dfrac{c}{n}, \ldots, \dfrac{c}{n}\right)$ we found in part (a). So the means are equal only when $x_1 = x_2 = x_3 = \cdots = x_n = \dfrac{c}{n}$.

15 Review

———————————————————— CONCEPT CHECK ————————————————————

1. (a) A function f of two variables is a rule that assigns to each ordered pair (x, y) of real numbers in its domain a unique real number denoted by $f(x, y)$.

(b) One way to visualize a function of two variables is by graphing it, resulting in the surface $z = f(x, y)$. Another method for visualizing a function of two variables is a contour map. The contour map consists of level curves of the function which are horizontal traces of the graph of the function projected onto the xy-plane. Also, we can use an arrow diagram such as Figure 1 in Section 15.1 [ET 14.1].

2. A function f of three variables is a rule that assigns to each ordered triple (x, y, z) in its domain a unique real number $f(x, y, z)$. We can visualize a function of three variables by examining its level surfaces $f(x, y, z) = k$, where k is a constant.

3. $\lim\limits_{(x,y)\to(a,b)} f(x, y) = L$ means the values of $f(x, y)$ approach the number L as the point (x, y) approaches the point (a, b) along any path that is within the domain of f. We can show that a limit at a point does not exist by finding two different paths approaching the point along which $f(x, y)$ has different limits.

4. (a) See Definition 15.2.4 [ET 14.2.4].

(b) If f is continuous on $\mathbb{R}^2$, its graph will appear as a surface without holes or breaks.

5. (a) See (2) and (3) in Section 15.3 [ET 14.3].

(b) See "Interpretations of Partial Derivatives" on page 948 [ET 912].

(c) To find f_x, regard y as a constant and differentiate $f(x, y)$ with respect to x. To find f_y, regard x as a constant and differentiate $f(x, y)$ with respect to y.

6. See the statement of Clairaut's Theorem on page 952 [ET 916].

7. (a) See (2) in Section 15.4 [ET 14.4].

(b) See (19) and the preceding discussion in Section 15.6 [ET 14.6].

8. See (3) and (4) and the accompanying discussion in Section 15.4 [ET 14.4]. We can interpret the linearization of f at (a, b) geometrically as the linear function whose graph is the tangent plane to the graph of f at (a, b). Thus it is the linear function which best approximates f near (a, b).

9. (a) See Definition 15.4.7 [ET 14.4.7].

(b) Use Theorem 15.4.8 [ET 14.4.8].

10. See (10) and the associated discussion in Section 15.4 [ET 14.4].

11. See (2) and (3) in Section 15.5 [ET 14.5].

12. See (7) and the preceding discussion in Section 15.5 [ET 14.5].

13. (a) See Definition 15.6.2 [ET 14.6.2]. We can interpret it as the rate of change of f at (x_0, y_0) in the direction of $\mathbf{u}$. Geometrically, if P is the point $(x_0, y_0, f(x_0, y_0))$ on the graph of f and C is the curve of intersection of the graph of f with the vertical plane that passes through P in the direction $\mathbf{u}$, the directional derivative of f at (x_0, y_0) in the direction of $\mathbf{u}$ is the slope of the tangent line to C at P. (See Figure 5 in Section 15.6 [ET 14.6].)

(b) See Theorem 15.6.3 [ET 14.6.3].

14. (a) See (8) and (13) in Section 15.6 [ET 14.6].

(b) $D_{\mathbf{u}} f(x, y) = \nabla f(x, y) \cdot \mathbf{u}$ or $D_{\mathbf{u}} f(x, y, z) = \nabla f(x, y, z) \cdot \mathbf{u}$

(c) The gradient vector of a function points in the direction of maximum rate of increase of the function. On a graph of the function, the gradient points in the direction of steepest ascent.

15. (a) f has a local maximum at (a, b) if $f(x, y) \leq f(a, b)$ when (x, y) is near (a, b).

(b) f has an absolute maximum at (a, b) if $f(x, y) \leq f(a, b)$ for all points (x, y) in the domain of f.

(c) f has a local minimum at (a, b) if $f(x, y) \geq f(a, b)$ when (x, y) is near (a, b).

(d) f has an absolute minimum at (a, b) if $f(x, y) \geq f(a, b)$ for all points (x, y) in the domain of f.

(e) f has a saddle point at (a, b) if $f(a, b)$ is a local maximum in one direction but a local minimum in another.

16. (a) By Theorem 15.7.2 [ET 14.7.2], if f has a local maximum at (a, b) and the first-order partial derivatives of f exist there, then $f_x(a, b) = 0$ and $f_y(a, b) = 0$.

(b) A critical point of f is a point (a, b) such that $f_x(a, b) = 0$ and $f_y(a, b) = 0$ or one of these partial derivatives does not exist.

17. See (3) in Section 15.7 [ET 14.7].

18. (a) See Figure 11 and the accompanying discussion in Section 15.7 [ET 14.7].

(b) See Theorem 15.7.8 [ET 14.7.8].

(c) See the procedure outlined in (9) in Section 15.7 [ET 14.7].

19. See the discussion beginning on page 1001 [ET 965]; see the discussion preceding Example 5 in Section 15.8 [ET 14.8].

─────────────────────────── TRUE-FALSE QUIZ ───────────────────────────

1. True. $f_y(a, b) = \lim\limits_{h \to 0} \dfrac{f(a, b + h) - f(a, b)}{h}$ from Equation 15.3.3 [ET 14.3.3]. Let $h = y - b$. As $h \to 0$, $y \to b$. Then by substituting, we get $f_y(a, b) = \lim\limits_{y \to b} \dfrac{f(a, y) - f(a, b)}{y - b}$.

3. False. $f_{xy} = \dfrac{\partial^2 f}{\partial y\, \partial x}$.

5. False. See Example 15.2.3 [ET 14.2.3].

7. True. If f has a local minimum and f is differentiable at (a, b) then by Theorem 15.7.2 [ET 14.7.2], $f_x(a, b) = 0$ and $f_y(a, b) = 0$, so $\nabla f(a, b) = \langle f_x(a, b), f_y(a, b) \rangle = \langle 0, 0 \rangle = \mathbf{0}$.

9. False. $\nabla f(x, y) = \langle 0, 1/y \rangle$.

11. True. $\nabla f = \langle \cos x, \cos y \rangle$, so $|\nabla f| = \sqrt{\cos^2 x + \cos^2 y}$. But $|\cos \theta| \leq 1$, so $|\nabla f| \leq \sqrt{2}$. Now $D_{\mathbf{u}} f(x, y) = \nabla f \cdot \mathbf{u} = |\nabla f|\, |\mathbf{u}| \cos \theta$, but $\mathbf{u}$ is a unit vector, so $|D_{\mathbf{u}} f(x, y)| \leq \sqrt{2} \cdot 1 \cdot 1 = \sqrt{2}$.

─────────────── EXERCISES ───────────────

1. The domain of $\sin^{-1} x$ is $-1 \leq x \leq 1$ while the domain of $\tan^{-1} y$ is all real numbers, so the domain of $f(x, y) = \sin^{-1} x + \tan^{-1} y$ is $\{(x, y) \mid -1 \leq x \leq 1\}$.

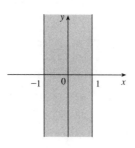

3. $z = f(x, y) = 1 - x^2 - y^2$, a paraboloid with vertex $(0, 0, 1)$.

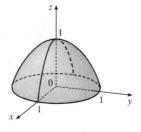

5. Let $k = e^{-c} = e^{-(x^2 + y^2)}$ be the level curves. Then $-\ln k = c = x^2 + y^2$, so we have a family of concentric circles.

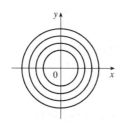

7.

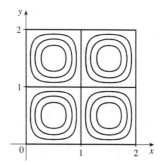

9. f is a rational function, so it is continuous on its domain. Since f is defined at $(1, 1)$, we use direct substitution to evaluate the limit: $\displaystyle\lim_{(x,y)\to(1,1)} \frac{2xy}{x^2 + 2y^2} = \frac{2(1)(1)}{1^2 + 2(1)^2} = \frac{2}{3}$.

11. (a) $T_x(6, 4) = \displaystyle\lim_{h \to 0} \frac{T(6 + h, 4) - T(6, 4)}{h}$, so we can approximate $T_x(6, 4)$ by considering $h = \pm 2$ and using the values given in the table: $T_x(6, 4) \approx \dfrac{T(8, 4) - T(6, 4)}{2} = \dfrac{86 - 80}{2} = 3$,

$T_x(6, 4) \approx \dfrac{T(4, 4) - T(6, 4)}{-2} = \dfrac{72 - 80}{-2} = 4$. Averaging these values, we estimate $T_x(6, 4)$ to be approximately $3.5°\text{C}/\text{m}$. Similarly, $T_y(6, 4) = \displaystyle\lim_{h \to 0} \frac{T(6, 4 + h) - T(6, 4)}{h}$, which we can approximate with $h = \pm 2$: $T_y(6, 4) \approx \dfrac{T(6, 6) - T(6, 4)}{2} = \dfrac{75 - 80}{2} = -2.5$,

$T_y(6, 4) \approx \dfrac{T(6, 2) - T(6, 4)}{-2} = \dfrac{87 - 80}{-2} = -3.5$. Averaging these values, we estimate $T_y(6, 4)$ to be approximately $-3.0°\text{C}/\text{m}$.

(b) Here $\mathbf{u} = \left\langle \frac{1}{\sqrt{2}}, \frac{1}{\sqrt{2}} \right\rangle$, so by Equation 15.6.9 [ET 14.6.9],

$D_{\mathbf{u}} T(6,4) = \nabla T(6,4) \cdot \mathbf{u} = T_x(6,4) \frac{1}{\sqrt{2}} + T_y(6,4) \frac{1}{\sqrt{2}}$. Using our estimates from part (a), we have

$D_{\mathbf{u}} T(6,4) \approx (3.5) \frac{1}{\sqrt{2}} + (-3.0) \frac{1}{\sqrt{2}} = \frac{1}{2\sqrt{2}} \approx 0.35$. This means that as we move through the point $(6,4)$ in the direction of $\mathbf{u}$, the temperature increases at a rate of approximately $0.35°C\,/m$.

Alternatively, we can use Definition 15.6.2 [ET 14.6.2]:

$D_{\mathbf{u}} T(6,4) = \lim_{h \to 0} \dfrac{T\left(6 + h\frac{1}{\sqrt{2}}, 4 + h\frac{1}{\sqrt{2}}\right) - T(6,4)}{h}$, which we can estimate with $h = \pm 2\sqrt{2}$. Then

$D_{\mathbf{u}} T(6,4) \approx \dfrac{T(8,6) - T(6,4)}{2\sqrt{2}} = \dfrac{80 - 80}{2\sqrt{2}} = 0$, $D_{\mathbf{u}} T(6,4) \approx \dfrac{T(4,2) - T(6,4)}{-2\sqrt{2}} = \dfrac{74 - 80}{-2\sqrt{2}} = \dfrac{3}{\sqrt{2}}$.

Averaging these values, we have $D_{\mathbf{u}} T(6,4) \approx \frac{3}{2\sqrt{2}} \approx 1.1°C\,/m$.

(c) $T_{xy}(x,y) = \dfrac{\partial}{\partial y} [T_x(x,y)] = \lim_{h \to 0} \dfrac{T_x(x, y+h) - T_x(x,y)}{h}$, so $T_{xy}(6,4) = \lim_{h \to 0} \dfrac{T_x(6, 4+h) - T_x(6,4)}{h}$

which we can estimate with $h = \pm 2$. We have $T_x(6,4) \approx 3.5$ from part (a), but we will also need values for $T_x(6,6)$ and $T_x(6,2)$. If we use $h = \pm 2$ and the values given in the table, we have

$T_x(6,6) \approx \dfrac{T(8,6) - T(6,6)}{2} = \dfrac{80 - 75}{2} = 2.5$, $T_x(6,6) \approx \dfrac{T(4,6) - T(6,6)}{-2} = \dfrac{68 - 75}{-2} = 3.5$.

Averaging these values, we estimate $T_x(6,6) \approx 3.0$. Similarly,

$T_x(6,2) \approx \dfrac{T(8,2) - T_x(6,2)}{2} = \dfrac{90 - 87}{2} = 1.5$, $T_x(6,2) \approx \dfrac{T(4,2) - T(6,2)}{-2} = \dfrac{74 - 87}{-2} = 6.5$.

Averaging these values, we estimate $T_x(6,2) \approx 4.0$. Finally, we estimate $T_{xy}(6,4)$:

$T_{xy}(6,4) \approx \dfrac{T_x(6,6) - T_x(6,4)}{2} = \dfrac{3.0 - 3.5}{2} = -0.25$,

$T_{xy}(6,4) \approx \dfrac{T_x(6,2) - T_x(6,4)}{-2} = \dfrac{4.0 - 3.5}{-2} = -0.25$. Averaging these values, we have

$T_{xy}(6,4) \approx -0.25$.

13. $f(x,y) = \sqrt{2x + y^2} \quad \Rightarrow \quad f_x = \frac{1}{2}(2x + y^2)^{-1/2}(2) = \dfrac{1}{\sqrt{2x + y^2}}$, $f_y = \frac{1}{2}(2x + y^2)^{-1/2}(2y) = \dfrac{y}{\sqrt{2x + y^2}}$

15. $g(u,v) = u\tan^{-1} v \quad \Rightarrow \quad g_u = \tan^{-1} v$, $g_v = \dfrac{u}{1 + v^2}$

17. $T(p,q,r) = p\ln(q + e^r) \quad \Rightarrow \quad T_p = \ln(q + e^r)$, $T_q = \dfrac{p}{q + e^r}$, $T_r = \dfrac{pe^r}{q + e^r}$

19. $f(x,y) = 4x^3 - xy^2 \quad \Rightarrow \quad f_x = 12x^2 - y^2$, $f_y = -2xy$, $f_{xx} = 24x$, $f_{yy} = -2x$, and $f_{xy} = f_{yx} = -2y$.

21. $f(x,y,z) = x^k y^l z^m \quad \Rightarrow \quad f_x = kx^{k-1}y^l z^m$, $f_y = lx^k y^{l-1} z^m$, $f_z = mx^k y^l z^{m-1}$,
$f_{xx} = k(k-1)x^{k-2}y^l z^m$, $f_{yy} = l(l-1)x^k y^{l-2} z^m$, $f_{zz} = m(m-1)x^k y^l z^{m-2}$, $f_{xy} = f_{yx} = klx^{k-1}y^{l-1} z^m$,
$f_{xz} = f_{zx} = kmx^{k-1}y^l z^{m-1}$, and $f_{yz} = f_{zy} = lmx^k y^{l-1} z^{m-1}$.

23. $u = x^y \quad \Rightarrow \quad u_x = yx^{y-1}$, $u_y = x^y \ln x$ and $(x/y)u_x + (\ln x)^{-1}u_y = x^y + x^y = 2u$.

25. (a) $z_x = 6x + 2 \quad \Rightarrow \quad z_x(1, -2) = 8$ and $z_y = -2y \quad \Rightarrow \quad z_y(1, -2) = 4$, so an equation of the tangent plane is
$z - 1 = 8(x - 1) + 4(y + 2)$ or $z = 8x + 4y + 1$.

(b) A normal vector to the tangent plane (and the surface) at $(1, -2, 1)$ is $\langle 8, 4, -1 \rangle$. Then parametric equations for the normal line there are $x = 1 + 8t$, $y = -2 + 4t$, $z = 1 - t$, and symmetric equations are

$$\frac{x - 1}{8} = \frac{y + 2}{4} = \frac{z - 1}{-1}.$$

27. (a) Let $F(x, y, z) = x^2 + 2y^2 - 3z^2$. Then $F_x = 2x$, $F_y = 4y$, $F_z = -6z$, so $F_x(2, -1, 1) = 4$, $F_y(2, -1, 1) = -4$, $F_z(2, -1, 1) = -6$. From Equation 15.6.19 [ET 14.6.19], an equation of the tangent plane is $4(x - 2) - 4(y + 1) - 6(z - 1) = 0$ or equivalently $2x - 2y - 3z = 3$.

(b) From Equations 15.6.20 [ET 14.6.20], symmetric equations for the normal line are $\dfrac{x - 2}{4} = \dfrac{y + 1}{-4} = \dfrac{z - 1}{-6}$.

29. (a) Let $F(x, y, z) = x + 2y + 3z - \sin(xyz)$. Then $F_x = 1 - yz\cos(xyz)$, $F_y = 2 - xz\cos(xyz)$, $F_z = 3 - xy\cos(xyz)$, so $F_x(2, -1, 0) = 1$, $F_y(2, -1, 0) = 2$, $F_z(2, -1, 0) = 5$. From Equation 15.6.19 [ET 14.6.19], an equation of the tangent plane is $1(x - 2) + 2(y + 1) + 5(z - 0) = 0$ or $x + 2y + 5z = 0$.

(b) From Equations 15.6.20 [ET 14.6.20], symmetric equations for the normal line are $\dfrac{x - 2}{1} = \dfrac{y + 1}{2} = \dfrac{z}{5}$.

31. $F(x, y, z) = x^2 + y^2 + z^2$, $\nabla F(x_0, y_0, z_0) = \langle 2x_0, 2y_0, 2z_0 \rangle = k \langle 2, 1, -3 \rangle$ or $x_0 = k$, $y_0 = \frac{1}{2}k$ and $z_0 = -\frac{3}{2}k$. But $x_0^2 + y_0^2 + z_0^2 = 1$, so $\frac{7}{2}k^2 = 1$ and $k = \pm\sqrt{\frac{2}{7}}$. Hence there are two such points: $\left(\pm\sqrt{\frac{2}{7}}, \pm\frac{1}{\sqrt{14}}, \mp\frac{3}{\sqrt{14}} \right)$.

33. $f(x, y, z) = x^3\sqrt{y^2 + z^2}$ $\Rightarrow$ $f_x(x, y, z) = 3x^2\sqrt{y^2 + z^2}$, $f_y(x, y, z) = \dfrac{yx^3}{\sqrt{y^2 + z^2}}$, and

$f_z(x, y, z) = \dfrac{zx^3}{\sqrt{y^2 + z^2}}$, so $f(2, 3, 4) = 8(5) = 40$, $f_x(2, 3, 4) = 3(4)\sqrt{25} = 60$, $f_y(2, 3, 4) = \dfrac{3(8)}{\sqrt{25}} = \dfrac{24}{5}$,

and $f_z(2, 3, 4) = \dfrac{4(8)}{\sqrt{25}} = \dfrac{32}{5}$. Then the linear approximation of f at $(2, 3, 4)$ is

$$f(x, y, z) \approx f(2, 3, 4) + f_x(2, 3, 4)(x - 2) + f_y(2, 3, 4)(y - 3) + f_z(2, 3, 4)(z - 4)$$
$$= 40 + 60(x - 2) + \tfrac{24}{5}(y - 3) + \tfrac{32}{5}(z - 4) = 60x + \tfrac{24}{5}y + \tfrac{32}{5}z - 120$$

Then

$$(1.98)^3\sqrt{(3.01)^2 + (3.97)^2} = f(1.98, 3.01, 3.97) \approx 60(1.98) + \tfrac{24}{5}(3.01) + \tfrac{32}{5}(3.97) - 120$$
$$= 38.656$$

35. $\dfrac{dw}{dt} = \dfrac{1}{2\sqrt{x}}(2e^{2t}) + \dfrac{2y}{z}(3t^2 + 4) + \dfrac{-y^2}{z^2}(2t) = e^t + \dfrac{2y}{z}(3t^2 + 4) - 2t\dfrac{y^2}{z^2}$

37. By the Chain Rule, $\dfrac{\partial z}{\partial s} = \dfrac{\partial z}{\partial x}\dfrac{\partial x}{\partial s} + \dfrac{\partial z}{\partial y}\dfrac{\partial y}{\partial s}$. When $s = 1$ and $t = 2$, $x = g(1, 2) = 3$ and $y = h(1, 2) = 6$, so

$\dfrac{\partial z}{\partial s} = f_x(3, 6)g_s(1, 2) + f_y(3, 6)h_s(1, 2) = (7)(-1) + (8)(-5) = -47$. Similarly, $\dfrac{\partial z}{\partial t} = \dfrac{\partial z}{\partial x}\dfrac{\partial x}{\partial t} + \dfrac{\partial z}{\partial y}\dfrac{\partial y}{\partial t}$, so

$\dfrac{\partial z}{\partial t} = f_x(3, 6)g_t(1, 2) + f_y(3, 6)h_t(1, 2) = (7)(4) + (8)(10) = 108$.

39. $\dfrac{\partial z}{\partial x} = 2xf'(x^2 - y^2)$, $\dfrac{\partial z}{\partial y} = 1 - 2yf'(x^2 - y^2)$ $\left[\text{where } f' = \dfrac{df}{d(x^2 - y^2)} \right]$. Then

$y\dfrac{\partial z}{\partial x} + x\dfrac{\partial z}{\partial y} = 2xyf'(x^2 - y^2) + x - 2xyf'(x^2 - y^2) = x$.

41. $\dfrac{\partial z}{\partial x} = \dfrac{\partial z}{\partial u}\,y + \dfrac{\partial z}{\partial v}\dfrac{-y}{x^2}$ and

$$\dfrac{\partial^2 z}{\partial x^2} = y\,\dfrac{\partial}{\partial x}\!\left(\dfrac{\partial z}{\partial u}\right) + \dfrac{2y}{x^3}\dfrac{\partial z}{\partial v} + \dfrac{-y}{x^2}\dfrac{\partial}{\partial x}\!\left(\dfrac{\partial z}{\partial v}\right)$$

$$= \dfrac{2y}{x^3}\dfrac{\partial z}{\partial v} + y\!\left(\dfrac{\partial^2 z}{\partial u^2}\,y + \dfrac{\partial^2 z}{\partial v\,\partial u}\dfrac{-y}{x^2}\right) + \dfrac{-y}{x^2}\!\left(\dfrac{\partial^2 z}{\partial v^2}\dfrac{-y}{x^2} + \dfrac{\partial^2 z}{\partial u\,\partial v}\,y\right)$$

$$= \dfrac{2y}{x^3}\dfrac{\partial z}{\partial v} + y^2\dfrac{\partial^2 z}{\partial u^2} - \dfrac{2y^2}{x^2}\dfrac{\partial^2 z}{\partial u\,\partial v} + \dfrac{y^2}{x^4}\dfrac{\partial^2 z}{\partial v^2}$$

Also $\dfrac{\partial z}{\partial y} = x\,\dfrac{\partial z}{\partial u} + \dfrac{1}{x}\dfrac{\partial z}{\partial v}$ and

$$\dfrac{\partial^2 z}{\partial y^2} = x\,\dfrac{\partial}{\partial y}\!\left(\dfrac{\partial z}{\partial u}\right) + \dfrac{1}{x}\dfrac{\partial}{\partial y}\!\left(\dfrac{\partial z}{\partial v}\right) = x\!\left(\dfrac{\partial^2 z}{\partial u^2}\,x + \dfrac{\partial^2 z}{\partial v\,\partial u}\dfrac{1}{x}\right) + \dfrac{1}{x}\!\left(\dfrac{\partial^2 z}{\partial v^2}\dfrac{1}{x} + \dfrac{\partial^2 z}{\partial u\,\partial v}\,x\right)$$

$$= x^2\dfrac{\partial^2 z}{\partial u^2} + 2\dfrac{\partial^2 z}{\partial u\,\partial v} + \dfrac{1}{x^2}\dfrac{\partial^2 z}{\partial v^2}$$

Thus

$$x^2\dfrac{\partial^2 z}{\partial x^2} - y^2\dfrac{\partial^2 z}{\partial y^2} = \dfrac{2y}{x}\dfrac{\partial z}{\partial v} + x^2 y^2\dfrac{\partial^2 z}{\partial u^2} - 2y^2\dfrac{\partial^2 z}{\partial u\,\partial v} + \dfrac{y^2}{x^2}\dfrac{\partial^2 z}{\partial v^2} - x^2 y^2\dfrac{\partial^2 z}{\partial u^2} - 2y^2\dfrac{\partial^2 z}{\partial u\,\partial v} - \dfrac{y^2}{x^2}\dfrac{\partial^2 z}{\partial v^2}$$

$$= \dfrac{2y}{x}\dfrac{\partial z}{\partial v} - 4y^2\dfrac{\partial^2 z}{\partial u\,\partial v} = 2v\dfrac{\partial z}{\partial v} - 4uv\dfrac{\partial^2 z}{\partial u\,\partial v}$$

since $y = xv = \dfrac{uv}{y}$ or $y^2 = uv$.

43. $\nabla f = \left\langle z^2\sqrt{y}\,e^{x\sqrt{y}},\ \dfrac{xz^2 e^{x\sqrt{y}}}{2\sqrt{y}},\ 2ze^{x\sqrt{y}}\right\rangle = ze^{x\sqrt{y}}\left\langle z\sqrt{y},\ \dfrac{xz}{2\sqrt{y}},\ 2\right\rangle$

45. $\nabla f = \langle 1/\sqrt{x},\,-2y\rangle,\ \nabla f(1,5) = \langle 1,-10\rangle,\ \mathbf{u} = \tfrac{1}{5}\langle 3,-4\rangle.$ Then $D_{\mathbf{u}}\,f(1,5) = \tfrac{43}{5}.$

47. $\nabla f = \left\langle 2xy,\, x^2 + 1/(2\sqrt{y})\right\rangle,\ |\nabla f(2,1)| = \left|\left\langle 4,\tfrac{9}{2}\right\rangle\right|.$ Thus the maximum rate of change of f at $(2,1)$ is $\dfrac{\sqrt{145}}{2}$ in the direction $\left\langle 4,\tfrac{9}{2}\right\rangle$.

49. First we draw a line passing through Homestead and the eye of the hurricane. We can approximate the directional derivative at Homestead in the direction of the eye of the hurricane by the average rate of change of wind speed between the points where this line intersects the contour lines closest to Homestead. In the direction of the eye of the hurricane, the wind speed changes from 45 to 50 knots. We estimate the distance between these two points to be approximately 8 miles, so the rate of change of wind speed in the direction given is approximately $\dfrac{50-45}{8} = \tfrac{5}{8} = 0.625$ knot/mi.

51. $f(x,y) = x^2 - xy + y^2 + 9x - 6y + 10 \ \Rightarrow\ f_x = 2x - y + 9,$
$f_y = -x + 2y - 6,\ f_{xx} = 2 = f_{yy},\ f_{xy} = -1.$ Then $f_x = 0$ and $f_y = 0$ imply $y = 1,\ x = -4.$ Thus the only critical point is $(-4,1)$ and $f_{xx}(-4,1) > 0,\ D(-4,1) = 3 > 0,$ so $f(-4,1) = -11$ is a local minimum.

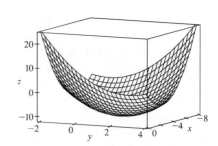

53. $f(x,y) = 3xy - x^2y - xy^2 \Rightarrow f_x = 3y - 2xy - y^2,$

$f_y = 3x - x^2 - 2xy$, $f_{xx} = -2y$, $f_{yy} = -2x$, $f_{xy} = 3 - 2x - 2y$. Then

$f_x = 0$ implies $y(3 - 2x - y) = 0$ so $y = 0$ or $y = 3 - 2x$. Substituting

into $f_y = 0$ implies $x(3 - x) = 0$ or $3x(-1 + x) = 0$. Hence the critical

points are $(0,0)$, $(3,0)$, $(0,3)$ and $(1,1)$.

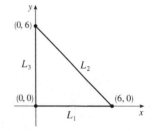

$D(0,0) = D(3,0) = D(0,3) = -9 < 0$ so $(0,0)$, $(3,0)$, and $(0,3)$ are

saddle points. $D(1,1) = 3 > 0$ and $f_{xx}(1,1) = -2 < 0$, so $f(1,1) = 1$

is a local maximum.

55. First solve inside D. Here $f_x = 4y^2 - 2xy^2 - y^3$,

$f_y = 8xy - 2x^2y - 3xy^2$. Then $f_x = 0$ implies $y = 0$ or $y = 4 - 2x$,

but $y = 0$ isn't inside D. Substituting $y = 4 - 2x$ into $f_y = 0$ implies

$x = 0$, $x = 2$ or $x = 1$, but $x = 0$ isn't inside D, and when $x = 2$, $y = 0$

but $(2,0)$ isn't inside D. Thus the only critical point inside D is $(1,2)$ and

$f(1,2) = 4$. Secondly we consider the boundary of D.

On L_1, $f(x,0) = 0$ and so $f = 0$ on L_1. On L_2, $x = -y + 6$ and

$f(-y + 6, y) = y^2(6 - y)(-2) = -2(6y^2 - y^3)$ which has

critical points at $y = 0$ and $y = 4$. Then $f(6,0) = 0$ while $f(2,4) = -64$. On L_3, $f(0,y) = 0$, so $f = 0$ on L_3.

Thus on D the absolute maximum of f is $f(1,2) = 4$ while the absolute minimum is $f(2,4) = -64$.

57. $f(x,y) = x^3 - 3x + y^4 - 2y^2$

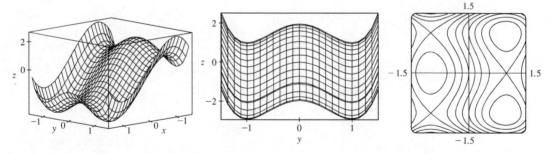

From the graphs, it appears that f has a local maximum $f(-1,0) \approx 2$, local minima $f(1, \pm 1) \approx -3$, and saddle

points at $(-1, \pm 1)$ and $(1,0)$.

To find the exact quantities, we calculate $f_x = 3x^2 - 3 = 0 \Leftrightarrow x = \pm 1$ and $f_y = 4y^3 - 4y = 0 \Leftrightarrow$

$y = 0, \pm 1$, giving the critical points estimated above. Also $f_{xx} = 6x$, $f_{xy} = 0$, $f_{yy} = 12y^2 - 4$, so using the

Second Derivatives Test, $D(-1,0) = 24 > 0$ and $f_{xx}(-1,0) = -6 < 0$ indicating a local maximum

$f(-1,0) = 2$; $D(1, \pm 1) = 48 > 0$ and $f_{xx}(1, \pm 1) = 6 > 0$ indicating local minima $f(1, \pm 1) = -3$; and

$D(-1, \pm 1) = -48$ and $D(1,0) = -24$, indicating saddle points.

59. $f(x,y) = x^2 y$, $g(x,y) = x^2 + y^2 = 1$ $\Rightarrow$ $\nabla f = \langle 2xy, x^2 \rangle = \lambda \nabla g = \langle 2\lambda x, 2\lambda y \rangle$. Then $2xy = 2\lambda x$ and

$x^2 = 2\lambda y$ imply $\lambda = x^2/(2y)$ and $\lambda = y$ if $x \neq 0$ and $y \neq 0$. Hence $x^2 = 2y^2$. Then $x^2 + y^2 = 1$ implies

$3y^2 = 1$ so $y = \pm\frac{1}{\sqrt{3}}$ and $x = \pm\sqrt{\frac{2}{3}}$. [Note if $x = 0$ then $x^2 = 2\lambda y$ implies $y = 0$ and $f(0,0) = 0$.] Thus the

possible points are $\left(\pm\sqrt{\frac{2}{3}}, \pm\frac{1}{\sqrt{3}}\right)$ and the absolute maxima are $f\left(\pm\sqrt{\frac{2}{3}}, \frac{1}{\sqrt{3}}\right) = \frac{2}{3\sqrt{3}}$ while the absolute minima

are $f\left(\pm\sqrt{\frac{2}{3}}, -\frac{1}{\sqrt{3}}\right) = -\frac{2}{3\sqrt{3}}$.

61. $f(x,y,z) = xyz$, $g(x,y,z) = x^2 + y^2 + z^2 = 3$. $\nabla f = \lambda \nabla g$ $\Rightarrow$ $\langle yz, xz, xy \rangle = \lambda \langle 2x, 2y, 2z \rangle$. If any of x, y,

or z is zero, then $x = y = z = 0$ which contradicts $x^2 + y^2 + z^2 = 3$. Then $\lambda = \dfrac{yz}{2x} = \dfrac{xz}{2y} = \dfrac{xy}{2z}$ $\Rightarrow$

$2y^2 z = 2x^2 z$ $\Rightarrow$ $y^2 = x^2$, and similarly $2yz^2 = 2x^2 y$ $\Rightarrow$ $z^2 = x^2$. Substituting into the constraint equation

gives $x^2 + x^2 + x^2 = 3$ $\Rightarrow$ $x^2 = 1 = y^2 = z^2$. Thus the possible points are

$(1,1,\pm 1)$, $(1,-1,\pm 1)$, $(-1,1,\pm 1)$, $(-1,-1,\pm 1)$. The absolute maximum is

$f(1,1,1) = f(1,-1,-1) = f(-1,1,-1) = f(-1,-1,1) = 1$ and the absolute minimum is

$f(1,1,-1) = f(1,-1,1) = f(-1,1,1) = f(-1,-1,-1) = -1$.

63. $f(x,y,z) = x^2 + y^2 + z^2$, $g(x,y,z) = xy^2 z^3 = 2$ $\Rightarrow$

$\nabla f = \langle 2x, 2y, 2z \rangle = \lambda \nabla g = \langle \lambda y^2 z^3, 2\lambda xyz^3, 3\lambda xy^2 z^2 \rangle$. Since $xy^2 z^3 = 2$, $x \neq 0$, $y \neq 0$ and $z \neq 0$, so

(1) $2x = \lambda y^2 z^3$, (2) $1 = \lambda xz^3$, (3) $2 = 3\lambda xy^2 z$. Then (2) and (3) imply $\dfrac{1}{xz^3} = \dfrac{2}{3xy^2 z}$ or $y^2 = \frac{2}{3}z^2$ so

$y = \pm z\sqrt{\frac{2}{3}}$. Similarly (1) and (3) imply $\dfrac{2x}{y^2 z^3} = \dfrac{2}{3xy^2 z}$ or $3x^2 = z^2$ so $x = \pm\frac{1}{\sqrt{3}}z$. But $xy^2 z^3 = 2$ so x and z

must have the same sign, that is, $x = \frac{1}{\sqrt{3}}z$. Thus $g(x,y,z) = 2$ implies $\frac{1}{\sqrt{3}}z\left(\frac{2}{3}z^2\right)z^3 = 2$ or $z = \pm 3^{1/4}$ and the

possible points are $(\pm 3^{-1/4}, 3^{-1/4}\sqrt{2}, \pm 3^{1/4})$, $(\pm 3^{-1/4}, -3^{-1/4}\sqrt{2}, \pm 3^{1/4})$. However at each of these points

f takes on the same value, $2\sqrt{3}$. But $(2,1,1)$ also satisfies $g(x,y,z) = 2$ and $f(2,1,1) = 6 > 2\sqrt{3}$. Thus f has

an absolute minimum value of $2\sqrt{3}$ and no absolute maximum subject to the constraint $xy^2 z^3 = 2$.

Alternate solution: $g(x,y,z) = xy^2 z^3 = 2$ implies $y^2 = \dfrac{2}{xz^3}$, so minimize $f(x,z) = x^2 + \dfrac{2}{xz^3} + z^2$. Then

$f_x = 2x - \dfrac{2}{x^2 z^3}$, $f_z = -\dfrac{6}{xz^4} + 2z$, $f_{xx} = 2 + \dfrac{4}{x^3 z^3}$, $f_{zz} = \dfrac{24}{xz^5} + 2$ and $f_{xz} = \dfrac{6}{x^2 z^4}$. Now $f_x = 0$ implies

$2x^3 z^3 - 2 = 0$ or $z = 1/x$. Substituting into $f_y = 0$ implies $-6x^3 + 2x^{-1} = 0$ or $x = \dfrac{1}{\sqrt[4]{3}}$, so the two critical

points are $\left(\pm\frac{1}{\sqrt[4]{3}}, \pm\sqrt[4]{3}\right)$. Then $D\left(\pm\frac{1}{\sqrt[4]{3}}, \pm\sqrt[4]{3}\right) = (2+4)\left(2+\frac{24}{3}\right) - \left(\frac{6}{\sqrt{3}}\right)^2 > 0$ and

$f_{xx}\left(\pm\frac{1}{\sqrt[4]{3}}, \pm\sqrt[4]{3}\right) = 6 > 0$, so each point is a minimum. Finally, $y^2 = \dfrac{2}{xz^3}$, so the four points closest to the

origin are $\left(\pm\frac{1}{\sqrt[4]{3}}, \frac{\sqrt{2}}{\sqrt[4]{3}}, \pm\sqrt[4]{3}\right)$, $\left(\pm\frac{1}{\sqrt[4]{3}}, -\frac{\sqrt{2}}{\sqrt[4]{3}}, \pm\sqrt[4]{3}\right)$.

65.

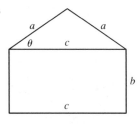

The area of the triangle is $\frac{1}{2}ca\sin\theta$ and the area of the rectangle is bc.
Thus, the area of the whole object is $f(a,b,c) = \frac{1}{2}ca\sin\theta + bc$. The
perimeter of the object is $g(a,b,c) = 2a + 2b + c = P$. To simplify
$\sin\theta$ in terms of a, b, and c notice that $a^2\sin^2\theta + \left(\frac{1}{2}c\right)^2 = a^2 \quad\Rightarrow$
$\sin\theta = \dfrac{1}{2a}\sqrt{4a^2 - c^2}$. Thus $f(a,b,c) = \dfrac{c}{4}\sqrt{4a^2 - c^2} + bc$.

(Instead of using θ, we could just have used the Pythagorean Theorem.) As a result, by Lagrange's method, we must

find a, b, c, and λ by solving $\nabla f = \lambda\nabla g$ which gives the following equations: (1) $ca(4a^2 - c^2)^{-1/2} = 2\lambda$,

(2) $c = 2\lambda$, (3) $\frac{1}{4}(4a^2 - c^2)^{1/2} - \frac{1}{4}c^2(4a^2 - c^2)^{-1/2} + b = \lambda$, and (4) $2a + 2b + c = P$. From (2), $\lambda = \frac{1}{2}c$ and

so (1) produces $ca(4a^2 - c^2)^{-1/2} = c \quad\Rightarrow\quad (4a^2 - c^2)^{1/2} = a \quad\Rightarrow\quad 4a^2 - c^2 = a^2 \quad\Rightarrow\quad$ (5) $c = \sqrt{3}\,a$.

Similarly, since $(4a^2 - c^2)^{1/2} = a$ and $\lambda = \frac{1}{2}c$, (3) gives $\dfrac{a}{4} - \dfrac{c^2}{4a} + b = \dfrac{c}{2}$, so from (5), $\dfrac{a}{4} - \dfrac{3a}{4} + b = \dfrac{\sqrt{3}\,a}{2}$

$\Rightarrow \quad -\dfrac{a}{2} - \dfrac{\sqrt{3}\,a}{2} = -b \quad\Rightarrow\quad$ (6) $b = \dfrac{a}{2}\left(1 + \sqrt{3}\right)$. Substituting (5) and (6) into (4) we get:

$2a + a\left(1 + \sqrt{3}\right) + \sqrt{3}\,a = P \quad\Rightarrow\quad 3a + 2\sqrt{3}\,a = P \quad\Rightarrow\quad a = \dfrac{P}{3 + 2\sqrt{3}} = \dfrac{2\sqrt{3} - 3}{3}P$ and thus

$b = \dfrac{\left(2\sqrt{3} - 3\right)\left(1 + \sqrt{3}\right)}{6}P = \dfrac{3 - \sqrt{3}}{6}P$ and $c = \left(2 - \sqrt{3}\right)P$.

☐ PROBLEMS PLUS

1. The areas of the smaller rectangles are $A_1 = xy$, $A_2 = (L - x)y$,

$A_3 = (L - x)(W - y)$, $A_4 = x(W - y)$. For $0 \leq x \leq L$,

$0 \leq y \leq W$, let

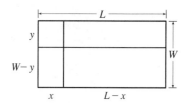

$$f(x, y) = A_1^2 + A_2^2 + A_3^2 + A_4^2$$

$$= x^2 y^2 + (L - x)^2 y^2 + (L - x)^2 (W - y)^2 + x^2 (W - y)^2$$

$$= [x^2 + (L - x)^2][y^2 + (W - y)^2]$$

Then we need to find the maximum and minimum values of $f(x, y)$. Here

$$f_x(x, y) = [2x - 2(L - x)][y^2 + (W - y)^2] = 0 \quad \Rightarrow \quad 4x - 2L = 0 \text{ or } x = \tfrac{1}{2}L, \text{ and}$$

$$f_y(x, y) = [x^2 + (L - x)^2][2y - 2(W - y)] = 0 \quad \Rightarrow \quad 4y - 2W = 0 \text{ or } y = W/2. \text{ Also}$$

$f_{xx} = 4[y^2 + (W - y)^2]$, $f_{yy} = 4[x^2 + (L - x)^2]$, and $f_{xy} = (4x - 2L)(4y - 2W)$. Then

$D = 16[y^2 + (W - y)^2][x^2 + (L - x)^2] - (4x - 2L)^2(4y - 2W)^2$. Thus when $x = \tfrac{1}{2}L$ and $y = \tfrac{1}{2}W$,

$D > 0$ and $f_{xx} = 2W^2 > 0$. Thus a minimum of f occurs at $\left(\tfrac{1}{2}L, \tfrac{1}{2}W\right)$ and this minimum value is

$f\left(\tfrac{1}{2}L, \tfrac{1}{2}W\right) = \tfrac{1}{4}L^2 W^2$. There are no other critical points, so the maximum must occur on the boundary. Now

along the width of the rectangle let $g(y) = f(0, y) = f(L, y) = L^2[y^2 + (W - y)^2]$, $0 \leq y \leq W$. Then

$g'(y) = L^2[2y - 2(W - y)] = 0 \quad \Leftrightarrow \quad y = \tfrac{1}{2}W$. And $g\left(\tfrac{1}{2}\right) = \tfrac{1}{2}L^2 W^2$. Checking the endpoints, we get

$g(0) = g(W) = L^2 W^2$. Along the length of the rectangle let $h(x) = f(x, 0) = f(x, W) = W^2[x^2 + (L - x)^2]$,

$0 \leq x \leq L$. By symmetry $h'(x) = 0 \quad \Leftrightarrow \quad x = \tfrac{1}{2}L$ and $h\left(\tfrac{1}{2}L\right) = \tfrac{1}{2}L^2 W^2$. At the endpoints we have

$h(0) = h(L) = L^2 W^2$. Therefore $L^2 W^2$ is the maximum value of f. This maximum value of f occurs when the

"cutting" lines correspond to sides of the rectangle.

3. (a) The area of a trapezoid is $\tfrac{1}{2}h(b_1 + b_2)$, where h is the height (the distance between the two parallel sides) and

b_1, b_2 are the lengths of the bases (the parallel sides). From the figure in the text, we see that $h = x \sin \theta$,

$b_1 = w - 2x$, and $b_2 = w - 2x + 2x \cos \theta$. Therefore the cross-sectional area of the rain gutter is

$$A(x, \theta) = \tfrac{1}{2}x \sin \theta \left[(w - 2x) + (w - 2x + 2x \cos \theta)\right] = (x \sin \theta)(w - 2x + x \cos \theta)$$

$$= wx \sin \theta - 2x^2 \sin \theta + x^2 \sin \theta \cos \theta, \ 0 < x \leq \tfrac{1}{2}w, 0 < \theta \leq \tfrac{\pi}{2}$$

We look for the critical points of A: $\partial A / \partial x = w \sin \theta - 4x \sin \theta + 2x \sin \theta \cos \theta$ and

$\partial A / \partial \theta = wx \cos \theta - 2x^2 \cos \theta + x^2(\cos^2 \theta - \sin^2 \theta)$, so $\partial A / \partial x = 0 \quad \Leftrightarrow \quad \sin \theta (w - 4x + 2x \cos \theta) = 0$

$$\Leftrightarrow \quad \cos\theta = \frac{4x - w}{2x} = 2 - \frac{w}{2x} \quad (0 < \theta \le \tfrac{\pi}{2} \quad \Rightarrow \quad \sin\theta > 0). \text{ If, in addition, } \partial A/\partial\theta = 0, \text{ then}$$

$$0 = wx\cos\theta - 2x^2\cos\theta + x^2(2\cos^2\theta - 1)$$

$$= wx\left(2 - \frac{w}{2x}\right) - 2x^2\left(2 - \frac{w}{2x}\right) + x^2\left[2\left(2 - \frac{w}{2x}\right)^2 - 1\right]$$

$$= 2wx - \tfrac{1}{2}w^2 - 4x^2 + wx + x^2\left[8 - \frac{4w}{x} + \frac{w^2}{2x^2} - 1\right] = -wx + 3x^2 = x(3x - w)$$

Since $x > 0$, we must have $x = \tfrac{1}{3}w$, in which case $\cos\theta = \tfrac{1}{2}$, so $\theta = \tfrac{\pi}{3}$, $\sin\theta = \tfrac{\sqrt{3}}{2}$, $k = \tfrac{\sqrt{3}}{6}w$, $b_1 = \tfrac{1}{3}w$,

$b_2 = \tfrac{2}{3}w$, and $A = \tfrac{\sqrt{3}}{12}w^2$. As in Example 15.7.6 [ET 14.7.6], we can argue from the physical nature of this

problem that we have found a local maximum of A. Now checking the boundary of A, let

$g(\theta) = A(w/2, \theta) = \tfrac{1}{2}w^2\sin\theta - \tfrac{1}{2}w^2\sin\theta + \tfrac{1}{4}w^2\sin\theta\cos\theta = \tfrac{1}{8}w^2\sin 2\theta$, $0 < \theta \le \tfrac{\pi}{2}$. Clearly g is

maximized when $\sin 2\theta = 1$ in which case $A = \tfrac{1}{8}w^2$. Also along the line $\theta = \tfrac{\pi}{2}$, let

$h(x) = A\left(x, \tfrac{\pi}{2}\right) = wx - 2x^2$, $0 < x < \tfrac{1}{2}w \quad \Rightarrow \quad h'(x) = w - 4x = 0 \quad \Leftrightarrow \quad x = \tfrac{1}{4}w$, and

$h\left(\tfrac{1}{4}w\right) = w\left(\tfrac{1}{4}w\right) - 2\left(\tfrac{1}{4}w\right)^2 = \tfrac{1}{8}w^2$. Since $\tfrac{1}{8}w^2 < \tfrac{\sqrt{3}}{12}w^2$, we conclude that the local maximum found earlier

was an absolute maximum.

(b) If the metal were bent into a semi-circular gutter of radius r, we would have $w = \pi r$ and

$$A = \tfrac{1}{2}\pi r^2 = \tfrac{1}{2}\pi\left(\frac{w}{\pi}\right)^2 = \frac{w^2}{2\pi}. \text{ Since } \frac{w^2}{2\pi} > \frac{\sqrt{3}\,w^2}{12}, \text{ it } would \text{ be better to bend the metal into a gutter with a}$$

semicircular cross-section.

5. Let $g(x, y) = xf\left(\dfrac{y}{x}\right)$. Then $g_x(x, y) = f\left(\dfrac{y}{x}\right) + xf'\left(\dfrac{y}{x}\right)\left(-\dfrac{y}{x^2}\right) = f\left(\dfrac{y}{x}\right) - \dfrac{y}{x}f'\left(\dfrac{y}{x}\right)$ and

$g_y(x, y) = xf'\left(\dfrac{y}{x}\right)\left(\dfrac{1}{x}\right) = f'\left(\dfrac{y}{x}\right)$. Thus the tangent plane at (x_0, y_0, z_0) on the surface has equation

$$z - x_0 f\left(\frac{y_0}{x_0}\right) = \left[f\left(\frac{y_0}{x_0}\right) - y_0 x_0^{-1} f'\left(\frac{y_0}{x_0}\right)\right](x - x_0) + f'\left(\frac{y_0}{x_0}\right)(y - y_0) \quad \Rightarrow$$

$\left[f\left(\dfrac{y_0}{x_0}\right) - y_0 x_0^{-1} f'\left(\dfrac{y_0}{x_0}\right)\right]x + \left[f'\left(\dfrac{y_0}{x_0}\right)\right]y - z = 0$. But any plane whose equation is of the form

$ax + by + cz = 0$ passes through the origin. Thus the origin is the common point of intersection.

7. (a) $x = r\cos\theta$, $y = r\sin\theta$, $z = z$. Then $\dfrac{\partial u}{\partial r} = \dfrac{\partial u}{\partial x}\dfrac{\partial x}{\partial r} + \dfrac{\partial u}{\partial y}\dfrac{\partial y}{\partial r} + \dfrac{\partial u}{\partial z}\dfrac{\partial z}{\partial r} = \dfrac{\partial u}{\partial x}\cos\theta + \dfrac{\partial u}{\partial y}\sin\theta$ and

$$\frac{\partial^2 u}{\partial r^2} = \cos\theta\left[\frac{\partial^2 u}{\partial x^2}\frac{\partial x}{\partial r} + \frac{\partial^2 u}{\partial y\,\partial x}\frac{\partial y}{\partial r} + \frac{\partial^2 u}{\partial z\,\partial x}\frac{\partial z}{\partial r}\right] + \sin\theta\left[\frac{\partial^2 u}{\partial y^2}\frac{\partial y}{\partial r} + \frac{\partial^2 u}{\partial x\,\partial y}\frac{\partial x}{\partial r} + \frac{\partial^2 u}{\partial z\,\partial y}\frac{\partial z}{\partial r}\right]$$

$$= \frac{\partial^2 u}{\partial x^2}\cos^2\theta + \frac{\partial^2 u}{\partial y^2}\sin^2\theta + 2\frac{\partial^2 u}{\partial y\,\partial x}\cos\theta\sin\theta$$

Similarly $\dfrac{\partial u}{\partial\theta} = -\dfrac{\partial u}{\partial x}r\sin\theta + \dfrac{\partial u}{\partial y}r\cos\theta$ and

$$\frac{\partial^2 u}{\partial \theta^2} = \frac{\partial^2 u}{\partial x^2} r^2 \sin^2 \theta + \frac{\partial^2 u}{\partial y^2} r^2 \cos^2 \theta - 2 \frac{\partial^2 u}{\partial y \, \partial x} r^2 \sin \theta \cos \theta - \frac{\partial u}{\partial x} r \cos \theta - \frac{\partial u}{\partial y} r \sin \theta. \text{ So}$$

$$\frac{\partial^2 u}{\partial r^2} + \frac{1}{r} \frac{\partial u}{\partial r} + \frac{1}{r^2} \frac{\partial^2 u}{\partial \theta^2} + \frac{\partial^2 u}{\partial z^2}$$

$$= \frac{\partial^2 u}{\partial x^2} \cos^2 \theta + \frac{\partial^2 u}{\partial y^2} \sin^2 \theta + 2 \frac{\partial^2 u}{\partial y \, \partial x} \cos \theta \sin \theta + \frac{\partial u}{\partial x} \frac{\cos \theta}{r} + \frac{\partial u}{\partial y} \frac{\sin \theta}{r}$$

$$+ \frac{\partial^2 u}{\partial x^2} \sin^2 \theta + \frac{\partial^2 u}{\partial y^2} \cos^2 \theta - 2 \frac{\partial^2 u}{\partial y \, \partial x} \sin \theta \cos \theta - \frac{\partial u}{\partial x} \frac{\cos \theta}{r} - \frac{\partial u}{\partial y} \frac{\sin \theta}{r} + \frac{\partial^2 u}{\partial z^2}$$

$$= \frac{\partial^2 u}{\partial x^2} + \frac{\partial^2 u}{\partial y^2} + \frac{\partial^2 u}{\partial z^2}$$

(b) $x = \rho \sin \phi \cos \theta$, $y = \rho \sin \phi \sin \theta$, $z = \rho \cos \phi$. Then

$$\frac{\partial u}{\partial \rho} = \frac{\partial u}{\partial x} \frac{\partial x}{\partial \rho} + \frac{\partial u}{\partial y} \frac{\partial y}{\partial \rho} + \frac{\partial u}{\partial z} \frac{\partial z}{\partial \rho} = \frac{\partial u}{\partial x} \sin \phi \cos \theta + \frac{\partial u}{\partial y} \sin \phi \sin \theta + \frac{\partial u}{\partial z} \cos \phi, \text{ and}$$

$$\frac{\partial^2 u}{\partial \rho^2} = \sin \phi \cos \theta \left[\frac{\partial^2 u}{\partial x^2} \frac{\partial x}{\partial \rho} + \frac{\partial^2 u}{\partial y \, \partial x} \frac{\partial y}{\partial \rho} + \frac{\partial^2 u}{\partial z \, \partial x} \frac{\partial z}{\partial \rho} \right]$$

$$+ \sin \phi \sin \theta \left[\frac{\partial^2 u}{\partial y^2} \frac{\partial y}{\partial \rho} + \frac{\partial^2 u}{\partial x \, \partial y} \frac{\partial x}{\partial \rho} + \frac{\partial^2 u}{\partial z \, \partial y} \frac{\partial z}{\partial \rho} \right]$$

$$+ \cos \phi \left[\frac{\partial^2 u}{\partial z^2} \frac{\partial z}{\partial \rho} + \frac{\partial^2 u}{\partial x \, \partial z} \frac{\partial x}{\partial \rho} + \frac{\partial^2 u}{\partial y \, \partial z} \frac{\partial y}{\partial \rho} \right]$$

$$= 2 \frac{\partial^2 u}{\partial y \, \partial x} \sin^2 \phi \sin \theta \cos \theta + 2 \frac{\partial^2 u}{\partial z \, \partial x} \sin \phi \cos \phi \cos \theta + 2 \frac{\partial^2 u}{\partial y \, \partial z} \sin \phi \cos \phi \sin \theta$$

$$+ \frac{\partial^2 u}{\partial x^2} \sin^2 \phi \cos^2 \theta + \frac{\partial^2 u}{\partial y^2} \sin^2 \phi \sin^2 \theta + \frac{\partial^2 u}{\partial z^2} \cos^2 \phi$$

Similarly $\dfrac{\partial u}{\partial \phi} = \dfrac{\partial u}{\partial x} \rho \cos \phi \cos \theta + \dfrac{\partial u}{\partial y} \rho \cos \phi \sin \theta - \dfrac{\partial u}{\partial z} \rho \sin \phi$, and

$$\frac{\partial^2 u}{\partial \phi^2} = 2 \frac{\partial^2 u}{\partial y \, \partial x} \rho^2 \cos^2 \phi \sin \theta \cos \theta - 2 \frac{\partial^2 u}{\partial x \, \partial z} \rho^2 \sin \phi \cos \phi \cos \theta$$

$$- 2 \frac{\partial^2 u}{\partial y \, \partial z} \rho^2 \sin \phi \cos \phi \sin \theta + \frac{\partial^2 u}{\partial x^2} \rho^2 \cos^2 \phi \cos^2 \theta + \frac{\partial^2 u}{\partial y^2} \rho^2 \cos^2 \phi \sin^2 \theta$$

$$+ \frac{\partial^2 u}{\partial z^2} \rho^2 \sin^2 \phi - \frac{\partial u}{\partial x} \rho \sin \phi \cos \theta - \frac{\partial u}{\partial y} \rho \sin \phi \sin \theta - \frac{\partial u}{\partial z} \rho \cos \phi$$

And $\dfrac{\partial u}{\partial \theta} = -\dfrac{\partial u}{\partial x} \rho \sin \phi \sin \theta + \dfrac{\partial u}{\partial y} \rho \sin \phi \cos \theta$, while

$$\frac{\partial^2 u}{\partial \theta^2} = -2 \frac{\partial^2 u}{\partial y \, \partial x} \rho^2 \sin^2 \phi \cos \theta \sin \theta + \frac{\partial^2 u}{\partial x^2} \rho^2 \sin^2 \phi \sin^2 \theta$$

$$+ \frac{\partial^2 u}{\partial y^2} \rho^2 \sin^2 \phi \cos^2 \theta - \frac{\partial u}{\partial x} \rho \sin \phi \cos \theta - \frac{\partial u}{\partial y} \rho \sin \phi \sin \theta$$

Therefore

$$\frac{\partial^2 u}{\partial \rho^2} + \frac{2}{\rho}\frac{\partial u}{\partial \rho} + \frac{\cot \phi}{\rho^2}\frac{\partial u}{\partial \phi} + \frac{1}{\rho^2}\frac{\partial^2 u}{\partial \phi^2} + \frac{1}{\rho^2 \sin^2 \phi}\frac{\partial^2 u}{\partial \theta^2}$$

$$= \frac{\partial^2 u}{\partial x^2}\left[(\sin^2 \phi \cos^2 \theta) + (\cos^2 \phi \cos^2 \theta) + \sin^2 \theta\right]$$

$$+ \frac{\partial^2 u}{\partial y^2}\left[(\sin^2 \phi \sin^2 \theta) + (\cos^2 \phi \sin^2 \theta) + \cos^2 \theta\right] + \frac{\partial^2 u}{\partial z^2}\left[\cos^2 \phi + \sin^2 \phi\right]$$

$$+ \frac{\partial u}{\partial x}\left[\frac{2\sin^2 \phi \cos \theta + \cos^2 \phi \cos \theta - \sin^2 \phi \cos \theta - \cos \theta}{\rho \sin \phi}\right]$$

$$+ \frac{\partial u}{\partial y}\left[\frac{2\sin^2 \phi \sin \theta + \cos^2 \phi \sin \theta - \sin^2 \phi \sin \theta - \sin \theta}{\rho \sin \phi}\right]$$

But $2\sin^2 \phi \cos \theta + \cos^2 \phi \cos \theta - \sin^2 \phi \cos \theta - \cos \theta = (\sin^2 \phi + \cos^2 \phi - 1)\cos \theta = 0$ and similarly the

coefficient of $\partial u/\partial y$ is 0. Also $\sin^2 \phi \cos^2 \theta + \cos^2 \phi \cos^2 \theta + \sin^2 \theta = \cos^2 \theta (\sin^2 \phi + \cos^2 \phi) + \sin^2 \theta = 1$,

and similarly the coefficient of $\partial^2 u/\partial y^2$ is 1. So Laplace's Equation in spherical coordinates is as stated.

9. Since we are minimizing the area of the ellipse, and the circle lies above

the x-axis, the ellipse will intersect the circle for only one value of y. This

y-value must satisfy both the equation of the circle and the equation of the

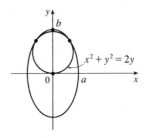

ellipse. Now $\dfrac{x^2}{a^2} + \dfrac{y^2}{b^2} = 1 \;\Rightarrow\; x^2 = \dfrac{a^2}{b^2}(b^2 - y^2)$. Substituting into

the equation of the circle gives $\dfrac{a^2}{b^2}(b^2 - y^2) + y^2 - 2y = 0 \;\Rightarrow$

$\left(\dfrac{b^2 - a^2}{b^2}\right)y^2 - 2y + a^2 = 0$. In order for there to be only one solution to this quadratic equation, the discriminant

must be 0, so $4 - 4a^2\dfrac{b^2 - a^2}{b^2} = 0 \;\Rightarrow\; b^2 - a^2 b^2 + a^4 = 0$. The area of the ellipse is $A(a, b) = \pi ab$, and we

minimize this function subject to the constraint $g(a, b) = b^2 - a^2 b^2 + a^4 = 0$.

Now $\nabla A = \lambda \nabla g \;\Leftrightarrow\; \pi b = \lambda(4a^3 - 2ab^2),\ \pi a = \lambda(2b - 2ba^2) \;\Rightarrow\;$ (1) $\lambda = \dfrac{\pi b}{2a(2a^2 - b^2)}$,

(2) $\lambda = \dfrac{\pi a}{2b(1 - a^2)}$, (3) $b^2 - a^2 b^2 + a^4 = 0$. Comparing (1) and (2) gives $\dfrac{\pi b}{2a(2a^2 - b^2)} = \dfrac{\pi a}{2b(1 - a^2)} \;\Rightarrow\;$

$2\pi b^2 = 4\pi a^4 \;\Leftrightarrow\; a^2 = \frac{1}{\sqrt{2}}b$. Substitute this into (3) to get $b = \frac{3}{\sqrt{2}} \;\Rightarrow\; a = \sqrt{\frac{3}{2}}$.

16 ☐ MULTIPLE INTEGRALS

16.1 Double Integrals over Rectangles

1. (a) The subrectangles are shown in the figure.

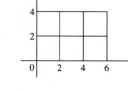

The surface is the graph of $f(x, y) = xy$ and $\Delta A = 4$,

so we estimate

$$V \approx \sum_{i=1}^{3} \sum_{j=1}^{2} f(x_i, y_j) \, \Delta A$$

$$= f(2,2) \, \Delta A + f(2,4) \, \Delta A + f(4,2) \, \Delta A + f(4,4) \, \Delta A + f(6,2) \, \Delta A + f(6,4) \, \Delta A$$

$$= 4(4) + 8(4) + 8(4) + 16(4) + 12(4) + 24(4) = 288$$

(b) $V \approx \displaystyle\sum_{i=1}^{3} \sum_{j=1}^{2} f(\overline{x}_i, \overline{y}_j) \, \Delta A$

$$= f(1,1) \, \Delta A + f(1,3) \, \Delta A + f(3,1) \, \Delta A + f(3,3) \, \Delta A + f(5,1) \, \Delta A + f(5,3) \, \Delta A$$

$$= 1(4) + 3(4) + 3(4) + 9(4) + 5(4) + 15(4) = 144$$

3. (a) The subrectangles are shown in the figure. Since $\Delta A = \pi^2/4$, we estimate

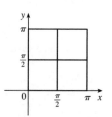

$$\iint_R \sin(x + y) \, dA \approx \sum_{i=1}^{2} \sum_{j=1}^{2} f(x_{ij}^*, y_{ij}^*) \, \Delta A$$

$$= f(0,0) \, \Delta A + f\left(0, \tfrac{\pi}{2}\right) \Delta A + f\left(\tfrac{\pi}{2}, 0\right) \Delta A + f\left(\tfrac{\pi}{2}, \tfrac{\pi}{2}\right) \Delta A$$

$$= 0\left(\tfrac{\pi^2}{4}\right) + 1\left(\tfrac{\pi^2}{4}\right) + 1\left(\tfrac{\pi^2}{4}\right) + 0\left(\tfrac{\pi^2}{4}\right) = \tfrac{\pi^2}{2} \approx 4.935$$

(b) $\iint_R \sin(x + y) \, dA \approx \displaystyle\sum_{i=1}^{2} \sum_{j=1}^{2} f(\overline{x}_i, \overline{y}_j) \, \Delta A$

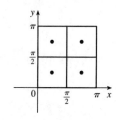

$$= f\left(\tfrac{\pi}{4}, \tfrac{\pi}{4}\right) \Delta A + f\left(\tfrac{\pi}{4}, \tfrac{3\pi}{4}\right) \Delta A$$

$$\quad + f\left(\tfrac{3\pi}{4}, \tfrac{\pi}{4}\right) \Delta A + f\left(\tfrac{3\pi}{4}, \tfrac{3\pi}{4}\right) \Delta A$$

$$= 1\left(\tfrac{\pi^2}{4}\right) + 0\left(\tfrac{\pi^2}{4}\right) + 0\left(\tfrac{\pi^2}{4}\right) + (-1)\left(\tfrac{\pi^2}{4}\right) = 0$$

5. (a) Each subrectangle and its midpoint are shown in the figure. The area of each

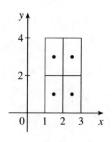

subrectangle is $\Delta A = 2$, so we evaluate f at each midpoint and estimate

$$\iint_R f(x, y) \, dA \approx \sum_{i=1}^{2} \sum_{j=1}^{2} f(\overline{x}_i, \overline{y}_j) \, \Delta A$$

$$= f(1.5, 1) \, \Delta A + f(1.5, 3) \, \Delta A$$

$$\quad + f(2.5, 1) \, \Delta A + f(2.5, 3) \, \Delta A$$

$$= 1(2) + (-8)(2) + 5(2) + (-1)(2) = -6$$

(b) The subrectangles are shown in the figure. In each subrectangle, the sample point farthest from the origin is the upper right corner, and the area of each subrectangle is $\Delta A = \frac{1}{2}$. Thus we estimate

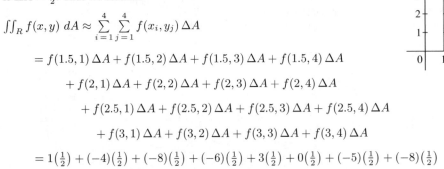

$$\iint_R f(x,y)\, dA \approx \sum_{i=1}^{4} \sum_{j=1}^{4} f(x_i, y_j)\, \Delta A$$

$$= f(1.5, 1)\, \Delta A + f(1.5, 2)\, \Delta A + f(1.5, 3)\, \Delta A + f(1.5, 4)\, \Delta A$$

$$+ f(2, 1)\, \Delta A + f(2, 2)\, \Delta A + f(2, 3)\, \Delta A + f(2, 4)\, \Delta A$$

$$+ f(2.5, 1)\, \Delta A + f(2.5, 2)\, \Delta A + f(2.5, 3)\, \Delta A + f(2.5, 4)\, \Delta A$$

$$+ f(3, 1)\, \Delta A + f(3, 2)\, \Delta A + f(3, 3)\, \Delta A + f(3, 4)\, \Delta A$$

$$= 1\left(\tfrac{1}{2}\right) + (-4)\left(\tfrac{1}{2}\right) + (-8)\left(\tfrac{1}{2}\right) + (-6)\left(\tfrac{1}{2}\right) + 3\left(\tfrac{1}{2}\right) + 0\left(\tfrac{1}{2}\right) + (-5)\left(\tfrac{1}{2}\right) + (-8)\left(\tfrac{1}{2}\right)$$

$$+ 5\left(\tfrac{1}{2}\right) + 3\left(\tfrac{1}{2}\right) + (-1)\left(\tfrac{1}{2}\right) + (-4)\left(\tfrac{1}{2}\right) + 8\left(\tfrac{1}{2}\right) + 6\left(\tfrac{1}{2}\right) + 3\left(\tfrac{1}{2}\right) + 0\left(\tfrac{1}{2}\right)$$

$$= -3.5$$

7. The values of $f(x,y) = \sqrt{52 - x^2 - y^2}$ get smaller as we move farther from the origin, so on any of the subrectangles in the problem, the function will have its largest value at the lower left corner of the subrectangle and its smallest value at the upper right corner, and any other value will lie between these two. So using these subrectangles we have $U < V < L$. (Note that this is true no matter how R is divided into subrectangles.)

9. (a) With $m = n = 2$, we have $\Delta A = 4$. Using the contour map to estimate the value of f at the center of each subrectangle, we have

$$\iint_R f(x,y)\, dA \approx \sum_{i=1}^{2} \sum_{j=1}^{2} f\left(\overline{x}_i, \overline{y}_j\right) \Delta A = \Delta A [f(1,1) + f(1,3) + f(3,1) + f(3,3)]$$

$$\approx 4(27 + 4 + 14 + 17) = 248$$

(b) $f_{\text{ave}} = \frac{1}{A(R)} \iint_R f(x,y)\, dA \approx \frac{1}{16}(248) = 15.5$

11. $z = 3 > 0$, so we can interpret the integral as the volume of the solid S that lies below the plane $z = 3$ and above the rectangle $[-2, 2] \times [1, 6]$. S is a rectangular solid, thus $\iint_R 3\, dA = 4 \cdot 5 \cdot 3 = 60$.

13. $z = f(x,y) = 4 - 2y \geq 0$ for $0 \leq y \leq 1$. Thus the integral represents the volume of that part of the rectangular solid $[0, 1] \times [0, 1] \times [0, 4]$ which lies below the plane $z = 4 - 2y$. So

$$\iint_R (4 - 2y)\, dA = (1)(1)(2) + \tfrac{1}{2}(1)(1)(2) = 3$$

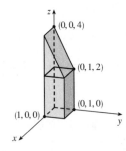

15. To calculate the estimates using a programmable calculator, we can use an algorithm similar to that of Exercise 5.1.7 [ET 5.1.7]. In Maple, we can define the function $f(x, y) = e^{-x^2-y^2}$ (calling it f), load the student package, and then use the command

$$\text{middlesum}(\text{middlesum}(f,x=0..1,m),$$
$$y=0..1,m);$$

to get the estimate with $n = m^2$ squares of equal size. Mathematica has no special Riemann sum command, but we can define f and then use nested Sum commands to calculate the estimates.

n	estimate
1	0.6065
4	0.5694
16	0.5606
64	0.5585
256	0.5579
1024	0.5578

17. If we divide R into mn subrectangles, $\iint_R k \, dA \approx \sum\limits_{i=1}^{m} \sum\limits_{j=1}^{n} f(x_{ij}^*, y_{ij}^*) \, \Delta A$ for any choice of sample points

(x_{ij}^*, y_{ij}^*). But $f(x_{ij}^*, y_{ij}^*) = k$ always and $\sum\limits_{i=1}^{m} \sum\limits_{j=1}^{n} \Delta A = $ area of $R = (b-a)(d-c)$. Thus, no matter how we

choose the sample points, $\sum\limits_{i=1}^{m} \sum\limits_{j=1}^{n} f(x_{ij}^*, y_{ij}^*) \, \Delta A = k \sum\limits_{i=1}^{m} \sum\limits_{j=1}^{n} \Delta A = k(b-a)(d-c)$ and so

$$\iint_R k \, dA = \lim_{m,n\to\infty} \sum_{i=1}^{m} \sum_{j=1}^{n} f(x_{ij}^*, y_{ij}^*) \, \Delta A = \lim_{m,n\to\infty} k \sum_{i=1}^{m} \sum_{j=1}^{n} \Delta A$$

$$= \lim_{m,n\to\infty} k(b-a)(d-c) = k(b-a)(d-c)$$

16.2 Iterated Integrals

ET 15.2

1. $\int_0^3 (2x + 3x^2y) \, dx = \left[x^2 + x^3y\right]_{x=0}^{x=3} = (9 + 27y) - (0 + 0) = 9 + 27y,$

$\int_0^4 (2x + 3x^2y) \, dy = \left[2xy + 3x^2 \dfrac{y^2}{2}\right]_{y=0}^{y=4} = \left(8x + 3x^2 \cdot \dfrac{16}{2}\right) - (0 + 0) = 8x + 24x^2$

3. $\int_1^3 \int_0^1 (1 + 4xy) \, dx \, dy = \int_1^3 \left[x + 2x^2y\right]_{x=0}^{x=1} dy = \int_1^3 (1 + 2y) \, dy = \left[y + y^2\right]_1^3 = (3 + 9) - (1 + 1) = 10$

5. $\int_0^2 \int_0^{\pi/2} x \sin y \, dy \, dx = \int_0^2 x \, dx \int_0^{\pi/2} \sin y \, dy$ [as in Example 5]

$= \left[\dfrac{x^2}{2}\right]_0^2 \left[-\cos y\right]_0^{\pi/2} = (2 - 0)(0 + 1) = 2.$

7. $\int_0^2 \int_0^1 (2x + y)^8 \, dx \, dy = \int_0^2 \left[\dfrac{1}{2} \dfrac{(2x + y)^9}{9}\right]_{x=0}^{x=1} dy$ [substitute $u = 2x + y \Rightarrow dx = \frac{1}{2}du$]

$= \dfrac{1}{18} \int_0^2 \left[(2 + y)^9 - (0 + y)^9\right] dy = \dfrac{1}{18} \left[\dfrac{(2+y)^{10}}{10} - \dfrac{y^{10}}{10}\right]_0^2$

$= \frac{1}{180}\left[(4^{10} - 2^{10}) - (2^{10} - 0^{10})\right] = \dfrac{1,046,528}{180} = \dfrac{261,632}{45}$

9. $\int_1^4 \int_1^2 \left(\frac{x}{y} + \frac{y}{x} \right) dy\, dx = \int_1^4 \left[x \ln |y| + \frac{1}{x} \cdot \frac{1}{2} y^2 \right]_{y=1}^{y=2} dx = \int_1^4 \left(x \ln 2 + \frac{3}{2x} \right) dx$

$= \left[\frac{1}{2} x^2 \ln 2 + \frac{3}{2} \ln |x| \right]_1^4 = 8 \ln 2 + \frac{3}{2} \ln 4 - \frac{1}{2} \ln 2$

$= \frac{15}{2} \ln 2 + 3 \ln 4^{1/2} = \frac{21}{2} \ln 2$

11. $\int_0^{\ln 2} \int_0^{\ln 5} e^{2x - y}\, dx\, dy = \left(\int_0^{\ln 5} e^{2x}\, dx \right) \left(\int_0^{\ln 2} e^{-y}\, dy \right) = \left[\frac{1}{2} e^{2x} \right]_0^{\ln 5} \left[-e^{-y} \right]_0^{\ln 2}$

$= \left(\frac{25}{2} - \frac{1}{2} \right) \left(-\frac{1}{2} + 1 \right) = 6$

13. $\iint_R (6x^2 y^3 - 5y^4)\, dA = \int_0^3 \int_0^1 (6x^2 y^3 - 5y^4)\, dy\, dx = \int_0^3 \left[\frac{3}{2} x^2 y^4 - y^5 \right]_{y=0}^{y=1} dx$

$= \int_0^3 \left(\frac{3}{2} x^2 - 1 \right) dx = \left[\frac{1}{2} x^3 - x \right]_0^3 = \frac{27}{2} - 3 = \frac{21}{2}$

15. $\iint_R \frac{xy^2}{x^2 + 1}\, dA = \int_0^1 \int_{-3}^3 \frac{xy^2}{x^2 + 1}\, dy\, dx = \int_0^1 \frac{x}{x^2 + 1}\, dx \int_{-3}^3 y^2\, dy$

$= \left[\frac{1}{2} \ln(x^2 + 1) \right]_0^1 \left[\frac{1}{3} y^3 \right]_{-3}^3 = \frac{1}{2} (\ln 2 - \ln 1) \cdot \frac{1}{3} (27 + 27) = 9 \ln 2$

17. $\int_0^{\pi/6} \int_0^{\pi/3} x \sin(x + y)\, dy\, dx$

$= \int_0^{\pi/6} \left[-x \cos(x + y) \right]_{y=0}^{y=\pi/3} dx = \int_0^{\pi/6} \left[x \cos x - x \cos\left(x + \frac{\pi}{3} \right) \right] dx$

$= x \left[\sin x - \sin\left(x + \frac{\pi}{3} \right) \right]_0^{\pi/6} - \int_0^{\pi/6} \left[\sin x - \sin\left(x + \frac{\pi}{3} \right) \right] dx$

[by integrating by parts separately for each term]

$= \frac{\pi}{6} \left[\frac{1}{2} - 1 \right] - \left[-\cos x + \cos\left(x + \frac{\pi}{3} \right) \right]_0^{\pi/6} = -\frac{\pi}{12} - \left[-\frac{\sqrt{3}}{2} + 0 - \left(-1 + \frac{1}{2} \right) \right]$

$= \frac{\sqrt{3} - 1}{2} - \frac{\pi}{12}$

19. $\iint_R xy e^{x^2 y}\, dA = \int_0^2 \int_0^1 xy e^{x^2 y}\, dx\, dy = \int_0^2 \left[\frac{1}{2} e^{x^2 y} \right]_{x=0}^{x=1} dy = \frac{1}{2} \int_0^2 (e^y - 1)\, dy$

$= \frac{1}{2} \left[e^y - y \right]_0^2 = \frac{1}{2} [(e^2 - 2) - (1 - 0)] = \frac{1}{2} (e^2 - 3)$

21. $z = f(x, y) = 4 - x - 2y \geq 0$ for $0 \leq x \leq 1$ and $0 \leq y \leq 1$.

So the solid is the region in the first octant which lies below the

plane $z = 4 - x - 2y$ and above $[0, 1] \times [0, 1]$.

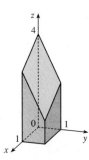

23. $V = \iint_R (12 - 3x - 2y)\, dA = \int_{-2}^3 \int_0^1 (12 - 3x - 2y)\, dx\, dy = \int_{-2}^3 \left[12x - \frac{3}{2} x^2 - 2xy \right]_{x=0}^{x=1} dy$

$= \int_{-2}^3 \left(\frac{21}{2} - 2y \right) dy = \left[\frac{21}{2} y - y^2 \right]_{-2}^3 = \frac{95}{2}$

25. $V = \int_{-2}^2 \int_{-1}^1 \left(1 - \frac{1}{4} x^2 - \frac{1}{9} y^2 \right) dx\, dy = 4 \int_0^2 \int_0^1 \left(1 - \frac{1}{4} x^2 - \frac{1}{9} y^2 \right) dx\, dy$

$= 4 \int_0^2 \left[x - \frac{1}{12} x^3 - \frac{1}{9} y^2 x \right]_{x=0}^{x=1} dy = 4 \int_0^2 \left(\frac{11}{12} - \frac{1}{9} y^2 \right) dy = 4 \left[\frac{11}{12} y - \frac{1}{27} y^3 \right]_0^2 = 4 \cdot \frac{83}{54} = \frac{166}{27}$

27. Here we need the volume of the solid lying under the surface $z = x\sqrt{x^2 + y}$ and above the square $R = [0, 1] \times [0, 1]$ in the xy-plane.

$$V \int_0^1 \int_0^1 x\sqrt{x^2 + y}\, dx\, dy = \int_0^1 \frac{1}{3}\left[(x^2 + y)^{3/2}\right]_{x=0}^{x=1} dy = \frac{1}{3}\int_0^1 \left[(1+y)^{3/2} - y^{3/2}\right] dy$$

$$= \frac{1}{3}\cdot\frac{2}{5}\left[(1+y)^{5/2} - y^{5/2}\right]_0^1 = \frac{4}{15}\left(2\sqrt{2} - 1\right)$$

29. In the first octant, $z \geq 0 \;\Rightarrow\; y \leq 3$, so

$$V = \int_0^3 \int_0^2 (9 - y^2)\, dx\, dy = \int_0^3 \left[9x - y^2 x\right]_{x=0}^{x=2} dy = \int_0^3 (18 - 2y^2)\, dy = \left[18y - \frac{2}{3}y^3\right]_0^3 = 36$$

31. In Maple, we can calculate the integral by defining the integrand as f and then using the command `int(int(f,x=0..1),y=0..1);`. In Mathematica, we can use the command `Integrate[Integrate[f,{x,0,1}],{y,0,1}]`. We find that $\iint_R x^5 y^3 e^{xy}\, dA = 21e - 57 \approx 0.0839$. We can use `plot3d` (in Maple) or `Plot3d` (in Mathematica) to graph the function.

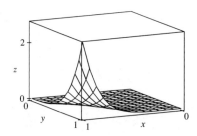

33. R is the rectangle $[-1, 1] \times [0, 5]$. Thus, $A(R) = 2 \cdot 5 = 10$ and

$$f_{ave} = \frac{1}{A(R)}\iint_R f(x, y)\, dA = \frac{1}{10}\int_0^5 \int_{-1}^1 x^2 y\, dx\, dy = \frac{1}{10}\int_0^5 \left[\frac{1}{3}x^3 y\right]_{x=-1}^{x=1} dy = \frac{1}{10}\int_0^5 \frac{2}{3}y\, dy$$

$$= \frac{1}{10}\left[\frac{1}{3}y^2\right]_0^5 = \frac{5}{6}$$

35. Let $f(x, y) = \dfrac{x - y}{(x + y)^3}$. Then a CAS gives $\int_0^1 \int_0^1 f(x, y)\, dy\, dx = \frac{1}{2}$ and $\int_0^1 \int_0^1 f(x, y)\, dx\, dy = -\frac{1}{2}$.

To explain the seeming violation of Fubini's Theorem, note that f has an infinite discontinuity at $(0, 0)$ and thus does not satisfy the conditions of Fubini's Theorem. In fact, both iterated integrals involve improper integrals which diverge at their lower limits of integration.

16.3 Double Integrals over General Regions   ET 15.3

1. $\int_0^1 \int_0^{x^2} (x + 2y)\, dy\, dx = \int_0^1 \left[xy + y^2\right]_{y=0}^{y=x^2} dx = \int_0^1 \left[x(x^2) + (x^2)^2 - 0 - 0\right] dx$

$$= \int_0^1 (x^3 + x^4)\, dx = \left[\frac{1}{4}x^4 + \frac{1}{5}x^5\right]_0^1 = \frac{9}{20}$$

3. $\int_0^1 \int_y^{e^y} \sqrt{x}\, dx\, dy = \int_0^1 \left[\frac{2}{3}x^{3/2}\right]_{x=y}^{x=e^y} dy = \frac{2}{3}\int_0^1 (e^{3y/2} - y^{3/2})\, dy = \frac{2}{3}\left[\frac{2}{3}e^{3y/2} - \frac{2}{5}y^{5/2}\right]_0^1$

$$= \frac{2}{3}\left(\frac{2}{3}e^{3/2} - \frac{2}{5} - \frac{2}{3}e^0 + 0\right) = \frac{4}{9}e^{3/2} - \frac{32}{45}$$

5. $\int_0^{\pi/2} \int_0^{\cos\theta} e^{\sin\theta}\, dr\, d\theta = \int_0^{\pi/2} \left[re^{\sin\theta}\right]_{r=0}^{r=\cos\theta} d\theta = \int_0^{\pi/2} (\cos\theta)\, e^{\sin\theta}\, d\theta = e^{\sin\theta}\Big]_0^{\pi/2}$

$$= e^{\sin(\pi/2)} - e^0 = e - 1$$

7. $\iint_D x^3 y^2\, dA = \int_0^2 \int_{-x}^x x^3 y^2\, dy\, dx = \int_0^2 \left[\frac{1}{3}x^3 y^3\right]_{y=-x}^{y=x} dx = \frac{1}{3}\int_0^2 2x^6\, dx$

$$= \frac{2}{3}\left[\frac{1}{7}x^7\right]_0^2 = \frac{2}{21}\left[2^7 - 0\right] = \frac{256}{21}$$

9. $\int_0^1 \int_0^{\sqrt{x}} \dfrac{2y}{x^2+1} \, dy \, dx = \int_0^1 \left[\dfrac{y^2}{x^2+1} \right]_{y=0}^{y=\sqrt{x}} dx = \int_0^1 \dfrac{x}{x^2+1} \, dx$

$$= \tfrac{1}{2} \ln \left| x^2 + 1 \right| \Big]_0^1 = \tfrac{1}{2}(\ln 2 - \ln 1) = \tfrac{1}{2} \ln 2$$

11. $\int_1^2 \int_y^{y^3} e^{x/y} \, dx \, dy = \int_1^2 \left[y e^{x/y} \right]_{x=y}^{x=y^3} dy = \int_1^2 \left(y e^{y^2} - ey \right) dy = \left[\tfrac{1}{2} e^{y^2} - \tfrac{1}{2} e y^2 \right]_1^2 = \tfrac{1}{2}(e^4 - 4e)$

13. $\int_0^1 \int_0^{x^2} x \cos y \, dy \, dx = \int_0^1 \left[x \sin y \right]_{y=0}^{y=x^2} dx = \int_0^1 x \sin x^2 \, dx = -\tfrac{1}{2} \cos x^2 \big]_0^1 = \tfrac{1}{2}(1 - \cos 1)$

15.

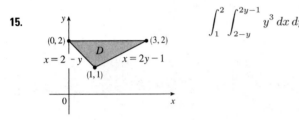

$\displaystyle \int_1^2 \int_{2-y}^{2y-1} y^3 \, dx \, dy = \int_1^2 \left[xy^3 \right]_{x=2-y}^{x=2y-1} dy$

$$= \int_1^2 \left[(2y-1) - (2-y) \right] y^3 \, dy$$

$$= \int_1^2 (3y^4 - 3y^3) \, dy = \left[\tfrac{3}{5} y^5 - \tfrac{3}{4} y^4 \right]_1^2$$

$$= \tfrac{96}{5} - 12 - \tfrac{3}{5} + \tfrac{3}{4} = \tfrac{147}{20}$$

17.

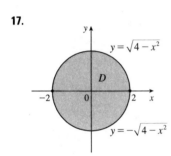

$\displaystyle \int_{-2}^2 \int_{-\sqrt{4-x^2}}^{\sqrt{4-x^2}} (2x-y) \, dy \, dx$

$$= \int_{-2}^2 \left[2xy - \tfrac{1}{2} y^2 \right]_{y=-\sqrt{4-x^2}}^{y=\sqrt{4-x^2}} dx$$

$$= \int_{-2}^2 \left[2x\sqrt{4-x^2} - \tfrac{1}{2}(4-x^2) + 2x\sqrt{4-x^2} + \tfrac{1}{2}(4-x^2) \right] dx$$

$$= \int_{-2}^2 4x\sqrt{4-x^2} \, dx = -\tfrac{4}{3}(4-x^2)^{3/2} \Big]_{-2}^2 = 0$$

(Or, note that $4x\sqrt{4-x^2}$ is an odd function, so $\int_{-2}^2 4x\sqrt{4-x^2} \, dx = 0$.)

19.

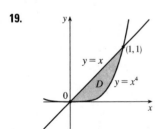

$V = \int_0^1 \int_{x^4}^x (x+2y) \, dy \, dx$

$$= \int_0^1 \left[xy + y^2 \right]_{y=x^4}^{y=x} dx = \int_0^1 (2x^2 - x^5 - x^8) \, dx$$

$$= \left[\tfrac{2}{3} x^3 - \tfrac{1}{6} x^6 - \tfrac{1}{9} x^9 \right]_0^1 = \tfrac{2}{3} - \tfrac{1}{6} - \tfrac{1}{9} = \tfrac{7}{18}$$

21.

$V = \int_1^2 \int_1^{7-3y} xy \, dx \, dy = \int_1^2 \left[\tfrac{1}{2} x^2 y \right]_{x=1}^{x=7-3y} dy$

$$= \tfrac{1}{2} \int_1^2 (48y - 42y^2 + 9y^3) \, dy$$

$$= \tfrac{1}{2} \left[24y^2 - 14y^3 + \tfrac{9}{4} y^4 \right]_1^2 = \tfrac{31}{8}$$

23.

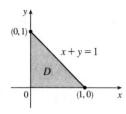

$V = \int_0^1 \int_0^{1-x} (1 - x - y)\, dy\, dx$

$= \int_0^1 \left[y - xy - \frac{y^2}{2} \right]_{y=0}^{y=1-x} dx$

$= \int_0^1 \left[(1-x)^2 - \frac{1}{2}(1-x)^2 \right] dx$

$= \int_0^1 \frac{1}{2}(1-x)^2\, dx = \left[-\frac{1}{6}(1-x)^3 \right]_0^1 = \frac{1}{6}$

25.

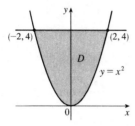

$V = \int_{-2}^2 \int_{x^2}^4 x^2\, dy\, dx$

$= \int_{-2}^2 x^2 \left[y \right]_{y=x^2}^{y=4} dx = \int_{-2}^2 (4x^2 - x^4)\, dx$

$= \left[\frac{4}{3}x^3 - \frac{1}{5}x^5 \right]_{-2}^2 = \frac{32}{3} - \frac{32}{5} + \frac{32}{3} - \frac{32}{5} = \frac{128}{15}$

27.

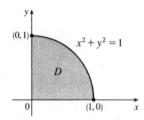

$V = \int_0^1 \int_0^{\sqrt{1-x^2}} y\, dy\, dx = \int_0^1 \left[\frac{y^2}{2} \right]_{y=0}^{y=\sqrt{1-x^2}} dx$

$= \int_0^1 \frac{1-x^2}{2}\, dx = \frac{1}{2} \left[x - \frac{1}{3}x^3 \right]_0^1 = \frac{1}{3}$

29.

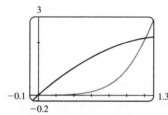

From the graph, it appears that the two curves intersect at $x = 0$ and at $x \approx 1.213$. Thus the desired integral is

$\iint_D x\, dA \approx \int_0^{1.213} \int_{x^4}^{3x - x^2} x\, dy\, dx = \int_0^{1.213} \left[xy \right]_{y=x^4}^{y=3x-x^2} dx$

$= \int_0^{1.213} (3x^2 - x^3 - x^5)\, dx = \left[x^3 - \frac{1}{4}x^4 - \frac{1}{6}x^6 \right]_0^{1.213}$

≈ 0.713

31. The two bounding curves $y = 1 - x^2$ and $y = x^2 - 1$ intersect at $(\pm 1, 0)$ with $1 - x^2 \geq x^2 - 1$ on $[-1, 1]$. Within this region, the plane $z = 2x + 2y + 10$ is above the plane $z = 2 - x - y$, so

$V = \int_{-1}^1 \int_{x^2-1}^{1-x^2} (2x + 2y + 10)\, dy\, dx - \int_{-1}^1 \int_{x^2-1}^{1-x^2} (2 - x - y)\, dy\, dx$

$= \int_{-1}^1 \int_{x^2-1}^{1-x^2} (2x + 2y + 10 - (2 - x - y))\, dy\, dx = \int_{-1}^1 \int_{x^2-1}^{1-x^2} (3x + 3y + 8)\, dy\, dx$

$= \int_{-1}^1 \left[3xy + \frac{3}{2}y^2 + 8y \right]_{y=x^2-1}^{y=1-x^2} dx$

$= \int_{-1}^1 \left[3x(1 - x^2) + \frac{3}{2}(1 - x^2)^2 + 8(1 - x^2) - 3x(x^2 - 1) - \frac{3}{2}(x^2 - 1)^2 - 8(x^2 - 1) \right] dx$

$= \int_{-1}^1 (-6x^3 - 16x^2 + 6x + 16)\, dx = \left[-\frac{3}{2}x^4 - \frac{16}{3}x^3 + 3x^2 + 16x \right]_{-1}^1$

$= -\frac{3}{2} - \frac{16}{3} + 3 + 16 + \frac{3}{2} - \frac{16}{3} - 3 + 16 = \frac{64}{3}$

33. The two bounding curves $y = x^3 - x$ and $y = x^2 + x$ intersect at the origin and at $x = 2$, with $x^2 + x > x^3 - x$ on $(0, 2)$. Using a CAS, we find that the volume is

$$V = \int_0^2 \int_{x^3 - x}^{x^2 + x} z \, dy \, dx = \int_0^2 \int_{x^3 - x}^{x^2 + x} (x^3 y^4 + xy^2) \, dy \, dx = \frac{13{,}984{,}735{,}616}{14{,}549{,}535}$$

35. The two surfaces intersect in the circle $x^2 + y^2 = 1$, $z = 0$ and the region of integration is the disk D: $x^2 + y^2 \le 1$. Using a CAS, the volume is $\iint_D (1 - x^2 - y^2) \, dA = \int_{-1}^1 \int_{-\sqrt{1-x^2}}^{\sqrt{1-x^2}} (1 - x^2 - y^2) \, dy \, dx = \frac{\pi}{2}$.

37.

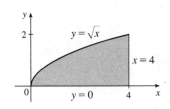

Because the region of integration is

$$D = \{(x,y) \mid 0 \le y \le \sqrt{x}, 0 \le x \le 4\}$$
$$= \{(x,y) \mid y^2 \le x \le 4, 0 \le y \le 2\}$$

we have

$\int_0^4 \int_0^{\sqrt{x}} f(x,y) \, dy \, dx = \iint_D f(x,y) \, dA = \int_0^2 \int_{y^2}^4 f(x,y) \, dx \, dy$.

39.

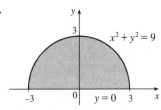

Because the region of integration is

$$D = \left\{ (x,y) \mid -\sqrt{9 - y^2} \le x \le \sqrt{9 - y^2}, 0 \le y \le 3 \right\}$$
$$= \{(x,y) \mid 0 \le y \le \sqrt{9 - x^2}, -3 \le x \le 3\}$$

we have

$$\int_0^3 \int_{-\sqrt{9-y^2}}^{\sqrt{9-y^2}} f(x,y) \, dx \, dy = \iint_D f(x,y) \, dA$$
$$= \int_{-3}^3 \int_0^{\sqrt{9-x^2}} f(x,y) \, dy \, dx$$

41.

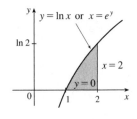

Because the region of integration is

$$D = \{(x,y) \mid 0 \le y \le \ln x, 1 \le x \le 2\}$$
$$= \{(x,y) \mid e^y \le x \le 2, 0 \le y \le \ln 2\}$$

we have

$$\int_1^2 \int_0^{\ln x} f(x,y) \, dy \, dx = \iint_D f(x,y) \, dA$$
$$= \int_0^{\ln 2} \int_{e^y}^2 f(x,y) \, dx \, dy$$

43.

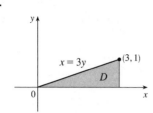

$$\int_0^1 \int_{3y}^3 e^{x^2} \, dx \, dy = \int_0^3 \int_0^{x/3} e^{x^2} \, dy \, dx$$

$$= \int_0^3 \left[e^{x^2} y \right]_{y=0}^{y=x/3} \, dx = \int_0^3 \left(\frac{x}{3} \right) e^{x^2} \, dx$$

$$= \frac{1}{6} \, e^{x^2} \Big]_0^3 = \frac{e^9 - 1}{6}$$

45.

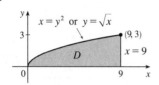

$$\int_0^3 \int_{y^2}^9 y \cos x^2 \, dx \, dy = \int_0^9 \int_0^{\sqrt{x}} y \cos x^2 \, dy \, dx$$

$$= \int_0^9 \cos x^2 \left[\frac{y^2}{2} \right]_{y=0}^{y=\sqrt{x}} \, dx = \int_0^9 \frac{1}{2} x \cos x^2 \, dx$$

$$= \frac{1}{4} \sin x^2 \Big]_0^9 = \frac{1}{4} \sin 81$$

47.

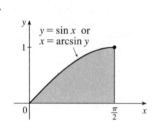

$$\int_0^1 \int_{\arcsin y}^{\pi/2} \cos x \sqrt{1 + \cos^2 x} \, dx \, dy$$

$$= \int_0^{\pi/2} \int_0^{\sin x} \cos x \sqrt{1 + \cos^2 x} \, dy \, dx$$

$$= \int_0^{\pi/2} \cos x \sqrt{1 + \cos^2 x} \left[y \right]_{y=0}^{y=\sin x} \, dx$$

$$= \int_0^{\pi/2} \cos x \sqrt{1 + \cos^2 x} \sin x \, dx$$

$$[\text{Let } u = \cos x, \, du = -\sin x \, dx, \, dx = du/(-\sin x)]$$

$$= \int_1^0 -u \sqrt{1 + u^2} \, du = -\frac{1}{3} \left(1 + u^2 \right)^{3/2} \Big]_1^0$$

$$= \frac{1}{3} \left(\sqrt{8} - 1 \right) = \frac{1}{3} \left(2\sqrt{2} - 1 \right)$$

49. $D = \{(x,y) \mid 0 \le x \le 1, \, -x + 1 \le y \le 1\} \cup \{(x,y) \mid -1 \le x \le 0, \, x + 1 \le y \le 1\}$

$$\cup \{(x,y) \mid 0 \le x \le 1, \, -1 \le y \le x - 1\} \cup \{(x,y) \mid -1 \le x \le 0, \, -1 \le y \le -x - 1\},$$

all type I.

$$\iint_D x^2 \, dA = \int_0^1 \int_{1-x}^1 x^2 \, dy \, dx + \int_{-1}^0 \int_{x+1}^1 x^2 \, dy \, dx + \int_0^1 \int_{-1}^{x-1} x^2 \, dy \, dx + \int_{-1}^0 \int_{-1}^{-x-1} x^2 \, dy \, dx$$

$$= 4 \int_0^1 \int_{1-x}^1 x^2 \, dy \, dx \qquad [\text{by symmetry of the regions and because } f(x,y) = x^2 \ge 0]$$

$$= 4 \int_0^1 x^3 \, dx = 4 \left[\frac{1}{4} x^4 \right]_0^1 = 1$$

51. For $D = [0, 1] \times [0, 1]$, $0 \le \sqrt{x^3 + y^3} \le \sqrt{2}$ and $A(D) = 1$, so $0 \le \iint_D \sqrt{x^3 + y^3} \, dA \le \sqrt{2}$.

53. Since $m \le f(x,y) \le M$, $\iint_D m \, dA \le \iint_D f(x,y) \, dA \le \iint_D M \, dA$ by (8) ⇒
$m \iint_D 1 \, dA \le \iint_D f(x,y) \, dA \le M \iint_D 1 \, dA$ by (7) ⇒ $mA(D) \le \iint_D f(x,y) \, dA \le MA(D)$ by (10).

55. $\iint_D (x^2 \tan x + y^3 + 4)\, dA = \iint_D x^2 \tan x\, dA + \iint_D y^3\, dA + \iint_D 4\, dA$. But $x^2 \tan x$ is an odd function of x and D is symmetric with respect to the y-axis, so $\iint_D x^2 \tan x\, dA = 0$. Similarly, y^3 is an odd function of y and D is symmetric with respect to the x-axis, so $\iint_D y^3\, dA = 0$. Thus

$$\iint_D (x^2 \tan x + y^3 + 4)\, dA = 4 \iint_D dA = 4(\text{area of } D) = 4 \cdot \pi \left(\sqrt{2}\right)^2 = 8\pi$$

57. Since $\sqrt{1 - x^2 - y^2} \ge 0$, we can interpret $\iint_D \sqrt{1 - x^2 - y^2}\, dA$ as the volume of the solid that lies below the graph of $z = \sqrt{1 - x^2 - y^2}$ and above the region D in the xy-plane. $z = \sqrt{1 - x^2 - y^2}$ is equivalent to $x^2 + y^2 + z^2 = 1$, $z \ge 0$ which meets the xy-plane in the circle $x^2 + y^2 = 1$, the boundary of D. Thus, the solid is an upper hemisphere of radius 1 which has volume $\frac{1}{2}\left[\frac{4}{3}\pi(1)^3\right] = \frac{2}{3}\pi$.

16.4 Double Integrals in Polar Coordinates ET 15.4

1. The region R is more easily described by polar coordinates: $R = \{(r, \theta) \mid 0 \le r \le 2, 0 \le \theta \le 2\pi\}$.
Thus $\iint_R f(x, y)\, dA = \int_0^{2\pi} \int_0^2 f(r\cos\theta, r\sin\theta)\, r\, dr\, d\theta$.

3. The region R is more easily described by rectangular coordinates: $R = \{(x, y) \mid -2 \le x \le 2, x \le y \le 2\}$.
Thus $\iint_R f(x, y)\, dA = \int_{-2}^2 \int_x^2 f(x, y)\, dy\, dx$.

5. The region R is more easily described by polar coordinates: $R = \{(r, \theta) \mid 2 \le r \le 5, 0 \le \theta \le 2\pi\}$.
Thus $\iint_R f(x, y)\, dA = \int_0^{2\pi} \int_2^5 f(r\cos\theta, r\sin\theta)\, r\, dr\, d\theta$.

7. The integral $\int_\pi^{2\pi} \int_4^7 r\, dr\, d\theta$ represents the area of the region
$R = \{(r, \theta) \mid 4 \le r \le 7, \pi \le \theta \le 2\pi\}$, the lower half of a ring.

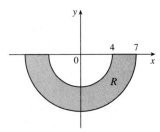

$$\int_\pi^{2\pi} \int_4^7 r\, dr\, d\theta = \left(\int_\pi^{2\pi} d\theta\right)\left(\int_4^7 r\, dr\right)$$

$$= \left[\theta\right]_\pi^{2\pi} \left[\tfrac{1}{2}r^2\right]_4^7 = \pi \cdot \frac{1}{2}(49 - 16) = \frac{33\pi}{2}$$

9. The disk D can be described in polar coordinates as $D = \{(r, \theta) \mid 0 \le r \le 3, 0 \le \theta \le 2\pi\}$. Then

$$\iint_D xy\, dA = \int_0^{2\pi} \int_0^3 (r\cos\theta)(r\sin\theta)\, r\, dr\, d\theta = \left(\int_0^{2\pi} \sin\theta\cos\theta\, d\theta\right)\left(\int_0^3 r^3\, dr\right)$$

$$= \left[\tfrac{1}{2}\sin^2\theta\right]_0^{2\pi} \left[\tfrac{1}{4}r^4\right]_0^3 = 0$$

11. $\iint_R \cos(x^2 + y^2)\, dA = \int_0^\pi \int_0^3 \cos(r^2)\, r\, dr\, d\theta = \left(\int_0^\pi d\theta\right)\left(\int_0^3 r\cos(r^2)\, dr\right)$

$$= \left[\theta\right]_0^\pi \left[\tfrac{1}{2}\sin(r^2)\right]_0^3 = \pi \cdot \tfrac{1}{2}(\sin 9 - \sin 0) = \tfrac{\pi}{2}\sin 9$$

13. $\iint_D e^{-x^2 - y^2}\, dA = \int_{-\pi/2}^{\pi/2} \int_0^2 e^{-r^2}\, r\, dr\, d\theta = \left(\int_{-\pi/2}^{\pi/2} d\theta\right)\left(\int_0^2 re^{-r^2}\, dr\right)$

$$= \left[\theta\right]_{-\pi/2}^{\pi/2} \left[-\tfrac{1}{2}e^{-r^2}\right]_0^2 = \pi\left(-\tfrac{1}{2}\right)(e^{-4} - e^0) = \tfrac{\pi}{2}(1 - e^{-4})$$

15. R is the region shown in the figure, and can be described

by $R = \{(r, \theta) \mid 0 \le \theta \le \pi/4,\, 1 \le r \le 2\}$. Thus

$$\iint_R \arctan(y/x)\, dA = \int_0^{\pi/4} \int_1^2 \arctan(\tan \theta)\, r\, dr\, d\theta$$

since $y/x = \tan \theta$. Also, $\arctan(\tan \theta) = \theta$ for $0 \le \theta \le \pi/4$,

so the integral becomes

$$\int_0^{\pi/4} \int_1^2 \theta\, r\, dr\, d\theta = \int_0^{\pi/4} \theta\, d\theta \int_1^2 r\, dr = \left[\tfrac{1}{2}\theta^2\right]_0^{\pi/4} \left[\tfrac{1}{2}r^2\right]_1^2 = \tfrac{\pi^2}{32} \cdot \tfrac{3}{2} = \tfrac{3}{64}\pi^2.$$

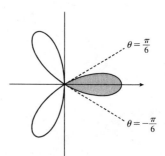

17. One loop is given by the region

$$D = \{(r, \theta)\mid -\pi/6 \le \theta \le \pi/6,\, 0 \le r \le \cos 3\theta\}, \text{ so the area is}$$

$$\iint_D dA = \int_{-\pi/6}^{\pi/6} \int_0^{\cos 3\theta} r\, dr\, d\theta = \int_{-\pi/6}^{\pi/6} \left[\frac{1}{2}r^2\right]_{r=0}^{r=\cos 3\theta} d\theta$$

$$= \int_{-\pi/6}^{\pi/6} \frac{1}{2}\cos^2 3\theta\, d\theta = 2\int_0^{\pi/6} \frac{1}{2}\left(\frac{1+\cos 6\theta}{2}\right) d\theta$$

$$= \frac{1}{2}\left[\theta + \frac{1}{6}\sin 6\theta\right]_0^{\pi/6} = \frac{\pi}{12}$$

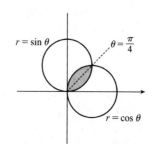

19. By symmetry,

$$A = 2\int_0^{\pi/4} \int_0^{\sin \theta} r\, dr\, d\theta = 2\int_0^{\pi/4} \left[\tfrac{1}{2}r^2\right]_{r=0}^{r=\sin \theta} d\theta$$

$$= \int_0^{\pi/4} \sin^2 \theta\, d\theta = \int_0^{\pi/4} \tfrac{1}{2}(1 - \cos 2\theta)\, d\theta$$

$$= \tfrac{1}{2}\left[\theta - \tfrac{1}{2}\sin 2\theta\right]_0^{\pi/4}$$

$$= \tfrac{1}{2}\left[\tfrac{\pi}{4} - \tfrac{1}{2}\sin \tfrac{\pi}{2} - 0 + \tfrac{1}{2}\sin 0\right] = \tfrac{1}{8}(\pi - 2)$$

21. $V = \iint\limits_{x^2+y^2 \le 9} (x^2+y^2)\, dA = \int_0^{2\pi} \int_0^3 (r^2)\, r\, dr\, d\theta = \int_0^{2\pi} d\theta \int_0^3 r^3\, dr = \left[\theta\right]_0^{2\pi} \left[\tfrac{1}{4}r^4\right]_0^3 = 2\pi\left(\tfrac{81}{4}\right) = \tfrac{81\pi}{2}$

23. By symmetry,

$$V = 2 \iint\limits_{x^2+y^2 \le a^2} \sqrt{a^2-x^2-y^2}\, dA = 2\int_0^{2\pi}\int_0^a \sqrt{a^2-r^2}\, r\, dr\, d\theta = 2\int_0^{2\pi} d\theta \int_0^a r\sqrt{a^2-r^2}\, dr$$

$$= 2\left[\theta\right]_0^{2\pi} \left[-\tfrac{1}{3}(a^2-r^2)^{3/2}\right]_0^a = 2(2\pi)\left(0 + \tfrac{1}{3}a^3\right) = \tfrac{4\pi}{3}a^3$$

25. The cone $z = \sqrt{x^2+y^2}$ intersects the sphere $x^2+y^2+z^2 = 1$ when $x^2+y^2+\left(\sqrt{x^2+y^2}\right)^2 = 1$

or $x^2+y^2 = \tfrac{1}{2}$. So

$$V = \iint\limits_{x^2+y^2 \le 1/2} \left(\sqrt{1-x^2-y^2} - \sqrt{x^2+y^2}\right) dA = \int_0^{2\pi} \int_0^{1/\sqrt{2}} \left(\sqrt{1-r^2} - r\right) r\, dr\, d\theta$$

$$= \int_0^{2\pi} d\theta \int_0^{1/\sqrt{2}} \left(r\sqrt{1-r^2} - r^2\right) dr = \left[\theta\right]_0^{2\pi} \left[-\tfrac{1}{3}(1-r^2)^{3/2} - \tfrac{1}{3}r^3\right]_0^{1/\sqrt{2}}$$

$$= 2\pi\left(-\tfrac{1}{3}\right)\left(\tfrac{1}{\sqrt{2}} - 1\right) = \tfrac{\pi}{3}\left(2 - \sqrt{2}\right)$$

27. The given solid is the region inside the cylinder $x^2 + y^2 = 4$ between the surfaces $z = \sqrt{64 - 4x^2 - 4y^2}$

and $z = -\sqrt{64 - 4x^2 - 4y^2}$. So

$$V = \iint\limits_{x^2+y^2 \le 4} \left[\sqrt{64 - 4x^2 - 4y^2} - \left(-\sqrt{64 - 4x^2 - 4y^2} \right) \right] dA$$

$$= \iint\limits_{x^2+y^2 \le 4} 2\sqrt{64 - 4x^2 - 4y^2} \, dA = 4 \int_0^{2\pi} \int_0^2 \sqrt{16 - r^2} \, r \, dr \, d\theta$$

$$= 4 \int_0^{2\pi} d\theta \int_0^2 r\sqrt{16 - r^2} \, dr = 4 \left[\theta \right]_0^{2\pi} \left[-\tfrac{1}{3}(16 - r^2)^{3/2} \right]_0^2$$

$$= 8\pi \left(-\tfrac{1}{3} \right) (12^{3/2} - 16^{2/3}) = \tfrac{8\pi}{3} \left(64 - 24\sqrt{3} \right)$$

29.

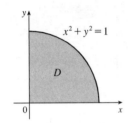

$x^2 + y^2 = 1$

D

$$\int_0^1 \int_0^{\sqrt{1-x^2}} e^{x^2+y^2} \, dy \, dx = \int_0^{\pi/2} \int_0^1 e^{r^2} r \, dr \, d\theta$$

$$= \int_0^{\pi/2} d\theta \int_0^1 r e^{r^2} \, dr$$

$$= \left[\theta \right]_0^{\pi/2} \left[\tfrac{1}{2} e^{r^2} \right]_0^1 = \tfrac{1}{4}\pi(e - 1)$$

31.

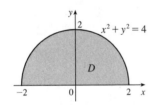

2

$x^2 + y^2 = 4$

D

-2 \quad 0 \quad 2 \quad x

$$\int_0^\pi \int_0^2 (r\cos\theta)^2 (r\sin\theta)^2 r \, dr \, d\theta = \int_0^\pi (\sin\theta\cos\theta)^2 d\theta \int_0^2 r^5 \, dr$$

$$= \int_0^\pi \left(\tfrac{1}{2}\sin 2\theta \right)^2 d\theta \int_0^2 r^5 \, dr$$

$$= \tfrac{1}{4} \left[\tfrac{1}{2}\theta - \tfrac{1}{8}\sin 4\theta \right]_0^\pi \left[\tfrac{1}{6}r^6 \right]_0^2$$

$$= \tfrac{1}{4} \left(\tfrac{\pi}{2} \right) \left(\tfrac{64}{6} \right) = \tfrac{4\pi}{3}$$

33. The surface of the water in the pool is a circular disk D with radius 20 ft. If we place D on coordinate axes with the
origin at the center of D and define $f(x, y)$ to be the depth of the water at (x, y), then the volume of water in the
pool is the volume of the solid that lies above $D = \{(x, y) \mid x^2 + y^2 \le 400\}$ and below the graph of $f(x, y)$. We
can associate north with the positive y-direction, so we are given that the depth is constant in the x-direction and the
depth increases linearly in the y-direction from $f(0, -20) = 2$ to $f(0, 20) = 7$. The trace in the yz-plane is a line
segment from $(0, -20, 2)$ to $(0, 20, 7)$. The slope of this line is $\frac{7-2}{20-(-20)} = \tfrac{1}{8}$, so an equation of the line is
$z - 7 = \tfrac{1}{8}(y - 20)$ $\Rightarrow$ $z = \tfrac{1}{8}y + \tfrac{9}{2}$. Since $f(x, y)$ is independent of x, $f(x, y) = \tfrac{1}{8}y + \tfrac{9}{2}$. Thus the volume is
given by $\iint_D f(x, y) \, dA$, which is most conveniently evaluated using polar coordinates. Then
$D = \{(r, \theta) \mid 0 \le r \le 20, 0 \le \theta \le 2\pi\}$ and substituting $x = r\cos\theta$, $y = r\sin\theta$ the integral becomes

$$\int_0^{2\pi} \int_0^{20} \left(\tfrac{1}{8}r\sin\theta + \tfrac{9}{2} \right) r \, dr \, d\theta = \int_0^{2\pi} \left[\tfrac{1}{24}r^3\sin\theta + \tfrac{9}{4}r^2 \right]_{r=0}^{r=20} d\theta = \int_0^{2\pi} \left(\tfrac{1000}{3}\sin\theta + 900 \right) d\theta$$

$$= \left[-\tfrac{1000}{3}\cos\theta + 900\theta \right]_0^{2\pi} = 1800\pi$$

Thus the pool contains $1800\pi \approx 5655$ ft³ of water.

35. $\displaystyle\int_{1/\sqrt{2}}^{1}\int_{\sqrt{1-x^2}}^{x} xy\,dy\,dx + \int_{1}^{\sqrt{2}}\int_{0}^{x} xy\,dy\,dx + \int_{\sqrt{2}}^{2}\int_{0}^{\sqrt{4-x^2}} xy\,dy\,dx$

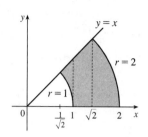

$\displaystyle = \int_{0}^{\pi/4}\int_{1}^{2} r^3\cos\theta\sin\theta\,dr\,d\theta = \int_{0}^{\pi/4}\left[\frac{r^4}{4}\cos\theta\sin\theta\right]_{r=1}^{r=2} d\theta$

$\displaystyle = \frac{15}{4}\int_{0}^{\pi/4}\sin\theta\cos\theta\,d\theta = \frac{15}{4}\left[\frac{\sin^2\theta}{2}\right]_{0}^{\pi/4} = \frac{15}{16}$

37. (a) We integrate by parts with $u = x$ and $dv = xe^{-x^2}\,dx$. Then $du = dx$ and $v = -\frac{1}{2}e^{-x^2}$, so

$\displaystyle\int_{0}^{\infty} x^2 e^{-x^2}\,dx = \lim_{t\to\infty}\int_{0}^{t} x^2 e^{-x^2}\,dx = \lim_{t\to\infty}\left(-\frac{1}{2}xe^{-x^2}\Big]_{0}^{t} + \int_{0}^{t}\frac{1}{2}e^{-x^2}\,dx\right)$

$\displaystyle\qquad = \lim_{t\to\infty}\left(-\frac{1}{2}te^{-t^2}\right) + \frac{1}{2}\int_{0}^{\infty} e^{-x^2}\,dx = 0 + \frac{1}{2}\int_{0}^{\infty} e^{-x^2}\,dx$ [by l'Hospital's Rule]

$\displaystyle\qquad = \frac{1}{4}\int_{-\infty}^{\infty} e^{-x^2}\,dx$ [since e^{-x^2} is an even function]

$\displaystyle\qquad = \frac{1}{4}\sqrt{\pi}$ [by Exercise 36(c)]

(b) Let $u = \sqrt{x}$. Then $u^2 = x \;\Rightarrow\; dx = 2u\,du \;\Rightarrow$

$\displaystyle\int_{0}^{\infty}\sqrt{x}\,e^{-x}\,dx = \lim_{t\to\infty}\int_{0}^{t}\sqrt{x}\,e^{-x}\,dx = \lim_{t\to\infty}\int_{0}^{\sqrt{t}} ue^{-u^2}\,2u\,du = 2\int_{0}^{\infty} u^2 e^{-u^2}\,du$

$\displaystyle\qquad = 2\left(\frac{1}{4}\sqrt{\pi}\right)$ [by part(a)] $= \frac{1}{2}\sqrt{\pi}$

16.5 Applications of Double Integrals   ET 15.5

1. $Q = \iint_D \sigma(x,y)\,dA = \int_1^3\int_0^2 (2xy + y^2)\,dy\,dx = \int_1^3\left[xy^2 + \frac{1}{3}y^3\right]_{y=0}^{y=2} dx$

$\displaystyle\quad = \int_1^3\left(4x + \frac{8}{3}\right) dx = \left[2x^2 + \frac{8}{3}x\right]_1^3 = 16 + \frac{16}{3} = \frac{64}{3}$ C

3. $m = \iint_D \rho(x,y)\,dA = \int_0^2\int_{-1}^1 xy^2\,dy\,dx = \int_0^2 x\,dx \int_{-1}^1 y^2\,dy = \left[\frac{1}{2}x^2\right]_0^2\left[\frac{1}{3}y^3\right]_{-1}^1 = 2\cdot\frac{2}{3} = \frac{4}{3}$,

$\overline{x} = \frac{1}{m}\iint_D x\rho(x,y)\,dA = \frac{3}{4}\int_0^2\int_{-1}^1 x^2 y^2\,dy\,dx = \frac{3}{4}\int_0^2 x^2\,dx\int_{-1}^1 y^2\,dy = \frac{3}{4}\left[\frac{1}{3}x^3\right]_0^2\left[\frac{1}{3}y^3\right]_{-1}^1 = \frac{3}{4}\cdot\frac{8}{3}\cdot\frac{2}{3} = \frac{4}{3}$,

$\overline{y} = \frac{1}{m}\iint_D y\rho(x,y)\,dA = \frac{3}{4}\int_0^2\int_{-1}^1 xy^3\,dy\,dx = \frac{3}{4}\int_0^2 x\,dx\int_{-1}^1 y^3\,dy = \frac{3}{4}\left[\frac{1}{2}x^2\right]_0^2\left[\frac{1}{4}y^4\right]_{-1}^1 = \frac{3}{4}\cdot 2\cdot 0 = 0$.

Hence, $(\overline{x},\overline{y}) = \left(\frac{4}{3}, 0\right)$.

5. $m = \int_0^2\int_{x/2}^{3-x} (x+y)\,dy\,dx = \int_0^2\left[xy + \frac{1}{2}y^2\right]_{y=x/2}^{y=3-x} dx = \int_0^2\left[x(3 - \frac{3}{2}x) + \frac{1}{2}(3-x)^2 - \frac{1}{8}x^2\right] dx$

$\displaystyle\quad = \int_0^2\left(-\frac{9}{8}x^2 + \frac{9}{2}\right) dx = \left[-\frac{9}{8}\left(\frac{1}{3}x^3\right) + \frac{9}{2}x\right]_0^2 = 6,$

$M_y = \int_0^2\int_{x/2}^{3-x} (x^2 + xy)\,dy\,dx = \int_0^2\left[x^2 y + \frac{1}{2}xy^2\right]_{y=x/2}^{y=3-x} dx = \int_0^2\left(\frac{9}{2}x - \frac{9}{8}x^3\right) dx = \frac{9}{2}$, and

$M_x = \int_0^2\int_{x/2}^{3-y} (xy + y^2)\,dy\,dx = \int_0^2\left[\frac{1}{2}xy^2 + \frac{1}{3}y^3\right]_{y=x/2}^{y=3-x} dx = \int_0^2\left(9 - \frac{9}{2}x\right) dx = 9.$

Hence $m = 6$, $(\overline{x},\overline{y}) = \left(\dfrac{M_y}{m}, \dfrac{M_x}{m}\right) = \left(\dfrac{3}{4}, \dfrac{3}{2}\right)$.

7. $m = \int_0^1 \int_0^{e^x} y \, dy \, dx = \int_0^1 \left[\frac{1}{2}y^2\right]_{y=0}^{y=e^x} dx = \frac{1}{2}\int_0^1 e^{2x} \, dx = \frac{1}{4}e^{2x}\big]_0^1 = \frac{1}{4}(e^2 - 1)$,

$M_y = \int_0^1 \int_0^{e^x} xy \, dy \, dx = \frac{1}{2}\int_0^1 xe^{2x} \, dx = \frac{1}{2}\left[\frac{1}{2}xe^{2x} - \frac{1}{4}e^{2x}\right]_0^1 = \frac{1}{8}(e^2 + 1)$, and

$M_x = \int_0^1 \int_0^{e^x} y^2 \, dy \, dx = \int_0^1 \left[\frac{1}{3}y^3\right]_{y=0}^{y=e^x} dx = \frac{1}{3}\int_0^1 e^{3x} \, dx = \frac{1}{3}\left[\frac{1}{3}e^{3x}\right]_0^1 = \frac{1}{9}(e^3 - 1)$.

Hence $m = \frac{1}{4}(e^2 - 1)$, $(\bar{x}, \bar{y}) = \left(\dfrac{\frac{1}{8}(e^2 + 1)}{\frac{1}{4}(e^2 - 1)}, \dfrac{\frac{1}{9}(e^3 - 1)}{\frac{1}{4}(e^2 - 1)}\right) = \left(\dfrac{e^2 + 1}{2(e^2 - 1)}, \dfrac{4(e^3 - 1)}{9(e^2 - 1)}\right)$.

9.

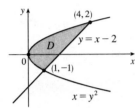

$m = \int_{-1}^{2} \int_{y^2}^{y+2} 3 \, dx \, dy = \int_{-1}^{2}(3y + 6 - 3y^2) \, dy = \frac{27}{2}$,

$M_y = \int_{-1}^{2} \int_{y^2}^{y+2} 3x \, dx \, dy = \int_{-1}^{2} \frac{3}{2}\left[(y+2)^2 - y^4\right] dy$

$= \left[\frac{1}{2}(y + 2)^3 - \frac{3}{10}y^5\right]_{-1}^{2} = \frac{108}{5}$

and

$M_x = \int_{-1}^{2} \int_{y^2}^{y+2} 3y \, dx \, dy = \int_{-1}^{2}(3y^2 + 6y - 3y^3) \, dy$

$= \left[y^3 + 3y^2 - \frac{3}{4}y^4\right]_{-1}^{2} = \frac{27}{4}$

Hence $m = \frac{27}{2}$, $(\bar{x}, \bar{y}) = \left(\frac{8}{5}, \frac{1}{2}\right)$.

11. $\rho(x, y) = ky = kr \sin \theta$, $m = \int_0^{\pi/2} \int_0^1 kr^2 \sin \theta \, dr \, d\theta = \frac{1}{3}k \int_0^{\pi/2} \sin \theta \, d\theta = \frac{1}{3}k\left[-\cos \theta\right]_0^{\pi/2} = \frac{1}{3}k$,

$M_y = \int_0^{\pi/2} \int_0^1 kr^3 \sin \theta \cos \theta \, dr \, d\theta = \frac{1}{4}k \int_0^{\pi/2} \sin \theta \cos \theta \, d\theta = \frac{1}{8}k\left[-\cos 2\theta\right]_0^{\pi/2} = \frac{1}{8}k$,

$M_x = \int_0^{\pi/2} \int_0^1 kr^3 \sin^2 \theta \, dr \, d\theta = \frac{1}{4}k \int_0^{\pi/2} \sin^2 \theta \, d\theta = \frac{1}{8}k\left[\theta + \sin 2\theta\right]_0^{\pi/2} = \frac{\pi}{16}k$.

Hence $(\bar{x}, \bar{y}) = \left(\frac{3}{8}, \frac{3\pi}{16}\right)$.

13. Placing the vertex opposite the hypotenuse at $(0, 0)$, $\rho(x, y) = k(x^2 + y^2)$. Then

$m = \int_0^a \int_0^{a-x} k(x^2 + y^2) \, dy \, dx = k \int_0^a \left[ax^2 - x^3 + \frac{1}{3}(a - x)^3\right] dx$

$= k\left[\frac{1}{3}ax^3 - \frac{1}{4}x^4 - \frac{1}{12}(a - x)^4\right]_0^a = \frac{1}{6}ka^4$

By symmetry,

$M_y = M_x = \int_0^a \int_0^{a-x} ky(x^2 + y^2) \, dy \, dx = k \int_0^a \left[\frac{1}{2}(a - x)^2 x^2 + \frac{1}{4}(a - x)^4\right] dx$

$= k\left[\frac{1}{6}a^2x^3 - \frac{1}{4}ax^4 + \frac{1}{10}x^5 - \frac{1}{20}(a - x)^5\right]_0^a = \frac{1}{15}ka^5$

Hence $(\bar{x}, \bar{y}) = \left(\frac{2}{5}a, \frac{2}{5}a\right)$.

15. $I_x = \iint_D y^2 \rho(x, y) \, dA = \int_0^1 \int_0^{e^x} y^2 \cdot y \, dy \, dx = \int_0^1 \left[\frac{1}{4}y^4\right]_{y=0}^{y=e^x} dx = \frac{1}{4}\int_0^1 e^{4x} \, dx$

$= \frac{1}{4}\left[\frac{1}{4}e^{4x}\right]_0^1 = \frac{1}{16}(e^4 - 1)$,

$I_y = \iint_D x^2 \rho(x, y) \, dA = \int_0^1 \int_0^{e^x} x^2 y \, dy \, dx = \int_0^1 x^2 \left[\frac{1}{2}y^2\right]_{y=0}^{y=e^x} dx = \frac{1}{2}\int_0^1 x^2 e^{2x} \, dx$

$= \frac{1}{2}\left[\left(\frac{1}{2}x^2 - \frac{1}{2}x + \frac{1}{4}\right)e^{2x}\right]_0^1$ [integrate by parts twice]

$= \frac{1}{8}(e^2 - 1)$,

and $I_0 = I_x + I_y = \frac{1}{16}(e^4 - 1) + \frac{1}{8}(e^2 - 1) = \frac{1}{16}(e^4 + 2e^2 - 3)$.

17. $I_x = \int_{-1}^{2} \int_{y^2}^{y+2} 3y^2 \, dx \, dy = \int_{-1}^{2} (3y^3 + 6y^2 - 3y^4) \, dy = \left[\frac{3}{4} y^4 + 2y^3 - \frac{3}{5} y^5 \right]_{-1}^{2} = \frac{189}{20}$,

$I_y = \int_{-1}^{2} \int_{y^2}^{y+2} 3x^2 \, dx \, dy = \int_{-1}^{2} \left[(y+2)^3 - y^6 \right] dy = \left[\frac{1}{4} (y+2)^4 - \frac{1}{7} y^7 \right]_{-1}^{2} = \frac{1269}{28}$, and

$I_0 = I_x + I_y = \frac{1917}{35}$.

19. Using a CAS, we find $m = \iint_D \rho(x, y) \, dA = \int_0^{\pi} \int_0^{\sin x} xy \, dy \, dx = \frac{\pi^2}{8}$. Then

$\bar{x} = \frac{1}{m} \iint_D x\rho(x, y) \, dA = \frac{8}{\pi^2} \int_0^{\pi} \int_0^{\sin x} x^2 y \, dy \, dx = \frac{2\pi}{3} - \frac{1}{\pi}$ and

$\bar{y} = \frac{1}{m} \iint_D y\rho(x, y) \, dA = \frac{8}{\pi^2} \int_0^{\pi} \int_0^{\sin x} xy^2 \, dy \, dx = \frac{16}{9\pi}$, so $(\bar{x}, \bar{y}) = \left(\frac{2\pi}{3} - \frac{1}{\pi}, \frac{16}{9\pi} \right)$.

The moments of inertia are $I_x = \iint_D y^2 \rho(x, y) \, dA = \int_0^{\pi} \int_0^{\sin x} xy^3 \, dy \, dx = \frac{3\pi^2}{64}$,

$I_y = \iint_D x^2 \rho(x, y) \, dA = \int_0^{\pi} \int_0^{\sin x} x^3 y \, dy \, dx = \frac{\pi^2}{16} (\pi^2 - 3)$, and $I_0 = I_x + I_y = \frac{\pi^2}{64} (4\pi^2 - 9)$.

21. $I_x = \int_0^a \int_0^a \rho y^2 \, dx \, dy = \rho \int_0^a dx \int_0^a y^2 \, dy = \rho \left[x \right]_0^a \left[\frac{1}{3} y^3 \right]_0^a = \rho a \left(\frac{1}{3} a^3 \right) = \frac{1}{3} \rho a^4 = I_y$ by symmetry, and

$m = \rho a^2$ since the lamina is homogeneous. Hence $\overline{\overline{x}}^2 = \frac{I_y}{m} \Rightarrow \overline{\overline{x}} = \left[\left(\frac{1}{3} \rho a^4 \right) / (\rho a^2) \right]^{1/2} = \frac{1}{\sqrt{3}} a$ and

$\overline{\overline{y}}^2 = \frac{I_x}{m} \Rightarrow \overline{\overline{y}} = \frac{1}{\sqrt{3}} a$.

23. (a) $f(x, y)$ is a joint density function, so we know $\iint_{\mathbb{R}^2} f(x, y) \, dA = 1$. Since $f(x, y) = 0$ outside the rectangle
$[0, 1] \times [0, 2]$, we can say

$$\iint_{\mathbb{R}^2} f(x, y) \, dA = \int_{-\infty}^{\infty} \int_{-\infty}^{\infty} f(x, y) \, dy \, dx = \int_0^1 \int_0^2 Cx(1 + y) \, dy \, dx$$

$$= C \int_0^1 x \left[y + \frac{1}{2} y^2 \right]_{y=0}^{y=2} dx = C \int_0^1 4x \, dx = C \left[2x^2 \right]_0^1 = 2C$$

Then $2C = 1 \Rightarrow C = \frac{1}{2}$.

(b) $P(X \le 1, Y \le 1) = \int_{-\infty}^{1} \int_{-\infty}^{1} f(x, y) \, dy \, dx = \int_0^1 \int_0^1 \frac{1}{2} x(1 + y) \, dy \, dx$

$$= \int_0^1 \frac{1}{2} x \left[y + \frac{1}{2} y^2 \right]_{y=0}^{y=1} dx = \int_0^1 \frac{1}{2} x \left(\frac{3}{2} \right) dx = \frac{3}{4} \left[\frac{1}{2} x^2 \right]_0^1 = \frac{3}{8} \text{ or } 0.375$$

(c) $P(X + Y \le 1) = P((X, Y) \in D)$ where D is the triangular region
shown in the figure. Thus

$P(X + Y \le 1) = \iint_D f(x, y) \, dA = \int_0^1 \int_0^{1-x} \frac{1}{2} x(1 + y) \, dy \, dx$

$= \int_0^1 \frac{1}{2} x \left[y + \frac{1}{2} y^2 \right]_{y=0}^{y=1-x} dx = \int_0^1 \frac{1}{2} x \left(\frac{1}{2} x^2 - 2x + \frac{3}{2} \right) dx$

$= \frac{1}{4} \int_0^1 (x^3 - 4x^2 + 3x) \, dx = \frac{1}{4} \left[\frac{x^4}{4} - 4\frac{x^3}{3} + 3\frac{x^2}{2} \right]_0^1$

$= \frac{5}{48} \approx 0.1042$

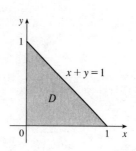

25. (a) $f(x, y) \geq 0$, so f is a joint density function if $\iint_{\mathbb{R}^2} f(x, y) \, dA = 1$. Here, $f(x, y) = 0$ outside the first quadrant, so

$$\iint_{\mathbb{R}^2} f(x, y) \, dA = \int_0^\infty \int_0^\infty 0.1 e^{-(0.5x + 0.2y)} \, dy \, dx = 0.1 \int_0^\infty \int_0^\infty e^{-0.5x} e^{-0.2y} \, dy \, dx$$

$$= 0.1 \int_0^\infty e^{-0.5x} \, dx \int_0^\infty e^{-0.2y} \, dy = 0.1 \lim_{t \to \infty} \int_0^t e^{-0.5x} \, dx \lim_{t \to \infty} \int_0^t e^{-0.2y} \, dy$$

$$= 0.1 \lim_{t \to \infty} \left[-2e^{-0.5x}\right]_0^t \lim_{t \to \infty} \left[-5e^{-0.2y}\right]_0^t$$

$$= 0.1 \lim_{t \to \infty} \left[-2(e^{-0.5t} - 1)\right] \lim_{t \to \infty} \left[-5(e^{-0.2t} - 1)\right]$$

$$= (0.1) \cdot (-2)(0 - 1) \cdot (-5)(0 - 1) = 1$$

Thus $f(x, y)$ is a joint density function.

(b) (i) No restriction is placed on X, so

$$P(Y \geq 1) = \int_{-\infty}^\infty \int_1^\infty f(x, y) \, dy \, dx = \int_0^\infty \int_1^\infty 0.1 e^{-(0.5x+0.2y)} \, dy \, dx$$

$$= 0.1 \int_0^\infty e^{-0.5x} \, dx \int_1^\infty e^{-0.2y} \, dy = 0.1 \lim_{t \to \infty} \int_0^t e^{-0.5x} \, dx \lim_{t \to \infty} \int_1^t e^{-0.2y} \, dy$$

$$= 0.1 \lim_{t \to \infty} \left[-2e^{-0.5x}\right]_0^t \lim_{t \to \infty} \left[-5e^{-0.2y}\right]_1^t$$

$$= 0.1 \lim_{t \to \infty} \left[-2(e^{-0.5t} - 1)\right] \lim_{t \to \infty} \left[-5(e^{-0.2t} - e^{-0.2})\right]$$

$$= (0.1) \cdot (-2)(0 - 1) \cdot (-5)(0 - e^{-0.2}) = e^{-0.2} \approx 0.8187$$

(ii) $P(X \leq 2, Y \leq 4) = \int_{-\infty}^2 \int_{-\infty}^4 f(x, y) \, dy \, dx = \int_0^2 \int_0^4 0.1 e^{-(0.5x+0.2y)} \, dy \, dx$

$$= 0.1 \int_0^2 e^{-0.5x} \, dx \int_0^4 e^{-0.2y} \, dy = 0.1 \left[-2e^{-0.5x}\right]_0^2 \left[-5e^{-0.2y}\right]_0^4$$

$$= (0.1) \cdot (-2)(e^{-1} - 1) \cdot (-5)(e^{-0.8} - 1)$$

$$= (e^{-1} - 1)(e^{-0.8} - 1) = 1 + e^{-1.8} - e^{-0.8} - e^{-1} \approx 0.3481$$

(c) The expected value of X is given by

$$\mu_1 = \iint_{\mathbb{R}^2} x f(x, y) \, dA = \int_0^\infty \int_0^\infty x \left[0.1 e^{-(0.5x+0.2y)}\right] dy \, dx$$

$$= 0.1 \int_0^\infty x e^{-0.5x} \, dx \int_0^\infty e^{-0.2y} \, dy = 0.1 \lim_{t \to \infty} \int_0^t x e^{-0.5x} \, dx \lim_{t \to \infty} \int_0^t e^{-0.2y} \, dy$$

To evaluate the first integral, we integrate by parts with $u = x$ and $dv = e^{-0.5x} \, dx$ (or we can use Formula 96 in the Table of Integrals):

$\int x e^{-0.5x} \, dx = -2x e^{-0.5x} - \int -2e^{-0.5x} \, dx = -2x e^{-0.5x} - 4e^{-0.5x} = -2(x + 2)e^{-0.5x}$. Thus

$$\mu_1 = 0.1 \lim_{t \to \infty} \left[-2(x + 2)e^{-0.5x}\right]_0^t \lim_{t \to \infty} \left[-5e^{-0.2y}\right]_0^t$$

$$= 0.1 \lim_{t \to \infty} (-2)\left[(t + 2)e^{-0.5t} - 2\right] \lim_{t \to \infty} (-5)\left[e^{-0.2t} - 1\right]$$

$$= 0.1(-2)\left(\lim_{t \to \infty} \frac{t + 2}{e^{0.5t}} - 2\right)(-5)(-1) = 2 \qquad \text{[by l'Hospital's Rule]}$$

The expected value of Y is given by

$$\mu_2 = \iint_{\mathbb{R}^2} y f(x, y) \, dA = \int_0^\infty \int_0^\infty y \left[0.1 e^{-(0.5 + 0.2y)}\right] dy \, dx$$

$$= 0.1 \int_0^\infty e^{-0.5x} \, dx \int_0^\infty y e^{-0.2y} \, dy = 0.1 \lim_{t \to \infty} \int_0^t e^{-0.5x} \, dx \lim_{t \to \infty} \int_0^t y e^{-0.2y} \, dy$$

To evaluate the second integral, we integrate by parts with $u = y$ and $dv = e^{-0.2y} \, dy$ (or again we can use Formula 96 in the Table of Integrals) which gives

$\int ye^{-0.2y} \, dy = -5ye^{-0.2y} + \int 5e^{-0.2y} \, dy = -5(y+5)e^{-0.2y}$. Then

$$\mu_2 = 0.1 \lim_{t \to \infty} \left[-2e^{-0.5x} \right]_0^t \lim_{t \to \infty} \left[-5(y+5)e^{-0.2y} \right]_0^t$$

$$= 0.1 \lim_{t \to \infty} \left[-2(e^{-0.5t} - 1) \right] \lim_{t \to \infty} \left(-5[(t+5)e^{-0.2t} - 5] \right)$$

$$= 0.1(-2)(-1) \cdot (-5) \left(\lim_{t \to \infty} \frac{t+5}{e^{0.2t}} - 5 \right) = 5 \qquad \text{[by l'Hospital's Rule]}$$

27. (a) The random variables X and Y are normally distributed with $\mu_1 = 45$, $\mu_2 = 20$, $\sigma_1 = 0.5$, and $\sigma_2 = 0.1$. The individual density functions for X and Y, then, are $f_1(x) = \dfrac{1}{0.5\sqrt{2\pi}} e^{-(x-45)^2/0.5}$ and

$f_2(y) = \dfrac{1}{0.1\sqrt{2\pi}} e^{-(y-20)^2/0.02}$. Since X and Y are independent, the joint density function is the product

$$f(x, y) = f_1(x)f_2(y) = \frac{1}{0.5\sqrt{2\pi}} e^{-(x-45)^2/0.5} \frac{1}{0.1\sqrt{2\pi}} e^{-(y-20)^2/0.02}$$

$$= \frac{10}{\pi} e^{-2(x-45)^2 - 50(y-20)^2}$$

Then
$$P(40 \leq X \leq 50,\ 20 \leq Y \leq 25) = \int_{40}^{50} \int_{20}^{25} f(x, y) \, dy \, dx$$

$$= \frac{10}{\pi} \int_{40}^{50} \int_{20}^{25} e^{-2(x-45)^2 - 50(y-20)^2} \, dy \, dx$$

Using a CAS or calculator to evaluate the integral, we get $P(40 \leq X \leq 50,\ 20 \leq Y \leq 25) \approx 0.500$.

(b) $P(4(X-45)^2 + 100(Y-20)^2 \leq 2) = \iint_D \frac{10}{\pi} e^{-2(x-45)^2 - 50(y-20)^2} \, dA$, where D is the region enclosed by the ellipse $4(x-45)^2 + 100(y-20)^2 = 2$. Solving for y gives $y = 20 \pm \frac{1}{10}\sqrt{2 - 4(x-45)^2}$, the upper and lower halves of the ellipse, and these two halves meet where $y = 20$ [since the ellipse is centered at $(45, 20)$] $\Rightarrow 4(x-45)^2 = 2 \Rightarrow x = 45 \pm \frac{1}{\sqrt{2}}$. Thus

$$\iint_D \frac{10}{\pi} e^{-2(x-45)^2 - 50(y-20)^2} \, dA = \frac{10}{\pi} \int_{45 - 1/\sqrt{2}}^{45 + 1/\sqrt{2}} \int_{20 - \frac{1}{10}\sqrt{2 - 4(x-45)^2}}^{20 + \frac{1}{10}\sqrt{2 - 4(x-45)^2}} e^{-2(x-45)^2 - 50(y-20)^2} \, dy \, dx.$$

Using a CAS or calculator to evaluate the integral, we get $P(4(X-45)^2 + 100(Y-20)^2 \leq 2) \approx 0.632$.

29. (a) If $f(P, A)$ is the probability that an individual at A will be infected by an individual at P, and $k \, dA$ is the number of infected individuals in an element of area dA, then $f(P, A)k \, dA$ is the number of infections that should result from exposure of the individual at A to infected people in the element of area dA. Integration over D gives the number of infections of the person at A due to all the infected people in D. In rectangular coordinates (with the origin at the city's center), the exposure of a person at A is

$$E = \iint_D kf(P, A) \, dA = k \iint_D \frac{20 - d(P, A)}{20} \, dA = k \iint_D \left[1 - \frac{\sqrt{(x - x_0)^2 + (y - y_0)^2}}{20} \right] dx \, dy.$$

(b) If $A = (0,0)$, then

$$E = k \iint_D \left[1 - \frac{1}{20}\sqrt{x^2 + y^2}\right] dx\, dy$$

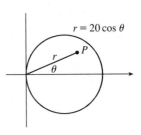
$r = 20\cos\theta$

$$= k \int_0^{2\pi} \int_0^{10} \left(1 - \frac{r}{20}\right) r\, dr\, d\theta = 2\pi k \left[\frac{r^2}{2} - \frac{r^3}{60}\right]_0^{10}$$

$$= 2\pi k\left(50 - \tfrac{50}{3}\right) = \tfrac{200}{3}\pi k \approx 209k$$

For A at the edge of the city, it is convenient to use a polar coordinate system centered at A. Then the polar equation for the circular boundary of the city becomes $r = 20\cos\theta$ instead of $r = 10$, and the distance from A to a point P in the city is again r (see the figure). So

$$E = k \int_{-\pi/2}^{\pi/2} \int_0^{20\cos\theta} \left(1 - \frac{r}{20}\right) r\, dr\, d\theta = k \int_{-\pi/2}^{\pi/2} \left[\frac{r^2}{2} - \frac{r^3}{60}\right]_{r=0}^{r=20\cos\theta} d\theta$$

$$= k \int_{-\pi/2}^{\pi/2} \left(200\cos^2\theta - \tfrac{400}{3}\cos^3\theta\right) d\theta = 200k \int_{-\pi/2}^{\pi/2} \left[\tfrac{1}{2} + \tfrac{1}{2}\cos 2\theta - \tfrac{2}{3}(1 - \sin^2\theta)\cos\theta\right] d\theta$$

$$= 200k \left[\tfrac{1}{2}\theta + \tfrac{1}{4}\sin 2\theta - \tfrac{2}{3}\sin\theta + \tfrac{2}{3}\cdot\tfrac{1}{3}\sin^3\theta\right]_{-\pi/2}^{\pi/2} = 200k\left[\tfrac{\pi}{4} + 0 - \tfrac{2}{3} + \tfrac{2}{9} + \tfrac{\pi}{4} + 0 - \tfrac{2}{3} + \tfrac{2}{9}\right]$$

$$= 200k\left(\tfrac{\pi}{2} - \tfrac{8}{9}\right) \approx 136k$$

Therefore the risk of infection is much lower at the edge of the city than in the middle, so it is better to live at the edge.

16.6 Surface Area   ET 15.6

1. Here $z = f(x, y) = 2 + 3x + 4y$ and D is the rectangle $[0, 5] \times [1, 4]$, so by Formula 2 the area of the surface is

$$A(S) = \iint_D \sqrt{[f_x(x, y)]^2 + [f_y(x, y)]^2 + 1}\, dA = \iint_D \sqrt{3^2 + 4^2 + 1}\, dA = \sqrt{26} \iint_D dA$$
$$= \sqrt{26}\, A(D) = \sqrt{26}\,(5)(3) = 15\sqrt{26}$$

3. $z = f(x, y) = 6 - 3x - 2y$ which intersects the xy-plane in the line $3x + 2y = 6$, so D is the triangular region given by $\left\{(x, y) \mid 0 \le x \le 2, 0 \le y \le 3 - \tfrac{3}{2}x\right\}$. Thus

$$A(S) = \iint_D \sqrt{(-3)^2 + (-2)^2 + 1}\, dA = \sqrt{14} \iint_D dA = \sqrt{14}\, A(D)$$
$$= \sqrt{14}\left(\tfrac{1}{2}\cdot 2 \cdot 3\right) = 3\sqrt{14}$$

5. $y^2 + z^2 = 9 \Rightarrow z = \sqrt{9 - y^2}$. $f_x = 0$, $f_y = -y(9 - y^2)^{-1/2} \Rightarrow$

$$A(S) = \int_0^4 \int_0^2 \sqrt{0^2 + [-y(9 - y^2)^{-1/2}]^2 + 1}\, dy\, dx = \int_0^4 \int_0^2 \sqrt{\frac{y^2}{9 - y^2} + 1}\, dy\, dx$$

$$= \int_0^4 \int_0^2 \frac{3}{\sqrt{9 - y^2}}\, dy\, dx = 3\int_0^4 \left[\sin^{-1}\frac{y}{3}\right]_{y=0}^{y=2} dx = 3\left[(\sin^{-1}(\tfrac{2}{3}))x\right]_0^4 = 12\sin^{-1}(\tfrac{2}{3})$$

7. $z = f(x, y) = y^2 - x^2$ with $1 \le x^2 + y^2 \le 4$. Then

$$A(S) = \iint_D \sqrt{1 + 4x^2 + 4y^2} \, dA = \int_0^{2\pi} \int_1^2 \sqrt{1 + 4r^2} \, r \, dr \, d\theta = \int_0^{2\pi} d\theta \int_1^2 r \sqrt{1 + 4r^2} \, dr$$

$$= \left[\theta \right]_0^{2\pi} \left[\tfrac{1}{12}(1 + 4r^2)^{3/2} \right]_1^2 = \tfrac{\pi}{6} \left(17\sqrt{17} - 5\sqrt{5} \right)$$

9. $z = f(x, y) = xy$ with $0 \le x^2 + y^2 \le 1$, so $f_x = y$, $f_y = x$ $\Rightarrow$

$$A(S) = \iint_D \sqrt{y^2 + x^2 + 1} \, dA = \int_0^{2\pi} \int_0^1 \sqrt{r^2 + 1} \, r \, dr \, d\theta = \int_0^{2\pi} \left[\tfrac{1}{3}(r^2 + 1)^{3/2} \right]_{r=0}^{r=1} d\theta$$

$$= \int_0^{2\pi} \tfrac{1}{3}(2\sqrt{2} - 1) \, d\theta = \tfrac{2\pi}{3}(2\sqrt{2} - 1)$$

11. $z = \sqrt{a^2 - x^2 - y^2}$, $z_x = -x(a^2 - x^2 - y^2)^{-1/2}$, $z_y = -y(a^2 - x^2 - y^2)^{-1/2}$,

$$A(S) = \iint_D \sqrt{\frac{x^2 + y^2}{a^2 - x^2 - y^2} + 1} \, dA$$

$$= \int_{-\pi/2}^{\pi/2} \int_0^{a\cos\theta} \sqrt{\frac{r^2}{a^2 - r^2} + 1} \, r \, dr \, d\theta$$

$$= \int_{-\pi/2}^{\pi/2} \int_0^{a\cos\theta} \frac{ar}{\sqrt{a^2 - r^2}} \, dr \, d\theta$$

$$= \int_{-\pi/2}^{\pi/2} \left[-a\sqrt{a^2 - r^2} \right]_{r=0}^{r=a\cos\theta} d\theta$$

$$= \int_{-\pi/2}^{\pi/2} -a\left(\sqrt{a^2 - a^2\cos^2\theta} - a \right) d\theta = 2a^2 \int_0^{\pi/2} \left(1 - \sqrt{1 - \cos^2\theta} \right) d\theta$$

$$= 2a^2 \int_0^{\pi/2} d\theta - 2a^2 \int_0^{\pi/2} \sqrt{\sin^2\theta} \, d\theta = a^2\pi - 2a^2 \int_0^{\pi/2} \sin\theta \, d\theta = a^2(\pi - 2)$$

$r = a\cos\theta$

13. $z = f(x, y) = e^{-x^2 - y^2}$, $f_x = -2xe^{-x^2 - y^2}$, $f_y = -2ye^{-x^2 - y^2}$. Then

$$A(S) = \iint_{x^2 + y^2 \le 4} \sqrt{(-2xe^{-x^2-y^2})^2 + (-2ye^{-x^2-y^2})^2 + 1} \, dA = \iint_{x^2 + y^2 \le 4} \sqrt{4(x^2 + y^2)e^{-2(x^2+y^2)} + 1} \, dA.$$

Converting to polar coordinates we have

$$A(S) = \int_0^{2\pi} \int_0^2 \sqrt{4r^2 e^{-2r^2} + 1} \, r \, dr \, d\theta = \int_0^{2\pi} d\theta \int_0^2 r \sqrt{4r^2 e^{-2r^2} + 1} \, dr$$

$$= 2\pi \int_0^2 r \sqrt{4r^2 e^{-2r^2} + 1} \, dr \approx 13.9783 \text{ using a calculator.}$$

15. (a) The midpoints of the four squares are $\left(\tfrac{1}{4}, \tfrac{1}{4}\right)$, $\left(\tfrac{1}{4}, \tfrac{3}{4}\right)$, $\left(\tfrac{3}{4}, \tfrac{1}{4}\right)$, and $\left(\tfrac{3}{4}, \tfrac{3}{4}\right)$. Here $f(x, y) = x^2 + y^2$, so the Midpoint Rule gives

$$A(S) = \iint_D \sqrt{[f_x(x, y)]^2 + [f_y(x, y)]^2 + 1} \, dA = \iint_D \sqrt{(2x)^2 + (2y)^2 + 1} \, dA$$

$$\approx \tfrac{1}{4} \left(\sqrt{\left[2\left(\tfrac{1}{4}\right)\right]^2 + \left[2\left(\tfrac{1}{4}\right)\right]^2 + 1} + \sqrt{\left[2\left(\tfrac{1}{4}\right)\right]^2 + \left[2\left(\tfrac{3}{4}\right)\right]^2 + 1} \right.$$

$$\left. + \sqrt{\left[2\left(\tfrac{3}{4}\right)\right]^2 + \left[2\left(\tfrac{1}{4}\right)\right]^2 + 1} + \sqrt{\left[2\left(\tfrac{3}{4}\right)\right]^2 + \left[2\left(\tfrac{3}{4}\right)\right]^2 + 1} \right)$$

$$= \tfrac{1}{4} \left(\sqrt{\tfrac{3}{2}} + 2\sqrt{\tfrac{7}{2}} + \sqrt{\tfrac{11}{2}} \right) \approx 1.8279$$

(b) A CAS estimates the integral to be

$A(S) = \iint_D \sqrt{1 + (2x)^2 + (2y)^2}\, dA = \int_0^1 \int_0^1 \sqrt{1 + 4x^2 + 4y^2}\, dy\, dx \approx 1.8616$. This agrees with the Midpoint estimate only in the first decimal place.

17. $z = 1 + 2x + 3y + 4y^2$, so

$$A(S) = \iint_D \sqrt{1 + \left(\frac{\partial z}{\partial x}\right)^2 + \left(\frac{\partial z}{\partial y}\right)^2}\, dA = \int_1^4 \int_0^1 \sqrt{1 + 4 + (3 + 8y)^2}\, dy\, dx$$

$$= \int_1^4 \int_0^1 \sqrt{14 + 48y + 64y^2}\, dy\, dx.$$

Using a CAS, we have

$\int_1^4 \int_0^1 \sqrt{14 + 48y + 64y^2}\, dy\, dx = \frac{45}{8}\sqrt{14} + \frac{15}{16}\ln\left(11\sqrt{5} + 3\sqrt{14}\sqrt{5}\right) - \frac{15}{16}\ln\left(3\sqrt{5} + \sqrt{14}\sqrt{5}\right)$

or $\frac{45}{8}\sqrt{14} + \frac{15}{16}\ln\dfrac{11\sqrt{5} + 3\sqrt{70}}{3\sqrt{5} + \sqrt{70}}$.

19. $f(x,y) = 1 + x^2y^2 \Rightarrow f_x = 2xy^2$, $f_y = 2x^2y$. We use a CAS (with precision reduced to five significant digits, to speed up the calculation) to estimate the integral

$$A(S) = \int_{-1}^1 \int_{-\sqrt{1-x^2}}^{\sqrt{1-x^2}} \sqrt{f_x^2 + f_y^2 + 1}\, dy\, dx = \int_{-1}^1 \int_{-\sqrt{1-x^2}}^{\sqrt{1-x^2}} \sqrt{4x^2y^4 + 4x^4y^2 + 1}\, dy\, dx,$$ and find that

$A(S) \approx 3.3213$.

21. Here $z = f(x,y) = ax + by + c$, $f_x(x,y) = a$, $f_y(x,y) = b$, so

$A(S) = \iint_D \sqrt{a^2 + b^2 + 1}\, dA = \sqrt{a^2 + b^2 + 1} \iint_D dA = \sqrt{a^2 + b^2 + 1}\, A(D)$.

23. If we project the surface onto the xz-plane, then the surface lies "above" the disk $x^2 + z^2 \leq 25$ in the xz-plane. We have $y = f(x,z) = x^2 + z^2$ and, adapting Formula 2, the area of the surface is

$$A(S) = \iint_{x^2 + z^2 \leq 25} \sqrt{[f_x(x,z)]^2 + [f_z(x,z)]^2 + 1}\, dA = \iint_{x^2 + z^2 \leq 25} \sqrt{4x^2 + 4z^2 + 1}\, dA$$

Converting to polar coordinates $x = r\cos\theta$, $z = r\sin\theta$ we have

$A(S) = \int_0^{2\pi} \int_0^5 \sqrt{4r^2 + 1}\, r\, dr\, d\theta = \int_0^{2\pi} d\theta \int_0^5 r(4r^2 + 1)^{1/2}\, dr = \left[\theta\right]_0^{2\pi} \left[\frac{1}{12}(4r^2 + 1)^{3/2}\right]_0^5$

$= \frac{\pi}{6}\left(101\sqrt{101} - 1\right)$

16.7 Triple Integrals   ET 15.7

1. $\iiint_B xyz^2\, dV = \int_0^1 \int_{-1}^2 \int_0^3 xyz^2\, dz\, dx\, dy = \int_0^1 \int_{-1}^2 xy\left[\frac{1}{3}z^3\right]_{z=0}^{z=3} dx\, dy = \int_0^1 \int_{-1}^2 9xy\, dx\, dy$

$= \int_0^1 \left[\frac{9}{2}x^2y\right]_{x=-1}^{x=2} dy = \int_0^1 \frac{27}{2}y\, dy = \frac{27}{4}y^2\Big]_0^1 = \frac{27}{4}$

3. $\int_0^1 \int_0^z \int_0^{x+z} 6xz\, dy\, dx\, dz = \int_0^1 \int_0^z \left[6xyz\right]_{y=0}^{y=x+z} dx\, dz = \int_0^1 \int_0^z 6xz(x + z)\, dx\, dz$

$= \int_0^1 \left[2x^3z + 3x^2z^2\right]_{x=0}^{x=z} dz = \int_0^1 (2z^4 + 3z^4)\, dz = \int_0^1 5z^4\, dz = z^5\Big]_0^1 = 1$

5. $\int_0^3 \int_0^1 \int_0^{\sqrt{1-z^2}} ze^y\, dx\, dz\, dy = \int_0^3 \int_0^1 \left[xze^y\right]_{x=0}^{x=\sqrt{1-z^2}} dz\, dy = \int_0^3 \int_0^1 ze^y\sqrt{1 - z^2}\, dz\, dy$

$= \int_0^3 \left[-\frac{1}{3}(1 - z^2)^{3/2}e^y\right]_{z=0}^{z=1} dy = \int_0^3 \frac{1}{3}e^y\, dy = \frac{1}{3}e^y\Big]_0^3 = \frac{1}{3}(e^3 - 1)$

7. $\iiint_E 2x\, dV = \int_0^2 \int_0^{\sqrt{4-y^2}} \int_0^y 2x\, dz\, dx\, dy = \int_0^2 \int_0^{\sqrt{4-y^2}} \left[2xz\right]_{z=0}^{z=y} dx\, dy = \int_0^2 \int_0^{\sqrt{4-y^2}} 2xy\, dx\, dy$

$= \int_0^2 \left[x^2y\right]_{x=0}^{x=\sqrt{4-y^2}} dy = \int_0^2 (4 - y^2)y\, dy = \left[2y^2 - \frac{1}{4}y^4\right]_0^2 = 4$

9. Here $E = \{(x, y, z) \mid 0 \le x \le 1, 0 \le y \le \sqrt{x}, 0 \le z \le 1 + x + y\}$, so

$$\iiint_E 6xy \, dV = \int_0^1 \int_0^{\sqrt{x}} \int_0^{1+x+y} 6xy \, dz \, dy \, dx = \int_0^1 \int_0^{\sqrt{x}} \left[6xyz\right]_{z=0}^{z=1+x+y} dy \, dx$$

$$= \int_0^1 \int_0^{\sqrt{x}} 6xy(1 + x + y) \, dy \, dx = \int_0^1 \left[3xy^2 + 3x^2y^2 + 2xy^3\right]_{y=0}^{y=\sqrt{x}} dx$$

$$= \int_0^1 \left(3x^2 + 3x^3 + 2x^{5/2}\right) dx = \left[x^3 + \tfrac{3}{4}x^4 + \tfrac{4}{7}x^{7/2}\right]_0^1 = \tfrac{65}{28}$$

11. Here E is the region that lies below the plane with x-, y-, and z-intercepts 1, 2, and 3 respectively, that is, below the plane $2z + 6x + 3y = 6$ and above the region in the xy-plane bounded by the lines $x = 0$, $y = 0$ and $6x + 3y = 6$. So

$$\iiint_E xy \, dV = \int_0^1 \int_0^{2-2x} \int_0^{3-3x-3y/2} xy \, dz \, dy \, dx = \int_0^1 \int_0^{2-2x} \left(3xy - 3x^2y - \tfrac{3}{2}xy^2\right) dy \, dx$$

$$= \int_0^1 \left[\tfrac{3}{2}xy^2 - \tfrac{3}{2}x^2y^2 - \tfrac{1}{2}xy^3\right]_{y=0}^{y=2-2x} dx = \int_0^1 \left(2x - 6x^2 + 6x^3 - 2x^4\right) dx$$

$$= \left[x^2 - 2x^3 + \tfrac{3}{2}x^4 - \tfrac{2}{5}x^5\right]_0^1 = \tfrac{1}{10}.$$

13.

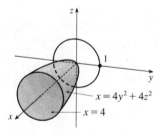

E is the region below the parabolic cylinder $z = 1 - y^2$ and above the square $[-1, 1] \times [-1, 1]$ in the xy-plane.

$$\iiint_E x^2 e^y \, dV = \int_{-1}^1 \int_{-1}^1 \int_0^{1-y^2} x^2 e^y \, dz \, dy \, dx$$

$$= \int_{-1}^1 \int_{-1}^1 x^2 e^y (1 - y^2) \, dy \, dx$$

$$= \int_{-1}^1 x^2 \, dx \int_{-1}^1 (e^y - y^2 e^y) \, dy$$

$$= \left[\tfrac{1}{3}x^3\right]_{-1}^1 \left[e^y - (y^2 - 2y + 2)e^y\right]_{-1}^1$$

[integrate by parts twice]

$$= \tfrac{1}{3}(2)[e - e - e^{-1} + 5e^{-1}] = \tfrac{8}{3e}$$

15.

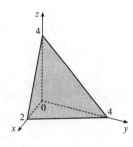

The projection E on the yz-plane is the disk $y^2 + z^2 \le 1$. Using polar coordinates $y = r\cos\theta$ and $z = r\sin\theta$, we get

$$\iiint_E x \, dV = \iint_D \left[\int_{4y^2+4z^2}^4 x \, dx\right] dA$$

$$= \tfrac{1}{2} \iint_D \left[4^2 - (4y^2 + 4z^2)^2\right] dA = 8 \int_0^{2\pi} \int_0^1 (1 - r^4) \, r \, dr \, d\theta$$

$$= 8 \int_0^{2\pi} d\theta \int_0^1 (r - r^5) \, dr = 8(2\pi)\left[\tfrac{1}{2}r^2 - \tfrac{1}{6}r^6\right]_0^1 = \tfrac{16\pi}{3}$$

17. The plane $2x + y + z = 4$ intersects the xy-plane when

$$2x + y + 0 = 4 \quad \Rightarrow \quad y = 4 - 2x, \text{ so}$$

$$E = \{(x, y, z) \mid 0 \le x \le 2, 0 \le y \le 4 - 2x, 0 \le z \le 4 - 2x - y\} \text{ and}$$

$$V = \int_0^2 \int_0^{4-2x} \int_0^{4-2x-y} dz \, dy \, dx = \int_0^2 \int_0^{4-2x} (4 - 2x - y) \, dy \, dx$$

$$= \int_0^2 \left[4y - 2xy - \tfrac{1}{2}y^2\right]_{y=0}^{y=4-2x} dx$$

$$= \int_0^2 \left[4(4 - 2x) - 2x(4 - 2x) - \tfrac{1}{2}(4 - 2x)^2\right] dx$$

$$= \int_0^2 (2x^2 - 8x + 8) \, dx = \left[\tfrac{2}{3}x^3 - 4x^2 + 8x\right]_0^2 = \tfrac{16}{3}$$

19. $V = \displaystyle\int_{-3}^{3}\int_{-\sqrt{9-x^2}}^{\sqrt{9-x^2}}\int_{1}^{5-y} dz\,dy\,dx = \int_{-3}^{3}\int_{-\sqrt{9-x^2}}^{\sqrt{9-x^2}} (5-y-1)\,dy\,dx = \int_{-3}^{3}\left[4y - \tfrac{1}{2}y^2\right]_{y=-\sqrt{9-x^2}}^{y=\sqrt{9-x^2}} dx$

$= \int_{-3}^{3} 8\sqrt{9-x^2}\,dx = 8\left[\tfrac{x}{2}\sqrt{9-x^2} + \tfrac{9}{2}\sin^{-1}\left(\tfrac{x}{3}\right)\right]_{-3}^{3}$ $\quad\left[\begin{array}{l}\text{using trigonometric substitution}\\ \text{or Formula 30 in the Table of Integrals}\end{array}\right]$

$= 8\left[\tfrac{9}{2}\sin^{-1}(1) - \tfrac{9}{2}\sin^{-1}(-1)\right] = 36\left(\tfrac{\pi}{2} - \left(-\tfrac{\pi}{2}\right)\right) = 36\pi$

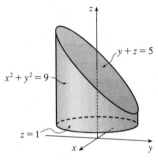

Alternatively, use polar coordinates to evaluate the double integral:

$\displaystyle\int_{-3}^{3}\int_{-\sqrt{9-x^2}}^{\sqrt{9-x^2}} (4-y)\,dy\,dx = \int_{0}^{2\pi}\int_{0}^{3} (4 - r\sin\theta)\,r\,dr\,d\theta$

$= \int_{0}^{2\pi}\left[2r^2 - \tfrac{1}{3}r^3\sin\theta\right]_{r=0}^{r=3} d\theta$

$= \int_{0}^{2\pi}(18 - 9\sin\theta)\,d\theta$

$= 18\theta + 9\cos\theta\Big]_{0}^{2\pi} = 36\pi$

21. (a) The wedge can be described as the region

$D = \{(x,y,z) \mid y^2 + z^2 \le 1, 0 \le x \le 1, 0 \le y \le x\}$

$= \{(x,y,z) \mid 0 \le x \le 1, 0 \le y \le x, 0 \le z \le \sqrt{1-y^2}\}$

So the integral expressing the volume of the wedge is

$\iiint_D dV = \int_0^1\int_0^x\int_0^{\sqrt{1-y^2}} dz\,dy\,dx.$

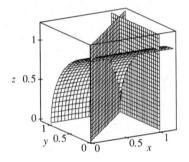

(b) A CAS gives $\int_0^1\int_0^x\int_0^{\sqrt{1-y^2}} dz\,dy\,dx = \tfrac{\pi}{4} - \tfrac{1}{3}.$

(Or use Formulas 30 and 87 from the Table of Integrals.)

23. Here $f(x,y,z) = \dfrac{1}{\ln(1+x+y+z)}$ and $\Delta V = 2\cdot 4\cdot 2 = 16$, so the Midpoint Rule gives

$$\iiint_B f(x,y,z)\,dV \approx \sum_{i=1}^{l}\sum_{j=1}^{m}\sum_{k=1}^{n} f\left(\overline{x}_i, \overline{y}_j, \overline{z}_k\right)\Delta V$$

$= 16\left[f(1,2,1) + f(1,2,3) + f(1,6,1) + f(1,6,3)\right.$

$\left. + f(3,2,1) + f(3,2,3) + f(3,6,1) + f(3,6,3)\right]$

$= 16\left[\dfrac{1}{\ln 5} + \dfrac{1}{\ln 7} + \dfrac{1}{\ln 9} + \dfrac{1}{\ln 11} + \dfrac{1}{\ln 7} + \dfrac{1}{\ln 9} + \dfrac{1}{\ln 11} + \dfrac{1}{\ln 13}\right] \approx 60.533$

25. $E = \{(x,y,z) \mid 0 \le x \le 1, 0 \le z \le 1 - x, 0 \le y \le 2 - 2z\}$,

the solid bounded by the three coordinate planes and the planes

$z = 1 - x, y = 2 - 2z.$

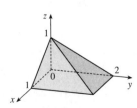

27.

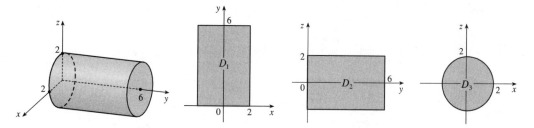

If D_1, D_2, D_3 are the projections of E on the xy-, yz-, and xz-planes, then

$$D_1 = \{(x, y) \mid -2 \le x \le 2, 0 \le y \le 6\}$$

$$D_2 = \{(y, z) \mid -2 \le z \le 2, 0 \le y \le 6\}$$

$$D_3 = \{(x, z) \mid x^2 + z^2 \le 4\}$$

Therefore

$$E = \left\{(x, y, z) \mid -\sqrt{4 - x^2} \le z \le \sqrt{4 - x^2}, \ -2 \le x \le 2, 0 \le y \le 6\right\}$$

$$= \left\{(x, y, z) \mid -\sqrt{4 - z^2} \le x \le \sqrt{4 - z^2}, \ -2 \le z \le 2, 0 \le y \le 6\right\}$$

$$\iiint_E f(x, y, z)\, dV = \int_{-2}^{2} \int_{0}^{6} \int_{-\sqrt{4-x^2}}^{\sqrt{4-x^2}} f(x, y, z)\, dz\, dy\, dx = \int_{0}^{6} \int_{-2}^{2} \int_{-\sqrt{4-x^2}}^{\sqrt{4-x^2}} f(x, y, z)\, dz\, dx\, dy$$

$$= \int_{0}^{6} \int_{-2}^{2} \int_{-\sqrt{4-z^2}}^{\sqrt{4-z^2}} f(x, y, z)\, dx\, dz\, dy = \int_{-2}^{2} \int_{0}^{6} \int_{-\sqrt{4-z^2}}^{\sqrt{4-z^2}} f(x, y, z)\, dx\, dy\, dz$$

$$= \int_{-2}^{2} \int_{-\sqrt{4-x^2}}^{\sqrt{4-x^2}} \int_{0}^{6} f(x, y, z)\, dy\, dz\, dx = \int_{-2}^{2} \int_{-\sqrt{4-z^2}}^{\sqrt{4-z^2}} \int_{0}^{6} f(x, y, z)\, dy\, dx\, dz$$

29.

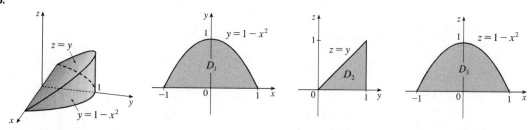

If D_1, D_2, and D_3 are the projections of E on the xy-, yz-, and xz-planes, then

$$D_1 = \left\{(x, y) \mid -1 \le x \le 1, 0 \le y \le 1 - x^2\right\} = \left\{(x, y) \mid 0 \le y \le 1, -\sqrt{1 - y} \le x \le \sqrt{1 - y}\right\},$$

$$D_2 = \{(y, z) \mid 0 \le y \le 1, 0 \le z \le y\} = \{(y, z) \mid 0 \le z \le 1, z \le y \le 1\}, \text{ and}$$

$$D_3 = \left\{(x, z) \mid -1 \le x \le 1, 0 \le z \le 1 - x^2\right\} = \left\{(x, z) \mid 0 \le z \le 1, -\sqrt{1 - z} \le x \le \sqrt{1 - z}\right\}$$

Therefore

$$E = \{(x,y,z) \mid -1 \le x \le 1, 0 \le y \le 1 - x^2, 0 \le z \le y\}$$
$$= \{(x,y,z) \mid 0 \le y \le 1, \ -\sqrt{1-y} \le x \le \sqrt{1-y}, 0 \le z \le y\}$$
$$= \{(x,y,z) \mid 0 \le y \le 1, 0 \le z \le y, \ -\sqrt{1-y} \le x \le \sqrt{1-y}\}$$
$$= \{(x,y,z) \mid 0 \le z \le 1, z \le y \le 1, \ -\sqrt{1-y} \le x \le \sqrt{1-y}\}$$
$$= \{(x,y,z) \mid -1 \le x \le 1, 0 \le z \le 1 - x^2, z \le y \le 1 - x^2\}$$
$$= \{(x,y,z) \mid 0 \le z \le 1, \ -\sqrt{1-z} \le x \le \sqrt{1-z}, z \le y \le 1 - x^2\}$$

Then

$$\iiint_E f(x,y,z)\,dV = \int_{-1}^{1} \int_{0}^{1-x^2} \int_{0}^{y} f(x,y,z)\,dz\,dy\,dx = \int_{0}^{1} \int_{-\sqrt{1-y}}^{\sqrt{1-y}} \int_{0}^{y} f(x,y,z)\,dz\,dx\,dy$$

$$= \int_{0}^{1} \int_{0}^{y} \int_{-\sqrt{1-y}}^{\sqrt{1-y}} f(x,y,z)\,dx\,dz\,dy = \int_{0}^{1} \int_{z}^{1} \int_{-\sqrt{1-y}}^{\sqrt{1-y}} f(x,y,z)\,dx\,dy\,dz$$

$$= \int_{-1}^{1} \int_{0}^{1-x^2} \int_{z}^{1-x^2} f(x,y,z)\,dy\,dz\,dx = \int_{0}^{1} \int_{-\sqrt{1-z}}^{\sqrt{1-z}} \int_{z}^{1-x^2} f(x,y,z)\,dy\,dx\,dz$$

31.

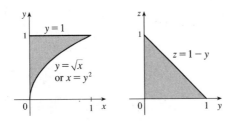 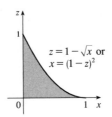

The diagrams show the projections of E on the xy-, yz-, and xz-planes. Therefore

$$\int_{0}^{1} \int_{\sqrt{x}}^{1} \int_{0}^{1-y} f(x,y,z)\,dz\,dy\,dx = \int_{0}^{1} \int_{0}^{y^2} \int_{0}^{1-y} f(x,y,z)\,dz\,dx\,dy$$

$$= \int_{0}^{1} \int_{0}^{1-z} \int_{0}^{y^2} f(x,y,z)\,dx\,dy\,dz$$

$$= \int_{0}^{1} \int_{0}^{1-y} \int_{0}^{y^2} f(x,y,z)\,dx\,dz\,dy$$

$$= \int_{0}^{1} \int_{0}^{1-\sqrt{x}} \int_{\sqrt{x}}^{1-z} f(x,y,z)\,dy\,dz\,dx$$

$$= \int_{0}^{1} \int_{0}^{(1-z)^2} \int_{\sqrt{x}}^{1-z} f(x,y,z)\,dy\,dx\,dz$$

33.

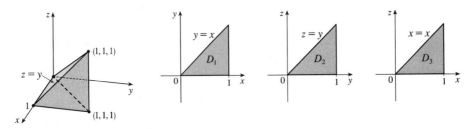

$$\int_{0}^{1} \int_{y}^{1} \int_{0}^{y} f(x,y,z)\,dz\,dx\,dy = \iiint_E f(x,y,z)\,dV \text{ where } E = \{(x,y,z) \mid 0 \le z \le y, y \le x \le 1, 0 \le y \le 1\}.$$

If D_1, D_2, and D_3 are the projections of E on the xy-, yz- and xz-planes then

$$D_1 = \{(x, y) \mid 0 \le y \le 1, y \le x \le 1\} = \{(x, y) \mid 0 \le x \le 1, 0 \le y \le x\},$$

$$D_2 = \{(y, z) \mid 0 \le y \le 1, 0 \le z \le y\} = \{(y, z) \mid 0 \le z \le 1, z \le y \le 1\}, \text{ and}$$

$$D_3 = \{(x, z) \mid 0 \le x \le 1, 0 \le z \le x\} = \{(x, z) \mid 0 \le z \le 1, z \le x \le 1\}.$$

Thus we also have

$$E = \{(x, y, z) \mid 0 \le x \le 1, 0 \le y \le x, 0 \le z \le y\} = \{(x, y, z) \mid 0 \le y \le 1, 0 \le z \le y, y \le x \le 1\}$$

$$= \{(x, y, z) \mid 0 \le z \le 1, z \le y \le 1, y \le x \le 1\} = \{(x, y, z) \mid 0 \le x \le 1, 0 \le z \le x, z \le y \le x\}$$

$$= \{(x, y, z) \mid 0 \le z \le 1, z \le x \le 1, z \le y \le x\}.$$

Then

$$\int_0^1 \int_y^1 \int_0^y f(x, y, z) \, dz \, dx \, dy = \int_0^1 \int_0^x \int_0^y f(x, y, z) \, dz \, dy \, dx = \int_0^1 \int_0^y \int_y^1 f(x, y, z) \, dx \, dz \, dy$$

$$= \int_0^1 \int_z^1 \int_y^1 f(x, y, z) \, dx \, dy \, dz = \int_0^1 \int_0^x \int_z^x f(x, y, z) \, dy \, dz \, dx$$

$$= \int_0^1 \int_z^1 \int_z^x f(x, y, z) \, dy \, dx \, dz$$

35. $m = \iiint_E \rho(x, y, z) \, dV = \int_0^1 \int_0^{\sqrt{x}} \int_0^{1+x+y} 2 \, dz \, dy \, dx$

$= \int_0^1 \int_0^{\sqrt{x}} 2(1 + x + y) \, dy \, dx = \int_0^1 \left[2y + 2xy + y^2 \right]_{y=0}^{y=\sqrt{x}} dx$

$= \int_0^1 \left(2\sqrt{x} + 2x^{3/2} + x \right) dx = \left[\frac{4}{3} x^{3/2} + \frac{4}{5} x^{5/2} + \frac{1}{2} x^2 \right]_0^1 = \frac{79}{30}$

$M_{yz} = \iiint_E x\rho(x, y, z) \, dV = \int_0^1 \int_0^{\sqrt{x}} \int_0^{1+x+y} 2x \, dz \, dy \, dx$

$= \int_0^1 \int_0^{\sqrt{x}} 2x(1 + x + y) \, dy \, dx = \int_0^1 \left[2xy + 2x^2y + xy^2 \right]_{y=0}^{y=\sqrt{x}} dx$

$= \int_0^1 (2x^{3/2} + 2x^{5/2} + x^2) \, dx = \left[\frac{4}{5} x^{5/2} + \frac{4}{7} x^{7/2} + \frac{1}{3} x^3 \right]_0^1 = \frac{179}{105}$

$M_{xz} = \iiint_E y\rho(x, y, z) \, dV = \int_0^1 \int_0^{\sqrt{x}} \int_0^{1+x+y} 2y \, dz \, dy \, dx$

$= \int_0^1 \int_0^{\sqrt{x}} 2y(1 + x + y) \, dy \, dx = \int_0^1 \left[y^2 + xy^2 + \frac{2}{3} y^3 \right]_{y=0}^{y=\sqrt{x}} dx$

$= \int_0^1 \left(x + x^2 + \frac{2}{3} x^{3/2} \right) dx = \left[\frac{1}{2} x^2 + \frac{1}{3} x^3 + \frac{4}{15} x^{5/2} \right]_0^1 = \frac{11}{10}$

$M_{xy} = \iiint_E z\rho(x, y, z) \, dV = \int_0^1 \int_0^{\sqrt{x}} \int_0^{1+x+y} 2z \, dz \, dy \, dx$

$= \int_0^1 \int_0^{\sqrt{x}} \left[z^2 \right]_{z=0}^{z=1+x+y} dy \, dx = \int_0^1 \int_0^{\sqrt{x}} (1 + x + y)^2 \, dy \, dx$

$= \int_0^1 \int_0^{\sqrt{x}} (1 + 2x + 2y + 2xy + x^2 + y^2) \, dy \, dx$

$= \int_0^1 \left[y + 2xy + y^2 + xy^2 + x^2y + \frac{1}{3} y^3 \right]_{y=0}^{y=\sqrt{x}} dx = \int_0^1 \left(\sqrt{x} + \frac{7}{3} x^{3/2} + x + x^2 + x^{5/2} \right) dx$

$= \left[\frac{2}{3} x^{3/2} + \frac{14}{15} x^{5/2} + \frac{1}{2} x^2 + \frac{1}{3} x^3 + \frac{2}{7} x^{7/2} \right]_0^1 = \frac{571}{210}$

Thus the mass is $\frac{79}{30}$ and the center of mass is $(\bar{x}, \bar{y}, \bar{z}) = \left(\dfrac{M_{yz}}{m}, \dfrac{M_{xz}}{m}, \dfrac{M_{xy}}{m} \right) = \left(\dfrac{358}{553}, \dfrac{33}{79}, \dfrac{571}{553} \right).$

37. $m = \int_0^a \int_0^a \int_0^a (x^2 + y^2 + z^2)\, dx\, dy\, dz = \int_0^a \int_0^a \left[\frac{1}{3}x^3 + xy^2 + xz^2\right]_{x=0}^{x=a} dy\, dz$

$\qquad = \int_0^a \int_0^a \left(\frac{1}{3}a^3 + ay^2 + az^2\right) dy\, dz = \int_0^a \left[\frac{1}{3}a^3 y + \frac{1}{3}ay^3 + ayz^2\right]_{y=0}^{y=a} dz$

$\qquad = \int_0^a \left(\frac{2}{3}a^4 + a^2 z^2\right) dz = \left[\frac{2}{3}a^4 z + \frac{1}{3}a^2 z^3\right]_0^a = \frac{2}{3}a^5 + \frac{1}{3}a^5 = a^5$

$\quad M_{yz} = \int_0^a \int_0^a \int_0^a \left[x^3 + x(y^2 + z^2)\right] dx\, dy\, dz = \int_0^a \int_0^a \left[\frac{1}{4}a^4 + \frac{1}{2}a^2(y^2 + z^2)\right] dy\, dz$

$\qquad = \int_0^a \left(\frac{1}{4}a^5 + \frac{1}{6}a^5 + \frac{1}{2}a^3 z^2\right) dz = \frac{1}{4}a^6 + \frac{1}{3}a^6 = \frac{7}{12}a^6$

$\qquad = M_{xz} = M_{xy}$ by symmetry of E and $\rho(x, y, z)$

Hence $(\bar{x}, \bar{y}, \bar{z}) = \left(\frac{7}{12}a, \frac{7}{12}a, \frac{7}{12}a\right)$.

39. (a) $m = \int_{-3}^3 \int_{-\sqrt{9-x^2}}^{\sqrt{9-x^2}} \int_1^{5-y} \sqrt{x^2 + y^2}\, dz\, dy\, dx$

(b) $(\bar{x}, \bar{y}, \bar{z}) = \left(\dfrac{M_{yz}}{m}, \dfrac{M_{xz}}{m}, \dfrac{M_{xy}}{m}\right)$ where

$\quad M_{yz} = \int_{-3}^3 \int_{-\sqrt{9-x^2}}^{\sqrt{9-x^2}} \int_1^{5-y} x\sqrt{x^2 + y^2}\, dz\, dy\, dx, \ M_{xz} = \int_{-3}^3 \int_{-\sqrt{9-x^2}}^{\sqrt{9-x^2}} \int_1^{5-y} y\sqrt{x^2 + y^2}\, dz\, dy\, dx$, and

$\quad M_{xy} = \int_{-3}^3 \int_{-\sqrt{9-x^2}}^{\sqrt{9-x^2}} \int_1^{5-y} z\sqrt{x^2 + y^2}\, dz\, dy\, dx$.

(c) $I_z = \int_{-3}^3 \int_{-\sqrt{9-x^2}}^{\sqrt{9-x^2}} \int_1^{5-y} (x^2 + y^2)\sqrt{x^2 + y^2}\, dz\, dy\, dx = \int_{-3}^3 \int_{-\sqrt{9-x^2}}^{\sqrt{9-x^2}} \int_1^{5-y} (x^2 + y^2)^{3/2}\, dz\, dy\, dx$

41. (a) $m = \int_0^1 \int_0^{\sqrt{1-x^2}} \int_0^y (1 + x + y + z)\, dz\, dy\, dx = \frac{3\pi}{32} + \frac{11}{24}$

(b) $(\bar{x}, \bar{y}, \bar{z}) = \left(m^{-1} \int_0^1 \int_0^{\sqrt{1-x^2}} \int_0^y x(1 + x + y + z)\, dz\, dy\, dx,\right.$

$\qquad\qquad\qquad m^{-1} \int_0^1 \int_0^{\sqrt{1-x^2}} \int_0^y y(1 + x + y + z)\, dz\, dy\, dx,$

$\qquad\qquad\qquad \left. m^{-1} \int_0^1 \int_0^{\sqrt{1-x^2}} \int_0^y z(1 + x + y + z)\, dz\, dy\, dx\right)$

$\qquad = \left(\dfrac{28}{9\pi + 44}, \dfrac{30\pi + 128}{45\pi + 220}, \dfrac{45\pi + 208}{135\pi + 660}\right)$

(c) $I_z = \int_0^1 \int_0^{\sqrt{1-x^2}} \int_0^y (x^2 + y^2)(1 + x + y + z)\, dz\, dy\, dx = \dfrac{68 + 15\pi}{240}$

43. $I_x = \int_0^L \int_0^L \int_0^L k(y^2 + z^2)\, dz\, dy\, dx = k\int_0^L \int_0^L \left(Ly^2 + \frac{1}{3}L^3\right) dy\, dx = k\int_0^L \frac{2}{3}L^4\, dx = \frac{2}{3}kL^5$.

By symmetry, $I_x = I_y = I_z = \frac{2}{3}kL^5$.

45. (a) $f(x, y, z)$ is a joint density function, so we know $\iiint_{\mathbb{R}^3} f(x, y, z)\, dV = 1$. Here we have

$$\iiint_{\mathbb{R}^3} f(x, y, z)\, dV = \int_{-\infty}^{\infty} \int_{-\infty}^{\infty} \int_{-\infty}^{\infty} f(x, y, z)\, dz\, dy\, dx = \int_0^2 \int_0^2 \int_0^2 Cxyz\, dz\, dy\, dx$$

$$= C \int_0^2 x\, dx \int_0^2 y\, dy \int_0^2 z\, dz = C\left[\frac{x^2}{2}\right]_0^2 \left[\frac{y^2}{2}\right]_0^2 \left[\frac{z^2}{2}\right]_0^2$$

$$= 8C$$

Then we must have $8C = 1 \;\Rightarrow\; C = \frac{1}{8}$.

(b) $P(X \le 1, Y \le 1, Z \le 1) = \int_{-\infty}^1 \int_{-\infty}^1 \int_{-\infty}^1 f(x, y, z)\, dz\, dy\, dx$

$$= \int_0^1 \int_0^1 \int_0^1 \tfrac{1}{8}xyz\, dz\, dy\, dx = \tfrac{1}{8} \int_0^1 x\, dx \int_0^1 y\, dy \int_0^1 z\, dz$$

$$= \frac{1}{8}\left[\frac{x^2}{2}\right]_0^1 \left[\frac{y^2}{2}\right]_0^1 \left[\frac{z^2}{2}\right]_0^1 = \tfrac{1}{8}\left(\tfrac{1}{2}\right)^3 = \tfrac{1}{64}$$

(c) $P(X + Y + Z \le 1) = P((X, Y, Z) \in E)$ where E is the solid region in the first octant bounded by the coordinate planes and the plane $x + y + z = 1$. The plane $x + y + z = 1$ meets the xy-plane in the line $x + y = 1$, so we have

$$P(X + Y + Z \le 1) = \iiint_E f(x, y, z)\, dV = \int_0^1 \int_0^{1-x} \int_0^{1-x-y} \tfrac{1}{8}xyz\, dz\, dy\, dx$$

$$= \tfrac{1}{8} \int_0^1 \int_0^{1-x} xy\left[\tfrac{1}{2}z^2\right]_{z=0}^{z=1-x-y}\, dy\, dx$$

$$= \tfrac{1}{16} \int_0^1 \int_0^{1-x} xy(1 - x - y)^2\, dy\, dx$$

$$= \tfrac{1}{16} \int_0^1 \int_0^{1-x} \left[(x^3 - 2x^2 + x)y + (2x^2 - 2x)y^2 + xy^3\right]\, dy\, dx$$

$$= \tfrac{1}{16} \int_0^1 \left[(x^3 - 2x^2 + x)\tfrac{1}{2}y^2 + (2x^2 - 2x)\tfrac{1}{3}y^3 + x\left(\tfrac{1}{4}y^4\right)\right]_{y=0}^{y=1-x}\, dx$$

$$= \tfrac{1}{192} \int_0^1 (x - 4x^2 + 6x^3 - 4x^4 + x^5)\, dx = \tfrac{1}{192}\left(\tfrac{1}{30}\right) = \tfrac{1}{5760}$$

47. $V(E) = L^3$,

$$f_{\text{ave}} = \frac{1}{L^3} \int_0^L \int_0^L \int_0^L xyz\, dx\, dy\, dz = \frac{1}{L^3} \int_0^L x\, dx \int_0^L y\, dy \int_0^L z\, dz$$

$$= \frac{1}{L^3}\left[\frac{x^2}{2}\right]_0^L \left[\frac{y^2}{2}\right]_0^L \left[\frac{z^2}{2}\right]_0^L = \frac{1}{L^3}\frac{L^2}{2}\frac{L^2}{2}\frac{L^2}{2} = \frac{L^3}{8}$$

49. The triple integral will attain its maximum when the integrand $1 - x^2 - 2y^2 - 3z^2$ is positive in the region E and negative everywhere else. For if E contains some region F where the integrand is negative, the integral could be increased by excluding F from E, and if E fails to contain some part G of the region where the integrand is positive, the integral could be increased by including G in E. So we require that $x^2 + 2y^2 + 3z^2 \le 1$. This describes the region bounded by the ellipsoid $x^2 + 2y^2 + 3z^2 = 1$.

16.8 Triple Integrals in Cylindrical and Spherical Coordinates ET 15.8

1. The region of integration is given in cylindrical coordinates by

$E = \{(r, \theta, z) \mid 0 \le \theta \le 2\pi, 0 \le r \le 4, r \le z \le 4\}$. This represents the solid region bounded below by the cone $z = r$ and above by the horizontal plane $z = 4$.

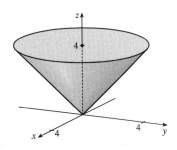

$$\int_0^4 \int_0^{2\pi} \int_r^4 r\, dz\, d\theta\, dr = \int_0^4 \int_0^{2\pi} [rz]_{z=r}^{z=4} d\theta\, dr$$

$$= \int_0^4 \int_0^{2\pi} r(4 - r)\, d\theta\, dr$$

$$= \int_0^4 (4r - r^2)\, dr \int_0^{2\pi} d\theta$$

$$= \left[2r^2 - \tfrac{1}{3}r^3\right]_0^4 \left[\theta\right]_0^{2\pi}$$

$$= \left(32 - \tfrac{64}{3}\right)(2\pi) = \tfrac{64\pi}{3}$$

3. The region of integration is given in spherical coordinates by

$E = \{(\rho, \theta, \phi) \mid 0 \le \rho \le 3, 0 \le \theta \le \pi/2, 0 \le \phi \le \pi/6\}$. This represents the solid region in the first octant bounded above by the sphere $\rho = 3$ and below by the cone $\phi = \pi/6$.

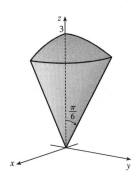

$$\int_0^{\pi/6} \int_0^{\pi/2} \int_0^3 \rho^2 \sin\phi\, d\rho\, d\theta\, d\phi = \int_0^{\pi/6} \sin\phi\, d\phi \int_0^{\pi/2} d\theta \int_0^3 \rho^2\, d\rho$$

$$= \left[-\cos\phi\right]_0^{\pi/6} \left[\theta\right]_0^{\pi/2} \left[\tfrac{1}{3}\rho^3\right]_0^3$$

$$= \left(1 - \frac{\sqrt{3}}{2}\right)\left(\frac{\pi}{2}\right)(9)$$

$$= \frac{9\pi}{4}\left(2 - \sqrt{3}\right)$$

5. The solid E is most conveniently described if we use cylindrical

coordinates: $E = \{(r, \theta, z) \mid 0 \le \theta \le \tfrac{\pi}{2}, 0 \le r \le 3, 0 \le z \le 2\}$. Then

$\iiint_E f(x, y, z)\, dV = \int_0^{\pi/2} \int_0^3 \int_0^2 f(r\cos\theta, r\sin\theta, z)\, r\, dz\, dr\, d\theta.$

7. In cylindrical coordinates, E is given by $\{(r, \theta, z) \mid 0 \le \theta \le 2\pi, 0 \le r \le 4, -5 \le z \le 4\}$. So

$\iiint_E \sqrt{x^2 + y^2}\, dV = \int_0^{2\pi} \int_0^4 \int_{-5}^4 \sqrt{r^2}\, r\, dz\, dr\, d\theta = \int_0^{2\pi} d\theta \int_0^4 r^2\, dr \int_{-5}^4 dz$

$= \left[\theta\right]_0^{2\pi} \left[\tfrac{1}{3}r^3\right]_0^4 \left[z\right]_{-5}^4 = (2\pi)\left(\tfrac{64}{3}\right)(9) = 384\pi$

9. In cylindrical coordinates E is bounded by the paraboloid $z = 1 + r^2$, the cylinder $r^2 = 5$ or $r = \sqrt{5}$, and the

xy-plane, so E is given by $\{(r, \theta, z) \mid 0 \le \theta \le 2\pi, 0 \le r \le \sqrt{5}, 0 \le z \le 1 + r^2\}$. Thus

$\iiint_E e^z\, dV = \int_0^{2\pi} \int_0^{\sqrt{5}} \int_0^{1+r^2} e^z\, r\, dz\, dr\, d\theta = \int_0^{2\pi} \int_0^{\sqrt{5}} r[e^z]_{z=0}^{z=1+r^2}\, dr\, d\theta = \int_0^{2\pi} \int_0^{\sqrt{5}} r(e^{1+r^2} - 1)\, dr\, d\theta$

$= \int_0^{2\pi} d\theta \int_0^{\sqrt{5}} \left(re^{1+r^2} - r\right) dr = 2\pi\left[\tfrac{1}{2}e^{1+r^2} - \tfrac{1}{2}r^2\right]_0^{\sqrt{5}} = \pi(e^6 - e - 5)$

11. In cylindrical coordinates, E is bounded by the cylinder $r = 1$, the plane $z = 0$, and the cone $z = 2r$. So
$E = \{(r, \theta, z) \mid 0 \le \theta \le 2\pi, 0 \le r \le 1, 0 \le z \le 2r\}$ and

$$\iiint_E x^2 \, dV = \int_0^{2\pi} \int_0^1 \int_0^{2r} r^2 \cos^2 \theta \, r \, dz \, dr \, d\theta = \int_0^{2\pi} \int_0^1 \left[r^3 \cos^2 \theta \, z \right]_{z=0}^{z=2r} dr \, d\theta$$

$$= \int_0^{2\pi} \int_0^1 2r^4 \cos^2 \theta \, dr \, d\theta = \int_0^{2\pi} \left[\tfrac{2}{5} r^5 \cos^2 \theta \right]_{r=0}^{r=1} d\theta = \tfrac{2}{5} \int_0^{2\pi} \cos^2 \theta \, d\theta$$

$$= \frac{2}{5} \int_0^{2\pi} \frac{1 + \cos 2\theta}{2} \, d\theta = \frac{1}{5} \left[\theta + \frac{1}{2} \sin 2\theta \right]_0^{2\pi} = \frac{2\pi}{5}$$

13. (a) The paraboloids intersect when $x^2 + y^2 = 36 - 3x^2 - 3y^2 \;\Rightarrow\; x^2 + y^2 = 9$, so the region of
integration is $D = \{(x, y) \mid x^2 + y^2 \le 9\}$. Then, in cylindrical coordinates,
$E = \{(r, \theta, z) \mid r^2 \le z \le 36 - 3r^2, 0 \le r \le 3, 0 \le \theta \le 2\pi\}$ and

$$V = \int_0^{2\pi} \int_0^3 \int_{r^2}^{36-3r^2} r \, dz \, dr \, d\theta = \int_0^{2\pi} \int_0^3 (36r - 4r^3) \, dr \, d\theta$$

$$= \int_0^{2\pi} \left[18r^2 - r^4 \right]_{r=0}^{r=3} d\theta = \int_0^{2\pi} 81 \, d\theta = 162\pi$$

(b) For constant density K, $m = KV = 162\pi K$ from part (a). Since the region is homogeneous and symmetric,
$M_{yz} = M_{xz} = 0$ and

$$M_{xy} = \int_0^{2\pi} \int_0^3 \int_{r^2}^{36-3r^2} (zK) r \, dz \, dr \, d\theta = K \int_0^{2\pi} \int_0^3 r \left[\tfrac{1}{2} z^2 \right]_{z=r^2}^{z=36-3r^2} dr \, d\theta$$

$$= \tfrac{K}{2} \int_0^{2\pi} \int_0^3 r((36 - 3r^2)^2 - r^4) \, dr \, d\theta = \tfrac{K}{2} \int_0^{2\pi} d\theta \int_0^3 (8r^5 - 216r^3 + 1296r) \, dr$$

$$= \tfrac{K}{2} (2\pi) \left[\tfrac{8}{6} r^6 - \tfrac{216}{4} r^4 + \tfrac{1296}{2} r^2 \right]_0^3 = \pi K (2430) = 2430\pi K$$

Thus $(\overline{x}, \overline{y}, \overline{z}) = \left(\dfrac{M_{yz}}{m}, \dfrac{M_{xz}}{m}, \dfrac{M_{xy}}{m} \right) = \left(0, 0, \dfrac{2430\pi K}{162\pi K} \right) = (0, 0, 15)$.

15. The paraboloid $z = 4x^2 + 4y^2$ intersects the plane $z = a$ when $a = 4x^2 + 4y^2$ or $x^2 + y^2 = \tfrac{1}{4} a$. So, in cylindrical
coordinates, $E = \{(r, \theta, z) \mid 0 \le r \le \tfrac{1}{2} \sqrt{a}, 0 \le \theta \le 2\pi, 4r^2 \le z \le a\}$. Thus

$$m = \int_0^{2\pi} \int_0^{\sqrt{a}/2} \int_{4r^2}^a Kr \, dz \, dr \, d\theta = K \int_0^{2\pi} \int_0^{\sqrt{a}/2} (ar - 4r^3) \, dr \, d\theta$$

$$= K \int_0^{2\pi} \left[\tfrac{1}{2} ar^2 - r^4 \right]_{r=0}^{r=\sqrt{a}/2} d\theta = K \int_0^{2\pi} \tfrac{1}{16} a^2 \, d\theta = \tfrac{1}{8} a^2 \pi K$$

Since the region is homogeneous and symmetric, $M_{yz} = M_{xz} = 0$ and

$$M_{xy} = \int_0^{2\pi} \int_0^{\sqrt{a}/2} \int_{4r^2}^a Krz \, dz \, dr \, d\theta = K \int_0^{2\pi} \int_0^{\sqrt{a}/2} \left(\tfrac{1}{2} a^2 r - 8r^5 \right) dr \, d\theta$$

$$= K \int_0^{2\pi} \left[\tfrac{1}{4} a^2 r^2 - \tfrac{4}{3} r^6 \right]_{r=0}^{r=\sqrt{a}/2} d\theta = K \int_0^{2\pi} \tfrac{1}{24} a^3 \, d\theta = \tfrac{1}{12} a^3 \pi K$$

Hence $(\overline{x}, \overline{y}, \overline{z}) = \left(0, 0, \tfrac{2}{3} a \right)$.

17. In spherical coordinates, B is represented by $\{(\rho, \theta, \phi) \mid 0 \le \rho \le 1, 0 \le \theta \le 2\pi, 0 \le \phi \le \pi\}$. Thus

$$\iiint_B (x^2 + y^2 + z^2) \, dV = \int_0^\pi \int_0^{2\pi} \int_0^1 (\rho^2) \rho^2 \sin \phi \, d\rho \, d\theta \, d\phi = \int_0^\pi \sin \phi \, d\phi \int_0^{2\pi} d\theta \int_0^1 \rho^4 \, d\rho$$

$$= \left[-\cos \phi \right]_0^\pi \left[\theta \right]_0^{2\pi} \left[\tfrac{1}{5} \rho^5 \right]_0^1 = (2)(2\pi) \left(\tfrac{1}{5} \right) = \tfrac{4\pi}{5}$$

19. In spherical coordinates, E is represented by $\{(\rho, \theta, \phi) \mid 1 \le \rho \le 2, 0 \le \theta \le \tfrac{\pi}{2}, 0 \le \phi \le \tfrac{\pi}{2} \}$. Thus

$$\iiint_E z \, dV = \int_0^{\pi/2} \int_0^{\pi/2} \int_1^2 (\rho \cos \phi) \rho^2 \sin \phi \, d\rho \, d\theta \, d\phi$$

$$= \int_0^{\pi/2} \cos \phi \sin \phi \, d\phi \int_0^{\pi/2} d\theta \int_1^2 \rho^3 \, d\rho = \left[\tfrac{1}{2} \sin^2 \phi \right]_0^{\pi/2} \left[\theta \right]_0^{\pi/2} \left[\tfrac{1}{4} \rho^4 \right]_1^2$$

$$= \left(\tfrac{1}{2} \right) \left(\tfrac{\pi}{2} \right) \left(\tfrac{15}{4} \right) = \tfrac{15\pi}{16}$$

21. $\iiint_E x^2 \, dV = \int_0^\pi \int_0^\pi \int_3^4 (\rho \sin\phi \cos\theta)^2 \, \rho^2 \sin\phi \, d\rho \, d\phi \, d\theta$

$\qquad = \int_0^\pi \cos^2\theta \, d\theta \int_0^\pi \sin^3\phi \, d\phi \int_3^4 \rho^4 \, d\rho$

$\qquad = \left[\frac{1}{2}\theta + \frac{1}{4}\sin 2\theta\right]_0^\pi \left[-\frac{1}{3}(2 + \sin^2\phi)\cos\phi\right]_0^\pi \left[\frac{1}{5}\rho^5\right]_3^4$

$\qquad = \left(\frac{\pi}{2}\right)\left(\frac{2}{3} + \frac{2}{3}\right)\frac{1}{5}(4^5 - 3^5) = \frac{1562}{15}\pi$

23. Since $\rho = 4\cos\phi$ implies $\rho^2 = 4\rho\cos\phi$, the equation is that of a sphere of radius 2 with center at $(0, 0, 2)$. Thus

$$V = \int_0^{2\pi} \int_0^{\pi/3} \int_0^{4\cos\phi} \rho^2 \sin\phi \, d\rho \, d\phi \, d\theta = \int_0^{2\pi} \int_0^{\pi/3} \left[\frac{1}{3}\rho^3\right]_{\rho=0}^{\rho=4\cos\phi} \sin\phi \, d\phi \, d\theta$$

$$= \int_0^{2\pi} \int_0^{\pi/3} \left(\frac{64}{3}\cos^3\phi\right) \sin\phi \, d\phi \, d\theta = \int_0^{2\pi} \left[-\frac{16}{3}\cos^4\phi\right]_{\phi=0}^{\phi=\pi/3} d\theta$$

$$= \int_0^{2\pi} -\frac{16}{3}\left(\frac{1}{16} - 1\right) d\theta = 5\theta \bigg]_0^{2\pi} = 10\pi$$

25. By the symmetry of the region, $M_{xy} = 0$ and $M_{yz} = 0$. Assuming constant density K,

$$m = \iiint_E KV = K \int_0^\pi \int_0^\pi \int_3^4 \rho^2 \sin\phi \, d\rho \, d\phi \, d\theta = K \int_0^\pi d\theta \int_0^\pi \sin\phi \, d\phi \int_3^4 \rho^2 \, d\rho$$

$$= K\pi \left[-\cos\phi\right]_0^\pi \left[\frac{1}{3}\rho^3\right]_3^4 = 2K\pi \cdot \frac{37}{3} = \frac{74}{3}\pi K$$

and

$$M_{xz} = \iiint_E yK \, dV = K \int_0^\pi \int_0^\pi \int_3^4 (\rho \sin\phi \sin\theta) \, \rho^2 \sin\phi \, d\rho \, d\phi \, d\theta$$

$$= K \int_0^\pi \sin\theta \, d\theta \int_0^\pi \sin^2\phi \, d\phi \int_3^4 \rho^3 \, d\rho$$

$$= K \left[-\cos\theta\right]_0^\pi \left[\frac{1}{2}\phi - \frac{1}{4}\sin 2\phi\right]_0^\pi \left[\frac{1}{4}\rho^4\right]_3^4$$

$$= K(2)\left(\frac{\pi}{2}\right)\frac{1}{4}(256 - 81) = \frac{175}{4}\pi K$$

Thus the centroid is $(\bar{x}, \bar{y}, \bar{z}) = \left(\dfrac{M_{yz}}{m}, \dfrac{M_{xz}}{m}, \dfrac{M_{xy}}{m}\right) = \left(0, \dfrac{175\pi K/4}{74\pi K/3}, 0\right) = \left(0, \dfrac{525}{296}, 0\right)$.

27. (a) The density function is $\rho(x, y, z) = K$, a constant, and by the symmetry of the problem $M_{xz} = M_{yz} = 0$.

Then $M_{xy} = \int_0^{2\pi} \int_0^{\pi/2} \int_0^a K\rho^3 \sin\phi \cos\phi \, d\rho \, d\phi \, d\theta = \frac{1}{2}\pi K a^4 \int_0^{\pi/2} \sin\phi \cos\phi \, d\phi = \frac{1}{8}\pi K a^4$. But the mass is

$K(\text{volume of the hemisphere}) = \frac{2}{3}\pi K a^3$, so the centroid is $\left(0, 0, \frac{3}{8}a\right)$.

(b) Place the center of the base at $(0, 0, 0)$; the density function is $\rho(x, y, z) = K$. By symmetry, the moments of inertia about any two such diameters will be equal, so we just need to find I_x:

$$I_x = \int_0^{2\pi} \int_0^{\pi/2} \int_0^a \left(K\rho^2 \sin\phi\right) \rho^2 \left(\sin^2\phi \sin^2\theta + \cos^2\phi\right) d\rho \, d\phi \, d\theta$$

$$= K \int_0^{2\pi} \int_0^{\pi/2} \left(\sin^3\phi \sin^2\theta + \sin\phi \cos^2\phi\right)\left(\frac{1}{5}a^5\right) d\phi \, d\theta$$

$$= \frac{1}{5}Ka^5 \int_0^{2\pi} \left[\sin^2\theta\left(-\cos\phi + \frac{1}{3}\cos^3\phi\right) + \left(-\frac{1}{3}\cos^3\phi\right)\right]_{\phi=0}^{\phi=\pi/2} d\theta$$

$$= \frac{1}{5}Ka^5 \int_0^{2\pi} \left[\frac{2}{3}\sin^2\theta + \frac{1}{3}\right] d\theta = \frac{1}{5}Ka^5 \left[\frac{2}{3}\left(\frac{1}{2}\theta - \frac{1}{4}\sin 2\theta\right) + \frac{1}{3}\theta\right]_0^{2\pi}$$

$$= \frac{1}{5}Ka^5 \left[\frac{2}{3}(\pi - 0) + \frac{1}{3}(2\pi - 0)\right] = \frac{4}{15}Ka^5\pi$$

29. In spherical coordinates $z = \sqrt{x^2 + y^2}$ becomes $\cos\phi = \sin\phi$ or $\phi = \frac{\pi}{4}$. Then

$$V = \int_0^{2\pi} \int_0^{\pi/4} \int_0^1 \rho^2 \sin\phi \, d\rho \, d\phi \, d\theta = \int_0^{2\pi} d\theta \int_0^{\pi/4} \sin\phi \, d\phi \int_0^1 \rho^2 \, d\rho = \frac{1}{3}\pi(2 - \sqrt{2}),$$

$$M_{xy} = \int_0^{2\pi} \int_0^{\pi/4} \int_0^1 \rho^3 \sin\phi \cos\phi \, d\rho \, d\phi \, d\theta = 2\pi \left[-\frac{1}{4}\cos 2\phi\right]_0^{\pi/4} \left(\frac{1}{4}\right) = \frac{\pi}{8} \text{ and by symmetry } M_{yz} = M_{xz} = 0.$$

Hence $(\bar{x}, \bar{y}, \bar{z}) = \left(0, 0, \dfrac{3}{8(2 - \sqrt{2})}\right)$.

31. In cylindrical coordinates the paraboloid is given by $z = r^2$ and the plane by $z = 2r\sin\theta$ and they intersect in the circle $r = 2\sin\theta$. Then $\iiint_E z \, dV = \int_0^\pi \int_0^{2\sin\theta} \int_{r^2}^{2r\sin\theta} rz \, dz \, dr \, d\theta = \frac{5\pi}{6}$ [using a CAS].

33. The region E of integration is the region above the paraboloid $z = x^2 + y^2$, or $z = r^2$, and below the paraboloid $z = 2 - x^2 - y^2$, or $z = 2 - r^2$. Also, we have $-1 \le x \le 1$ with $-\sqrt{1 - x^2} \le y \le \sqrt{1 - x^2}$ which describes the unit circle in the xy-plane. Thus,

$$\int_{-1}^1 \int_{-\sqrt{1-x^2}}^{\sqrt{1-x^2}} \int_{x^2+y^2}^{2-x^2-y^2} (x^2 + y^2)^{3/2} \, dz \, dy \, dx = \int_0^{2\pi} \int_0^1 \int_{r^2}^{2-r^2} (r^2)^{3/2} r \, dz \, dr \, d\theta$$

$$= \int_0^{2\pi} \int_0^1 \left[r^4 z\right]_{z=r^2}^{z=2-r^2} dr \, d\theta = \int_0^{2\pi} \int_0^1 (2r^4 - r^6 - r^6) \, dr \, d\theta = \int_0^{2\pi} \left(\frac{2}{5} - \frac{2}{7}\right) d\theta = \frac{8\pi}{35}$$

35. The region of integration E is the top half of the sphere $x^2 + y^2 + z^2 = 9$. So

$$\int_{-3}^3 \int_{-\sqrt{9-x^2}}^{\sqrt{9-x^2}} \int_0^{\sqrt{9-x^2-y^2}} z\sqrt{x^2 + y^2 + z^2} \, dz \, dy \, dx = \iiint_E z\sqrt{x^2 + y^2 + z^2} \, dV$$

$$= \int_0^{2\pi} \int_0^{\pi/2} \int_0^3 (\rho^2 \cos\phi)(\rho^2 \sin\phi) \, d\rho \, d\phi \, d\theta = \int_0^{2\pi} d\theta \int_0^{\pi/2} \cos\phi \sin\phi \, d\phi \int_0^3 \rho^4 \, d\rho$$

$$= \left[\theta\right]_0^{2\pi} \left[\frac{1}{2}\sin^2\phi\right]_0^{\pi/2} \left[\frac{1}{5}\rho^5\right]_0^3 = (2\pi)\left(\frac{1}{2}\right)\left(\frac{243}{5}\right) = \frac{243}{5}\pi$$

37. If E is the solid enclosed by the surface $\rho = 1 + \frac{1}{5}\sin 6\theta \sin 5\phi$, it can be described in spherical coordinates as

$E = \left\{(\rho, \theta, \phi) \mid 0 \le \rho \le 1 + \frac{1}{5}\sin 6\theta \sin 5\phi, 0 \le \theta \le 2\pi, 0 \le \phi \le \pi\right\}$. Its volume is given by

$$V(E) = \iiint_E dV = \int_0^\pi \int_0^{2\pi} \int_0^{1 + (\sin 6\theta \sin 5\phi)/5} \rho^2 \sin\phi \, d\rho \, d\theta \, d\phi = \frac{136\pi}{99} \text{ [using a CAS]}.$$

39. (a) From the diagram, $z = r\cot\phi_0$ to $z = \sqrt{a^2 - r^2}$, $r = 0$ to

$r = a\sin\phi_0$ (or use $a^2 - r^2 = r^2\cot^2\phi_0$). Thus

$$V = \int_0^{2\pi} \int_0^{a\sin\phi_0} \int_{r\cot\phi_0}^{\sqrt{a^2-r^2}} r \, dz \, dr \, d\theta$$

$$= 2\pi \int_0^{a\sin\phi_0} \left(r\sqrt{a^2 - r^2} - r^2\cot\phi_0\right) dr$$

$$= \frac{2\pi}{3} \left[-(a^2 - r^2)^{3/2} - r^3\cot\phi_0\right]_0^{a\sin\phi_0}$$

$$= \frac{2\pi}{3} \left[-\left(a^2 - a^2\sin^2\phi_0\right)^{3/2} - a^3\sin^3\phi_0\cot\phi_0 + a^3\right]$$

$$= \frac{2}{3}\pi a^3 \left[1 - \left(\cos^3\phi_0 + \sin^2\phi_0\cos\phi_0\right)\right] = \frac{2}{3}\pi a^3(1 - \cos\phi_0)$$

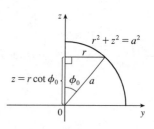

(b) The wedge in question is the shaded area rotated from $\theta = \theta_1$ to $\theta = \theta_2$.

Letting

V_{ij} = volume of the region bounded by the sphere of radius ρ_i

and the cone with angle ϕ_j $(\theta = \theta_1$ to $\theta_2)$

and letting V be the volume of the wedge, we have

$V = (V_{22} - V_{21}) - (V_{12} - V_{11})$

$\quad = \frac{1}{3}(\theta_2 - \theta_1)\left[\rho_2^3(1 - \cos\phi_2) - \rho_2^3(1 - \cos\phi_1) - \rho_1^3(1 - \cos\phi_2) + \rho_1^3(1 - \cos\phi_1)\right]$

$\quad = \frac{1}{3}(\theta_2 - \theta_1)\left[(\rho_2^3 - \rho_1^3)(1 - \cos\phi_2) - (\rho_2^3 - \rho_1^3)(1 - \cos\phi_1)\right]$

$\quad = \frac{1}{3}(\theta_2 - \theta_1)\left[(\rho_2^3 - \rho_1^3)(\cos\phi_1 - \cos\phi_2)\right]$

Or: Show that $V = \displaystyle\int_{\theta_1}^{\theta_2}\int_{\rho_1\sin\phi_1}^{\rho_2\sin\phi_2}\int_{r\cot\phi_2}^{r\cot\phi_1} r\,dz\,dr\,d\theta$.

(c) By the Mean Value Theorem with $f(\rho) = \rho^3$ there exists some $\tilde{\rho}$ with $\rho_1 \le \tilde{\rho} \le \rho_2$ such that

$f(\rho_2) - f(\rho_1) = f'(\tilde{\rho})(\rho_2 - \rho_1)$ or $\rho_1^3 - \rho_2^3 = 3\tilde{\rho}^2\Delta\rho$. Similarly there exists ϕ with $\phi_1 \le \tilde{\phi} \le \phi_2$ such that

$\cos\phi_2 - \cos\phi_1 = \left(-\sin\tilde{\phi}\right)\Delta\phi$. Substituting into the result from (b) gives

$\Delta V = (\tilde{\rho}^2\,\Delta\rho)(\theta_2 - \theta_1)(\sin\tilde{\phi})\,\Delta\phi = \tilde{\rho}^2\sin\tilde{\phi}\,\Delta\rho\,\Delta\phi\,\Delta\theta$.

16.9 Change of Variables in Multiple Integrals ET 15.9

1. $x = u + 4v$, $y = 3u - 2v$.

The Jacobian is $\dfrac{\partial(x, y)}{\partial(u, v)} = \begin{vmatrix} \partial x/\partial u & \partial x/\partial v \\ \partial y/\partial u & \partial y/\partial v \end{vmatrix} = \begin{vmatrix} 1 & 4 \\ 3 & -2 \end{vmatrix} = 1(-2) - 4(3) = -14.$

3. $\dfrac{\partial(x, y)}{\partial(u, v)} = \begin{vmatrix} \dfrac{\partial x}{\partial u} & \dfrac{\partial x}{\partial v} \\ \dfrac{\partial y}{\partial u} & \dfrac{\partial y}{\partial v} \end{vmatrix} = \begin{vmatrix} \dfrac{v}{(u+v)^2} & -\dfrac{u}{(u+v)^2} \\ -\dfrac{v}{(u-v)^2} & \dfrac{u}{(u-v)^2} \end{vmatrix} = \dfrac{uv}{(u+v)^2(u-v)^2} - \dfrac{uv}{(u+v)^2(u-v)^2} = 0$

5. $\dfrac{\partial(x, y, z)}{\partial(u, v, w)} = \begin{vmatrix} \partial x/\partial u & \partial x/\partial v & \partial x/\partial w \\ \partial y/\partial u & \partial y/\partial v & \partial y/\partial w \\ \partial z/\partial u & \partial z/\partial v & \partial z/\partial w \end{vmatrix} = \begin{vmatrix} v & u & 0 \\ 0 & w & v \\ w & 0 & u \end{vmatrix}$

$\quad = v\begin{vmatrix} w & v \\ 0 & u \end{vmatrix} - u\begin{vmatrix} 0 & v \\ w & u \end{vmatrix} + 0\begin{vmatrix} 0 & w \\ w & 0 \end{vmatrix} = v(uw - 0) - u(0 - vw) = 2uvw$

7. The transformation maps the boundary of S to the boundary of the image R, so we first look at side S_1 in the

uv-plane. S_1 is described by $v = 0$ $(0 \le u \le 3)$, so $x = 2u + 3v = 2u$ and $y = u - v = u$. Eliminating u,

we have $x = 2y$, $0 \le x \le 6$. S_2 is the line segment $u = 3$, $0 \le v \le 2$, so $x = 6 + 3v$ and $y = 3 - v$. Then

$v = 3 - y \implies x = 6 + 3(3 - y) = 15 - 3y, 6 \le x \le 12$. S_3 is the line segment $v = 2, 0 \le u \le 3$, so $x = 2u + 6$ and $y = u - 2$, giving $u = y + 2 \implies x = 2y + 10, 6 \le x \le 12$. Finally, S_4 is the segment $u = 0$, $0 \le v \le 2$, so $x = 3v$ and $y = -v \implies x = -3y, 0 \le x \le 6$. The image of set S is the region R shown in the xy-plane, a parallelogram bounded by these four segments.

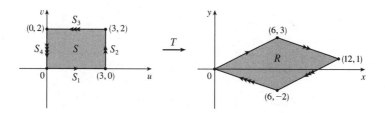

9. S_1 is the line segment $u = v, 0 \le u \le 1$, so $y = v = u$ and $x = u^2 = y^2$. Since $0 \le u \le 1$, the image is the portion of the parabola $x = y^2, 0 \le y \le 1$. S_2 is the segment $v = 1, 0 \le u \le 1$, thus $y = v = 1$ and $x = u^2$, so $0 \le x \le 1$. The image is the line segment $y = 1, 0 \le x \le 1$. S_3 is the segment $u = 0, 0 \le v \le 1$, so $x = u^2 = 0$ and $y = v \implies 0 \le y \le 1$. The image is the segment $x = 0, 0 \le y \le 1$. Thus, the image of S is the region R in the first quadrant bounded by the parabola $x = y^2$, the y-axis, and the line $y = 1$.

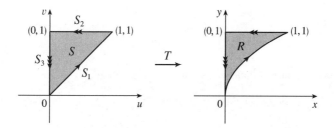

11. $\dfrac{\partial(x, y)}{\partial(u, v)} = \begin{vmatrix} 2 & 1 \\ 1 & 2 \end{vmatrix} = 3$ and $x - 3y = (2u + v) - 3(u + 2v) = -u - 5v$. To find the region S in the uv-plane that corresponds to R we first find the corresponding boundary under the given transformation. The line through $(0, 0)$ and $(2, 1)$ is $y = \frac{1}{2}x$ which is the image of $u + 2v = \frac{1}{2}(2u + v) \implies v = 0$; the line through $(2, 1)$ and $(1, 2)$ is $x + y = 3$ which is the image of $(2u + v) + (u + 2v) = 3 \implies u + v = 1$; the line through $(0, 0)$ and $(1, 2)$ is $y = 2x$ which is the image of $u + 2v = 2(2u + v) \implies u = 0$. Thus S is the triangle $0 \le v \le 1 - u, 0 \le u \le 1$ in the uv-plane and

$$\iint_R (x - 3y) \, dA = \int_0^1 \int_0^{1-u} (-u - 5v) \, |3| \, dv \, du$$

$$= -3 \int_0^1 \left[uv + \tfrac{5}{2}v^2 \right]_{v=0}^{v=1-u} du = -3 \int_0^1 \left(u - u^2 + \tfrac{5}{2}(1 - u)^2 \right) du$$

$$= -3 \left[\tfrac{1}{2}u^2 - \tfrac{1}{3}u^3 - \tfrac{5}{6}(1 - u)^3 \right]_0^1 = -3 \left(\tfrac{1}{2} - \tfrac{1}{3} + \tfrac{5}{6} \right) = -3$$

13. $\dfrac{\partial(x, y)}{\partial(u, v)} = \begin{vmatrix} 2 & 0 \\ 0 & 3 \end{vmatrix} = 6$, $x^2 = 4u^2$ and the planar ellipse $9x^2 + 4y^2 \le 36$ is the image of the disk $u^2 + v^2 \le 1$. Thus

$$\iint_R x^2\, dA = \iint_{u^2+v^2\le 1} (4u^2)(6)\, du\, dv = \int_0^{2\pi} \int_0^1 (24r^2 \cos^2 \theta)\, r\, dr\, d\theta$$

$$= 24 \int_0^{2\pi} \cos^2 \theta\, d\theta \int_0^1 r^3\, dr = 24\Big[\tfrac{1}{2}x + \tfrac{1}{4}\sin 2x\Big]_0^{2\pi} \Big[\tfrac{1}{4}r^4\Big]_0^1$$

$$= 24(\pi)\big(\tfrac{1}{4}\big) = 6\pi$$

15. $\dfrac{\partial(x, y)}{\partial(u, v)} = \begin{vmatrix} 1/v & -u/v^2 \\ 0 & 1 \end{vmatrix} = \dfrac{1}{v}$, $xy = u$, $y = x$ is the image of the parabola $v^2 = u$, $y = 3x$ is the image of the

parabola $v^2 = 3u$, and the hyperbolas $xy = 1$, $xy = 3$ are the images of the lines $u = 1$ and $u = 3$ respectively.
Thus

$$\iint_R xy\, dA = \int_1^3 \int_{\sqrt{u}}^{\sqrt{3u}} u\Big(\frac{1}{v}\Big)\, dv\, du = \int_1^3 u\Big(\ln \sqrt{3u} - \ln \sqrt{u}\Big)\, du$$

$$= \int_1^3 u \ln \sqrt{3}\, du = 4 \ln \sqrt{3} = 2 \ln 3$$

17. (a) $\dfrac{\partial(x, y, z)}{\partial(u, v, w)} = \begin{vmatrix} a & 0 & 0 \\ 0 & b & 0 \\ 0 & 0 & c \end{vmatrix} = abc$ and since $u = \dfrac{x}{a}$, $v = \dfrac{y}{b}$, $w = \dfrac{z}{c}$ the solid enclosed by the ellipsoid is the

image of the ball $u^2 + v^2 + w^2 \le 1$. So

$$\iiint_E dV = \iiint_{u^2+v^2+w^2 \le 1} abc\, du\, dv\, dw = (abc)(\text{volume of the ball}) = \tfrac{4}{3}\pi abc$$

(b) If we approximate the surface of Earth by the ellipsoid $\dfrac{x^2}{6378^2} + \dfrac{y^2}{6378^2} + \dfrac{z^2}{6356^2} = 1$, then we can estimate

the volume of Earth by finding the volume of the solid E enclosed by the ellipsoid. From part (a), this is

$$\iiint_E dV = \tfrac{4}{3}\pi (6378)(6378)(6356) \approx 1.083 \times 10^{12}\ \text{km}^3.$$

19. Letting $u = x - 2y$ and $v = 3x - y$, we have $x = \tfrac{1}{5}(2v - u)$ and $y = \tfrac{1}{5}(v - 3u)$. Then

$\dfrac{\partial(x, y)}{\partial(u, v)} = \begin{vmatrix} -1/5 & 2/5 \\ -3/5 & 1/5 \end{vmatrix} = \dfrac{1}{5}$ and R is the image of the rectangle enclosed by the lines $u = 0$, $u = 4$, $v = 1$, and

$v = 8$. Thus $\displaystyle\iint_R \frac{x - 2y}{3x - y}\, dA = \int_0^4 \int_1^8 \frac{u}{v} \left| \frac{1}{5} \right| dv\, du = \frac{1}{5} \int_0^4 u\, du \int_1^8 \frac{1}{v}\, dv = \frac{1}{5}\Big[\tfrac{1}{2}u^2\Big]_0^4 \big[\ln |v|\big]_1^8 = \frac{8}{5} \ln 8.$

21. Letting $u = y - x$, $v = y + x$, we have $y = \tfrac{1}{2}(u + v)$, $x = \tfrac{1}{2}(v - u)$. Then $\dfrac{\partial(x, y)}{\partial(u, v)} = \begin{vmatrix} -1/2 & 1/2 \\ 1/2 & 1/2 \end{vmatrix} = -\dfrac{1}{2}$ and

R is the image of the trapezoidal region with vertices $(-1, 1)$, $(-2, 2)$, $(2, 2)$, and $(1, 1)$. Thus

$$\iint_R \cos \frac{y - x}{y + x}\, dA = \int_1^2 \int_{-v}^{v} \cos \frac{u}{v} \left| -\frac{1}{2} \right| du\, dv = \frac{1}{2} \int_1^2 \Big[v \sin \frac{u}{v} \Big]_{u=-v}^{u=v} dv$$

$$= \tfrac{1}{2} \int_1^2 2v \sin(1)\, dv = \tfrac{3}{2} \sin 1$$

23. Let $u = x + y$ and $v = -x + y$. Then $u + v = 2y$ $\Rightarrow$ $y = \frac{1}{2}(u + v)$ and

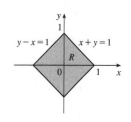

$$u - v = 2x \quad \Rightarrow \quad x = \frac{1}{2}(u - v). \ \frac{\partial(x, y)}{\partial(u, v)} = \begin{vmatrix} 1/2 & -1/2 \\ 1/2 & 1/2 \end{vmatrix} = \frac{1}{2}.$$

Now $|u| = |x + y| \leq |x| + |y| \leq 1$ $\Rightarrow$ $-1 \leq u \leq 1$, and

$|v| = |-x + y| \leq |x| + |y| \leq 1 \Rightarrow$ $-1 \leq v \leq 1$. R is the image of

the square region with vertices $(1, 1)$, $(1, -1)$, $(-1, -1)$, and $(-1, 1)$.

So $\iint_R e^{x+y} \, dA = \frac{1}{2} \int_{-1}^{1} \int_{-1}^{1} e^u \, du \, dv = \frac{1}{2} \left[e^u \right]_{-1}^{1} \left[v \right]_{-1}^{1} = e - e^{-1}$.

16 Review   ET 15

─────────────────────────────── CONCEPT CHECK ───────────────────────────────

1. (a) A double Riemann sum of f is $\sum_{i=1}^{m} \sum_{j=1}^{n} f\left(x_{ij}^*, y_{ij}^*\right) \Delta A$, where ΔA is the area of each subrectangle and

$\left(x_{ij}^*, y_{ij}^*\right)$ is a sample point in each subrectangle. If $f(x, y) \geq 0$, this sum represents an approximation to the

volume of the solid that lies above the rectangle R and below the graph of f.

(b) $\iint_R f(x, y) \, dA = \lim_{m, n \to \infty} \sum_{i=1}^{m} \sum_{j=1}^{n} f\left(x_{ij}^*, y_{ij}^*\right) \Delta A$

(c) If $f(x, y) \geq 0$, $\iint_R f(x, y) \, dA$ represents the volume of the solid that lies above the rectangle R and below the

surface $z = f(x, y)$. If f takes on both positive and negative values, $\iint_R f(x, y) \, dA$ is the difference of the

volume above R but below the surface $z = f(x, y)$ and the volume below R but above the surface $z = f(x, y)$.

(d) We usually evaluate $\iint_R f(x, y) \, dA$ as an iterated integral according to Fubini's Theorem

(see Theorem 16.2.4 [ET 15.2.4]).

(e) The Midpoint Rule for Double Integrals says that we approximate the double integral $\iint_R f(x, y) \, dA$ by the

double Riemann sum $\sum_{i=1}^{m} \sum_{j=1}^{n} f\left(\overline{x}_i, \overline{y}_j\right) \Delta A$ where the sample points $\left(\overline{x}_i, \overline{y}_j\right)$ are the centers of the

subrectangles.

(f) $f_{\text{ave}} = \dfrac{1}{A(R)} \iint_R f(x, y) \, dA$ where $A(R)$ is the area of R.

2. (a) See (1) and (2) and the accompanying discussion in Section 16.3 [ET 15.3].

(b) See (3) and the accompanying discussion in Section 16.3 [ET 15.3].

(c) See (5) and the preceding discussion in Section 16.3 [ET 15.3].

(d) See (6)–(11) in Section 16.3 [ET 15.3].

3. We may want to change from rectangular to polar coordinates in a double integral if the region R of

integration is more easily described in polar coordinates. To accomplish this, we use

$\iint_R f(x, y) \, dA = \int_{\alpha}^{\beta} \int_{a}^{b} f(r \cos \theta, r \sin \theta) \, r \, dr \, d\theta$ where R is given by $0 \leq a \leq r \leq b, \alpha \leq \theta \leq \beta$.

4. (a) $m = \iint_D \rho(x, y) \, dA$

(b) $M_x = \iint_D y\rho(x, y) \, dA$, $M_y = \iint_D x\rho(x, y) \, dA$

(c) The center of mass is $(\overline{x}, \overline{y})$ where $\overline{x} = \dfrac{M_y}{m}$ and $\overline{y} = \dfrac{M_x}{m}$.

(d) $I_x = \iint_D y^2\rho(x, y) \, dA$, $I_y = \iint_D x^2\rho(x, y) \, dA$, $I_0 = \iint_D (x^2 + y^2)\rho(x, y) \, dA$

5. (a) $P(a \leq X \leq b, c \leq Y \leq d) = \int_a^b \int_c^d f(x,y) \, dy \, dx$

(b) $f(x,y) \geq 0$ and $\iint_{\mathbb{R}^2} f(x,y) \, dA = 1$.

(c) The expected value of X is $\mu_1 = \iint_{\mathbb{R}^2} x f(x,y) \, dA$; the expected value of Y is $\mu_2 = \iint_{\mathbb{R}^2} y f(x,y) \, dA$.

6. $A(S) = \iint_D \sqrt{[f_x(x,y)]^2 + [f_y(x,y)]^2 + 1} \, dA$

7. (a) $\iiint_B f(x,y,z) \, dV = \lim\limits_{l,m,n \to \infty} \sum\limits_{i=1}^{l} \sum\limits_{j=1}^{m} \sum\limits_{k=1}^{n} f(x_{ijk}^*, y_{ijk}^*, z_{ijk}^*) \, \Delta V$

(b) We usually evaluate $\iiint_B f(x,y,z) \, dV$ as an iterated integral according to Fubini's Theorem for Triple Integrals (see Theorem 16.7.4 [ET 15.7.4]).

(c) See the paragraph following Example 16.7.1 [ET 15.7.1].

(d) See (5) and (6) and the accompanying discussion in Section 16.7 [ET 15.7].

(e) See (10) and the accompanying discussion in Section 16.7 [ET 15.7].

(f) See (11) and the preceding discussion in Section 16.7 [ET 15.7].

8. (a) $m = \iiint_E \rho(x,y,z) \, dV$

(b) $M_{yz} = \iiint_E x \rho(x,y,z) \, dV$, $M_{xz} = \iiint_E y \rho(x,y,z) \, dV$, $M_{xy} = \iiint_E z \rho(x,y,z) \, dV$.

(c) The center of mass is $(\bar{x}, \bar{y}, \bar{z})$ where $\bar{x} = \dfrac{M_{yz}}{m}$, $\bar{y} = \dfrac{M_{xz}}{m}$, and $\bar{z} = \dfrac{M_{xy}}{m}$.

(d) $I_x = \iiint_E (y^2 + z^2) \rho(x,y,z) \, dV$, $I_y = \iiint_E (x^2 + z^2) \rho(x,y,z) \, dV$, $I_z = \iiint_E (x^2 + y^2) \rho(x,y,z) \, dV$.

9. (a) See Formula 16.8.2 [ET 15.8.2] and the accompanying discussion.

(b) See Formula 16.8.4 [ET 15.8.4] and the accompanying discussion.

(c) We may want to change from rectangular to cylindrical or spherical coordinates in a triple integral if the region E of integration is more easily described in cylindrical or spherical coordinates or if the triple integral is easier to evaluate using cylindrical or spherical coordinates.

10. (a) $\dfrac{\partial (x,y)}{\partial (u,v)} = \begin{vmatrix} \partial x/\partial u & \partial x/\partial v \\ \partial y/\partial u & \partial y/\partial v \end{vmatrix} = \dfrac{\partial x}{\partial u} \dfrac{\partial y}{\partial v} - \dfrac{\partial x}{\partial v} \dfrac{\partial y}{\partial u}$

(b) See (9) and the accompanying discussion in Section 16.9 [ET 15.9].

(c) See (13) and the accompanying discussion in Section 16.9 [ET 15.9].

─────────────── TRUE-FALSE QUIZ ───────────────

1. This is true by Fubini's Theorem.

3. True. See the discussion following Example 4 on page 1029 [ET page 993].

5. True:

$\iint_D \sqrt{4 - x^2 - y^2} \, dA =$ the volume under the surface $x^2 + y^2 + z^2 = 4$ and above the xy-plane

$= \frac{1}{2}$ (the volume of the sphere $x^2 + y^2 + z^2 = 4$) $= \frac{1}{2} \cdot \frac{4}{3}\pi(2)^3 = \frac{16}{3}\pi$

7. The volume enclosed by the cone $z = \sqrt{x^2 + y^2}$ and the plane $z = 2$ is, in cylindrical coordinates,

$V = \int_0^{2\pi} \int_0^2 \int_r^2 r \, dz \, dr \, d\theta \neq \int_0^{2\pi} \int_0^2 \int_r^2 dz \, dr \, d\theta$, so the assertion is false.

——————————————— EXERCISES ———————————————

1. As shown in the contour map, we divide R into 9 equally sized subsquares, each with area $\Delta A = 1$. Then we approximate $\iint_R f(x, y)\, dA$ by a Riemann sum with $m = n = 3$ and the sample points the upper right corners of each square, so

$$\iint_R f(x, y)\, dA \approx \sum_{i=1}^{3} \sum_{j=1}^{3} f(x_i, y_j)\, \Delta A$$

$$= \Delta A\, [f(1, 1) + f(1, 2) + f(1, 3) + f(2, 1) + f(2, 2)$$

$$+ f(2, 3) + f(3, 1) + f(3, 2) + f(3, 3)]$$

Using the contour lines to estimate the function values, we have

$$\iint_R f(x, y)\, dA \approx 1[2.7 + 4.7 + 8.0 + 4.7 + 6.7 + 10.0 + 6.7 + 8.6 + 11.9] \approx 64.0$$

3. $\int_1^2 \int_0^2 (y + 2xe^y)\, dx\, dy = \int_1^2 \left[xy + x^2 e^y \right]_{x=0}^{x=2} dy = \int_1^2 (2y + 4e^y)\, dy = \left[y^2 + 4e^y \right]_1^2$
$$= 4 + 4e^2 - 1 - 4e = 4e^2 - 4e + 3$$

5. $\int_0^1 \int_0^x \cos(x^2)\, dy\, dx = \int_0^1 \left[\cos(x^2) y \right]_{y=0}^{y=x} dx = \int_0^1 x\cos(x^2)\, dx = \frac{1}{2} \sin(x^2) \Big]_0^1 = \frac{1}{2} \sin 1$

7. $\int_0^\pi \int_0^1 \int_0^{\sqrt{1-y^2}} y \sin x\, dz\, dy\, dx = \int_0^\pi \int_0^1 \left[(y \sin x) z \right]_{z=0}^{z=\sqrt{1-y^2}} dy\, dx = \int_0^\pi \int_0^1 y \sqrt{1-y^2} \sin x\, dy\, dx$
$$= \int_0^\pi \left[-\frac{1}{3}(1-y^2)^{3/2} \sin x \right]_{y=0}^{y=1} dx = \int_0^\pi \frac{1}{3} \sin x\, dx = -\frac{1}{3} \cos x \Big]_0^\pi = \frac{2}{3}$$

9. The region R is more easily described by polar coordinates: $R = \{(r, \theta) \mid 2 \le r \le 4, 0 \le \theta \le \pi\}$. Thus
$$\iint_R f(x, y)\, dA = \int_0^\pi \int_2^4 f(r\cos\theta, r\sin\theta)\, r\, dr\, d\theta.$$

11.

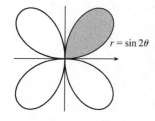

$r = \sin 2\theta$

The region whose area is given by $\int_0^{\pi/2} \int_0^{\sin 2\theta} r\, dr\, d\theta$ is

$\{(r, \theta) \mid 0 \le \theta \le \frac{\pi}{2}, 0 \le r \le \sin 2\theta\}$, which is the region contained in the loop in the first quadrant of the four-leaved rose $r = \sin 2\theta$.

13.

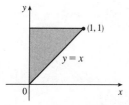

$\int_0^1 \int_x^1 \cos(y^2)\, dy\, dx = \int_0^1 \int_0^y \cos(y^2)\, dx\, dy$

$$= \int_0^1 \cos(y^2) [x]_{x=0}^{x=y}\, dy = \int_0^1 y\cos(y^2)\, dy$$

$$= \left[\frac{1}{2} \sin(y^2) \right]_0^1 = \frac{1}{2} \sin 1$$

15. $\iint_R ye^{xy}\, dA = \int_0^3 \int_0^2 ye^{xy}\, dx\, dy = \int_0^3 \left[e^{xy} \right]_{x=0}^{x=2} dy = \int_0^3 (e^{2y} - 1)\, dy = \left[\frac{1}{2}e^{2y} - y \right]_0^3$
$$= \frac{1}{2}e^6 - 3 - \frac{1}{2} = \frac{1}{2}e^6 - \frac{7}{2}$$

17.

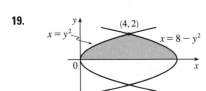

$$\iint_D \frac{y}{1+x^2}\, dA = \int_0^1 \int_0^{\sqrt{x}} \frac{y}{1+x^2}\, dy\, dx = \int_0^1 \frac{1}{1+x^2} \left[\tfrac{1}{2}y^2\right]_{y=0}^{y=\sqrt{x}} dx$$

$$= \tfrac{1}{2}\int_0^1 \frac{x}{1+x^2}\, dx = \left[\tfrac{1}{4}\ln(1+x^2)\right]_0^1 = \tfrac{1}{4}\ln 2$$

19.

$$\iint_D y\, dA = \int_0^2 \int_{y^2}^{8-y^2} y\, dx\, dy$$

$$= \int_0^2 y\left[x\right]_{x=y^2}^{x=8-y^2} dy = \int_0^2 y(8 - y^2 - y^2)\, dy$$

$$= \int_0^2 (8y - 2y^3)\, dy = \left[4y^2 - \tfrac{1}{2}y^4\right]_0^2 = 8$$

21.

$$\iint_D (x^2 + y^2)^{3/2}\, dA = \int_0^{\pi/3} \int_0^3 (r^2)^{3/2} r\, dr\, d\theta$$

$$= \int_0^{\pi/3} d\theta \int_0^3 r^4\, dr = \left[\theta\right]_0^{\pi/3} \left[\tfrac{1}{5}r^5\right]_0^3$$

$$= \frac{\pi}{3}\frac{3^5}{5} = \frac{81\pi}{5}$$

23. $\iiint_E xy\, dV = \int_0^3 \int_0^x \int_0^{x+y} xy\, dz\, dy\, dx = \int_0^3 \int_0^x xy\left[z\right]_{z=0}^{z=x+y} dy\, dx = \int_0^3 \int_0^x xy(x+y)\, dy\, dx$

$= \int_0^3 \int_0^x (x^2y + xy^2)\, dy\, dx = \int_0^3 \left[\tfrac{1}{2}x^2y^2 + \tfrac{1}{3}xy^3\right]_{y=0}^{y=x} dx = \int_0^3 \left(\tfrac{1}{2}x^4 + \tfrac{1}{3}x^4\right) dx$

$= \tfrac{5}{6}\int_0^3 x^4\, dx = \left[\tfrac{1}{6}x^5\right]_0^3 = \tfrac{81}{2} = 40.5$

25. $\iiint_E y^2 z^2\, dV = \int_{-1}^1 \int_{-\sqrt{1-y^2}}^{\sqrt{1-y^2}} \int_0^{1-y^2-z^2} y^2 z^2\, dx\, dz\, dy = \int_{-1}^1 \int_{-\sqrt{1-y^2}}^{\sqrt{1-y^2}} y^2 z^2 (1 - y^2 - z^2)\, dz\, dy$

$= \int_0^{2\pi} \int_0^1 (r^2 \cos^2\theta)(r^2 \sin^2\theta)(1 - r^2) r\, dr\, d\theta = \int_0^{2\pi} \int_0^1 \tfrac{1}{4}\sin^2 2\theta (r^5 - r^7)\, dr\, d\theta$

$= \int_0^{2\pi} \tfrac{1}{8}(1 - \cos 4\theta)\left[\tfrac{1}{6}r^6 - \tfrac{1}{8}r^8\right]_{r=0}^{r=1} d\theta = \tfrac{1}{192}\left[\theta - \tfrac{1}{4}\sin 4\theta\right]_0^{2\pi} = \tfrac{2\pi}{192} = \tfrac{\pi}{96}$

27. $\iiint_E yz\, dV = \int_{-2}^2 \int_0^{\sqrt{4-x^2}} \int_0^y yz\, dz\, dy\, dx = \int_{-2}^2 \int_0^{\sqrt{4-x^2}} \tfrac{1}{2}y^3\, dy\, dx = \int_0^\pi \int_0^2 \tfrac{1}{2}r^3 (\sin^3\theta) r\, dr\, d\theta$

$= \tfrac{16}{5}\int_0^\pi \sin^3\theta\, d\theta = \tfrac{16}{5}\left[-\cos\theta + \tfrac{1}{3}\cos^3\theta\right]_0^\pi = \tfrac{64}{15}$

29. $V = \int_0^2 \int_1^4 (x^2 + 4y^2)\, dy\, dx = \int_0^2 \left[x^2 y + \tfrac{4}{3}y^3\right]_{y=1}^{y=4} dx = \int_0^2 (3x^2 + 84)\, dx = 176$

31.

$$V = \int_0^2 \int_0^y \int_0^{(2-y)/2} dz\, dx\, dy$$

$$= \int_0^2 \int_0^y \left(1 - \tfrac{1}{2}y\right) dx\, dy$$

$$= \int_0^2 \left(y - \tfrac{1}{2}y^2\right) dy = \tfrac{2}{3}$$

33. Using the wedge above the plane $z = 0$ and below the plane $z = mx$ and noting that we have the same volume for $m < 0$ as for $m > 0$ (so use $m > 0$), we have

$$V = 2 \int_0^{a/3} \int_0^{\sqrt{a^2 - 9y^2}} mx \, dx \, dy = 2 \int_0^{a/3} \tfrac{1}{2} m (a^2 - 9y^2) \, dy = m \left[a^2 y - 3y^3 \right]_0^{a/3}$$

$$= m \left(\tfrac{1}{3} a^3 - \tfrac{1}{9} a^3 \right) = \tfrac{2}{9} m a^3$$

35. (a) $m = \int_0^1 \int_0^{1-y^2} y \, dx \, dy = \int_0^1 (y - y^3) \, dy = \tfrac{1}{2} - \tfrac{1}{4} = \tfrac{1}{4}$

(b) $M_y = \int_0^1 \int_0^{1-y^2} xy \, dx \, dy = \int_0^1 \tfrac{1}{2} y (1 - y^2)^2 \, dy = -\tfrac{1}{12} (1 - y^2)^3 \big]_0^1 = \tfrac{1}{12}$,

$M_x = \int_0^1 \int_0^{1-y^2} y^2 \, dx \, dy = \int_0^1 (y^2 - y^4) \, dy = \tfrac{2}{15}$. Hence $(\overline{x}, \overline{y}) = \left(\tfrac{1}{3}, \tfrac{8}{15} \right)$.

(c) $I_x = \int_0^1 \int_0^{1-y^2} y^3 \, dx \, dy = \int_0^1 (y^3 - y^5) \, dy = \tfrac{1}{12}$,

$I_y = \int_0^1 \int_0^{1-y^2} yx^2 \, dx \, dy = \int_0^1 \tfrac{1}{3} y (1 - y^2)^3 \, dy = -\tfrac{1}{24} (1 - y^2)^4 \big]_0^1 = \tfrac{1}{24}$,

$I_0 = I_x + I_y = \tfrac{1}{8}, \overline{\overline{y}}^2 = \tfrac{1/12}{1/4} = \tfrac{1}{3} \ \Rightarrow \ \overline{\overline{y}} = \tfrac{1}{\sqrt{3}}$, and $\overline{\overline{x}}^2 = \tfrac{1/24}{1/4} = \tfrac{1}{6} \ \Rightarrow \ \overline{\overline{x}} = \tfrac{1}{\sqrt{6}}$.

37. (a) The equation of the cone with the suggested orientation is $(h - z) = \tfrac{h}{a} \sqrt{x^2 + y^2}, 0 \le z \le h$. Then $V = \tfrac{1}{3} \pi a^2 h$ is the volume of one frustum of a cone; by symmetry $M_{yz} = M_{xz} = 0$; and

$$M_{xy} = \iint_{x^2 + y^2 \le a^2} \int_0^{h - (h/a)\sqrt{x^2 + y^2}} z \, dz \, dA = \int_0^{2\pi} \int_0^a \int_0^{(h/a)(a-r)} rz \, dz \, dr \, d\theta$$

$$= \pi \int_0^a r \frac{h^2}{a^2} (a - r)^2 \, dr = \frac{\pi h^2}{a^2} \int_0^a (a^2 r - 2ar^2 + r^3) \, dr = \frac{\pi h^2}{a^2} \left(\frac{a^4}{2} - \frac{2a^4}{3} + \frac{a^4}{4} \right) = \frac{\pi h^2 a^2}{12}$$

Hence the centroid is $(\overline{x}, \overline{y}, \overline{z}) = \left(0, 0, \tfrac{1}{4} h \right)$.

(b) $I_z = \int_0^{2\pi} \int_0^a \int_0^{(h/a)(a-r)} r^3 \, dz \, dr \, d\theta = 2\pi \int_0^a \frac{h}{a} (ar^3 - r^4) \, dr = \frac{2\pi h}{a} \left(\frac{a^5}{4} - \frac{a^5}{5} \right) = \frac{\pi a^4 h}{10}$

39. Let D represent the given triangle; then D can be described as the area enclosed by the x- and y-axes and the line $y = 2 - 2x$, or equivalently $D = \{(x, y) \mid 0 \le x \le 1, 0 \le y \le 2 - 2x\}$. We want to find the surface area of the part of the graph of $z = x^2 + y$ that lies over D, so using Equation 16.6.3 [ET 15.6.3] we have

$$A(S) = \iint_D \sqrt{1 + \left(\frac{\partial z}{\partial x} \right)^2 + \left(\frac{\partial z}{\partial y} \right)^2} \, dA = \iint_D \sqrt{1 + (2x)^2 + (1)^2} \, dA$$

$$= \int_0^1 \int_0^{2-2x} \sqrt{2 + 4x^2} \, dy \, dx = \int_0^1 \sqrt{2 + 4x^2} \left[y \right]_{y=0}^{y=2-2x} \, dx = \int_0^1 (2 - 2x) \sqrt{2 + 4x^2} \, dx$$

$$= \int_0^1 2 \sqrt{2 + 4x^2} \, dx - \int_0^1 2x \sqrt{2 + 4x^2} \, dx$$

Using Formula 21 in the Table of Integrals with $a = \sqrt{2}, u = 2x$, and $du = 2 \, dx$, we have

$\int 2 \sqrt{2 + 4x^2} \, dx = x \sqrt{2 + 4x^2} + \ln \left(2x + \sqrt{2 + 4x^2} \right)$. If we substitute $u = 2 + 4x^2$ in the second integral,

then $du = 8x \, dx$ and $\int 2x \sqrt{2 + 4x^2} \, dx = \tfrac{1}{4} \int \sqrt{u} \, du = \tfrac{1}{4} \cdot \tfrac{2}{3} u^{3/2} = \tfrac{1}{6} (2 + 4x^2)^{3/2}$. Thus

$$A(S) = \left[x \sqrt{2 + 4x^2} + \ln \left(2x + \sqrt{2 + 4x^2} \right) - \tfrac{1}{6} (2 + 4x^2)^{3/2} \right]_0^1$$

$$= \sqrt{6} + \ln \left(2 + \sqrt{6} \right) - \tfrac{1}{6} (6)^{3/2} - \ln \sqrt{2} + \tfrac{\sqrt{2}}{3} = \ln \frac{2 + \sqrt{6}}{\sqrt{2}} + \tfrac{\sqrt{2}}{3}$$

$$= \ln \left(\sqrt{2} + \sqrt{3} \right) + \tfrac{\sqrt{2}}{3} \approx 1.6176$$

41.

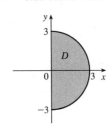

$$\int_0^3 \int_{-\sqrt{9-x^2}}^{\sqrt{9-x^2}} (x^3 + xy^2)\, dy\, dx = \int_0^3 \int_{-\sqrt{9-x^2}}^{\sqrt{9-x^2}} x(x^2 + y^2)\, dy\, dx$$

$$= \int_{-\pi/2}^{\pi/2} \int_0^3 (r\cos\theta)(r^2)\, r\, dr\, d\theta$$

$$= \int_{-\pi/2}^{\pi/2} \cos\theta\, d\theta \int_0^3 r^4\, dr$$

$$= \big[\sin\theta\big]_{-\pi/2}^{\pi/2} \big[\tfrac{1}{5}r^5\big]_0^3 = 2\cdot\tfrac{1}{5}(243) = \tfrac{486}{5} = 97.2$$

43. From the graph, it appears that $1 - x^2 = e^x$ at $x \approx -0.71$ and at

$x = 0$, with $1 - x^2 > e^x$ on $(-0.71, 0)$. So the desired integral is

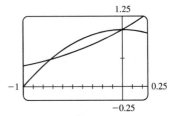

$$\iint_D y^2\, dA \approx \int_{-0.71}^0 \int_{e^x}^{1-x^2} y^2\, dy\, dx$$

$$= \tfrac{1}{3}\int_{-0.71}^0 [(1-x^2)^3 - e^{3x}]\, dx$$

$$= \tfrac{1}{3}\big[x - x^3 + \tfrac{3}{5}x^5 - \tfrac{1}{7}x^7 - \tfrac{1}{3}e^{3x}\big]_{-0.71}^0 \approx 0.0512$$

45. (a) $f(x, y)$ is a joint density function, so we know that $\iint_{\mathbb{R}^2} f(x, y)\, dA = 1$. Since $f(x, y) = 0$ outside the

rectangle $[0, 3] \times [0, 2]$, we can say

$$\iint_{\mathbb{R}^2} f(x, y)\, dA = \int_{-\infty}^{\infty}\int_{-\infty}^{\infty} f(x, y)\, dy\, dx = \int_0^3\int_0^2 C(x + y)\, dy\, dx$$

$$= C\int_0^3 \big[xy + \tfrac{1}{2}y^2\big]_{y=0}^{y=2}\, dx = C\int_0^3 (2x + 2)\, dx = C\big[x^2 + 2x\big]_0^3 = 15C$$

Then $15C = 1 \;\Rightarrow\; C = \tfrac{1}{15}$.

(b) $P(X \le 2, Y \ge 1) = \int_{-\infty}^{2}\int_1^{\infty} f(x, y)\, dy\, dx = \int_0^2\int_1^2 \tfrac{1}{15}(x, y)\, dy\, dx = \tfrac{1}{15}\int_0^2 \big[xy + \tfrac{1}{2}y^2\big]_{y=1}^{y=2}\, dx$

$\qquad = \tfrac{1}{15}\int_0^2 \big(x + \tfrac{3}{2}\big)\, dx = \tfrac{1}{15}\big[\tfrac{1}{2}x^2 + \tfrac{3}{2}x\big]_0^2 = \tfrac{1}{3}$

(c) $P(X + Y \le 1) = P((X, Y) \in D)$ where D is the triangular region

shown in the figure. Thus

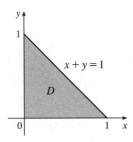

$$P(X + Y \le 1) = \iint_D f(x, y)\, dA = \int_0^1\int_0^{1-x} \tfrac{1}{15}(x + y)\, dy\, dx$$

$$= \tfrac{1}{15}\int_0^1 \big[xy + \tfrac{1}{2}y^2\big]_{y=0}^{y=1-x}\, dx$$

$$= \tfrac{1}{15}\int_0^1 \big[x(1 - x) + \tfrac{1}{2}(1 - x)^2\big]\, dx$$

$$= \tfrac{1}{30}\int_0^1 (1 - x^2)\, dx = \tfrac{1}{30}\big[x - \tfrac{1}{3}x^3\big]_0^1 = \tfrac{1}{45}$$

47.

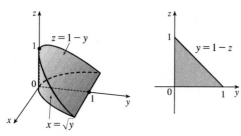

$$\int_{-1}^{1} \int_{x^2}^{1} \int_{0}^{1-y} f(x, y, z) \, dz \, dy \, dx = \int_{0}^{1} \int_{0}^{1-z} \int_{-\sqrt{y}}^{\sqrt{y}} f(x, y, z) \, dx \, dy \, dz$$

49. Since $u = x - y$ and $v = x + y$, $x = \frac{1}{2}(u + v)$ and $y = \frac{1}{2}(v - u)$.

Thus $\dfrac{\partial(x, y)}{\partial(u, v)} = \begin{vmatrix} 1/2 & 1/2 \\ -1/2 & 1/2 \end{vmatrix} = \dfrac{1}{2}$ and $\displaystyle\iint_R \dfrac{x - y}{x + y} \, dA = \int_2^4 \int_{-2}^{0} \dfrac{u}{v} \left(\dfrac{1}{2}\right) du \, dv = -\int_2^4 \dfrac{dv}{v} = -\ln 2$.

51. Let $u = y - x$ and $v = y + x$ so $x = y - u = (v - x) - u \;\Rightarrow\; x = \frac{1}{2}(v - u)$ and

$y = v - \frac{1}{2}(v - u) = \frac{1}{2}(v + u)$. $\left| \dfrac{\partial(x, y)}{\partial(u, v)} \right| = \left| \dfrac{\partial x}{\partial u}\dfrac{\partial y}{\partial v} - \dfrac{\partial x}{\partial v}\dfrac{\partial y}{\partial u} \right| = \left| -\frac{1}{2}\left(\frac{1}{2}\right) - \frac{1}{2}\left(\frac{1}{2}\right) \right| = \left| -\frac{1}{2} \right| = \frac{1}{2}$. R is the

image under this transformation of the square with vertices $(u, v) = (0, 0)$, $(-2, 0)$, $(0, 2)$, and $(-2, 2)$. So

$$\iint_R xy \, dA = \int_0^2 \int_{-2}^{0} \dfrac{v^2 - u^2}{4}\left(\dfrac{1}{2}\right) du \, dv = \frac{1}{8}\int_0^2 \left[v^2 u - \frac{1}{3}u^3\right]_{u=-2}^{u=0} dv = \frac{1}{8}\int_0^2 \left(2v^2 - \frac{8}{3}\right) dv$$

$$= \frac{1}{8}\left[\frac{2}{3}v^3 - \frac{8}{3}v\right]_0^2 = 0$$

This result could have been anticipated by symmetry, since the integrand is an odd function of y and R is symmetric
about the x-axis.

53. For each r such that D_r lies within the domain, $A(D_r) = \pi r^2$, and by the Mean Value Theorem for Double

Integrals there exists (x_r, y_r) in D_r such that $f(x_r, y_r) = \dfrac{1}{\pi r^2} \displaystyle\iint_{D_r} f(x, y) \, dA$. But $\displaystyle\lim_{r \to 0^+} (x_r, y_r) = (a, b)$,

so $\displaystyle\lim_{r \to 0^+} \dfrac{1}{\pi r^2} \iint_{D_r} f(x, y) \, dA = \lim_{r \to 0^+} f(x_r, y_r) = f(a, b)$ by the continuity of f.

□ PROBLEMS PLUS

1.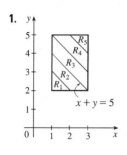

$x+y=5$

Let $R = \bigcup_{i=1}^{5} R_i$, where

$$R_i = \{(x, y) \mid x + y \geq i + 2, x + y < i + 3, 1 \leq x \leq 3, 2 \leq y \leq 5\}.$$

$\iint_R [\![x + y]\!]\, dA = \sum_{i=1}^{5} \iint_{R_i} [\![x + y]\!]\, dA = \sum_{i=1}^{5} [\![x + y]\!] \iint_{R_i} dA$, since

$[\![x + y]\!] = \text{constant} = i + 2$ for $(x, y) \in R_i$. Therefore

$$\iint_R [\![x + y]\!]\, dA = \sum_{i=1}^{5} (i + 2)\, [A(R_i)]$$

$$= 3A(R_1) + 4A(R_2) + 5A(R_3) + 6A(R_4) + 7A(R_5)$$

$$= 3\left(\tfrac{1}{2}\right) + 4\left(\tfrac{3}{2}\right) + 5(2) + 6\left(\tfrac{3}{2}\right) + 7\left(\tfrac{1}{2}\right) = 30$$

3. $f_{\text{ave}} = \dfrac{1}{b - a} \displaystyle\int_a^b f(x)\, dx = \dfrac{1}{1 - 0} \int_0^1 \left[\int_x^1 \cos(t^2)\, dt \right] dx$

$= \int_0^1 \int_x^1 \cos(t^2)\, dt\, dx$

$= \int_0^1 \int_0^t \cos(t^2)\, dx\, dt$ [changing the order of integration]

$= \int_0^1 t \cos(t^2)\, dt = \tfrac{1}{2} \sin(t^2) \big]_0^1 = \tfrac{1}{2} \sin 1$

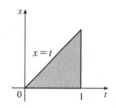

$x = t$

5. Since $|xy| < 1$, except at $(1, 1)$, the formula for the sum of a geometric series gives $\dfrac{1}{1 - xy} = \displaystyle\sum_{n=0}^{\infty} (xy)^n$, so

$$\int_0^1 \int_0^1 \frac{1}{1 - xy}\, dx\, dy = \int_0^1 \int_0^1 \sum_{n=0}^{\infty} (xy)^n\, dx\, dy = \sum_{n=0}^{\infty} \int_0^1 \int_0^1 (xy)^n\, dx\, dy = \sum_{n=0}^{\infty} \left[\int_0^1 x^n\, dx \right]\left[\int_0^1 y^n\, dy \right]$$

$$= \sum_{n=0}^{\infty} \frac{1}{n + 1} \cdot \frac{1}{n + 1} = \sum_{n=0}^{\infty} \frac{1}{(n + 1)^2} = \frac{1}{1^2} + \frac{1}{2^2} + \frac{1}{3^2} + \cdots = \sum_{n=1}^{\infty} \frac{1}{n^2}$$

7. (a) Since $|xyz| < 1$ except at $(1, 1, 1)$, the formula for the sum of a geometric series gives $\dfrac{1}{1 - xyz} = \displaystyle\sum_{n=0}^{\infty} (xyz)^n$,

so

$$\int_0^1 \int_0^1 \int_0^1 \frac{1}{1 - xyz}\, dx\, dy\, dz = \int_0^1 \int_0^1 \int_0^1 \sum_{n=0}^{\infty} (xyz)^n\, dx\, dy\, dz = \sum_{n=0}^{\infty} \int_0^1 \int_0^1 \int_0^1 (xyz)^n\, dx\, dy\, dz$$

$$= \sum_{n=0}^{\infty} \left[\int_0^1 x^n\, dx \right]\left[\int_0^1 y^n\, dy \right]\left[\int_0^1 z^n\, dz \right]$$

$$= \sum_{n=0}^{\infty} \frac{1}{n + 1} \cdot \frac{1}{n + 1} \cdot \frac{1}{n + 1}$$

$$= \sum_{n=0}^{\infty} \frac{1}{(n + 1)^3} = \frac{1}{1^3} + \frac{1}{2^3} + \frac{1}{3^3} + \cdots = \sum_{n=1}^{\infty} \frac{1}{n^3}$$

(b) Since $|-xyz| < 1$, except at $(1, 1, 1)$, the formula for the sum of a geometric series gives

$$\frac{1}{1 + xyz} = \sum_{n=0}^{\infty} (-xyz)^n, \text{ so}$$

$$\int_0^1 \int_0^1 \int_0^1 \frac{1}{1 + xyz} \, dx \, dy \, dz = \int_0^1 \int_0^1 \int_0^1 \sum_{n=0}^{\infty} (-xyz)^n \, dx \, dy \, dz$$

$$= \sum_{n=0}^{\infty} \int_0^1 \int_0^1 \int_0^1 (-xyz)^n \, dx \, dy \, dz$$

$$= \sum_{n=0}^{\infty} (-1)^n \left[\int_0^1 x^n \, dx \right] \left[\int_0^1 y^n \, dy \right] \left[\int_0^1 z^n \, dz \right]$$

$$= \sum_{n=0}^{\infty} (-1)^n \frac{1}{n+1} \cdot \frac{1}{n+1} \cdot \frac{1}{n+1}$$

$$= \sum_{n=0}^{\infty} \frac{(-1)^n}{(n+1)^3} = \frac{1}{1^3} - \frac{1}{2^3} + \frac{1}{3^3} - \cdots = \sum_{n=1}^{\infty} \frac{(-1)^{n-1}}{n^3}$$

To evaluate this sum, we first write out a few terms: $s = 1 - \dfrac{1}{2^3} + \dfrac{1}{3^3} - \dfrac{1}{4^3} + \dfrac{1}{5^3} - \dfrac{1}{6^3} \approx 0.8998$. Notice that

$a_7 = \dfrac{1}{7^3} < 0.003$. By the Alternating Series Estimation Theorem from Section 12.5 [ET 11.5], we have

$|s - s_6| \le a_7 < 0.003$. This error of 0.003 will not affect the second decimal place, so we have $s \approx 0.90$.

9. $\int_0^x \int_0^y \int_0^z f(t) \, dt \, dz \, dy = \iiint_E f(t) \, dV$, where

$E = \{(t, z, y) \mid 0 \le t \le z, 0 \le z \le y, 0 \le y \le x\}.$

If we let D be the projection of E on the yt-plane then

$D = \{(y, t) \mid 0 \le t \le x, t \le y \le x\}.$ And we see from the diagram

that $E = \{(t, z, y) \mid t \le z \le y, t \le y \le x, 0 \le t \le x\}.$ So

$$\int_0^x \int_0^y \int_0^z f(t) \, dt \, dz \, dy = \int_0^x \int_t^x \int_t^y f(t) \, dz \, dy \, dt = \int_0^x \left[\int_t^x (y - t) f(t) \, dy \right] dt$$

$$= \int_0^x \left[\left(\tfrac{1}{2} y^2 - ty \right) f(t) \right]_{y=t}^{y=x} dt = \int_0^x \left[\tfrac{1}{2} x^2 - tx - \tfrac{1}{2} t^2 + t^2 \right] f(t) \, dt$$

$$= \int_0^x \left[\tfrac{1}{2} x^2 - tx + \tfrac{1}{2} t^2 \right] f(t) \, dt = \int_0^x \left(\tfrac{1}{2} x^2 - 2tx + t^2 \right) f(t) \, dt$$

$$= \tfrac{1}{2} \int_0^x (x - t)^2 f(t) \, dt$$

17 □ VECTOR CALCULUS

17.1 Vector Fields

1. $\mathbf{F}(x,y) = \frac{1}{2}(\mathbf{i}+\mathbf{j})$

All vectors in this field are identical, with length $\frac{1}{\sqrt{2}}$ and direction parallel to the line $y = x$.

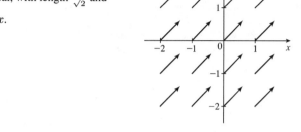

3. $\mathbf{F}(x,y) = y\,\mathbf{i} + \frac{1}{2}\mathbf{j}$. The length of the vector $y\,\mathbf{i} + \frac{1}{2}\mathbf{j}$ is $\sqrt{y^2 + \frac{1}{4}}$. Vectors are tangent to parabolas opening about the x-axis.

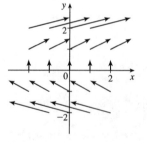

5. $\mathbf{F}(x,y) = \dfrac{y\,\mathbf{i}+x\,\mathbf{j}}{\sqrt{x^2+y^2}}$

The length of the vector $\dfrac{y\,\mathbf{i}+x\,\mathbf{j}}{\sqrt{x^2+y^2}}$ is 1.

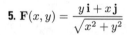

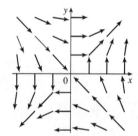

7. $\mathbf{F}(x,y,z) = \mathbf{j}$

All vectors in this field are parallel to the y-axis and have length 1.

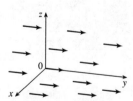

9. $\mathbf{F}(x,y,z) = y\,\mathbf{j}$

The length of $\mathbf{F}(x,y,z)$ is $|y|$. No vectors emanate from the xz-plane since $y = 0$ there. In each plane $y = b$, all the vectors are identical.

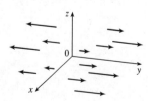

11. $\mathbf{F}(x, y) = \langle y, x \rangle$ corresponds to graph II. In the first quadrant all the vectors have positive x- and y-components, in the second quadrant all vectors have positive x-components and negative y-components, in the third quadrant all vectors have negative x- and y-components, and in the fourth quadrant all vectors have negative x-components and positive y-components. In addition, the vectors get shorter as we approach the origin.

13. $\mathbf{F}(x, y) = \langle x - 2, x + 1 \rangle$ corresponds to graph I since the vectors are independent of y (vectors along vertical lines are identical) and, as we move to the right, both the x- and the y-components get larger.

15. $\mathbf{F}(x, y, z) = \mathbf{i} + 2\mathbf{j} + 3\mathbf{k}$ corresponds to graph IV, since all vectors have identical length and direction.

17. $\mathbf{F}(x, y, z) = x\mathbf{i} + y\mathbf{j} + 3\mathbf{k}$ corresponds to graph III; the projection of each vector onto the xy-plane is $x\mathbf{i} + y\mathbf{j}$, which points away from the origin, and the vectors point generally upward because their z-components are all 3.

19.

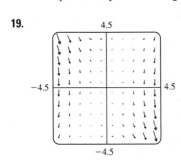

The vector field seems to have very short vectors near the line $y = 2x$.

For $\mathbf{F}(x, y) = \langle 0, 0 \rangle$ we must have $y^2 - 2xy = 0$ and $3xy - 6x^2 = 0$.

The first equation holds if $y = 0$ or $y = 2x$, and the second holds if $x = 0$ or $y = 2x$. So both equations hold [and thus $\mathbf{F}(x, y) = \mathbf{0}$] along the line $y = 2x$.

21. $\nabla f(x, y) = f_x(x, y)\mathbf{i} + f_y(x, y)\mathbf{j} = \dfrac{1}{x + 2y}\mathbf{i} + \dfrac{2}{x + 2y}\mathbf{j}$

23. $\nabla f(x, y, z) = f_x(x, y, z)\mathbf{i} + f_y(x, y, z)\mathbf{j} + f_z(x, y, z)\mathbf{k}$

$$= \frac{x}{\sqrt{x^2 + y^2 + z^2}}\mathbf{i} + \frac{y}{\sqrt{x^2 + y^2 + z^2}}\mathbf{j} + \frac{z}{\sqrt{x^2 + y^2 + z^2}}\mathbf{k}$$

25. $f(x, y) = xy - 2x \quad \Rightarrow$

$\nabla f(x, y) = (y - 2)\mathbf{i} + x\mathbf{j}$.

The length of $\nabla f(x, y)$ is $\sqrt{(y - 2)^2 + x^2}$ and $\nabla f(x, y)$ terminates on the line $y = x + 2$ at the point $(x + y - 2, x + y)$.

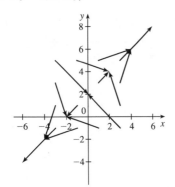

27. We graph ∇f along with a contour map of f.

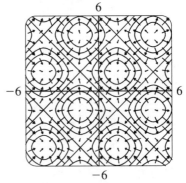

The graph shows that the gradient vectors are perpendicular to the level curves. Also, the gradient vectors point in the direction in which f is increasing and are longer where the level curves are closer together.

29. $f(x, y) = xy \quad \Rightarrow \quad \nabla f(x, y) = y\,\mathbf{i} + x\,\mathbf{j}$. In the first quadrant, both components of each vector are positive, while in the third quadrant both components are negative. However, in the second quadrant each vector's x-component is positive while its y-component is negative (and vice versa in the fourth quadrant). Thus, ∇f is graph IV.

31. $f(x, y) = x^2 + y^2 \quad \Rightarrow \quad \nabla f(x, y) = 2x\,\mathbf{i} + 2y\,\mathbf{j}$. Thus, each vector $\nabla f(x, y)$ has the same direction and twice the length of the position vector of the point (x, y), so the vectors all point directly away from the origin and their lengths increase as we move away from the origin. Hence, ∇f is graph II.

33. (a) We sketch the vector field $\mathbf{F}(x, y) = x\,\mathbf{i} - y\,\mathbf{j}$ along with several approximate flow lines. The flow lines appear to be hyperbolas with shape similar to the graph of $y = \pm 1/x$, so we might guess that the flow lines have equations $y = C/x$.

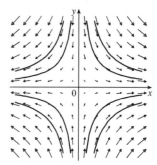

(b) If $x = x(t)$ and $y = y(t)$ are parametric equations of a flow line, then the velocity vector of the flow line at the point (x, y) is $x'(t)\,\mathbf{i} + y'(t)\,\mathbf{j}$. Since the velocity vectors coincide with the vectors in the vector field, we have $x'(t)\,\mathbf{i} + y'(t)\,\mathbf{j} = x\,\mathbf{i} - y\,\mathbf{j} \quad \Rightarrow \quad dx/dt = x,\ dy/dt = -y$. To solve these differential equations, we know $dx/dt = x \quad \Rightarrow \quad dx/x = dt \quad \Rightarrow \quad \ln|x| = t + C \quad \Rightarrow \quad x = \pm e^{t+C} = Ae^t$ for some constant A, and $dy/dt = -y \quad \Rightarrow \quad dy/y = -dt \quad \Rightarrow \quad \ln|y| = -t + K \quad \Rightarrow \quad y = \pm e^{-t+K} = Be^{-t}$ for some constant B. Therefore $xy = Ae^t Be^{-t} = AB = $ constant. If the flow line passes through $(1, 1)$ then $(1)(1) = $ constant $= 1 \quad \Rightarrow \quad xy = 1 \quad \Rightarrow \quad y = 1/x,\ x > 0$.

17.2 Line Integrals   ET 16.2

1. $x = t^2$ and $y = t$, $0 \le t \le 2$, so by Formula 3

$$\int_C y\, ds = \int_0^2 t \sqrt{\left(\frac{dx}{dt}\right)^2 + \left(\frac{dy}{dt}\right)^2}\, dt = \int_0^2 t \sqrt{(2t)^2 + (1)^2}\, dt$$

$$= \int_0^2 t\sqrt{4t^2 + 1}\, dt = \tfrac{1}{12}\left(4t^2 + 1\right)^{3/2}\Big]_0^2 = \tfrac{1}{12}\left(17\sqrt{17} - 1\right)$$

3. Parametric equations for C are $x = 4\cos t$, $y = 4\sin t$, $-\frac{\pi}{2} \le t \le \frac{\pi}{2}$. Then

$$\int_C xy^4\, ds = \int_{-\pi/2}^{\pi/2} (4\cos t)(4\sin t)^4 \sqrt{(-4\sin t)^2 + (4\cos t)^2}\, dt$$

$$= \int_{-\pi/2}^{\pi/2} 4^5 \cos t \sin^4 t \sqrt{16\left(\sin^2 t + \cos^2 t\right)}\, dt$$

$$= 4^5 \int_{-\pi/2}^{\pi/2} \left(\sin^4 t \cos t\right)(4)\, dt = (4)^6 \left[\tfrac{1}{5}\sin^5 t\right]_{-\pi/2}^{\pi/2} = \tfrac{2 \cdot 4^6}{5} = 1638.4$$

5. If we choose x as the parameter, parametric equations for C are $x = x$, $y = x^2$ for $1 \le x \le 3$ and

$$\int_C (xy + \ln x)\, dy = \int_1^3 \left(x \cdot x^2 + \ln x\right) 2x\, dx = \int_1^3 2\left(x^4 + x\ln x\right) dx$$

$$= 2\left[\tfrac{1}{5}x^5 + \tfrac{1}{2}x^2 \ln x - \tfrac{1}{4}x^2\right]_1^3 \quad \text{(by integrating by parts in the second term)}$$

$$= 2\left(\tfrac{243}{5} + \tfrac{9}{2}\ln 3 - \tfrac{9}{4} - \tfrac{1}{5} + \tfrac{1}{4}\right) = \tfrac{464}{5} + 9\ln 3$$

7.

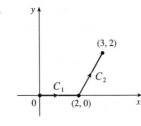

$$C = C_1 + C_2$$

On C_1: $x = x, y = 0 \Rightarrow dy = 0\,dx, 0 \leq x \leq 2$.

On C_2: $x = x, y = 2x - 4 \Rightarrow dy = 2\,dx, 2 \leq x \leq 3$.

Then

$$\int_C xy\,dx + (x - y)\,dy = \int_{C_1} xy\,dx + (x - y)\,dy + \int_{C_2} xy\,dx + (x - y)\,dy$$

$$= \int_0^2 (0 + 0)\,dx + \int_2^3 [(2x^2 - 4x) + (-x + 4)(2)]\,dx$$

$$= \int_2^3 (2x^2 - 6x + 8)\,dx = \tfrac{17}{3}$$

9. $x = 4\sin t, y = 4\cos t, z = 3t, 0 \leq t \leq \frac{\pi}{2}$. Then by Formula 9,

$$\int_C xy^3\,ds = \int_0^{\pi/2} (4\sin t)(4\cos t)^3 \sqrt{\left(\frac{dx}{dt}\right)^2 + \left(\frac{dy}{dt}\right)^2 + \left(\frac{dz}{dt}\right)^2}\,dt$$

$$= \int_0^{\pi/2} 4^4 \cos^3 t \sin t \sqrt{(4\cos t)^2 + (-4\sin t)^2 + (3)^2}\,dt$$

$$= \int_0^{\pi/2} 256 \cos^3 t \sin t \sqrt{16(\cos^2 t + \sin^2 t) + 9}\,dt$$

$$= 1280 \int_0^{\pi/2} \cos^3 t \sin t\,dt = -320 \cos^4 t\big]_0^{\pi/2} = 320$$

11. Parametric equations for C are $x = t, y = 2t, z = 3t, 0 \leq t \leq 1$. Then

$$\int_C xe^{yz}\,ds = \int_0^1 te^{(2t)(3t)} \sqrt{1^2 + 2^2 + 3^2}\,dt = \sqrt{14} \int_0^1 te^{6t^2}\,dt$$

$$= \sqrt{14} \left[\tfrac{1}{12} e^{6t^2}\right]_0^1 = \tfrac{\sqrt{14}}{12}(e^6 - 1)$$

13. $\int_C x^2 y\sqrt{z}\,dz = \int_0^1 (t^3)^2 (t)\sqrt{t^2} \cdot 2t\,dt = \int_0^1 2t^9\,dt = \tfrac{1}{5}t^{10}\big]_0^1 = \tfrac{1}{5}$

15.

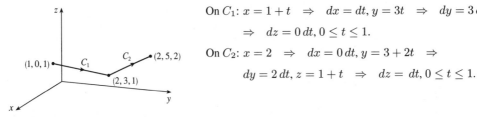

On C_1: $x = 1 + t \Rightarrow dx = dt, y = 3t \Rightarrow dy = 3\,dt, z = 1$

$\Rightarrow dz = 0\,dt, 0 \leq t \leq 1$.

On C_2: $x = 2 \Rightarrow dx = 0\,dt, y = 3 + 2t \Rightarrow$

$dy = 2\,dt, z = 1 + t \Rightarrow dz = dt, 0 \leq t \leq 1$.

Then $\int_C (x + yz)\,dx + 2x\,dy + xyz\,dz$

$$= \int_{C_1} (x + yz)\,dx + 2x\,dy + xyz\,dz + \int_{C_2} (x + yz)\,dx + 2x\,dy + xyz\,dz$$

$$= \int_0^1 (1 + t + (3t)(1))\,dt + 2(1 + t) \cdot 3\,dt + (1 + t)(3t)(1) \cdot 0\,dt$$

$$\quad + \int_0^1 (2 + (3 + 2t)(1 + t)) \cdot 0\,dt + 2(2) \cdot 2\,dt + (2)(3 + 2t)(1 + t)\,dt$$

$$= \int_0^1 (10t + 7)\,dt + \int_0^1 (4t^2 + 10t + 14)\,dt$$

$$= \left[5t^2 + 7t\right]_0^1 + \left[\tfrac{4}{3}t^3 + 5t^2 + 14t\right]_0^1 = 12 + \tfrac{61}{3} = \tfrac{97}{3}$$

17. (a) Along the line $x = -3$, the vectors of $\mathbf{F}$ have positive y-components, so since the path goes upward, the integrand $\mathbf{F} \cdot \mathbf{T}$ is always positive. Therefore $\int_{C_1} \mathbf{F} \cdot d\mathbf{r} = \int_{C_1} \mathbf{F} \cdot \mathbf{T}\,ds$ is positive.

(b) All of the (nonzero) field vectors along the circle with radius 3 are pointed in the clockwise direction, that is, opposite the direction to the path. So $\mathbf{F} \cdot \mathbf{T}$ is negative, and therefore $\int_{C_2} \mathbf{F} \cdot d\mathbf{r} = \int_{C_2} \mathbf{F} \cdot \mathbf{T} \, ds$ is negative.

19. $r(t) = t^2 \mathbf{i} - t^3 \mathbf{j}$, so $\mathbf{F}(\mathbf{r}(t)) = (t^2)^2 (-t^3)^3 \mathbf{i} - (-t^3) \sqrt{t^2} \mathbf{j} = -t^{13} \mathbf{i} + t^4 \mathbf{j}$ and $\mathbf{r}'(t) = 2t \mathbf{i} - 3t^2 \mathbf{j}$.

Thus $\int_C \mathbf{F} \cdot d\mathbf{r} = \int_0^1 \mathbf{F}(\mathbf{r}(t)) \cdot \mathbf{r}'(t) \, dt = \int_0^1 \left(-2t^{14} - 3t^6\right) dt = \left[-\frac{2}{15} t^{15} - \frac{3}{7} t^7\right]_0^1 = -\frac{59}{105}$.

21. $\int_C \mathbf{F} \cdot d\mathbf{r} = \int_0^1 \langle \sin t^3, \cos (-t^2), t^4 \rangle \cdot \langle 3t^2, -2t, 1 \rangle \, dt$

$\qquad = \int_0^1 \left(3t^2 \sin t^3 - 2t \cos t^2 + t^4\right) dt = \left[-\cos t^3 - \sin t^2 + \frac{1}{5} t^5\right]_0^1 = \frac{6}{5} - \cos 1 - \sin 1$

23. We graph $\mathbf{F}(x, y) = (x - y) \mathbf{i} + xy \mathbf{j}$ and the curve C. We see that most of the vectors starting on C point in roughly the same direction as C, so for these portions of C the tangential component $\mathbf{F} \cdot \mathbf{T}$ is positive. Although some vectors in the third quadrant which start on C point in roughly the opposite direction, and hence give negative tangential components, it seems reasonable that the effect of these portions of C is outweighed by the positive tangential components. Thus, we would expect $\int_C \mathbf{F} \cdot d\mathbf{r} = \int_C \mathbf{F} \cdot \mathbf{T} \, ds$ to be positive.

To verify, we evaluate $\int_C \mathbf{F} \cdot d\mathbf{r}$. The curve C can be represented by $\mathbf{r}(t) = 2\cos t \, \mathbf{i} + 2\sin t \, \mathbf{j}$, $0 \le t \le \frac{3\pi}{2}$, so $\mathbf{F}(\mathbf{r}(t)) = (2\cos t - 2\sin t) \mathbf{i} + 4\cos t \sin t \, \mathbf{j}$ and $\mathbf{r}'(t) = -2\sin t \, \mathbf{i} + 2\cos t \, \mathbf{j}$. Then

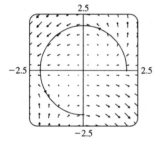

$\int_C \mathbf{F} \cdot d\mathbf{r} = \int_0^{3\pi/2} \mathbf{F}(\mathbf{r}(t)) \cdot \mathbf{r}'(t) \, dt$

$\qquad = \int_0^{3\pi/2} \left[-2\sin t(2\cos t - 2\sin t) + 2\cos t(4\cos t \sin t)\right] dt$

$\qquad = 4 \int_0^{3\pi/2} (\sin^2 t - \sin t \cos t + 2\sin t \cos^2 t) \, dt$

$\qquad = 3\pi + \frac{2}{3}$ [using a CAS]

25. (a) $\int_C \mathbf{F} \cdot d\mathbf{r} = \int_0^1 \langle e^{t^2-1}, t^5 \rangle \cdot \langle 2t, 3t^2 \rangle \, dt = \int_0^1 \left(2te^{t^2-1} + 3t^7\right) dt = \left[e^{t^2-1} + \frac{3}{8} t^8\right]_0^1 = \frac{11}{8} - 1/e$

(b) $\mathbf{r}(0) = \mathbf{0}$, $\mathbf{F}(\mathbf{r}(0)) = \langle e^{-1}, 0 \rangle$;

$\mathbf{r}\left(\frac{1}{\sqrt{2}}\right) = \langle \frac{1}{2}, \frac{1}{2\sqrt{2}} \rangle$, $\mathbf{F}\left(\mathbf{r}\left(\frac{1}{\sqrt{2}}\right)\right) = \langle e^{-1/2}, \frac{1}{4\sqrt{2}} \rangle$;

$\mathbf{r}(1) = \langle 1, 1 \rangle$, $\mathbf{F}(\mathbf{r}(1)) = \langle 1, 1 \rangle$.

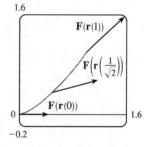

In order to generate the graph with Maple, we use the PLOT command (not to be confused with the plot command) to define each of the vectors. For example,

```
v1:=PLOT(CURVES([[0,0],[evalf(1/exp(1)),0]]));
```

generates the vector from the vector field at the point $(0, 0)$ (but without an arrowhead) and gives it the name v1. To show everything on the same screen, we use the display command. In Mathematica, we use ListPlot (with the PlotJoined - > True option) to generate the vectors, and then Show to show everything on the same screen.

27. The part of the astroid that lies in the quadrant is parametrized by $x = \cos^3 t$, $y = \sin^3 t$, $0 \le t \le \frac{\pi}{2}$.

Now $\dfrac{dx}{dt} = 3\cos^2 t \,(-\sin t)$ and $\dfrac{dy}{dt} = 3\sin^2 t \cos t$, so

$$\sqrt{\left(\frac{dx}{dt}\right)^2 + \left(\frac{dy}{dt}\right)^2} = \sqrt{9\cos^4 t \sin^2 t + 9\sin^4 t \cos^2 t} = 3\cos t \sin t \sqrt{\cos^2 t + \sin^2 t} = 3\cos t \sin t.$$

Therefore $\int_C x^3 y^5 \, ds = \int_0^{\pi/2} \cos^9 t \sin^{15} t \,(3\cos t \sin t)\, dt = \frac{945}{16,777,216}\pi.$

29. A calculator or CAS gives $\int_C x \sin y \, ds = \int_1^2 \ln t \sin\left(e^{-t}\right)\sqrt{(1/t)^2 + (-e^{-t})^2}\, dt \approx 0.052.$

31. We use the parametrization $x = 2\cos t$, $y = 2\sin t$, $-\frac{\pi}{2} \le t \le \frac{\pi}{2}$.

Then $ds = \sqrt{\left(\dfrac{dx}{dt}\right)^2 + \left(\dfrac{dy}{dt}\right)^2}\, dt = \sqrt{(-2\sin t)^2 + (2\cos t)^2}\, dt = 2\, dt$, so

$m = \int_C k \, ds = 2k \int_{-\pi/2}^{\pi/2} dt = 2k(\pi)$, $\bar{x} = \frac{1}{2\pi k}\int_C xk \, ds = \frac{1}{2\pi}\int_{-\pi/2}^{\pi/2}(2\cos t)2\, dt = \frac{1}{2\pi}\left[4\sin t\right]_{-\pi/2}^{\pi/2} = \frac{4}{\pi},$

$\bar{y} = \frac{1}{2\pi k}\int_C yk \, ds = \frac{1}{2\pi}\int_{-\pi/2}^{\pi/2}(2\sin t)2\, dt = 0.$ Hence $(\bar{x}, \bar{y}) = \left(\frac{4}{\pi}, 0\right).$

33. (a) $\bar{x} = \dfrac{1}{m}\displaystyle\int_C x\rho(x,y,z)\, ds$, $\bar{y} = \dfrac{1}{m}\displaystyle\int_C y\rho(x,y,z)\, ds$, $\bar{z} = \dfrac{1}{m}\displaystyle\int_C z\rho(x,y,z)\, ds$ where $m = \int_C \rho(x,y,z)\, ds$.

(b) $m = \int_C k \, ds = k\int_0^{2\pi}\sqrt{4\sin^2 t + 4\cos^2 t + 9}\, dt = k\sqrt{13}\int_0^{2\pi} dt = 2\pi k \sqrt{13},$

$\bar{x} = \dfrac{1}{2\pi k \sqrt{13}}\displaystyle\int_0^{2\pi} k2\sqrt{13}\sin t\, dt = 0$, $\bar{y} = \dfrac{1}{2\pi k \sqrt{13}}\displaystyle\int_0^{2\pi} k2\sqrt{13}\cos t\, dt = 0,$

$\bar{z} = \dfrac{1}{2\pi k \sqrt{13}}\displaystyle\int_0^{2\pi}\left(k\sqrt{13}\right)(3t)\, dt = \dfrac{3}{2\pi}\left(2\pi^2\right) = 3\pi.$ Hence $(\bar{x}, \bar{y}, \bar{z}) = (0, 0, 3\pi).$

35. From Example 3, $\rho(x,y) = k(1-y)$, $x = \cos t$, $y = \sin t$, and $ds = dt$, $0 \le t \le \pi$ $\Rightarrow$

$$I_x = \int_C y^2 \rho(x,y)\, ds = \int_0^\pi \sin^2 t \,[k(1 - \sin t)]\, dt = k\int_0^\pi (\sin^2 t - \sin^3 t)\, dt$$

$$= \tfrac{1}{2}k\int_0^\pi (1 - \cos 2t)\, dt - k\int_0^\pi (1 - \cos^2 t)\sin t\, dt \qquad \begin{bmatrix}\text{Let } u = \cos t, \, du = -\sin t\, dt \\ \text{in the second integral}\end{bmatrix}$$

$$= k\left[\tfrac{\pi}{2} + \int_1^{-1}(1 - u^2)\, du\right] = k\left(\tfrac{\pi}{2} - \tfrac{4}{3}\right)$$

$$I_y = \int_C x^2 \rho(x,y)\, ds = k\int_0^\pi \cos^2 t\,(1 - \sin t)\, dt = \tfrac{k}{2}\int_0^\pi (1 + \cos 2t)\, dt - k\int_0^\pi \cos^2 t \sin t\, dt$$

$$= k\left(\tfrac{\pi}{2} - \tfrac{2}{3}\right), \text{ using the same substitution as above.}$$

37. $W = \int_C \mathbf{F}\cdot d\mathbf{r} = \int_0^{2\pi}\langle t - \sin t, 3 - \cos t\rangle \cdot \langle 1 - \cos t, \sin t\rangle\, dt$

$= \int_0^{2\pi}(t - t\cos t - \sin t + \sin t\cos t + 3\sin t - \sin t\cos t)\, dt$

$= \int_0^{2\pi}(t - t\cos t + 2\sin t)\, dt = \left[\tfrac{1}{2}t^2 - (t\sin t + \cos t) - 2\cos t\right]_0^{2\pi} \qquad \begin{bmatrix}\text{by integrating by parts} \\ \text{in the second term}\end{bmatrix}$

$= 2\pi^2$

39. $\mathbf{r}(t) = \langle 1 + 2t, 4t, 2t\rangle$, $0 \le t \le 1$,

$$W = \int_C \mathbf{F}\cdot d\mathbf{r} = \int_0^1 \langle 6t, 1 + 4t, 1 + 6t\rangle \cdot \langle 2, 4, 2\rangle\, dt = \int_0^1 (12t + 4(1 + 4t) + 2(1 + 6t))\, dt$$

$$= \int_0^1 (40t + 6)\, dt = \left[20t^2 + 6t\right]_0^1 = 26$$

41. Let $\mathbf{F} = 185\,\mathbf{k}$. To parametrize the staircase, let

$$x = 20\cos t, \; y = 20\sin t, \; z = \frac{90}{6\pi}t = \frac{15}{\pi}t, \; 0 \le t \le 6\pi \quad \Rightarrow$$

$$W = \int_C \mathbf{F} \cdot d\mathbf{r} = \int_0^{6\pi} \langle 0, 0, 185\rangle \cdot \langle -20\sin t, 20\cos t, \tfrac{15}{\pi}\rangle \, dt = (185)\tfrac{15}{\pi}\int_0^{6\pi} dt = (185)(90)$$

$$\approx 1.67 \times 10^4 \text{ ft-lb}$$

43. (a) $\mathbf{r}(t) = \langle \cos t, \sin t\rangle$, $0 \le t \le 2\pi$, and let $\mathbf{F} = \langle a, b\rangle$. Then

$$\begin{aligned}
W = \int_C \mathbf{F} \cdot d\mathbf{r} &= \int_0^{2\pi} \langle a, b\rangle \cdot \langle -\sin t, \cos t\rangle \, dt \\
&= \int_0^{2\pi} (-a\sin t + b\cos t)\, dt = \big[a\cos t + b\sin t\big]_0^{2\pi} \\
&= a + 0 - a + 0 = 0
\end{aligned}$$

(b) Yes. $\mathbf{F}(x,y) = k\mathbf{x} = \langle kx, ky\rangle$ and

$$\begin{aligned}
W = \int_C \mathbf{F} \cdot d\mathbf{r} &= \int_0^{2\pi} \langle k\cos t, k\sin t\rangle \cdot \langle -\sin t, \cos t\rangle \, dt \\
&= \int_0^{2\pi} (-k\sin t\cos t + k\sin t\cos t)\, dt = \int_0^{2\pi} 0\, dt = 0
\end{aligned}$$

45. The work done in moving the object is $\int_C \mathbf{F} \cdot d\mathbf{r} = \int_C \mathbf{F} \cdot \mathbf{T}\, ds$. We can approximate this integral by dividing C into 7 segments of equal length $\Delta s = 2$ and approximating $\mathbf{F} \cdot \mathbf{T}$, that is, the tangential component of force, at a point (x_i^*, y_i^*) on each segment. Since C is composed of straight line segments, $\mathbf{F} \cdot \mathbf{T}$ is the scalar projection of each force vector onto C. If we choose (x_i^*, y_i^*) to be the point on the segment closest to the origin, then the work done is $\int_C \mathbf{F} \cdot \mathbf{T}\, ds \approx \sum_{i=1}^{7} [\mathbf{F}(x_i^*, y_i^*) \cdot \mathbf{T}(x_i^*, y_i^*)]\, \Delta s = [2 + 2 + 2 + 2 + 1 + 1 + 1](2) = 22$. Thus, we estimate the work done to be approximately 22 J.

17.3 The Fundamental Theorem for Line Integrals   ET 16.3

1. C appears to be a smooth curve, and since ∇f is continuous, we know f is differentiable. Then Theorem 2 says that the value of $\int_C \nabla f \cdot d\mathbf{r}$ is simply the difference of the values of f at the terminal and initial points of C. From the graph, this is $50 - 10 = 40$.

3. $\partial(6x + 5y)/\partial y = 5 = \partial(5x + 4y)/\partial x$ and the domain of $\mathbf{F}$ is $\mathbb{R}^2$ which is open and simply-connected, so by Theorem 6 $\mathbf{F}$ is conservative. Thus, there exists a function f such that $\nabla f = \mathbf{F}$, that is, $f_x(x,y) = 6x + 5y$ and $f_y(x,y) = 5x + 4y$. But $f_x(x,y) = 6x + 5y$ implies $f(x,y) = 3x^2 + 5xy + g(y)$ and differentiating both sides of this equation with respect to y gives $f_y(x,y) = 5x + g'(y)$. Thus $5x + 4y = 5x + g'(y)$ so $g'(y) = 4y$ and $g(y) = 2y^2 + K$ where K is a constant. Hence $f(x,y) = 3x^2 + 5xy + 2y^2 + K$ is a potential function for $\mathbf{F}$.

5. $\partial(xe^y)/\partial y = xe^y$, $\partial(ye^x)/\partial x = ye^x$. Since these are not equal, $\mathbf{F}$ is not conservative.

7. $\partial(2x\cos y - y\cos x)/\partial y = -2x\sin y - \cos x = \partial(-x^2\sin y - \sin x)/\partial x$ and the domain of $\mathbf{F}$ is $\mathbb{R}^2$. Hence $\mathbf{F}$ is conservative so there exists a function f such that $\nabla f = \mathbf{F}$. Then $f_x(x,y) = 2x\cos y - y\cos x$ implies $f(x,y) = x^2\cos y - y\sin x + g(y)$ and $f_y(x,y) = -x^2\sin y - \sin x + g'(y)$. But $f_y(x,y) = -x^2\sin y - \sin x$ so $g'(y) = 0 \quad \Rightarrow \quad g(y) = K$. Then $f(x,y) = x^2\cos y - y\sin x + K$ is a potential function for $\mathbf{F}$.

9. $\partial(ye^x + \sin y)/\partial y = e^x + \cos y = \partial(e^x + x \cos y)/\partial x$ and the domain of $\mathbf{F}$ is $\mathbb{R}^2$. Hence $\mathbf{F}$ is conservative so there exists a function f such that $\nabla f = \mathbf{F}$. Then $f_x(x, y) = ye^x + \sin y$ implies $f(x, y) = ye^x + x \sin y + g(y)$ and $f_y(x, y) = e^x + x \cos y + g'(y)$. But $f_y(x, y) = e^x + x \cos y$ so $g(y) = K$ and $f(x, y) = ye^x + x \sin y + K$ is a potential function for $\mathbf{F}$.

11. (a) $\mathbf{F}$ has continuous first-order partial derivatives and $\dfrac{\partial}{\partial y} 2xy = 2x = \dfrac{\partial}{\partial x}(x^2)$ on $\mathbb{R}^2$, which is open and simply-connected. Thus, $\mathbf{F}$ is conservative by Theorem 6. Then we know that the line integral of $\mathbf{F}$ is independent of path; in particular, the value of $\int_C \mathbf{F} \cdot d\mathbf{r}$ depends only on the endpoints of C. Since all three curves have the same initial and terminal points, $\int_C \mathbf{F} \cdot d\mathbf{r}$ will have the same value for each curve.

(b) We first find a potential function f, so that $\nabla f = \mathbf{F}$. We know $f_x(x, y) = 2xy$ and $f_y(x, y) = x^2$. Integrating $f_x(x, y)$ with respect to x, we have $f(x, y) = x^2y + g(y)$. Differentiating both sides with respect to y gives $f_y(x, y) = x^2 + g'(y)$, so we must have $x^2 + g'(y) = x^2 \;\Rightarrow\; g'(y) = 0 \;\Rightarrow\; g(y) = K$, a constant. Thus $f(x, y) = x^2y + K$. All three curves start at $(1, 2)$ and end at $(3, 2)$, so by Theorem 2, $\int_C \mathbf{F} \cdot d\mathbf{r} = f(3, 2) - f(1, 2) = 18 - 2 = 16$ for each curve.

13. (a) $f_x(x, y) = x^3y^4$ implies $f(x, y) = \frac{1}{4}x^4y^4 + g(y)$ and $f_y(x, y) = x^4y^3 + g'(y)$. But $f_y(x, y) = x^4y^3$ so $g'(y) = 0 \;\Rightarrow\; g(y) = K$, a constant. We can take $K = 0$, so $f(x, y) = \frac{1}{4}x^4y^4$.

(b) The initial point of C is $\mathbf{r}(0) = (0, 1)$ and the terminal point is $\mathbf{r}(1) = (1, 2)$, so $\int_C \mathbf{F} \cdot d\mathbf{r} = f(1, 2) - f(0, 1) = 4 - 0 = 4$.

15. (a) $f_x(x, y, z) = yz$ implies $f(x, y, z) = xyz + g(y, z)$ and so $f_y(x, y, z) = xz + g_y(y, z)$. But $f_y(x, y, z) = xz$ so $g_y(y, z) = 0 \;\Rightarrow\; g(y, z) = h(z)$. Thus $f(x, y, z) = xyz + h(z)$ and $f_z(x, y, z) = xy + h'(z)$. But $f_z(x, y, z) = xy + 2z$, so $h'(z) = 2z \;\Rightarrow\; h(z) = z^2 + K$. Hence $f(x, y, z) = xyz + z^2$ (taking $K = 0$).

(b) $\int_C \mathbf{F} \cdot d\mathbf{r} = f(4, 6, 3) - f(1, 0, -2) = 81 - 4 = 77$.

17. (a) $f_x(x, y, z) = y^2 \cos z$ implies $f(x, y, z) = xy^2 \cos z + g(y, z)$ and so $f_y(x, y, z) = 2xy \cos z + g_y(y, z)$. But $f_y(x, y, z) = 2xy \cos z$ so $g_y(y, z) = 0 \;\Rightarrow\; g(y, z) = h(z)$. Thus $f(x, y, z) = xy^2 \cos z + h(z)$ and $f_z(x, y, z) = -xy^2 \sin z + h'(z)$. But $f_z(x, y, z) = -xy^2 \sin z$, so $h'(z) = 0 \;\Rightarrow\; h(z) = K$. Hence $f(x, y, z) = xy^2 \cos z$ (taking $K = 0$).

(b) $\mathbf{r}(0) = \langle 0, 0, 0 \rangle$, $\mathbf{r}(\pi) = \langle \pi^2, 0, \pi \rangle$ so $\int_C \mathbf{F} \cdot d\mathbf{r} = f(\pi^2, 0, \pi) - f(0, 0, 0) = 0 - 0 = 0$.

19. Here $\mathbf{F}(x, y) = \tan y \,\mathbf{i} + x \sec^2 y \,\mathbf{j}$. Then $f(x, y) = x \tan y$ is a potential function for $\mathbf{F}$, that is, $\nabla f = \mathbf{F}$ so $\mathbf{F}$ is conservative and thus its line integral is independent of path. Hence $\int_C \tan y \, dx + x \sec^2 y \, dy = \int_C \mathbf{F} \cdot d\mathbf{r} = f\left(2, \frac{\pi}{4}\right) - f(1, 0) = 2 \tan \frac{\pi}{4} - \tan 0 = 2$.

21. $\mathbf{F}(x, y) = 2y^{3/2} \,\mathbf{i} + 3x \sqrt{y} \,\mathbf{j}$, $W = \int_C \mathbf{F} \cdot d\mathbf{r}$. Since $\partial(2y^{3/2})/\partial y = 3\sqrt{y} = \partial(3x\sqrt{y})/\partial x$, there exists a function f such that $\nabla f = \mathbf{F}$. In fact, $f_x(x, y) = 2y^{3/2} \;\Rightarrow\; f(x, y) = 2xy^{3/2} + g(y) \;\Rightarrow\;$ $f_y(x, y) = 3xy^{1/2} + g'(y)$. But $f_y(x, y) = 3x\sqrt{y}$ so $g'(y) = 0$ or $g(y) = K$. We can take $K = 0 \;\Rightarrow\;$ $f(x, y) = 2xy^{3/2}$. Thus $W = \int_C \mathbf{F} \cdot d\mathbf{r} = f(2, 4) - f(1, 1) = 2(2)(8) - 2(1) = 30$.

23. We know that if the vector field (call it **F**) is conservative, then around any closed path C, $\int_C \mathbf{F} \cdot d\mathbf{r} = 0$. But take C to be some circle centered at the origin, oriented counterclockwise. All of the field vectors along C oppose motion along C, so the integral around C will be negative. Therefore the field is not conservative.

25. From the graph, it appears that **F** is not conservative. For example,

any closed curve containing the point $(2, 1)$ seems to have many field

vectors pointing counterclockwise along it, and none pointing

clockwise. So along this path the integral $\int \mathbf{F} \cdot d\mathbf{r} \neq 0$. To confirm

our guess, we calculate

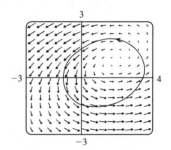

$$\frac{\partial}{\partial y} \left(\frac{x - 2y}{\sqrt{1 + x^2 + y^2}} \right) = (x - 2y) \left[\frac{-y}{(1 + x^2 + y^2)^{3/2}} \right] - \frac{2}{\sqrt{1 + x^2 + y^2}} = \frac{-2 - 2x^2 - xy}{(1 + x^2 + y^2)^{3/2}},$$

$$\frac{\partial}{\partial x} \left(\frac{x - 2}{\sqrt{1 + x^2 + y^2}} \right) = (x - 2) \left[\frac{-x}{(1 + x^2 + y^2)^{3/2}} \right] + \frac{1}{\sqrt{1 + x^2 + y^2}} = \frac{1 + y^2 + 2x}{(1 + x^2 + y^2)^{3/2}}.$$

These are not equal, so the field is not conservative, by Theorem 5.

27. Since **F** is conservative, there exists a function f such that $\mathbf{F} = \nabla f$, that is, $P = f_x, Q = f_y$, and $R = f_z$. Since P, Q and R have continuous first order partial derivatives, Clairaut's Theorem says that

$\partial P / \partial y = f_{xy} = f_{yx} = \partial Q / \partial x, \partial P / \partial z = f_{xz} = f_{zx} = \partial R / \partial x,$ and $\partial Q / \partial z = f_{yz} = f_{zy} = \partial R / \partial y.$

29. $D = \{(x, y) \mid x > 0, y > 0\} =$ the first quadrant (excluding the axes).

(a) D is open because around every point in D we can put a disk that lies in D.

(b) D is connected because the straight line segment joining any two points in D lies in D.

(c) D is simply-connected because it's connected and has no holes.

31. $D = \{(x, y) \mid 1 < x^2 + y^2 < 4\} =$ the annular region between the circles with center $(0, 0)$ and radii 1 and 2.

(a) D is open.

(b) D is connected.

(c) D is not simply-connected. For example, $x^2 + y^2 = (1.5)^2$ is simple and closed and lies within D but encloses points that are not in D. (Or we can say, D has a hole, so is not simply-connected.)

33. (a) $P = -\dfrac{y}{x^2 + y^2}, \dfrac{\partial P}{\partial y} = \dfrac{y^2 - x^2}{(x^2 + y^2)^2}$ and $Q = \dfrac{x}{x^2 + y^2}, \dfrac{\partial Q}{\partial x} = \dfrac{y^2 - x^2}{(x^2 + y^2)^2}.$ Thus $\dfrac{\partial P}{\partial y} = \dfrac{\partial Q}{\partial x}.$

(b) C_1: $x = \cos t, y = \sin t, 0 \le t \le \pi, C_2$: $x = \cos t, y = \sin t, t = 2\pi$ to $t = \pi$. Then

$$\int_{C_1} \mathbf{F} \cdot d\mathbf{r} = \int_0^\pi \frac{(-\sin t)(-\sin t) + (\cos t)(\cos t)}{\cos^2 t + \sin^2 t} \, dt = \int_0^\pi dt = \pi \text{ and } \int_{C_2} \mathbf{F} \cdot d\mathbf{r} = \int_{2\pi}^\pi dt = -\pi$$

Since these aren't equal, the line integral of **F** isn't independent of path. (Or notice that

$\int_{C_3} \mathbf{F} \cdot d\mathbf{r} = \int_0^{2\pi} dt = 2\pi$ where C_3 is the circle $x^2 + y^2 = 1$, and apply the contrapositive of Theorem 3.)

This doesn't contradict Theorem 6, since the domain of **F**, which is $\mathbb{R}^2$ except the origin, isn't

simply-connected.

17.4 Green's Theorem

1. (a)

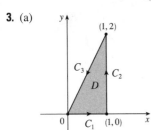

C_1: $x = t$ $\Rightarrow$ $dx = dt, y = 0$ $\Rightarrow$ $dy = 0\,dt, 0 \le t \le 2$.

C_2: $x = 2$ $\Rightarrow$ $dx = 0\,dt, y = t$ $\Rightarrow$ $dy = dt, 0 \le t \le 3$.

C_3: $x = 2 - t$ $\Rightarrow$ $dx = -dt, y = 3$ $\Rightarrow$ $dy = 0\,dt, 0 \le t \le 2$.

C_4: $x = 0$ $\Rightarrow$ $dx = 0\,dt, y = 3 - t$ $\Rightarrow$ $dy = -dt, 0 \le t \le 3$.

Thus $\oint_C xy^2\,dx + x^3\,dy = \oint_{C_1 + C_2 + C_3 + C_4} xy^2\,dx + x^3\,dy$

$$= \int_0^2 0\,dt + \int_0^3 8\,dt + \int_0^2 -9(2 - t)\,dt + \int_0^3 0\,dt$$

$$= 0 + 24 - 18 + 0 = 6$$

(b) $\oint_C xy^2\,dx + x^3\,dy = \iint_D \left[\frac{\partial}{\partial x}(x^3) - \frac{\partial}{\partial y}(xy^2)\right] dA = \int_0^2 \int_0^3 (3x^2 - 2xy)\,dy\,dx$

$$= \int_0^2 (9x^2 - 9x)\,dx = 24 - 18 = 6$$

3. (a)

C_1: $x = t$ $\Rightarrow$ $dx = dt, y = 0$ $\Rightarrow$ $dy = 0\,dt, 0 \le t \le 1$.

C_2: $x = 1$ $\Rightarrow$ $dx = 0\,dt, y = t$ $\Rightarrow$ $dy = dt, 0 \le t \le 2$.

C_3: $x = 1 - t$ $\Rightarrow$ $dx = -dt, y = 2 - 2t$ $\Rightarrow$ $dy = -2\,dt, 0 \le t \le 1$.

Thus $\oint_C xy\,dx + x^2 y^3\,dy = \oint_{C_1 + C_2 + C_3} xy\,dx + x^2 y^3\,dy$

$$= \int_0^1 0\,dt + \int_0^2 t^3\,dt + \int_0^1 \left[-(1 - t)(2 - 2t) - 2(1 - t)^2(2 - 2t)^3\right] dt$$

$$= 0 + \left[\tfrac{1}{4}t^4\right]_0^2 + \left[\tfrac{2}{3}(1 - t)^3 + \tfrac{8}{3}(1 - t)^6\right]_0^1 = 4 - \tfrac{10}{3} = \tfrac{2}{3}$$

(b) $\oint_C xy\,dx + x^2 y^3\,dy = \iint_D \left[\frac{\partial}{\partial x}(x^2 y^3) - \frac{\partial}{\partial y}(xy)\right] dA = \int_0^1 \int_0^{2x} (2xy^3 - x)\,dy\,dx$

$$= \int_0^1 \left[\tfrac{1}{2}xy^4 - xy\right]_{y=0}^{y=2x} dx = \int_0^1 (8x^5 - 2x^2)\,dx = \tfrac{4}{3} - \tfrac{2}{3} = \tfrac{2}{3}$$

5. We can parametrize C as $x = \cos\theta$, $y = \sin\theta$, $0 \le \theta \le 2\pi$. Then the line integral is

$\oint_C P\,dx + Q\,dy = \int_0^{2\pi} \cos^4\theta \sin^5\theta\,(-\sin\theta)\,d\theta + \int_0^{2\pi}(-\cos^7\theta \sin^6\theta)\cos\theta\,d\theta = -\frac{29\pi}{1024}$, according to a CAS. The double integral is

$$\iint_D \left(\frac{\partial Q}{\partial x} - \frac{\partial P}{\partial y}\right) dA = \int_{-1}^1 \int_{-\sqrt{1 - x^2}}^{\sqrt{1 - x^2}} (-7x^6 y^6 - 5x^4 y^4)\,dy\,dx = -\frac{29\pi}{1024},$$

verifying Green's Theorem in this case.

7. The region D enclosed by C is $[0, 1] \times [0, 1]$, so

$$\int_C e^y\,dx + 2xe^y\,dy = \iint_D \left[\frac{\partial}{\partial x}(2xe^y) - \frac{\partial}{\partial y}(e^y)\right] dA = \int_0^1 \int_0^1 (2e^y - e^y)\,dy\,dx$$

$$= \int_0^1 dx \int_0^1 e^y\,dy = (1)(e^1 - e^0) = e - 1$$

9. $\int_C \left(y + e^{\sqrt{x}}\right) dx + (2x + \cos y^2)\,dy = \iint_D \left[\frac{\partial}{\partial x}(2x + \cos y^2) - \frac{\partial}{\partial y}\left(y + e^{\sqrt{x}}\right)\right] dA$

$$= \int_0^1 \int_{y^2}^{\sqrt{y}}(2 - 1)\,dx\,dy = \int_0^1 (y^{1/2} - y^2)\,dy = \tfrac{1}{3}$$

11. $\int_C y^3 \, dx - x^3 \, dy = \iint_D \left[\frac{\partial}{\partial x} \left(-x^3 \right) - \frac{\partial}{\partial y} \left(y^3 \right) \right] dA = \iint_D (-3x^2 - 3y^2) \, dA = \int_0^{2\pi} \int_0^2 (-3r^2) \, r \, dr \, d\theta$

$$= -3 \int_0^{2\pi} d\theta \int_0^2 r^3 \, dr = -3(2\pi)(4) = -24\pi$$

13. $\mathbf{F}(x, y) = \langle \sqrt{x} + y^3, x^2 + \sqrt{y} \rangle$ and the region D enclosed by C is given by

$\{(x, y) \mid 0 \le x \le \pi, 0 \le y \le \sin x\}$. C is traversed clockwise, so $-C$ gives the positive orientation.

$$\int_C \mathbf{F} \cdot d\mathbf{r} = -\int_{-C} \left(\sqrt{x} + y^3 \right) dx + \left(x^2 + \sqrt{y} \right) dy = -\iint_D \left[\frac{\partial}{\partial x} \left(x^2 + \sqrt{y} \right) - \frac{\partial}{\partial y} \left(\sqrt{x} + y^3 \right) \right] dA$$

$$= -\int_0^{\pi} \int_0^{\sin x} (2x - 3y^2) \, dy \, dx = -\int_0^{\pi} \left[2xy - y^3 \right]_{y=0}^{y=\sin x} dx$$

$$= -\int_0^{\pi} (2x \sin x - \sin^3 x) \, dx = -\int_0^{\pi} (2x \sin x - (1 - \cos^2 x) \sin x) \, dx$$

$$= -\left[2 \sin x - 2x \cos x + \cos x - \tfrac{1}{3} \cos^3 x \right]_0^{\pi} \qquad \text{[integrate by parts in the first term]}$$

$$= -\left(2\pi - 2 + \tfrac{2}{3} \right) = \tfrac{4}{3} - 2\pi$$

15. $\mathbf{F}(x, y) = \langle e^x + x^2 y, e^y - xy^2 \rangle$ and the region D enclosed by C is the disk $x^2 + y^2 \le 25$.

C is traversed clockwise, so $-C$ gives the positive orientation.

$$\int_C \mathbf{F} \cdot d\mathbf{r} = -\int_{-C} (e^x + x^2 y) \, dx + (e^y - xy^2) \, dy$$

$$= -\iint_D \left[\frac{\partial}{\partial x} \left(e^y - xy^2 \right) - \frac{\partial}{\partial y} \left(e^x + x^2 y \right) \right] dA = -\iint_D (-y^2 - x^2) \, dA$$

$$= \iint_D (x^2 + y^2) \, dA = \int_0^{2\pi} \int_0^5 (r^2) \, r \, dr \, d\theta = \int_0^{2\pi} d\theta \int_0^5 r^3 \, dr = 2\pi \left[\tfrac{1}{4} r^4 \right]_0^5 = \tfrac{625}{2} \pi$$

17. By Green's Theorem, $W = \int_C \mathbf{F} \cdot d\mathbf{r} = \int_C x(x + y) \, dx + xy^2 \, dy = \iint_D (y^2 - x) \, dy \, dx$ where C is the path

described in the question and D is the triangle bounded by C. So

$$W = \int_0^1 \int_0^{1-x} (y^2 - x) \, dy \, dx = \int_0^1 \left[\tfrac{1}{3} y^3 - xy \right]_{y=0}^{y=1-x} dx = \int_0^1 \left(\tfrac{1}{3} (1 - x)^3 - x(1 - x) \right) dx$$

$$= \left[-\tfrac{1}{12} (1 - x)^4 - \tfrac{1}{2} x^2 + \tfrac{1}{3} x^3 \right]_0^1 = \left(-\tfrac{1}{2} + \tfrac{1}{3} \right) - \left(-\tfrac{1}{12} \right) = -\tfrac{1}{12}$$

19. Let C_1 be the arch of the cycloid from $(0, 0)$ to $(2\pi, 0)$, which corresponds to $0 \le t \le 2\pi$, and let C_2 be the

segment from $(2\pi, 0)$ to $(0, 0)$, so C_2 is given by $x = 2\pi - t$, $y = 0$, $0 \le t \le 2\pi$. Then $C = C_1 \cup C_2$ is traversed

clockwise, so $-C$ is oriented positively. Thus $-C$ encloses the area under one arch of the cycloid and from (5) we

have

$$A = -\oint_{-C} y \, dx = \int_{C_1} y \, dx + \int_{C_2} y \, dx = \int_0^{2\pi} (1 - \cos t)(1 - \cos t) \, dt + \int_0^{2\pi} 0 \, (-dt)$$

$$= \int_0^{2\pi} (1 - 2 \cos t + \cos^2 t) \, dt + 0 = \left[t - 2 \sin t + \tfrac{1}{2} t + \tfrac{1}{4} \sin 2t \right]_0^{2\pi} = 3\pi$$

21. (a) Using Equation 17.2.8 [ET 16.2.8], we write parametric equations of the line segment as $x = (1 - t)x_1 + tx_2$,

$y = (1 - t)y_1 + ty_2$, $0 \le t \le 1$. Then $dx = (x_2 - x_1) \, dt$ and $dy = (y_2 - y_1) \, dt$, so

$$\int_C x \, dy - y \, dx = \int_0^1 \left[(1 - t)x_1 + tx_2 \right] (y_2 - y_1) \, dt + \left[(1 - t)y_1 + ty_2 \right] (x_2 - x_1) \, dt$$

$$= \int_0^1 (x_1(y_2 - y_1) - y_1(x_2 - x_1) + t[(y_2 - y_1)(x_2 - x_1) - (x_2 - x_1)(y_2 - y_1)]) \, dt$$

$$= \int_0^1 (x_1 y_2 - x_2 y_1) \, dt = x_1 y_2 - x_2 y_1$$

(b) We apply Green's Theorem to the path $C = C_1 \cup C_2 \cup \cdots \cup C_n$, where C_i is the line segment that joins (x_i, y_i) to (x_{i+1}, y_{i+1}) for $i = 1, 2, \ldots, n-1$, and C_n is the line segment that joins (x_n, y_n) to (x_1, y_1). From (5), $\frac{1}{2} \int_C x\,dy - y\,dx = \iint_D dA$, where D is the polygon bounded by C. Therefore

$$\text{area of polygon} = A(D) = \iint_D dA = \tfrac{1}{2} \int_C x\,dy - y\,dx$$

$$= \tfrac{1}{2} \left(\int_{C_1} x\,dy - y\,dx + \int_{C_2} x\,dy - y\,dx + \cdots + \int_{C_{n-1}} x\,dy - y\,dx + \int_{C_n} x\,dy - y\,dx \right)$$

To evaluate these integrals we use the formula from (a) to get

$$A(D) = \tfrac{1}{2}[(x_1 y_2 - x_2 y_1) + (x_2 y_3 - x_3 y_2) + \cdots + (x_{n-1} y_n - x_n y_{n-1}) + (x_n y_1 - x_1 y_n)].$$

(c) $A = \tfrac{1}{2}[(0 \cdot 1 - 2 \cdot 0) + (2 \cdot 3 - 1 \cdot 1) + (1 \cdot 2 - 0 \cdot 3) + (0 \cdot 1 - (-1) \cdot 2) + (-1 \cdot 0 - 0 \cdot 1)]$

$= \tfrac{1}{2}(0 + 5 + 2 + 2) = \tfrac{9}{2}$

23. Here $A = \tfrac{1}{2}(1)(1) = \tfrac{1}{2}$ and $C = C_1 + C_2 + C_3$, where C_1: $x = x$, $y = 0$, $0 \le x \le 1$;

C_2: $x = x$, $y = 1 - x$, $x = 1$ to $x = 0$; and C_3: $x = 0$, $y = 1$ to $y = 0$. Then

$\bar{x} = \tfrac{1}{2A} \int_C x^2\,dy = \int_{C_1} x^2\,dy + \int_{C_2} x^2\,dy + \int_{C_3} x^2\,dy = 0 + \int_1^0 (x^2)(-dx) + 0 = \tfrac{1}{3}$. Similarly,

$\bar{y} = -\tfrac{1}{2A} \int_C y^2\,dx = \int_{C_1} y^2\,dx + \int_{C_2} y^2\,dx + \int_{C_3} y^2\,dx = 0 + \int_1^0 (1 - x)^2(-dx) + 0 = \tfrac{1}{3}$.

Therefore $(\bar{x}, \bar{y}) = \left(\tfrac{1}{3}, \tfrac{1}{3} \right)$.

25. By Green's Theorem, $-\tfrac{1}{3}\rho \oint_C y^3\,dx = -\tfrac{1}{3}\rho \iint_D (-3y^2)\,dA = \iint_D y^2 \rho\,dA = I_x$ and

$\tfrac{1}{3}\rho \oint_C x^3\,dy = \tfrac{1}{3}\rho \iint_D (3x^2)\,dA = \iint_D x^2 \rho\,dA = I_y$.

27. Since C is a simple closed path which doesn't pass through or enclose the origin, there exists an open region that doesn't contain the origin but does contain D. Thus $P = -y/(x^2 + y^2)$ and $Q = x/(x^2 + y^2)$ have continuous partial derivatives on this open region containing D and we can apply Green's Theorem. But by Exercise 17.3.33(a) [ET 16.3.33(a)], $\partial P/\partial y = \partial Q/\partial x$, so $\oint_C \mathbf{F} \cdot d\mathbf{r} = \iint_D 0\,dA = 0$.

29. Using the first part of (5), we have that $\iint_R dx\,dy = A(R) = \int_{\partial R} x\,dy$. But $x = g(u, v)$, and

$dy = \dfrac{\partial h}{\partial u}\,du + \dfrac{\partial h}{\partial v}\,dv$, and we orient ∂S by taking the positive direction to be that which corresponds, under the mapping, to the positive direction along ∂R, so

$$\int_{\partial R} x\,dy = \int_{\partial S} g(u, v)\left(\frac{\partial h}{\partial u}\,du + \frac{\partial h}{\partial v}\,dv \right) = \int_{\partial S} g(u, v)\frac{\partial h}{\partial u}\,du + g(u, v)\frac{\partial h}{\partial v}\,dv$$

$$= \pm \iint_S \left[\frac{\partial}{\partial u}\left(g(u, v)\frac{\partial h}{\partial v} \right) - \frac{\partial}{\partial v}\left(g(u, v)\frac{\partial h}{\partial u} \right) \right] dA \quad \text{[using Green's Theorem in the } uv\text{-plane]}$$

$$= \pm \iint_S \left(\frac{\partial g}{\partial u}\frac{\partial h}{\partial v} + g(u, v)\frac{\partial^2 h}{\partial u\,\partial v} - \frac{\partial g}{\partial v}\frac{\partial h}{\partial u} - g(u, v)\frac{\partial^2 h}{\partial v\,\partial u} \right) dA \quad \text{[using the Chain Rule]}$$

$$= \pm \iint_S \left(\frac{\partial x}{\partial u}\frac{\partial y}{\partial v} - \frac{\partial x}{\partial v}\frac{\partial y}{\partial u} \right) dA \quad \text{[by the equality of mixed partials]} \quad = \pm \iint_S \frac{\partial(x, y)}{\partial(u, v)}\,du\,dv$$

The sign is chosen to be positive if the orientation that we gave to ∂S corresponds to the usual positive orientation, and it is negative otherwise. In either case, since $A(R)$ is positive, the sign chosen must be the same as the sign of

$\dfrac{\partial(x, y)}{\partial(u, v)}$. Therefore $A(R) = \iint_R dx\,dy = \iint_S \left| \dfrac{\partial(x, y)}{\partial(u, v)} \right| du\,dv$.

17.5 Curl and Divergence

1. (a) curl $\mathbf{F} = \nabla \times \mathbf{F} = \begin{vmatrix} \mathbf{i} & \mathbf{j} & \mathbf{k} \\ \partial/\partial x & \partial/\partial y & \partial/\partial z \\ xyz & 0 & -x^2y \end{vmatrix} = (-x^2 - 0)\,\mathbf{i} - (-2xy - xy)\,\mathbf{j} + (0 - xz)\,\mathbf{k}$

$\qquad = -x^2\,\mathbf{i} + 3xy\,\mathbf{j} - xz\,\mathbf{k}$

(b) div $\mathbf{F} = \nabla \cdot \mathbf{F} = \dfrac{\partial}{\partial x}\,(xyz) + \dfrac{\partial}{\partial y}\,(0) + \dfrac{\partial}{\partial z}\,(-x^2y) = yz + 0 + 0 = yz$

3. (a) curl $\mathbf{F} = \nabla \times \mathbf{F} = \begin{vmatrix} \mathbf{i} & \mathbf{j} & \mathbf{k} \\ \partial/\partial x & \partial/\partial y & \partial/\partial z \\ 1 & x + yz & xy - \sqrt{z} \end{vmatrix} = (x - y)\,\mathbf{i} - (y - 0)\,\mathbf{j} + (1 - 0)\,\mathbf{k}$

$\qquad = (x - y)\,\mathbf{i} - y\,\mathbf{j} + \mathbf{k}$

(b) div $\mathbf{F} = \nabla \cdot \mathbf{F} = \dfrac{\partial}{\partial x}\,(1) + \dfrac{\partial}{\partial y}\,(x + yz) + \dfrac{\partial}{\partial z}\,(xy - \sqrt{z}) = z - \dfrac{1}{2\sqrt{z}}$

5. (a) curl $\mathbf{F} = \nabla \times \mathbf{F} = \begin{vmatrix} \mathbf{i} & \mathbf{j} & \mathbf{k} \\ \partial/\partial x & \partial/\partial y & \partial/\partial z \\ e^x \sin y & e^x \cos y & z \end{vmatrix} = (0 - 0)\,\mathbf{i} - (0 - 0)\,\mathbf{j} + (e^x \cos y - e^x \cos y)\,\mathbf{k} = \mathbf{0}$

(b) div $\mathbf{F} = \nabla \cdot \mathbf{F} = \dfrac{\partial}{\partial x}\,(e^x \sin y) + \dfrac{\partial}{\partial y}\,(e^x \cos y) + \dfrac{\partial}{\partial z}\,(z) = e^x \sin y - e^x \sin y + 1 = 1$

7. (a) curl $\mathbf{F} = \nabla \times \mathbf{F} = \begin{vmatrix} \mathbf{i} & \mathbf{j} & \mathbf{k} \\ \partial/\partial x & \partial/\partial y & \partial/\partial z \\ \ln x & \ln(xy) & \ln(xyz) \end{vmatrix} = \left(\dfrac{xz}{xyz} - 0\right)\mathbf{i} - \left(\dfrac{yz}{xyz} - 0\right)\mathbf{j} + \left(\dfrac{y}{xy} - 0\right)\mathbf{k}$

$\qquad = \left\langle \dfrac{1}{y}, -\dfrac{1}{x}, \dfrac{1}{x} \right\rangle$

(b) div $\mathbf{F} = \nabla \cdot \mathbf{F} = \dfrac{\partial}{\partial x}\,(\ln x) + \dfrac{\partial}{\partial y}\,(\ln(xy)) + \dfrac{\partial}{\partial z}\,(\ln(xyz)) = \dfrac{1}{x} + \dfrac{x}{xy} + \dfrac{xy}{xyz} = \dfrac{1}{x} + \dfrac{1}{y} + \dfrac{1}{z}$

9. If the vector field is $\mathbf{F} = P\,\mathbf{i} + Q\,\mathbf{j} + R\,\mathbf{k}$, then we know $R = 0$. In addition, the x-component of each vector of $\mathbf{F}$

is 0, so $P = 0$, hence $\dfrac{\partial P}{\partial x} = \dfrac{\partial P}{\partial y} = \dfrac{\partial P}{\partial z} = \dfrac{\partial R}{\partial x} = \dfrac{\partial R}{\partial y} = \dfrac{\partial R}{\partial z} = 0$. Q decreases as y increases, so $\dfrac{\partial Q}{\partial y} < 0$, but

Q doesn't change in the x- or z-directions, so $\dfrac{\partial Q}{\partial x} = \dfrac{\partial Q}{\partial z} = 0$.

(a) div $\mathbf{F} = \dfrac{\partial P}{\partial x} + \dfrac{\partial Q}{\partial y} + \dfrac{\partial R}{\partial z} = 0 + \dfrac{\partial Q}{\partial y} + 0 < 0$

(b) curl $\mathbf{F} = \left(\dfrac{\partial R}{\partial y} - \dfrac{\partial Q}{\partial z}\right)\mathbf{i} + \left(\dfrac{\partial P}{\partial z} - \dfrac{\partial R}{\partial x}\right)\mathbf{j} + \left(\dfrac{\partial Q}{\partial x} - \dfrac{\partial P}{\partial y}\right)\mathbf{k} = (0 - 0)\,\mathbf{i} + (0 - 0)\,\mathbf{j} + (0 - 0)\,\mathbf{k} = \mathbf{0}$

11. If the vector field is $\mathbf{F} = P\,\mathbf{i} + Q\,\mathbf{j} + R\,\mathbf{k}$, then we know $R = 0$. In addition, the y-component of each vector of $\mathbf{F}$

is 0, so $Q = 0$, hence $\dfrac{\partial Q}{\partial x} = \dfrac{\partial Q}{\partial y} = \dfrac{\partial Q}{\partial z} = \dfrac{\partial R}{\partial x} = \dfrac{\partial R}{\partial y} = \dfrac{\partial R}{\partial z} = 0$. P increases as y increases, so $\dfrac{\partial P}{\partial y} > 0$, but P

doesn't change in the x- or z-directions, so $\dfrac{\partial P}{\partial x} = \dfrac{\partial P}{\partial z} = 0$.

(a) div $\mathbf{F} = \dfrac{\partial P}{\partial x} + \dfrac{\partial Q}{\partial y} + \dfrac{\partial R}{\partial z} = 0 + 0 + 0 = 0$

(b) $\operatorname{curl} \mathbf{F} = \left(\dfrac{\partial R}{\partial y} - \dfrac{\partial Q}{\partial z} \right) \mathbf{i} + \left(\dfrac{\partial P}{\partial z} - \dfrac{\partial R}{\partial x} \right) \mathbf{j} + \left(\dfrac{\partial Q}{\partial x} - \dfrac{\partial P}{\partial y} \right) \mathbf{k}$

$\qquad = (0 - 0)\,\mathbf{i} + (0 - 0)\,\mathbf{j} + \left(0 - \dfrac{\partial P}{\partial y} \right) \mathbf{k} = -\dfrac{\partial P}{\partial y}\mathbf{k}$

Since $\dfrac{\partial P}{\partial y} > 0$, $-\dfrac{\partial P}{\partial y}\mathbf{k}$ is a vector pointing in the negative z-direction.

13. $\operatorname{curl} \mathbf{F} = \nabla \times \mathbf{F} = \begin{vmatrix} \mathbf{i} & \mathbf{j} & \mathbf{k} \\ \partial/\partial x & \partial/\partial y & \partial/\partial z \\ yz & xz & xy \end{vmatrix} = (x - x)\,\mathbf{i} - (y - y)\,\mathbf{j} + (z - z)\,\mathbf{k} = \mathbf{0}$

and $\mathbf{F}$ is defined on all of $\mathbb{R}^3$ with component functions which have continuous partial derivatives, so by Theorem 4,

$\mathbf{F}$ is conservative. Thus, there exists a function f such that $\mathbf{F} = \nabla f$. Then $f_x(x, y, z) = yz$ implies

$f(x, y, z) = xyz + g(y, z)$ and $f_y(x, y, z) = xz + g_y(y, z)$. But $f_y(x, y, z) = xz$, so $g(y, z) = h(z)$ and

$f(x, y, z) = xyz + h(z)$. Thus $f_z(x, y, z) = xy + h'(z)$ but $f_z(x, y, z) = xy$ so $h(z) = K$, a constant. Hence a

potential function for $\mathbf{F}$ is $f(x, y, z) = xyz + K$.

15. $\operatorname{curl} \mathbf{F} = \nabla \times \mathbf{F} = \begin{vmatrix} \mathbf{i} & \mathbf{j} & \mathbf{k} \\ \partial/\partial x & \partial/\partial y & \partial/\partial z \\ 2xy & x^2 + 2yz & y^2 \end{vmatrix} = (2y - 2y)\,\mathbf{i} - (0 - 0)\,\mathbf{j} + (2x - 2x)\,\mathbf{k} = \mathbf{0}$, $\mathbf{F}$ is defined on all

of $\mathbb{R}^3$, and the partial derivatives of the component functions are continuous, so $\mathbf{F}$ is conservative. Thus there exists

a function f such that $\nabla f = \mathbf{F}$. Then $f_x(x, y, z) = 2xy$ implies $f(x, y, z) = x^2 y + g(y, z)$ and

$f_y(x, y, z) = x^2 + g_y(y, z)$. But $f_y(x, y, z) = x^2 + 2yz$, so $g(y, z) = y^2 z + h(z)$ and

$f(x, y, z) = x^2 y + y^2 z + h(z)$. Thus $f_z(x, y, z) = y^2 + h'(z)$ but $f_z(x, y, z) = y^2$ so $h(z) = K$ and

$f(x, y, z) = x^2 y + y^2 z + K$.

17. $\operatorname{curl} \mathbf{F} = \nabla \times \mathbf{F} = \begin{vmatrix} \mathbf{i} & \mathbf{j} & \mathbf{k} \\ \partial/\partial x & \partial/\partial y & \partial/\partial z \\ ye^{-x} & e^{-x} & 2z \end{vmatrix} = (0 - 0)\,\mathbf{i} - (0 - 0)\,\mathbf{j} + (-e^{-x} - e^{-x})\,\mathbf{k} = -2e^{-x}\mathbf{k} \neq \mathbf{0}$,

so $\mathbf{F}$ is not conservative.

19. No. Assume there is such a $\mathbf{G}$. Then $\operatorname{div}(\operatorname{curl} \mathbf{G}) = y^2 + z^2 + x^2 \neq 0$, which contradicts Theorem 11.

21. $\operatorname{curl} \mathbf{F} = \begin{vmatrix} \mathbf{i} & \mathbf{j} & \mathbf{k} \\ \partial/\partial x & \partial/\partial y & \partial/\partial z \\ f(x) & g(y) & h(z) \end{vmatrix} = (0 - 0)\,\mathbf{i} + (0 - 0)\,\mathbf{j} + (0 - 0)\,\mathbf{k} = \mathbf{0}$.

Hence $\mathbf{F} = f(x)\,\mathbf{i} + g(y)\,\mathbf{j} + h(z)\,\mathbf{k}$ is irrotational.

For Exercises 23–29, let $\mathbf{F}(x, y, z) = P_1\,\mathbf{i} + Q_1\,\mathbf{j} + R_1\,\mathbf{k}$ and $\mathbf{G}(x, y, z) = P_2\,\mathbf{i} + Q_2\,\mathbf{j} + R_2\,\mathbf{k}$.

23. $\operatorname{div}(\mathbf{F} + \mathbf{G}) = \dfrac{\partial(P_1 + P_2)}{\partial x} + \dfrac{\partial(Q_1 + Q_2)}{\partial y} + \dfrac{\partial(R_1 + R_2)}{\partial z}$

$$= \left(\frac{\partial P_1}{\partial x} + \frac{\partial Q_1}{\partial y} + \frac{\partial R_1}{\partial z} \right) + \left(\frac{\partial P_2}{\partial x} + \frac{\partial Q_2}{\partial y} + \frac{\partial R_3}{\partial z} \right) = \operatorname{div}\mathbf{F} + \operatorname{div}\mathbf{G}$$

25. $\operatorname{div}(f\mathbf{F}) = \dfrac{\partial(fP_1)}{\partial x} + \dfrac{\partial(fQ_1)}{\partial y} + \dfrac{\partial(fR_1)}{\partial z}$

$$= \left(f\frac{\partial P_1}{\partial x} + P_1\frac{\partial f}{\partial x} \right) + \left(f\frac{\partial Q_1}{\partial y} + Q_1\frac{\partial f}{\partial y} \right) + \left(f\frac{\partial R_1}{\partial z} + R_1\frac{\partial f}{\partial z} \right)$$

$$= f\left(\frac{\partial P_1}{\partial x} + \frac{\partial Q_1}{\partial y} + \frac{\partial R_1}{\partial z} \right) + \langle P_1, Q_1, R_1 \rangle \cdot \left\langle \frac{\partial f}{\partial x}, \frac{\partial f}{\partial y}, \frac{\partial f}{\partial z} \right\rangle = f\operatorname{div}\mathbf{F} + \mathbf{F}\cdot\nabla f$$

27. $\operatorname{div}(\mathbf{F} \times \mathbf{G}) = \nabla \cdot (\mathbf{F} \times \mathbf{G}) = \begin{vmatrix} \partial/\partial x & \partial/\partial y & \partial/\partial z \\ P_1 & Q_1 & R_1 \\ P_2 & Q_2 & R_2 \end{vmatrix} = \dfrac{\partial}{\partial x}\begin{vmatrix} Q_1 & R_1 \\ Q_2 & R_2 \end{vmatrix} - \dfrac{\partial}{\partial y}\begin{vmatrix} P_1 & R_1 \\ P_2 & R_2 \end{vmatrix} + \dfrac{\partial}{\partial z}\begin{vmatrix} P_1 & Q_1 \\ P_2 & Q_2 \end{vmatrix}$

$$= \left[Q_1\frac{\partial R_2}{\partial x} + R_2\frac{\partial Q_1}{\partial x} - Q_2\frac{\partial R_1}{\partial x} - R_1\frac{\partial Q_2}{\partial x} \right]$$

$$- \left[P_1\frac{\partial R_2}{\partial y} + R_2\frac{\partial P_1}{\partial y} - P_2\frac{\partial R_1}{\partial y} - R_1\frac{\partial P_2}{\partial y} \right]$$

$$+ \left[P_1\frac{\partial Q_2}{\partial z} + Q_2\frac{\partial P_1}{\partial z} - P_2\frac{\partial Q_1}{\partial z} - Q_1\frac{\partial P_2}{\partial z} \right]$$

$$= \left[P_2\left(\frac{\partial R_1}{\partial y} - \frac{\partial Q_1}{\partial z} \right) + Q_2\left(\frac{\partial P_1}{\partial z} - \frac{\partial R_1}{\partial x} \right) + R_2\left(\frac{\partial Q_1}{\partial x} - \frac{\partial P_1}{\partial y} \right) \right]$$

$$- \left[P_1\left(\frac{\partial R_2}{\partial y} - \frac{\partial Q_2}{\partial z} \right) + Q_1\left(\frac{\partial P_2}{\partial z} - \frac{\partial R_2}{\partial x} \right) + R_1\left(\frac{\partial Q_2}{\partial x} - \frac{\partial P_2}{\partial y} \right) \right]$$

$$= \mathbf{G}\cdot\operatorname{curl}\mathbf{F} - \mathbf{F}\cdot\operatorname{curl}\mathbf{G}$$

29. $\operatorname{curl}\operatorname{curl}\mathbf{F} = \nabla \times (\nabla \times \mathbf{F}) = \begin{vmatrix} \mathbf{i} & \mathbf{j} & \mathbf{k} \\ \partial/\partial x & \partial/\partial y & \partial/\partial z \\ \partial R_1/\partial y - \partial Q_1/\partial z & \partial P_1/\partial z - \partial R_1/\partial x & \partial Q_1/\partial x - \partial P_1/\partial y \end{vmatrix}$

$$= \left(\frac{\partial^2 Q_1}{\partial y\partial x} - \frac{\partial^2 P_1}{\partial y^2} - \frac{\partial^2 P_1}{\partial z^2} + \frac{\partial^2 R_1}{\partial z\partial x} \right)\mathbf{i} + \left(\frac{\partial^2 R_1}{\partial z\partial y} - \frac{\partial^2 Q_1}{\partial z^2} - \frac{\partial^2 Q_1}{\partial x^2} + \frac{\partial^2 P_1}{\partial x\partial y} \right)\mathbf{j}$$

$$+ \left(\frac{\partial^2 P_1}{\partial x\partial z} - \frac{\partial^2 R_1}{\partial x^2} - \frac{\partial^2 R_1}{\partial y^2} + \frac{\partial^2 Q_1}{\partial y\partial z} \right)\mathbf{k}$$

Now let's consider grad div $\mathbf{F} - \nabla^2\mathbf{F}$ and compare with the above.

(Note that $\nabla^2 \mathbf{F}$ is defined on page 1130 [ET 1094].)

$$\operatorname{grad} \operatorname{div} \mathbf{F} - \nabla^2 \mathbf{F} = \left[\left(\frac{\partial^2 P_1}{\partial x^2} + \frac{\partial^2 Q_1}{\partial x \partial y} + \frac{\partial^2 R_1}{\partial x \partial z} \right) \mathbf{i} + \left(\frac{\partial^2 P_1}{\partial y \partial x} + \frac{\partial^2 Q_1}{\partial y^2} + \frac{\partial^2 R_1}{\partial y \partial z} \right) \mathbf{j} \right.$$

$$\left. + \left(\frac{\partial^2 P_1}{\partial z \partial x} + \frac{\partial^2 Q_1}{\partial z \partial y} + \frac{\partial^2 R_1}{\partial z^2} \right) \mathbf{k} \right]$$

$$- \left[\left(\frac{\partial^2 P_1}{\partial x^2} + \frac{\partial^2 P_1}{\partial y^2} + \frac{\partial^2 P_1}{\partial z^2} \right) \mathbf{i} + \left(\frac{\partial^2 Q_1}{\partial x^2} + \frac{\partial^2 Q_1}{\partial y^2} + \frac{\partial^2 Q_1}{\partial z^2} \right) \mathbf{j} \right.$$

$$\left. + \left(\frac{\partial^2 R_1}{\partial x^2} + \frac{\partial^2 R_1}{\partial y^2} + \frac{\partial^2 R_1}{\partial z^2} \right) \mathbf{k} \right]$$

$$= \left(\frac{\partial^2 Q_1}{\partial x \partial y} + \frac{\partial^2 R_1}{\partial x \partial z} - \frac{\partial^2 P_1}{\partial y^2} - \frac{\partial^2 P_1}{\partial z^2} \right) \mathbf{i} + \left(\frac{\partial^2 P_1}{\partial y \partial x} + \frac{\partial^2 R_1}{\partial y \partial z} - \frac{\partial^2 Q_1}{\partial x^2} - \frac{\partial^2 Q_1}{\partial z^2} \right) \mathbf{j}$$

$$+ \left(\frac{\partial^2 P_1}{\partial z \partial x} + \frac{\partial^2 Q_1}{\partial z \partial y} - \frac{\partial^2 R_1}{\partial x^2} - \frac{\partial^2 R_2}{\partial y^2} \right) \mathbf{k}$$

Then applying Clairaut's Theorem to reverse the order of differentiation in the second partial derivatives as needed and comparing, we have $\operatorname{curl} \operatorname{curl} \mathbf{F} = \operatorname{grad} \operatorname{div} \mathbf{F} - \nabla^2 \mathbf{F}$ as desired.

31. (a) $\nabla r = \nabla \sqrt{x^2 + y^2 + z^2} = \dfrac{x}{\sqrt{x^2 + y^2 + z^2}} \mathbf{i} + \dfrac{y}{\sqrt{x^2 + y^2 + z^2}} \mathbf{j} + \dfrac{z}{\sqrt{x^2 + y^2 + z^2}} \mathbf{k}$

$$= \frac{x \mathbf{i} + y \mathbf{j} + z \mathbf{k}}{\sqrt{x^2 + y^2 + z^2}} = \frac{\mathbf{r}}{r}$$

(b) $\nabla \times \mathbf{r} = \begin{vmatrix} \mathbf{i} & \mathbf{j} & \mathbf{k} \\ \dfrac{\partial}{\partial x} & \dfrac{\partial}{\partial y} & \dfrac{\partial}{\partial z} \\ x & y & z \end{vmatrix}$

$$= \left[\frac{\partial}{\partial y} (z) - \frac{\partial}{\partial z} (y) \right] \mathbf{i} + \left[\frac{\partial}{\partial z} (x) - \frac{\partial}{\partial x} (z) \right] \mathbf{j} + \left[\frac{\partial}{\partial x} (y) - \frac{\partial}{\partial y} (x) \right] \mathbf{k} = \mathbf{0}$$

(c) $\nabla \left(\dfrac{1}{r} \right) = \nabla \left(\dfrac{1}{\sqrt{x^2 + y^2 + z^2}} \right)$

$$= \frac{-\dfrac{1}{2\sqrt{x^2 + y^2 + z^2}} (2x)}{x^2 + y^2 + z^2} \mathbf{i} - \frac{\dfrac{1}{2\sqrt{x^2 + y^2 + z^2}} (2y)}{x^2 + y^2 + z^2} \mathbf{j} - \frac{\dfrac{1}{2\sqrt{x^2 + y^2 + z^2}} (2z)}{x^2 + y^2 + z^2} \mathbf{k}$$

$$= -\frac{x \mathbf{i} + y \mathbf{j} + z \mathbf{k}}{(x^2 + y^2 + z^2)^{3/2}} = -\frac{\mathbf{r}}{r^3}$$

(d) $\nabla \ln r = \nabla \ln (x^2 + y^2 + z^2)^{1/2} = \tfrac{1}{2} \nabla \ln (x^2 + y^2 + z^2)$

$$= \frac{x}{x^2 + y^2 + z^2} \mathbf{i} + \frac{y}{x^2 + y^2 + z^2} \mathbf{j} + \frac{z}{x^2 + y^2 + z^2} \mathbf{k} = \frac{x \mathbf{i} + y \mathbf{j} + z \mathbf{k}}{x^2 + y^2 + z^2} = \frac{\mathbf{r}}{r^2}$$

33. By (13), $\oint_C f(\nabla g) \cdot \mathbf{n}\, ds = \iint_D \operatorname{div}(f\nabla g)\, dA = \iint_D [f \operatorname{div}(\nabla g) + \nabla g \cdot \nabla f]\, dA$ by Exercise 25. But
$\operatorname{div}(\nabla g) = \nabla^2 g$. Hence $\iint_D f \nabla^2 g\, dA = \oint_C f(\nabla g) \cdot \mathbf{n}\, ds - \iint_D \nabla g \cdot \nabla f\, dA.$

35. (a) We know that $\omega = v/d$, and from the diagram $\sin\theta = d/r \;\Rightarrow\; v = d\omega = (\sin\theta)r\omega = |\mathbf{w} \times \mathbf{r}|$. But $\mathbf{v}$ is
perpendicular to both $\mathbf{w}$ and $\mathbf{r}$, so that $\mathbf{v} = \mathbf{w} \times \mathbf{r}$.

(b) From (a), $\mathbf{v} = \mathbf{w} \times \mathbf{r} = \begin{vmatrix} \mathbf{i} & \mathbf{j} & \mathbf{k} \\ 0 & 0 & \omega \\ x & y & z \end{vmatrix} = (0 \cdot z - \omega y)\mathbf{i} + (\omega x - 0 \cdot z)\mathbf{j} + (0 \cdot y - x \cdot 0)\mathbf{k} = -\omega y\,\mathbf{i} + \omega x\,\mathbf{j}$

(c) $\operatorname{curl}\mathbf{v} = \nabla \times \mathbf{v} = \begin{vmatrix} \mathbf{i} & \mathbf{j} & \mathbf{k} \\ \partial/\partial x & \partial/\partial y & \partial/\partial z \\ -\omega y & \omega x & 0 \end{vmatrix}$

$$= \left[\frac{\partial}{\partial y}(0) - \frac{\partial}{\partial z}(\omega x)\right]\mathbf{i} + \left[\frac{\partial}{\partial z}(-\omega y) - \frac{\partial}{\partial x}(0)\right]\mathbf{j} + \left[\frac{\partial}{\partial x}(\omega x) - \frac{\partial}{\partial y}(-\omega y)\right]\mathbf{k}$$

$$= [\omega - (-\omega)]\mathbf{k} = 2\omega\,\mathbf{k} = 2\mathbf{w}$$

37. For any continuous function f on $\mathbb{R}^3$, define a vector field $\mathbf{G}(x, y, z) = \langle g(x, y, z), 0, 0 \rangle$ where
$g(x, y, z) = \int_0^x f(t, y, z)\, dt$. Then

$$\operatorname{div}\mathbf{G} = \frac{\partial}{\partial x}(g(x, y, z)) + \frac{\partial}{\partial y}(0) + \frac{\partial}{\partial z}(0) = \frac{\partial}{\partial x}\int_0^x f(t, y, z)\, dt = f(x, y, z)$$ by the Fundamental Theorem of

Calculus. Thus every continuous function f on $\mathbb{R}^3$ is the divergence of some vector field.

17.6 Parametric Surfaces and Their Areas ET 16.6

1. $\mathbf{r}(u, v) = u\cos v\,\mathbf{i} + u\sin v\,\mathbf{j} + u^2\,\mathbf{k}$, so the corresponding parametric equations for the surface are
$x = u\cos v, y = u\sin v, z = u^2$. For any point (x, y, z) on the surface, we have
$x^2 + y^2 = u^2\cos^2 v + u^2\sin^2 v = u^2 = z$. Since no restrictions are placed on the parameters, the surface is
$z = x^2 + y^2$, which we recognize as a circular paraboloid opening upward whose axis is the z-axis.

3. $\mathbf{r}(x, \theta) = \langle x, \cos\theta, \sin\theta \rangle$, so the corresponding parametric equations for the surface are
$x = x, y = \cos\theta, z = \sin\theta$. For any point (x, y, z) on the surface, we have $y^2 + z^2 = \cos^2\theta + \sin^2\theta = 1$, so any
vertical trace in $x = k$ is the circle $y^2 + z^2 = 1, x = k$. Since $x = x$ with no restriction, the surface is a circular
cylinder with radius 1 whose axis is the x-axis.

5. $\mathbf{r}(u, v) = \langle u^2 + 1, v^3 + 1, u + v \rangle, -1 \le u \le 1, -1 \le v \le 1$.

The surface has parametric equations $x = u^2 + 1, y = v^3 + 1$,
$z = u + v, -1 \le u \le 1, -1 \le v \le 1$. If we keep u constant at
u_0, $x = u_0^2 + 1$, a constant, so the corresponding grid curves must
be the curves parallel to the yz-plane. If v is constant, we have
$y = v_0^3 + 1$, a constant, so these grid curves are the curves parallel
to the xz-plane.

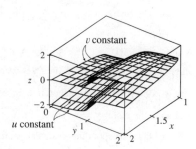

7. $\mathbf{r}(u, v) = \langle \cos^3 u \cos^3 v, \sin^3 u \cos^3 v, \sin^3 v \rangle$.

The surface has parametric equations $x = \cos^3 u \cos^3 v$,

$y = \sin^3 u \cos^3 v$, $z = \sin^3 v$, $0 \le u \le \pi$, $0 \le v \le 2\pi$. Note

that if $v = v_0$ is constant then $z = \sin^3 v_0$ is constant, so the

corresponding grid curves must be the curves parallel to the

xy-plane. The vertically oriented grid curves, then, correspond

to $u = u_0$ being held constant, giving $x = \cos^3 u_0 \cos^3 v$,

$y = \sin^3 u_0 \cos^3 v$, $z = \sin^3 v$. These curves lie in vertical

planes that contain the z-axis.

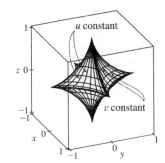

9. $x = \cos u \sin 2v$, $y = \sin u \sin 2v$, $z = \sin v$.

The complete graph of the surface is given by the parametric

domain $0 \le u \le \pi$, $0 \le v \le 2\pi$. Note that if $v = v_0$ is

constant, the parametric equations become $x = \cos u \sin 2v_0$,

$y = \sin u \sin 2v_0$, $z = \sin v_0$ which represent a circle of radius

$\sin 2v_0$ in the plane $z = \sin v_0$. So the circular grid curves we

see lying horizontally are the grid curves which have

v constant. The vertical grid curves, then, correspond to $u = u_0$

being held constant, giving $x = \cos u_0 \sin 2v$ and

$y = \sin u_0 \sin 2v$ with $z = \sin v$ which has a "figure-eight"

shape.

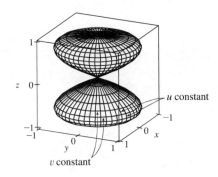

11. $\mathbf{r}(u, v) = \cos v \, \mathbf{i} + \sin v \, \mathbf{j} + u \, \mathbf{k}$. The parametric equations for the surface are $x = \cos v$, $y = \sin v$, $z = u$. Then

$x^2 + y^2 = \cos^2 v + \sin^2 v = 1$ and $z = u$ with no restriction on u, so we have a circular cylinder, graph IV. The

grid curves with u constant are the horizontal circles we see in the plane $z = u$. If v is constant, both x and y are

constant with z free to vary, so the corresponding grid curves are the lines on the cylinder parallel to the z-axis.

13. $\mathbf{r}(u, v) = u \cos v \, \mathbf{i} + u \sin v \, \mathbf{j} + v \, \mathbf{k}$. The parametric equations for the surface are $x = u \cos v$, $y = u \sin v$, $z = v$.

We look at the grid curves first; if we fix v, then x and y parametrize a straight line in the plane $z = v$ which

intersects the z-axis. If u is held constant, the projection onto the xy-plane is circular; with $z = v$, each grid curve is

a helix. The surface is a spiraling ramp, graph I.

15. $x = (u - \sin u) \cos v$, $y = (1 - \cos u) \sin v$, $z = u$. If u is held constant, x and y give an equation of an ellipse in

the plane $z = u$, thus the grid curves are horizontally oriented ellipses. Note that when $u = 0$, the "ellipse" is the

single point $(0, 0, 0)$, and when $u = \pi$, we have $y = 0$ while x ranges from $-\pi$ to π, a line segment parallel to the

x-axis in the plane $z = \pi$. This is the upper "seam" we see in graph II. When v is held constant, $z = u$ is free to

vary, so the corresponding grid curves are the curves we see running up and down along the surface.

17. From Example 3, parametric equations for the plane through the point $(1, 2, -3)$ that contains the vectors $\mathbf{a} = \langle 1, 1, -1 \rangle$ and $\mathbf{b} = \langle 1, -1, 1 \rangle$ are $x = 1 + u(1) + v(1) = 1 + u + v$, $y = 2 + u(1) + v(-1) = 2 + u - v$, $z = -3 + u(-1) + v(1) = -3 - u + v$.

19. Solving the equation for y gives $y^2 = 1 - x^2 + z^2 \Rightarrow y = \sqrt{1 - x^2 + z^2}$. (We choose the positive root since we want the part of the hyperboloid that corresponds to $y \geq 0$.) If we let x and z be the parameters, parametric equations are $x = x$, $z = z$, $y = \sqrt{1 - x^2 + z^2}$.

21. Since the cone intersects the sphere in the circle $x^2 + y^2 = 2$, $z = \sqrt{2}$ and we want the portion of the sphere above this, we can parametrize the surface as $x = x$, $y = y$, $z = \sqrt{4 - x^2 - y^2}$ where $x^2 + y^2 \leq 2$.
Alternate solution: Using spherical coordinates, $x = 2 \sin \phi \cos \theta$, $y = 2 \sin \phi \sin \theta$, $z = 2 \cos \phi$ where $0 \leq \phi \leq \frac{\pi}{4}$ and $0 \leq \theta \leq 2\pi$.

23. Parametric equations are $x = x$, $y = 4 \cos \theta$, $z = 4 \sin \theta$, $0 \leq x \leq 5$, $0 \leq \theta \leq 2\pi$.

25. The surface appears to be a portion of a circular cylinder of radius 3 with axis the x-axis. An equation of the cylinder is $y^2 + z^2 = 9$, and we can impose the restrictions $0 \leq x \leq 5$, $y \leq 0$ to obtain the portion shown. To graph the surface on a CAS, we can use parametric equations $x = u$, $y = 3 \cos v$, $z = 3 \sin v$ with the parameter domain $0 \leq u \leq 5$, $\frac{\pi}{2} \leq v \leq \frac{3\pi}{2}$. Alternatively, we can regard x and z as parameters. Then parametric equations are $x = x$, $z = z$, $y = -\sqrt{9 - z^2}$, where $0 \leq x \leq 5$ and $-3 \leq z \leq 3$.

27. Using Equations 3, we have the parametrization $x = x$, $y = e^{-x} \cos \theta$, $z = e^{-x} \sin \theta$, $0 \leq x \leq 3$, $0 \leq \theta \leq 2\pi$.

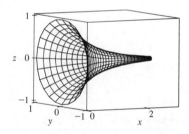

29. (a) Replacing $\cos u$ by $\sin u$ and $\sin u$ by $\cos u$ gives parametric equations $x = (2 + \sin v) \sin u$, $y = (2 + \sin v) \cos u$, $z = u + \cos v$. From the graph, it appears that the direction of the spiral is reversed. We can verify this observation by noting that the projection of the spiral grid curves onto the xy-plane, given by $x = (2 + \sin v) \sin u$, $y = (2 + \sin v) \cos u$, $z = 0$, draws a circle in the clockwise direction for each value of v. The original equations, on the other hand, give circular projections drawn in the counterclockwise direction. The equation for z is identical in both surfaces, so as z increases, these grid curves spiral up in opposite directions for the two surfaces.

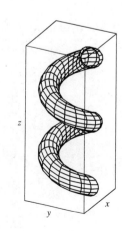

(b) Replacing $\cos u$ by $\cos 2u$ and $\sin u$ by $\sin 2u$ gives parametric equations

$x = (2 + \sin v)\cos 2u$, $y = (2 + \sin v)\sin 2u$, $z = u + \cos v$. From the graph, it appears that the number of coils in the surface doubles within the same parametric domain. We can verify this observation by noting that the projection of the spiral grid curves onto the xy-plane, given by $x = (2 + \sin v)\cos 2u$,

$y = (2 + \sin v)\sin 2u$, $z = 0$ (where v is constant), complete circular revolutions for $0 \le u \le \pi$ while the original surface requires $0 \le u \le 2\pi$ for a complete revolution. Thus, the new surface winds around twice as fast as the original surface, and since the equation for z is identical in both surfaces, we observe twice as many circular coils in the same z-interval.

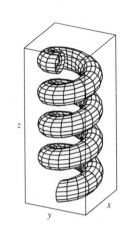

31. $\mathbf{r}(u, v) = (u + v)\,\mathbf{i} + 3u^2\,\mathbf{j} + (u - v)\,\mathbf{k}$.

$\mathbf{r}_u = \mathbf{i} + 6u\,\mathbf{j} + \mathbf{k}$ and $\mathbf{r}_v = \mathbf{i} - \mathbf{k}$, so

$\mathbf{r}_u \times \mathbf{r}_v = -6u\,\mathbf{i} + 2\,\mathbf{j} - 6u\,\mathbf{k}$. Since the point $(2, 3, 0)$

corresponds to $u = 1$, $v = 1$, a normal vector to the surface at

$(2, 3, 0)$ is $-6\,\mathbf{i} + 2\,\mathbf{j} - 6\,\mathbf{k}$, and an equation of the tangent plane is

$-6x + 2y - 6z = -6$ or $3x - y + 3z = 3$.

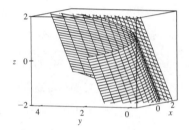

33. $\mathbf{r}(u, v) = u^2\,\mathbf{i} + 2u\sin v\,\mathbf{j} + u\cos v\,\mathbf{k}$ $\Rightarrow$ $\mathbf{r}(1, 0) = (1, 0, 1)$.

$\mathbf{r}_u = 2u\,\mathbf{i} + 2\sin v\,\mathbf{j} + \cos v\,\mathbf{k}$ and $\mathbf{r}_v = 2u\cos v\,\mathbf{j} - u\sin v\,\mathbf{k}$,

so a normal vector to the surface at the point $(1, 0, 1)$ is

$\mathbf{r}_u(1, 0) \times \mathbf{r}_v(1, 0) = (2\,\mathbf{i} + \mathbf{k}) \times (2\,\mathbf{j}) = -2\,\mathbf{i} + 4\,\mathbf{k}$.

Thus an equation of the tangent plane at $(1, 0, 1)$ is

$-2(x - 1) + 0(y - 0) + 4(z - 1) = 0$ or $-x + 2z = 1$.

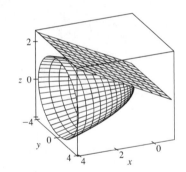

35. Here $z = f(x, y) = 4 - x - 2y$ and D is the disk $x^2 + y^2 \le 4$. Thus, by Formula 9,

$$A(S) = \iint_D \sqrt{1 + (-1)^2 + (-2)^2}\, dA = \sqrt{6} \iint_D dA = \sqrt{6}\, A(D) = 4\sqrt{6}\,\pi$$

37. $z = f(x, y) = xy$ with $0 \le x^2 + y^2 \le 1$, so $f_x = y$, $f_y = x$ $\Rightarrow$

$$A(S) = \iint_D \sqrt{1 + y^2 + x^2}\, dA = \int_0^{2\pi} \int_0^1 \sqrt{r^2 + 1}\, r\, dr\, d\theta = \int_0^{2\pi} \left[\frac{1}{3}(r^2 + 1)^{3/2} \right]_{r=0}^{r=1} d\theta$$

$$= \int_0^{2\pi} \frac{1}{3}\left(2\sqrt{2} - 1\right) d\theta = \frac{2\pi}{3}\left(2\sqrt{2} - 1\right)$$

39. $z = f(x, y) = y^2 - x^2$ with $1 \le x^2 + y^2 \le 4$. Then

$$A(S) = \iint_D \sqrt{1 + 4x^2 + 4y^2}\, dA = \int_0^{2\pi} \int_1^2 \sqrt{1 + 4r^2}\, r\, dr\, d\theta = \int_0^{2\pi} d\theta \int_1^2 \sqrt{1 + 4r^2}\, r\, dr$$

$$= \left[\theta\right]_0^{2\pi} \left[\frac{1}{12}(1 + 4r^2)^{3/2}\right]_1^2 = \frac{\pi}{6}\left(17\sqrt{17} - 5\sqrt{5}\right)$$

41. A parametric representation of the surface is $x = x$, $y = 4x + z^2$, $z = z$ with $0 \le x \le 1$, $0 \le z \le 1$.

Hence $\mathbf{r}_x \times \mathbf{r}_z = (\mathbf{i} + 4\mathbf{j}) \times (2z\,\mathbf{j} + \mathbf{k}) = 4\,\mathbf{i} - \mathbf{j} + 2z\,\mathbf{k}$.

Note: In general, if $y = f(x, z)$ then $\mathbf{r}_x \times \mathbf{r}_z = \dfrac{\partial f}{\partial x}\,\mathbf{i} - \mathbf{j} + \dfrac{\partial f}{\partial z}\,\mathbf{k}$ and

$$A(S) = \iint_D \sqrt{1 + \left(\frac{\partial f}{\partial x}\right)^2 + \left(\frac{\partial f}{\partial z}\right)^2}\, dA.\ \text{Then}$$

$$A(S) = \int_0^1 \int_0^1 \sqrt{17 + 4z^2}\, dx\, dz = \int_0^1 \sqrt{17 + 4z^2}\, dz$$

$$= \tfrac{1}{2}\left(z\sqrt{17 + 4z^2} + \tfrac{17}{2}\ln\left|2z + \sqrt{4z^2 + 17}\right|\right)\Big]_0^1 = \tfrac{\sqrt{21}}{2} + \tfrac{17}{4}\left[\ln\left(2 + \sqrt{21}\right) - \ln\sqrt{17}\right]$$

43. Let $A(S_1)$ be the surface area of that portion of the surface which lies above the plane $z = 0$. Then

$A(S) = 2A(S_1)$. Following Example 10, a parametric representation of S_1 is $x = a\sin\phi\cos\theta$, $y = a\sin\phi\sin\theta$,

$z = a\cos\phi$ and $|\mathbf{r}_\phi \times \mathbf{r}_\theta| = a^2\sin\phi$. For D, $0 \le \phi \le \tfrac{\pi}{2}$ and for each fixed ϕ, $\left(x - \tfrac{1}{2}a\right)^2 + y^2 \le \left(\tfrac{1}{2}a\right)^2$ or

$\left[a\sin\phi\cos\theta - \tfrac{1}{2}a\right]^2 + a^2\sin^2\phi\sin^2\theta \le (a/2)^2$ implies $a^2\sin^2\phi - a^2\sin\phi\cos\theta \le 0$ or

$\sin\phi\,(\sin\phi - \cos\theta) \le 0$. But $0 \le \phi \le \tfrac{\pi}{2}$, so $\cos\theta \ge \sin\phi$ or $\sin\left(\tfrac{\pi}{2} + \theta\right) \ge \sin\phi$ or $\phi - \tfrac{\pi}{2} \le \theta \le \tfrac{\pi}{2} - \phi$.

Hence $D = \left\{(\phi, \theta) \mid 0 \le \phi \le \tfrac{\pi}{2}, \phi - \tfrac{\pi}{2} \le \theta \le \tfrac{\pi}{2} - \phi\right\}$. Then

$$A(S_1) = \int_0^{\pi/2} \int_{\phi - (\pi/2)}^{(\pi/2) - \phi} a^2\sin\phi\, d\theta\, d\phi = a^2 \int_0^{\pi/2} (\pi - 2\phi)\sin\phi\, d\phi$$

$$= a^2\left[(-\pi\cos\phi) - 2(-\phi\cos\phi + \sin\phi)\right]_0^{\pi/2} = a^2(\pi - 2)$$

Thus $A(S) = 2a^2(\pi - 2)$.

Alternate solution: Working on S_1 we could parametrize the portion of the sphere by $x = x$, $y = y$,

$z = \sqrt{a^2 - x^2 - y^2}$. Then $|\mathbf{r}_x \times \mathbf{r}_y| = \sqrt{1 + \dfrac{x^2}{a^2 - x^2 - y^2} + \dfrac{y^2}{a^2 - x^2 - y^2}} = \dfrac{a}{\sqrt{a^2 - x^2 - y^2}}$ and

$$A(S_1) = \iint_{0 \le (x - (a/2))^2 + y^2 \le (a/2)^2} \frac{a}{\sqrt{a^2 - x^2 - y^2}}\, dA = \int_{-\pi/2}^{\pi/2} \int_0^{a\cos\theta} \frac{a}{\sqrt{a^2 - r^2}}\, r\, dr\, d\theta$$

$$= \int_{-\pi/2}^{\pi/2} -a(a^2 - r^2)^{1/2}\Big]_{r = 0}^{r = a\cos\theta}\, d\theta = \int_{-\pi/2}^{\pi/2} a^2[1 - (1 - \cos^2\theta)^{1/2}]\, d\theta$$

$$= \int_{-\pi/2}^{\pi/2} a^2(1 - |\sin\theta|)\, d\theta = 2a^2 \int_0^{\pi/2}(1 - \sin\theta)\, d\theta = 2a^2\left(\tfrac{\pi}{2} - 1\right)$$

Thus $A(S) = 4a^2\left(\tfrac{\pi}{2} - 1\right) = 2a^2(\pi - 2)$.

Notes:

(1) Perhaps working in spherical coordinates is the most obvious approach here. However, you must be careful in setting up D.

(2) In the alternate solution, you can avoid having to use $|\sin\theta|$ by working in the first octant and then multiplying by 4. However, if you set up S_1 as above and arrived at $A(S_1) = a^2\pi$, you now see your error.

45. $\mathbf{r}_u = \langle v, 1, 1 \rangle$, $\mathbf{r}_v = \langle u, 1, -1 \rangle$ and $\mathbf{r}_u \times \mathbf{r}_v = \langle -2, u + v, v - u \rangle$. Then

$$A(S) = \iint_{u^2 + v^2 \le 1} \sqrt{4 + 2u^2 + 2v^2} \, dA = \int_0^{2\pi} \int_0^1 r \sqrt{4 + 2r^2} \, dr \, d\theta = \int_0^{2\pi} d\theta \int_0^1 r \sqrt{4 + 2r^2} \, dr$$

$$= 2\pi \left[\tfrac{1}{6}(4 + 2r^2)^{3/2} \right]_0^1 = \tfrac{\pi}{3}\left(6\sqrt{6} - 8 \right) = \pi\left(2\sqrt{6} - \tfrac{8}{3} \right)$$

47. $z = f(x, y) = e^{-x^2 - y^2}$ with $x^2 + y^2 \le 4$.

$$A(S) = \iint_D \sqrt{1 + \left(-2xe^{-x^2 - y^2} \right)^2 + \left(-2ye^{-x^2 - y^2} \right)^2} \, dA$$

$$= \iint_D \sqrt{1 + 4(x^2 + y^2)e^{-2(x^2 + y^2)}} \, dA$$

$$= \int_0^{2\pi} \int_0^2 \sqrt{1 + 4r^2 e^{-2r^2}} \, r \, dr \, d\theta = \int_0^{2\pi} d\theta \int_0^2 r \sqrt{1 + 4r^2 e^{-2r^2}} \, dr$$

$$= 2\pi \int_0^2 r \sqrt{1 + 4r^2 e^{-2r^2}} \, dr \approx 13.9783$$

49. (a) The midpoints of the four squares are $\left(\tfrac{1}{4}, \tfrac{1}{4} \right)$, $\left(\tfrac{1}{4}, \tfrac{3}{4} \right)$, $\left(\tfrac{3}{4}, \tfrac{1}{4} \right)$, and $\left(\tfrac{3}{4}, \tfrac{3}{4} \right)$; the derivatives of the function $f(x, y) = x^2 + y^2$ are $f_x(x, y) = 2x$ and $f_y(x, y) = 2y$, so the Midpoint Rule gives

$$A(S) = \int_0^1 \int_0^1 \sqrt{[f_x(x, y)]^2 + [f_y(x, y)]^2 + 1} \, dy \, dx$$

$$\approx \tfrac{1}{4}\left(\sqrt{\left[2(\tfrac{1}{4}) \right]^2 + \left[2(\tfrac{1}{4}) \right]^2 + 1} + \sqrt{\left[2(\tfrac{1}{4}) \right]^2 + \left[2(\tfrac{3}{4}) \right]^2 + 1} \right.$$

$$\left. + \sqrt{\left[2(\tfrac{3}{4}) \right]^2 + \left[2(\tfrac{1}{4}) \right]^2 + 1} + \sqrt{\left[2(\tfrac{3}{4}) \right]^2 + \left[2(\tfrac{3}{4}) \right]^2 + 1} \right)$$

$$= \tfrac{1}{4}\left(\sqrt{\tfrac{3}{2}} + 2\sqrt{\tfrac{7}{2}} + \sqrt{\tfrac{11}{2}} \right) \approx 1.8279$$

(b) A CAS estimates the integral to be

$$A(S) = \int_0^1 \int_0^1 \sqrt{f_x^2 + f_y^2 + 1} \, dy \, dx = \int_0^1 \int_0^1 \sqrt{4x^2 + 4y^2 + 1} \, dy \, dx \approx 1.8616. \text{ This agrees with the}$$

Midpoint estimate only in the first decimal place.

51. $z = 1 + 2x + 3y + 4y^2$, so

$$A(S) = \iint_D \sqrt{1 + \left(\frac{\partial z}{\partial x} \right)^2 + \left(\frac{\partial z}{\partial y} \right)^2} \, dA = \int_1^4 \int_0^1 \sqrt{1 + 4 + (3 + 8y)^2} \, dy \, dx$$

$$= \int_1^4 \int_0^1 \sqrt{14 + 48y + 64y^2} \, dy \, dx.$$

Using a CAS, we have

$$\int_1^4 \int_0^1 \sqrt{14 + 48y + 64y^2} \, dy \, dx = \tfrac{45}{8}\sqrt{14} + \tfrac{15}{16}\ln\left(11\sqrt{5} + 3\sqrt{14}\sqrt{5} \right) - \tfrac{15}{16}\ln\left(3\sqrt{5} + \sqrt{14}\sqrt{5} \right)$$

or $\tfrac{45}{8}\sqrt{14} + \tfrac{15}{16}\ln\frac{11\sqrt{5} + 3\sqrt{70}}{3\sqrt{5} + \sqrt{70}}$.

53. (a) $x = a \sin u \cos v$, $y = b \sin u \sin v$, $z = c \cos u$ $\Rightarrow$

(b)
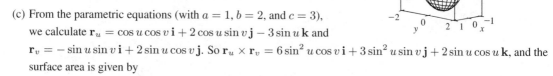

$$\frac{x^2}{a^2} + \frac{y^2}{b^2} + \frac{z^2}{c^2} = (\sin u \cos v)^2 + (\sin u \sin v)^2 + (\cos u)^2$$
$$= \sin^2 u + \cos^2 u = 1$$

and since the ranges of u and v are sufficient to generate the entire graph, the parametric equations represent an ellipsoid.

(c) From the parametric equations (with $a = 1$, $b = 2$, and $c = 3$), we calculate $\mathbf{r}_u = \cos u \cos v \, \mathbf{i} + 2 \cos u \sin v \, \mathbf{j} - 3 \sin u \, \mathbf{k}$ and
$\mathbf{r}_v = -\sin u \sin v \, \mathbf{i} + 2 \sin u \cos v \, \mathbf{j}$. So $\mathbf{r}_u \times \mathbf{r}_v = 6 \sin^2 u \cos v \, \mathbf{i} + 3 \sin^2 u \sin v \, \mathbf{j} + 2 \sin u \cos u \, \mathbf{k}$, and the surface area is given by

$$A(S) = \int_0^{2\pi} \int_0^{\pi} |\mathbf{r}_u \times \mathbf{r}_v| \, du \, dv$$
$$= \int_0^{2\pi} \int_0^{\pi} \sqrt{36 \sin^4 u \cos^2 v + 9 \sin^4 u \sin^2 v + 4 \cos^2 u \sin^2 u} \, du \, dv$$

55. $\mathbf{r}(u,v) = \langle \cos^3 u \cos^3 v, \sin^3 u \cos^3 v, \sin^3 v \rangle$, so $\mathbf{r}_u = \langle -3 \cos^2 u \sin u \cos^3 v, 3 \sin^2 u \cos u \cos^3 v, 0 \rangle$,
$\mathbf{r}_v = \langle -3 \cos^3 u \cos^2 v \sin v, -3 \sin^3 u \cos^2 v \sin v, 3 \sin^2 v \cos v \rangle$, and
$\mathbf{r}_u \times \mathbf{r}_v = \langle 9 \cos u \sin^2 u \cos^4 v \sin^2 v, 9 \cos^2 u \sin u \cos^4 v \sin^2 v, 9 \cos^2 u \sin^2 u \cos^5 v \sin v \rangle$. Then

$$|\mathbf{r}_u \times \mathbf{r}_v| = 9 \sqrt{\cos^2 u \sin^4 u \cos^8 v \sin^4 v + \cos^4 u \sin^2 u \cos^8 v \sin^4 v + \cos^4 u \sin^4 u \cos^{10} v \sin^2 v}$$

$$= 9 \sqrt{\cos^2 u \sin^2 u \cos^8 v \sin^2 v \, (\sin^2 v + \cos^2 u \sin^2 u \cos^2 v)}$$

$$= 9 \cos^4 v \, |\cos u \sin u \sin v| \sqrt{\sin^2 v + \cos^2 u \sin^2 u \cos^2 v}$$

Using a CAS, we have

$$A(S) = \int_0^{\pi} \int_0^{2\pi} 9 \cos^4 v \, |\cos u \sin u \sin v| \sqrt{\sin^2 v + \cos^2 u \sin^2 u \cos^2 v} \, dv \, du \approx 4.4506.$$

17.7 Surface Integrals

ET 16.7

1. Each face of the cube has surface area $2^2 = 4$, and the points P_{ij}^* are the points where the cube intersects the coordinate axes. Here, $f(x, y, z) = \sqrt{x^2 + 2y^2 + 3z^2}$, so by Definition 1,

$$\iint_S f(x, y, z) \, dS \approx [f(1,0,0)](4) + [f(-1,0,0)](4) + [f(0,1,0)](4) + [f(0,-1,0)](4)$$
$$+ [f(0,0,1)](4) + [f(0,0,-1)](4)$$
$$= 4 \left(1 + 1 + 2\sqrt{2} + 2\sqrt{3} \right) = 8 \left(1 + \sqrt{2} + \sqrt{3} \right) \approx 33.170$$

3. We can use the xz- and yz-planes to divide H into four patches of equal size, each with surface area equal to $\frac{1}{8}$ the surface area of a sphere with radius $\sqrt{50}$, so $\Delta S = \frac{1}{8}(4)\pi \left(\sqrt{50} \right)^2 = 25\pi$. Then $(\pm 3, \pm 4, 5)$ are sample points in the four patches, and using a Riemann sum as in Definition 1, we have

$$\iint_H f(x, y, z) \, dS \approx f(3, 4, 5) \, \Delta S + f(3, -4, 5) \, \Delta S + f(-3, 4, 5) \, \Delta S + f(-3, -4, 5) \, \Delta S$$
$$= (7 + 8 + 9 + 12)(25\pi) = 900\pi \approx 2827$$

5. $z = 1 + 2x + 3y$ so $\dfrac{\partial z}{\partial x} = 2$ and $\dfrac{\partial z}{\partial y} = 3$. Then by Formula 2,

$$\iint_S x^2 yz\, dS = \iint_D x^2 yz\, \sqrt{\left(\dfrac{\partial z}{\partial x}\right)^2 + \left(\dfrac{\partial z}{\partial y}\right)^2 + 1}\, dA$$

$$= \int_0^3 \int_0^2 x^2 y(1 + 2x + 3y)\, \sqrt{4 + 9 + 1}\, dy\, dx$$

$$= \sqrt{14} \int_0^3 \int_0^2 (x^2 y + 2x^3 y + 3x^2 y^2)\, dy\, dx$$

$$= \sqrt{14} \int_0^3 \left[\tfrac{1}{2}x^2 y^2 + x^3 y^2 + x^2 y^3\right]_{y=0}^{y=2}\, dx$$

$$= \sqrt{14} \int_0^3 (10x^2 + 4x^3)\, dx = \sqrt{14} \left[\tfrac{10}{3}x^3 + x^4\right]_0^3 = 171\sqrt{14}$$

7. S is the part of the plane $z = 1 - x - y$ over the region $D = \{(x, y) \mid 0 \le x \le 1, 0 \le y \le 1 - x\}$. Thus

$$\iint_S yz\, dS = \iint_D y(1 - x - y)\, \sqrt{(-1)^2 + (-1)^2 + 1}\, dA$$

$$= \sqrt{3} \int_0^1 \int_0^{1-x} (y - xy - y^2)\, dy\, dx = \sqrt{3} \int_0^1 \left[\tfrac{1}{2}y^2 - \tfrac{1}{2}xy^2 - \tfrac{1}{3}y^3\right]_{y=0}^{y=1-x}\, dx$$

$$= \sqrt{3} \int_0^1 \tfrac{1}{6}(1 - x)^3\, dx = -\tfrac{\sqrt{3}}{24}(1 - x)^4 \Big]_0^1 = \tfrac{\sqrt{3}}{24}$$

9. S is the portion of the cone $z^2 = x^2 + y^2$ for $1 \le z \le 3$, or equivalently, S is the part of the surface $z = \sqrt{x^2 + y^2}$ over the region $D = \{(x, y) \mid 1 \le x^2 + y^2 \le 9\}$. Thus

$$\iint_S x^2 z^2\, dS = \iint_D x^2(x^2 + y^2)\, \sqrt{\left(\dfrac{x}{\sqrt{x^2 + y^2}}\right)^2 + \left(\dfrac{y}{\sqrt{x^2 + y^2}}\right)^2 + 1}\, dA$$

$$= \iint_D x^2(x^2 + y^2)\, \sqrt{\dfrac{x^2 + y^2}{x^2 + y^2} + 1}\, dA = \iint_D \sqrt{2}\, x^2(x^2 + y^2)\, dA$$

$$= \sqrt{2} \int_0^{2\pi} \int_1^3 (r\cos\theta)^2(r^2)\, r\, dr\, d\theta = \sqrt{2} \int_0^{2\pi} \cos^2\theta\, d\theta \int_1^3 r^5\, dr$$

$$= \sqrt{2} \left[\tfrac{1}{2}\theta + \tfrac{1}{4}\sin 2\theta\right]_0^{2\pi} \left[\tfrac{1}{6}r^6\right]_1^3 = \sqrt{2}\,(\pi) \cdot \tfrac{1}{6}(3^6 - 1) = \dfrac{364\sqrt{2}}{3}\pi$$

11. Using x and z as parameters, we have $\mathbf{r}(x, z) = x\,\mathbf{i} + (x^2 + z^2)\,\mathbf{j} + z\,\mathbf{k}$, $x^2 + z^2 \le 4$. Then
$\mathbf{r}_x \times \mathbf{r}_z = (\mathbf{i} + 2x\,\mathbf{j}) \times (2z\,\mathbf{j} + \mathbf{k}) = 2x\,\mathbf{i} - \mathbf{j} + 2z\,\mathbf{k}$ and $|\mathbf{r}_x \times \mathbf{r}_z| = \sqrt{4x^2 + 1 + 4z^2} = \sqrt{1 + 4(x^2 + z^2)}$.
Thus

$$\iint_S y\, dS = \iint_{x^2 + z^2 \le 4} (x^2 + z^2)\sqrt{1 + 4(x^2 + z^2)}\, dA = \int_0^{2\pi} \int_0^2 r^2\, \sqrt{1 + 4r^2}\, r\, dr\, d\theta$$

$$= \int_0^{2\pi} d\theta \int_0^2 r^2\, \sqrt{1 + 4r^2}\, r\, dr = 2\pi \int_0^2 r^2\, \sqrt{1 + 4r^2}\, r\, dr$$

$$[\text{let } u = 1 + 4r^2 \quad \Rightarrow \quad r^2 = \tfrac{1}{4}(u - 1) \text{ and } \tfrac{1}{8}du = r\, dr]$$

$$= 2\pi \int_1^{17} \tfrac{1}{4}(u - 1)\sqrt{u} \cdot \tfrac{1}{8}du = \tfrac{1}{16}\pi \int_1^{17} (u^{3/2} - u^{1/2})\, du$$

$$= \tfrac{1}{16}\pi \left[\tfrac{2}{5}u^{5/2} - \tfrac{2}{3}u^{3/2}\right]_1^{17} = \tfrac{1}{16}\pi \left[\tfrac{2}{5}(17)^{5/2} - \tfrac{2}{3}(17)^{3/2} - \tfrac{2}{5} + \tfrac{2}{3}\right] = \dfrac{\pi}{60}\left(391\sqrt{17} + 1\right)$$

13. Using spherical coordinates and Example 17.6.10 [ET 16.6.10] we have
$\mathbf{r}(\phi, \theta) = 2\sin\phi\cos\theta\,\mathbf{i} + 2\sin\phi\sin\theta\,\mathbf{j} + 2\cos\phi\,\mathbf{k}$ and $|\mathbf{r}_\phi \times \mathbf{r}_\theta| = 4\sin\phi$. Then
$\iint_S (x^2 z + y^2 z)\, dS = \int_0^{2\pi} \int_0^{\pi/2} (4\sin^2\phi)(2\cos\phi)(4\sin\phi)\, d\phi\, d\theta = 16\pi \sin^4\phi\,\big]_0^{\pi/2} = 16\pi$.

15. Using cylindrical coordinates, we have $\mathbf{r}(\theta, z) = 3\cos\theta\,\mathbf{i} + 3\sin\theta\,\mathbf{j} + z\,\mathbf{k}$, $0 \le \theta \le 2\pi$, $0 \le z \le 2$, and $|\mathbf{r}_\theta \times \mathbf{r}_z| = 3$.

$$\iint_S (x^2 y + z^2)\, dS = \int_0^{2\pi}\int_0^2 (27\cos^2\theta\sin\theta + z^2)\,3\,dz\,d\theta = \int_0^{2\pi}(162\cos^2\theta\sin\theta + 8)\,d\theta = 16\pi$$

17. $\mathbf{r}(u, v) = u^2\,\mathbf{i} + u\sin v\,\mathbf{j} + u\cos v\,\mathbf{k}$, $0 \le u \le 1$, $0 \le v \le \pi/2$ and

$\mathbf{r}_u \times \mathbf{r}_v = (2u\,\mathbf{i} + \sin v\,\mathbf{j} + \cos v\,\mathbf{k}) \times (u\cos v\,\mathbf{j} - u\sin v\,\mathbf{k}) = -u\,\mathbf{i} + 2u^2\sin v\,\mathbf{j} + 2u^2\cos v\,\mathbf{k}$ and

$|\mathbf{r}_u \times \mathbf{r}_v| = \sqrt{u^2 + 4u^4\sin^2 v + 4u^4\cos^2 v} = \sqrt{u^2 + 4u^4(\sin^2 v + \cos^2 v)} = u\sqrt{1 + 4u^2}$ (since $u \ge 0$). Then

$$\iint_S yz\,dS = \int_0^{\pi/2}\int_0^1 (u\sin v)(u\cos v)\cdot u\sqrt{1 + 4u^2}\,du\,dv = \int_0^1 u^3\sqrt{1 + 4u^2}\,du\int_0^{\pi/2}\sin v\cos v\,dv$$

$$[\text{let } t = 1 + 4u^2 \Rightarrow u^2 = \tfrac14(t - 1) \text{ and } \tfrac18\,dt = u\,du]$$

$$= \int_1^5 \tfrac18\cdot\tfrac14(t - 1)\sqrt{t}\,dt\int_0^{\pi/2}\sin v\cos v\,dv = \tfrac{1}{32}\int_1^5\left(t^{3/2} - \sqrt{t}\right)dt\int_0^{\pi/2}\sin v\cos v\,dv$$

$$= \tfrac{1}{32}\left[\tfrac25 t^{5/2} - \tfrac23 t^{3/2}\right]_1^5\left[\tfrac12\sin^2 v\right]_0^{\pi/2} = \tfrac{1}{32}\left(\tfrac25(5)^{5/2} - \tfrac23(5)^{3/2} - \tfrac25 + \tfrac23\right)\cdot\tfrac12(1 - 0)$$

$$= \tfrac{5}{48}\sqrt5 + \tfrac{1}{240}$$

19. $\mathbf{F}(x, y, z) = xy\,\mathbf{i} + yz\,\mathbf{j} + zx\,\mathbf{k}$, $z = g(x, y) = 4 - x^2 - y^2$, and D is the square $[0, 1] \times [0, 1]$, so by Equation 8

$$\iint_S \mathbf{F}\cdot d\mathbf{S} = \iint_D [-xy(-2x) - yz(-2y) + zx]\,dA$$

$$= \int_0^1\int_0^1 [2x^2 y + 2y^2(4 - x^2 - y^2) + x(4 - x^2 - y^2)]\,dy\,dx$$

$$= \int_0^1 \left(\tfrac13 x^2 + \tfrac{11}{3}x - x^3 + \tfrac{34}{15}\right)dx = \tfrac{713}{180}$$

21. $\mathbf{F}(x, y, z) = xze^y\,\mathbf{i} - xze^y\,\mathbf{j} + z\,\mathbf{k}$, $z = g(x, y) = 1 - x - y$, and $D = \{(x, y) \mid 0 \le x \le 1, 0 \le y \le 1 - x\}$. Since S has downward orientation, we have

$$\iint_S \mathbf{F}\cdot d\mathbf{S} = -\iint_D [-xze^y(-1) - (-xze^y)(-1) + z]\,dA = -\int_0^1\int_0^{1-x}(1 - x - y)\,dy\,dx$$

$$= -\int_0^1\left(\tfrac12 x^2 - x + \tfrac12\right)dx = -\tfrac16$$

23. $\mathbf{F}(x, y, z) = x\,\mathbf{i} - z\,\mathbf{j} + y\,\mathbf{k}$, $z = g(x, y) = \sqrt{4 - x^2 - y^2}$ and D is the quarter disk $\{(x, y) \mid 0 \le x \le 2, 0 \le y \le \sqrt{4 - x^2}\}$. S has downward orientation, so by Formula 8,

$$\iint_S \mathbf{F}\cdot d\mathbf{S} = -\iint_D\left[-x\cdot\tfrac12(4 - x^2 - y^2)^{-1/2}(-2x) - (-z)\cdot\tfrac12(4 - x^2 - y^2)^{-1/2}(-2y) + y\right]dA$$

$$= -\iint_D\left(\frac{x^2}{\sqrt{4 - x^2 - y^2}} - \sqrt{4 - x^2 - y^2}\cdot\frac{y}{\sqrt{4 - x^2 - y^2}} + y\right)dA$$

$$= -\iint_D x^2(4 - (x^2 + y^2))^{-1/2}\,dA = -\int_0^{\pi/2}\int_0^2 (r\cos\theta)^2(4 - r^2)^{-1/2}\,r\,dr\,d\theta$$

$$= -\int_0^{\pi/2}\cos^2\theta\,d\theta\int_0^2 r^3(4 - r^2)^{-1/2}\,dr$$

$$[\text{let } u = 4 - r^2 \Rightarrow r^2 = 4 - u \text{ and } -\tfrac12\,du = r\,dr]$$

$$= -\int_0^{\pi/2}\left(\tfrac12 + \tfrac12\cos 2\theta\right)d\theta\int_4^0 -\tfrac12(4 - u)(u)^{-1/2}\,du$$

$$= -\left[\tfrac12\theta + \tfrac14\sin 2\theta\right]_0^{\pi/2}\left(-\tfrac12\right)\left[8\sqrt u - \tfrac23 u^{3/2}\right]_4^0$$

$$= -\tfrac{\pi}{4}\left(-\tfrac12\right)\left(-16 + \tfrac{16}{3}\right) = -\frac43\pi$$

25. Let S_1 be the paraboloid $y = x^2 + z^2$, $0 \leq y \leq 1$ and S_2 the disk $x^2 + z^2 \leq 1$, $y = 1$. Since S is a closed surface, we use the outward orientation. On S_1: $\mathbf{F}(\mathbf{r}(x,z)) = (x^2 + z^2)\mathbf{j} - z\mathbf{k}$ and $\mathbf{r}_x \times \mathbf{r}_z = 2x\mathbf{i} - \mathbf{j} + 2z\mathbf{k}$ (since the $\mathbf{j}$-component must be negative on S_1). Then

$$\iint_{S_1} \mathbf{F} \cdot d\mathbf{S} = \iint_{x^2+z^2 \leq 1} [-(x^2 + z^2) - 2z^2]\, dA = -\int_0^{2\pi} \int_0^1 (r^2 + 2r^2 \cos^2 \theta)\, r\, dr\, d\theta$$

$$= -\int_0^{2\pi} \tfrac{1}{4}(1 + 2\cos^2 \theta)\, d\theta = -\left(\tfrac{\pi}{2} + \tfrac{\pi}{2}\right) = -\pi$$

On S_2: $\mathbf{F}(\mathbf{r}(x,z)) = \mathbf{j} - z\mathbf{k}$ and $\mathbf{r}_z \times \mathbf{r}_x = \mathbf{j}$. Then $\iint_{S_2} \mathbf{F} \cdot d\mathbf{S} = \iint_{x^2+z^2 \leq 1} (1)\, dA = \pi$.

Hence $\iint_S \mathbf{F} \cdot d\mathbf{S} = -\pi + \pi = 0$.

27. Here S consists of the six faces of the cube as labeled in the figure. On S_1:

$\mathbf{F} = \mathbf{i} + 2y\mathbf{j} + 3z\mathbf{k}$, $\mathbf{r}_y \times \mathbf{r}_z = \mathbf{i}$ and $\iint_{S_1} \mathbf{F} \cdot d\mathbf{S} = \int_{-1}^1 \int_{-1}^1 dy\, dz = 4$;

S_2: $\mathbf{F} = x\mathbf{i} + 2\mathbf{j} + 3z\mathbf{k}$, $\mathbf{r}_z \times \mathbf{r}_x = \mathbf{j}$ and $\iint_{S_2} \mathbf{F} \cdot d\mathbf{S} = \int_{-1}^1 \int_{-1}^1 2\, dx\, dz = 8$;

S_3: $\mathbf{F} = x\mathbf{i} + 2y\mathbf{j} + 3\mathbf{k}$, $\mathbf{r}_x \times \mathbf{r}_y = \mathbf{k}$ and $\iint_{S_3} \mathbf{F} \cdot d\mathbf{S} = \int_{-1}^1 \int_{-1}^1 3\, dx\, dy = 12$;

S_4: $\mathbf{F} = -\mathbf{i} + 2y\mathbf{j} + 3z\mathbf{k}$, $\mathbf{r}_z \times \mathbf{r}_y = -\mathbf{i}$ and $\iint_{S_4} \mathbf{F} \cdot d\mathbf{S} = 4$;

S_5: $\mathbf{F} = x\mathbf{i} - 2\mathbf{j} + 3z\mathbf{k}$, $\mathbf{r}_x \times \mathbf{r}_z = -\mathbf{j}$ and $\iint_{S_5} \mathbf{F} \cdot d\mathbf{S} = 8$;

S_6: $\mathbf{F} = x\mathbf{i} + 2y\mathbf{j} - 3\mathbf{k}$, $\mathbf{r}_y \times \mathbf{r}_x = -\mathbf{k}$ and $\iint_{S_6} \mathbf{F} \cdot d\mathbf{S} = \int_{-1}^1 \int_{-1}^1 3\, dx\, dy = 12$.

Hence $\iint_S \mathbf{F} \cdot d\mathbf{S} = \sum_{i=1}^6 \iint_{S_i} \mathbf{F} \cdot d\mathbf{S} = 48$.

29. $z = xy$ $\Rightarrow$ $\partial z/\partial x = y$, $\partial z/\partial y = x$, so by Formula 2, a CAS gives

$\iint_S xyz\, dS = \int_0^1 \int_0^1 xy\,(xy)\sqrt{y^2 + x^2 + 1}\, dx\, dy \approx 0.1642$.

31. We use Formula 2 with $z = 3 - 2x^2 - y^2$ $\Rightarrow$ $\partial z/\partial x = -4x$, $\partial z/\partial y = -2y$. The boundaries of the region

$3 - 2x^2 - y^2 \geq 0$ are $-\sqrt{\tfrac{3}{2}} \leq x \leq \sqrt{\tfrac{3}{2}}$ and $-\sqrt{3 - 2x^2} \leq y \leq \sqrt{3 - 2x^2}$, so we use a CAS (with precision reduced to seven or fewer digits; otherwise the calculation takes a very long time) to calculate

$$\iint_S x^2 y^2 z^2\, dS = \int_{-\sqrt{3/2}}^{\sqrt{3/2}} \int_{-\sqrt{3-2x^2}}^{\sqrt{3-2x^2}} x^2 y^2 (3 - 2x^2 - y^2)^2 \sqrt{16x^2 + 4y^2 + 1}\, dy\, dx \approx 3.4895$$

33. If S is given by $y = h(x, z)$, then S is also the level surface $f(x, y, z) = y - h(x, z) = 0$.

$\mathbf{n} = \dfrac{\nabla f(x, y, z)}{|\nabla f(x, y, z)|} = \dfrac{-h_x \mathbf{i} + \mathbf{j} - h_z \mathbf{k}}{\sqrt{h_x^2 + 1 + h_z^2}}$, and $-\mathbf{n}$ is the unit normal that points to the left. Now we proceed as in

the derivation of (8), using Formula 2 to evaluate

$$\iint_S \mathbf{F} \cdot d\mathbf{S} = \iint_S \mathbf{F} \cdot \mathbf{n}\, dS$$

$$= \iint_D (P\mathbf{i} + Q\mathbf{j} + R\mathbf{k}) \frac{\dfrac{\partial h}{\partial x}\mathbf{i} - \mathbf{j} + \dfrac{\partial h}{\partial z}\mathbf{k}}{\sqrt{\left(\dfrac{\partial h}{\partial x}\right)^2 + 1 + \left(\dfrac{\partial h}{\partial z}\right)^2}} \sqrt{\left(\dfrac{\partial h}{\partial x}\right)^2 + 1 + \left(\dfrac{\partial h}{\partial z}\right)^2}\, dA$$

where D is the projection of $f(x, y, z)$ onto the xz-plane. Therefore

$$\iint_S \mathbf{F} \cdot d\mathbf{S} = \iint_D \left(P\frac{\partial h}{\partial x} - Q + R\frac{\partial h}{\partial z}\right) dA$$

35. $m = \iint_S K \, dS = K \cdot 4\pi\left(\frac{1}{2}a^2\right) = 2\pi a^2 K$; by symmetry $M_{xz} = M_{yz} = 0$, and

$M_{xy} = \iint_S zK \, dS = K \int_0^{2\pi} \int_0^{\pi/2} (a\cos\phi)(a^2\sin\phi) \, d\phi \, d\theta = 2\pi Ka^3 \left[-\frac{1}{4}\cos 2\phi\right]_0^{\pi/2} = \pi Ka^3$.

Hence $(\overline{x}, \overline{y}, \overline{z}) = \left(0, 0, \frac{1}{2}a\right)$.

37. (a) $I_z = \iint_S (x^2 + y^2)\rho(x, y, z) \, dS$

(b) $I_z = \iint_S (x^2 + y^2)\left(10 - \sqrt{x^2 + y^2}\right) dS = \iint_{1 \le x^2 + y^2 \le 16} (x^2 + y^2)\left(10 - \sqrt{x^2 + y^2}\right)\sqrt{2} \, dA$

$= \int_0^{2\pi} \int_1^4 \sqrt{2}\,(10r^3 - r^4) \, dr \, d\theta = 2\sqrt{2}\,\pi\left(\frac{4329}{10}\right) = \frac{4329}{5}\sqrt{2}\,\pi$

39. $\rho(x, y, z) = 1200$, $\mathbf{V} = y\,\mathbf{i} + \mathbf{j} + z\,\mathbf{k}$, $\mathbf{F} = \rho\mathbf{V} = (1200)(y\,\mathbf{i} + \mathbf{j} + z\,\mathbf{k})$. S is given by

$\mathbf{r}(x, y) = x\,\mathbf{i} + y\,\mathbf{j} + \left[9 - \frac{1}{4}(x^2 + y^2)\right]\mathbf{k}$, $0 \le x^2 + y^2 \le 36$ and $\mathbf{r}_x \times \mathbf{r}_y = \frac{1}{2}x\,\mathbf{i} + \frac{1}{2}y\,\mathbf{j} + \mathbf{k}$.

Thus the rate of flow is given by

$$\iint_S \mathbf{F} \cdot d\mathbf{S} = \iint_{0 \le x^2 + y^2 \le 36} (1200)\left(\tfrac{1}{2}xy + \tfrac{1}{2}y + \left[9 - \tfrac{1}{4}(x^2 + y^2)\right]\right) dA$$

$$= 1200 \int_0^6 \int_0^{2\pi} \left[\tfrac{1}{2}r^2\sin\theta\cos\theta + \tfrac{1}{2}r\sin\theta + 9 - \tfrac{1}{4}r^2\right] r \, d\theta \, dr$$

$$= 1200 \int_0^6 2\pi\left(9r - \tfrac{1}{4}r^3\right) dr = (1200)(2\pi)(81) = 194{,}400\pi$$

41. S consists of the hemisphere S_1 given by $z = \sqrt{a^2 - x^2 - y^2}$ and the disk S_2 given by

$0 \le x^2 + y^2 \le a^2$, $z = 0$. On S_1: $\mathbf{E} = a\sin\phi\cos\theta\,\mathbf{i} + a\sin\phi\sin\theta\,\mathbf{j} + 2a\cos\phi\,\mathbf{k}$,

$\mathbf{T}_\phi \times \mathbf{T}_\theta = a^2\sin^2\phi\cos\theta\,\mathbf{i} + a^2\sin^2\phi\sin\theta\,\mathbf{j} + a^2\sin\phi\cos\phi\,\mathbf{k}$. Thus

$$\iint_{S_1} \mathbf{E} \cdot d\mathbf{S} = \int_0^{2\pi} \int_0^{\pi/2} (a^3\sin^3\phi + 2a^3\sin\phi\cos^2\phi) \, d\phi \, d\theta$$

$$= \int_0^{2\pi} \int_0^{\pi/2} (a^3\sin\phi + a^3\sin\phi\cos^2\phi) \, d\phi \, d\theta = (2\pi)a^3\left(1 + \tfrac{1}{3}\right) = \tfrac{8}{3}\pi a^3$$

On S_2: $\mathbf{E} = x\,\mathbf{i} + y\,\mathbf{j}$, and $\mathbf{r}_y \times \mathbf{r}_x = -\mathbf{k}$ so $\iint_{S_2} \mathbf{E} \cdot d\mathbf{S} = 0$.

Hence the total charge is $q = \varepsilon_0 \iint_S \mathbf{E} \cdot d\mathbf{S} = \frac{8}{3}\pi a^3 \varepsilon_0$.

43. $K\nabla u = 6.5(4y\,\mathbf{j} + 4z\,\mathbf{k})$. S is given by $\mathbf{r}(x, \theta) = x\,\mathbf{i} + \sqrt{6}\,\cos\theta\,\mathbf{j} + \sqrt{6}\,\sin\theta\,\mathbf{k}$ and since we want the inward

heat flow, we use $\mathbf{r}_x \times \mathbf{r}_\theta = -\sqrt{6}\,\cos\theta\,\mathbf{j} - \sqrt{6}\,\sin\theta\,\mathbf{k}$. Then the rate of heat flow inward is given by

$\iint_S (-K\nabla u) \cdot d\mathbf{S} = \int_0^{2\pi} \int_0^4 -(6.5)(-24) \, dx \, d\theta = (2\pi)(156)(4) = 1248\pi$.

17.8 Stokes' Theorem

ET 16.8

1. Both H and P are oriented piecewise-smooth surfaces that are bounded by the simple, closed, smooth curve $x^2 + y^2 = 4$, $z = 0$ (which we can take to be oriented positively for both surfaces). Then H and P satisfy the hypotheses of Stokes' Theorem, so by (3) we know $\iint_H \text{curl } \mathbf{F} \cdot d\mathbf{S} = \int_C \mathbf{F} \cdot d\mathbf{r} = \iint_P \text{curl } \mathbf{F} \cdot d\mathbf{S}$ (where C is the boundary curve).

3. The boundary curve C is the circle $x^2 + y^2 = 4$, $z = 0$ oriented in the counterclockwise direction. The vector equation is $\mathbf{r}(t) = 2\cos t\,\mathbf{i} + 2\sin t\,\mathbf{j}$, $0 \le t \le 2\pi$, so $\mathbf{r}'(t) = -2\sin t\,\mathbf{i} + 2\cos t\,\mathbf{j}$ and

$\mathbf{F}(\mathbf{r}(t)) = (2\cos t)^2 e^{(2\sin t)(0)}\,\mathbf{i} + (2\sin t)^2 e^{(2\cos t)(0)}\,\mathbf{j} + (0)^2 e^{(2\cos t)(2\sin t)}\,\mathbf{k} = 4\cos^2 t\,\mathbf{i} + 4\sin^2 t\,\mathbf{j}$. Then,

by Stokes' Theorem,

$$\iint_S \text{curl } \mathbf{F} \cdot d\mathbf{S} = \int_C \mathbf{F} \cdot d\mathbf{r} = \int_0^{2\pi} \mathbf{F}(\mathbf{r}(t)) \cdot \mathbf{r}'(t) \, dt = \int_0^{2\pi} (-8\cos^2 t\sin t + 8\sin^2 t\cos t) \, dt$$

$$= 8\left[\tfrac{1}{3}\cos^3 t + \tfrac{1}{3}\sin^3 t\right]_0^{2\pi} = 0$$

5. C is the square in the plane $z = -1$. By (3), $\iint_{S_1} \text{curl } \mathbf{F} \cdot d\mathbf{S} = \oint_C \mathbf{F} \cdot d\mathbf{r} = \iint_{S_2} \text{curl } \mathbf{F} \cdot d\mathbf{S}$ where

S_1 is the original cube without the bottom and S_2 is the bottom face of the cube.

$\text{curl } \mathbf{F} = x^2 z \mathbf{i} + (xy - 2xyz) \mathbf{j} + (y - xz) \mathbf{k}$. For S_2, we choose $\mathbf{n} = \mathbf{k}$ so that C has the same orientation for

both surfaces. Then $\text{curl } \mathbf{F} \cdot \mathbf{n} = y - xz = x + y$ on S_2, where $z = -1$. Thus

$\iint_{S_2} \text{curl } \mathbf{F} \cdot d\mathbf{S} = \int_{-1}^{1} \int_{-1}^{1} (x + y) \, dx \, dy = 0$ so $\iint_{S_1} \text{curl } \mathbf{F} \cdot d\mathbf{S} = 0$.

7. $\text{curl } \mathbf{F} = -2z \mathbf{i} - 2x \mathbf{j} - 2y \mathbf{k}$ and we take the surface S to be the planar region enclosed by C, so S is the portion

of the plane $x + y + z = 1$ over $D = \{(x, y) \mid 0 \le x \le 1, 0 \le y \le 1 - x\}$. Since C is oriented counterclockwise,

we orient S upward. Using Equation 17.7.8 [ET 16.7.8], we have $z = g(x, y) = 1 - x - y$, $P = -2z$, $Q = -2x$,

$R = -2y$, and

$$\int_C \mathbf{F} \cdot d\mathbf{r} = \iint_S \text{curl } \mathbf{F} \cdot d\mathbf{S} = \iint_D [-(-2z)(-1) - (-2x)(-1) + (-2y)] \, dA$$

$$= \int_0^1 \int_0^{1-x} (-2) \, dy \, dx = -2 \int_0^1 (1 - x) \, dx = -1$$

9. $\text{curl } \mathbf{F} = (xe^{xy} - 2x) \mathbf{i} - (ye^{xy} - y) \mathbf{j} + (2z - z) \mathbf{k}$ and we take S to be the disk $x^2 + y^2 \le 16$, $z = 5$. Since C is

oriented counterclockwise (from above), we orient S upward. Then $\mathbf{n} = \mathbf{k}$ and $\text{curl } \mathbf{F} \cdot \mathbf{n} = 2z - z$ on S, where

$z = 5$. Thus

$$\oint \mathbf{F} \cdot d\mathbf{r} = \iint_S \text{curl } \mathbf{F} \cdot \mathbf{n} \, dS = \iint_S (2z - z) \, dS = \iint_S (10 - 5) \, dS = 5(\text{area of } S) = 5(\pi \cdot 4^2) = 80\pi$$

11. (a) The curve of intersection is an ellipse in the plane $x + y + z = 1$ with unit normal $\mathbf{n} = \frac{1}{\sqrt{3}} (\mathbf{i} + \mathbf{j} + \mathbf{k})$,

$\text{curl } \mathbf{F} = x^2 \mathbf{j} + y^2 \mathbf{k}$ and $\text{curl } \mathbf{F} \cdot \mathbf{n} = \frac{1}{\sqrt{3}} (x^2 + y^2)$. Then

$$\oint_C \mathbf{F} \cdot d\mathbf{r} = \iint_S \frac{1}{\sqrt{3}} (x^2 + y^2) \, dS = \iint_{x^2 + y^2 \le 9} (x^2 + y^2) \, dx \, dy$$

$$= \int_0^{2\pi} \int_0^3 r^3 \, dr \, d\theta = 2\pi \left(\frac{81}{4} \right) = \frac{81\pi}{2}$$

(b)

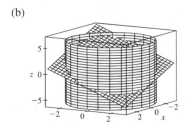

(c) One possible parametrization is $x = 3 \cos t$, $y = 3 \sin t$,

$z = 1 - 3 \cos t - 3 \sin t$, $0 \le t \le 2\pi$.

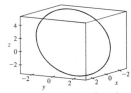

13. The boundary curve C is the circle $x^2 + y^2 = 1$, $z = 1$ oriented in the counterclockwise direction as viewed from

above. We can parametrize C by $\mathbf{r}(t) = \cos t \, \mathbf{i} + \sin t \, \mathbf{j} + \mathbf{k}$, $0 \le t \le 2\pi$, and then $\mathbf{r}'(t) = -\sin t \, \mathbf{i} + \cos t \, \mathbf{j}$. Thus

$\mathbf{F}(\mathbf{r}(t)) = \sin^2 t \, \mathbf{i} + \cos t \, \mathbf{j} + \mathbf{k}$, $\mathbf{F}(\mathbf{r}(t)) \cdot \mathbf{r}'(t) = \cos^2 t - \sin^3 t$, and

$$\oint_C \mathbf{F} \cdot d\mathbf{r} = \int_0^{2\pi} (\cos^2 t - \sin^3 t) \, dt = \int_0^{2\pi} \tfrac{1}{2} (1 + \cos 2t) \, dt - \int_0^{2\pi} (1 - \cos^2 t) \sin t \, dt$$

$$= \tfrac{1}{2} \left[t + \tfrac{1}{2} \sin 2t \right]_0^{2\pi} - \left[-\cos t + \tfrac{1}{3} \cos^3 t \right]_0^{2\pi} = \pi$$

Now $\text{curl } \mathbf{F} = (1 - 2y) \mathbf{k}$, and the projection D of S on the xy-plane is the disk $x^2 + y^2 \le 1$, so by

Equation 17.7.8 [ET 16.7.8] with $z = g(x, y) = x^2 + y^2$ we have

$$\iint_S \text{curl } \mathbf{F} \cdot d\mathbf{S} = \iint_D (1 - 2y) \, dA = \int_0^{2\pi} \int_0^1 (1 - 2r \sin \theta) \, r \, dr \, d\theta = \int_0^{2\pi} \left(\tfrac{1}{2} - \tfrac{2}{3} \sin \theta \right) d\theta = \pi$$

15. The boundary curve C is the circle $x^2 + z^2 = 1$, $y = 0$ oriented in the counterclockwise direction as viewed from the positive y-axis. Then C can be described by $\mathbf{r}(t) = \cos t\,\mathbf{i} - \sin t\,\mathbf{k}$, $0 \le t \le 2\pi$, and $\mathbf{r}'(t) = -\sin t\,\mathbf{i} - \cos t\,\mathbf{k}$. Thus $\mathbf{F}(\mathbf{r}(t)) = -\sin t\,\mathbf{j} + \cos t\,\mathbf{k}$, $\mathbf{F}(\mathbf{r}(t)) \cdot \mathbf{r}'(t) = -\cos^2 t$, and

$$\oint_C \mathbf{F} \cdot d\mathbf{r} = \int_0^{2\pi} -\cos^2 t\,dt = -\tfrac{1}{2}t - \tfrac{1}{4}\sin 2t\Big]_0^{2\pi} = -\pi$$

Now curl $\mathbf{F} = -\mathbf{i} - \mathbf{j} - \mathbf{k}$, and S can be parametrized (see Example 17.6.10 [ET 16.6.10]) by $\mathbf{r}(\phi, \theta) = \sin\phi\cos\theta\,\mathbf{i} + \sin\phi\sin\theta\,\mathbf{j} + \cos\phi\,\mathbf{k}$, $0 \le \theta \le \pi$, $0 \le \phi \le \pi$. Then $\mathbf{r}_\phi \times \mathbf{r}_\theta = \sin^2\phi\cos\theta\,\mathbf{i} + \sin^2\phi\sin\theta\,\mathbf{j} + \sin\phi\cos\phi\,\mathbf{k}$ and

$$\iint_S \text{curl }\mathbf{F} \cdot d\mathbf{S} = \iint_{x^2+z^2\le 1} \text{curl }\mathbf{F} \cdot (\mathbf{r}_\phi \times \mathbf{r}_\theta)\,dA$$

$$= \int_0^\pi \int_0^\pi (-\sin^2\phi\cos\theta - \sin^2\phi\sin\theta - \sin\phi\cos\phi)\,d\theta\,d\phi$$

$$= \int_0^\pi (-2\sin^2\phi - \pi\sin\phi\cos\phi)\,d\phi = \left[\tfrac{1}{2}\sin 2\phi - \phi - \tfrac{\pi}{2}\sin^2\phi\right]_0^\pi = -\pi$$

17. $\text{curl }\mathbf{F} = \begin{vmatrix} \mathbf{i} & \mathbf{j} & \mathbf{k} \\ \partial/\partial x & \partial/\partial y & \partial/\partial z \\ x^x + z^2 & y^y + x^2 & z^z + y^2 \end{vmatrix} = 2y\,\mathbf{i} + 2z\,\mathbf{j} + 2x\,\mathbf{k}$ and $W = \int_C \mathbf{F} \cdot d\mathbf{r} = \iint_S \text{curl }\mathbf{F} \cdot d\mathbf{S}$.

To parametrize the surface, let $x = 2\cos\theta\sin\phi$, $y = 2\sin\theta\sin\phi$, $z = 2\cos\phi$, so that $\mathbf{r}(\phi, \theta) = 2\sin\phi\cos\theta\,\mathbf{i} + 2\sin\phi\sin\theta\,\mathbf{j} + 2\cos\phi\,\mathbf{k}$, $0 \le \phi \le \tfrac{\pi}{2}$, $0 \le \theta \le \tfrac{\pi}{2}$, and $\mathbf{r}_\phi \times \mathbf{r}_\theta = 4\sin^2\phi\cos\theta\,\mathbf{i} + 4\sin^2\phi\sin\theta\,\mathbf{j} + 4\sin\phi\cos\phi\,\mathbf{k}$. Then $\text{curl }\mathbf{F}(\mathbf{r}(\phi, \theta)) = 4\sin\phi\sin\theta\,\mathbf{i} + 4\cos\phi\,\mathbf{j} + 4\sin\phi\cos\theta\,\mathbf{k}$, and $\text{curl }\mathbf{F} \cdot (\mathbf{r}_\phi \times \mathbf{r}_\theta) = 16\sin^3\phi\sin\theta\cos\theta + 16\cos\phi\sin^2\phi\sin\theta + 16\sin^2\phi\cos\phi\cos\theta$. Therefore

$$\iint_S \text{curl }\mathbf{F} \cdot d\mathbf{S} = \iint_D \text{curl }\mathbf{F} \cdot (\mathbf{r}_\phi \times \mathbf{r}_\theta)\,dA$$

$$= 16\left[\int_0^{\pi/2}\sin\theta\cos\theta\,d\theta\right]\left[\int_0^{\pi/2}\sin^3\phi\,d\phi\right] + 16\left[\int_0^{\pi/2}\sin\theta\,d\theta\right]\left[\int_0^{\pi/2}\sin^2\phi\cos\phi\,d\phi\right]$$

$$+ 16\left[\int_0^{\pi/2}\cos\theta\,d\theta\right]\left[\int_0^{\pi/2}\sin^2\phi\cos\phi\,d\phi\right]$$

$$= 8\left[-\cos\phi + \tfrac{1}{3}\cos^3\phi\right]_0^{\pi/2} + 16(1)\left[\tfrac{1}{3}\sin^3\phi\right]_0^{\pi/2} + 16(1)\left[\tfrac{1}{3}\sin^3\phi\right]_0^{\pi/2}$$

$$= 8\left[0 + 1 + 0 - \tfrac{1}{3}\right] + 16\left(\tfrac{1}{3}\right) + 16\left(\tfrac{1}{3}\right) = \tfrac{16}{3} + \tfrac{16}{3} + \tfrac{16}{3} = 16$$

19. Assume S is centered at the origin with radius a and let H_1 and H_2 be the upper and lower hemispheres, respectively, of S. Then $\iint_S \text{curl }\mathbf{F} \cdot d\mathbf{S} = \iint_{H_1} \text{curl }\mathbf{F} \cdot d\mathbf{S} + \iint_{H_2} \text{curl }\mathbf{F} \cdot d\mathbf{S} = \oint_{C_1} \mathbf{F} \cdot d\mathbf{r} + \oint_{C_2} \mathbf{F} \cdot d\mathbf{r}$ by Stokes' Theorem. But C_1 is the circle $x^2 + y^2 = a^2$ oriented in the counterclockwise direction while C_2 is the same circle oriented in the clockwise direction. Hence $\oint_{C_2} \mathbf{F} \cdot d\mathbf{r} = -\oint_{C_1} \mathbf{F} \cdot d\mathbf{r}$ so $\iint_S \text{curl }\mathbf{F} \cdot d\mathbf{S} = 0$ as desired.

17.9 The Divergence Theorem

1. The vectors that end near P_1 are longer than the vectors that start near P_1, so the net flow is inward near P_1 and div $\mathbf{F}(P_1)$ is negative. The vectors that end near P_2 are shorter than the vectors that start near P_2, so the net flow is outward near P_2 and div $\mathbf{F}(P_2)$ is positive.

3. div $\mathbf{F} = 3 + x + 2x = 3 + 3x$, so

$\iiint_E \operatorname{div} \mathbf{F} \, dV = \int_0^1 \int_0^1 \int_0^1 (3x + 3) \, dx \, dy \, dz = \frac{9}{2}$ (notice the triple integral is

three times the volume of the cube plus three times $\overline{x}$).

To compute $\iint_S \mathbf{F} \cdot d\mathbf{S}$, on S_1: $\mathbf{n} = \mathbf{i}$, $\mathbf{F} = 3\mathbf{i} + y\mathbf{j} + 2z\,\mathbf{k}$, and

$\iint_{S_1} \mathbf{F} \cdot d\mathbf{S} = \iint_{S_1} 3 \, dS = 3$;

S_2: $\mathbf{F} = 3x\,\mathbf{i} + x\,\mathbf{j} + 2xz\,\mathbf{k}$, $\mathbf{n} = \mathbf{j}$ and $\iint_{S_2} \mathbf{F} \cdot d\mathbf{S} = \iint_{S_2} x \, dS = \frac{1}{2}$;

S_3: $\mathbf{F} = 3x\,\mathbf{i} + xy\,\mathbf{j} + 2x\,\mathbf{k}$, $\mathbf{n} = \mathbf{k}$ and $\iint_{S_3} \mathbf{F} \cdot d\mathbf{S} = \iint_{S_3} 2x \, dS = 1$;

S_4: $\mathbf{F} = 0$, $\iint_{S_4} \mathbf{F} \cdot d\mathbf{S} = 0$; S_5: $\mathbf{F} = 3x\,\mathbf{i} + 2x\,\mathbf{k}$, $\mathbf{n} = -\mathbf{j}$ and $\iint_{S_5} \mathbf{F} \cdot d\mathbf{S} = \iint_{S_5} 0 \, dS = 0$;

S_6: $\mathbf{F} = 3x\,\mathbf{i} + xy\,\mathbf{j}$, $\mathbf{n} = -\mathbf{k}$ and $\iint_{S_6} \mathbf{F} \cdot d\mathbf{S} = \iint_{S_6} 0 \, dS = 0$. Thus $\iint_S \mathbf{F} \cdot d\mathbf{S} = \frac{9}{2}$.

5. div $\mathbf{F} = x + y + z$, so

$\iiint_E \operatorname{div} \mathbf{F} \, dV = \int_0^{2\pi} \int_0^1 \int_0^1 (r \cos\theta + r \sin\theta + z)\, r \, dz \, dr \, d\theta = \int_0^{2\pi} \int_0^1 \left(r^2 \cos\theta + r^2 \sin\theta + \frac{1}{2}r\right) dr \, d\theta$

$= \int_0^{2\pi} \left(\frac{1}{3}\cos\theta + \frac{1}{3}\sin\theta + \frac{1}{4}\right) d\theta = \frac{1}{4}(2\pi) = \frac{\pi}{2}$

Let S_1 be the top of the cylinder, S_2 the bottom, and S_3 the vertical edge. On S_1, $z = 1$, $\mathbf{n} = \mathbf{k}$, and

$\mathbf{F} = xy\,\mathbf{i} + y\,\mathbf{j} + x\,\mathbf{k}$, so

$\iint_{S_1} \mathbf{F} \cdot d\mathbf{S} = \iint_{S_1} \mathbf{F} \cdot \mathbf{n} \, dS = \iint_{S_1} x \, dS = \int_0^{2\pi} \int_0^1 (r \cos\theta)\, r \, dr \, d\theta = \left[\sin\theta\right]_0^{2\pi} \left[\frac{1}{3}r^3\right]_0^1 = 0$. On S_2, $z = 0$,

$\mathbf{n} = -\mathbf{k}$, and $\mathbf{F} = xy\,\mathbf{i}$ so $\iint_{S_2} \mathbf{F} \cdot d\mathbf{S} = \iint_{S_2} 0 \, dS = 0$. S_3 is given by $\mathbf{r}(\theta, z) = \cos\theta\,\mathbf{i} + \sin\theta\,\mathbf{j} + z\,\mathbf{k}$,

$0 \le \theta \le 2\pi, 0 \le z \le 1$. Then $\mathbf{r}_\theta \times \mathbf{r}_z = \cos\theta\,\mathbf{i} + \sin\theta\,\mathbf{j}$ and

$\iint_{S_3} \mathbf{F} \cdot d\mathbf{S} = \iint_D \mathbf{F} \cdot (\mathbf{r}_\theta \times \mathbf{r}_z)\, dA = \int_0^{2\pi} \int_0^1 (\cos^2\theta \sin\theta + z \sin^2\theta)\, dz \, d\theta$

$= \int_0^{2\pi} \left(\cos^2\theta \sin\theta + \frac{1}{2}\sin^2\theta\right) d\theta = \left[-\frac{1}{3}\cos^3\theta + \frac{1}{4}\left(\theta - \frac{1}{2}\sin 2\theta\right)\right]_0^{2\pi} = \frac{\pi}{2}$

Thus $\iint_S \mathbf{F} \cdot d\mathbf{S} = 0 + 0 + \frac{\pi}{2} = \frac{\pi}{2}$.

7. div $\mathbf{F} = \frac{\partial}{\partial x}(e^x \sin y) + \frac{\partial}{\partial y}(e^x \cos y) + \frac{\partial}{\partial z}(yz^2) = e^x \sin y - e^x \sin y + 2yz = 2yz$, so by the Divergence Theorem,

$\iint_S \mathbf{F} \cdot d\mathbf{S} = \iiint_E \operatorname{div} \mathbf{F} \, dV = \int_0^1 \int_0^1 \int_0^2 2yz \, dz \, dy \, dx = 2 \int_0^1 dx \int_0^1 y \, dy \int_0^1 z \, dz$

$= 2 \left[x\right]_0^1 \left[\frac{1}{2}y^2\right]_0^1 \left[\frac{1}{2}z^2\right]_0^2 = 2$

9. div $\mathbf{F} = 3y^2 + 0 + 3z^2$, so using cylindrical coordinates with $y = r \cos\theta$, $z = r \sin\theta$, $x = x$ we have

$\iint_S \mathbf{F} \cdot d\mathbf{S} = \iiint_E (3y^2 + 3z^2)\, dV = \int_0^{2\pi} \int_0^1 \int_{-1}^2 (3r^2 \cos^2\theta + 3r^2 \sin^2\theta)\, r \, dx \, dr \, d\theta$

$= 3 \int_0^{2\pi} d\theta \int_0^1 r^3 \, dr \int_{-1}^2 dx = 3(2\pi)\left(\frac{1}{4}\right)(3) = \frac{9\pi}{2}$

11. div $\mathbf{F} = y \sin z + 0 - y \sin z = 0$, so by the Divergence Theorem, $\iint_S \mathbf{F} \cdot d\mathbf{S} = \iiint_E 0 \, dV = 0$.

13. div $\mathbf{F} = y^2 + 0 + x^2 = x^2 + y^2$ so

$$\iint_S \mathbf{F} \cdot d\mathbf{S} = \iiint_E (x^2 + y^2)\, dV = \int_0^{2\pi} \int_0^2 \int_{r^2}^4 r^2 \cdot r\, dz\, dr\, d\theta = \int_0^{2\pi} \int_0^2 r^3(4 - r^2)\, dr\, d\theta$$

$$= \int_0^{2\pi} d\theta \int_0^2 (4r^3 - r^5)\, dr = 2\pi \left[r^4 - \tfrac{1}{6}r^6 \right]_0^2 = \tfrac{32}{3}\pi$$

15. div $\mathbf{F} = 12x^2 z + 12y^2 z + 12z^3$ so

$$\iint_S \mathbf{F} \cdot d\mathbf{S} = \iiint_E 12z(x^2 + y^2 + z^2)\, dV = \int_0^{2\pi} \int_0^\pi \int_0^R 12(\rho\cos\phi)(\rho^2)\rho^2 \sin\phi\, d\rho\, d\phi\, d\theta$$

$$= 12 \int_0^{2\pi} d\theta \int_0^\pi \sin\phi\cos\phi\, d\phi \int_0^R \rho^5\, d\rho = 12(2\pi)\left[\tfrac{1}{2}\sin^2\phi\right]_0^\pi \left[\tfrac{1}{6}\rho^6\right]_0^R = 0$$

17. $\iint_S \mathbf{F} \cdot d\mathbf{S} = \iiint_E \sqrt{3 - x^2}\, dV = \int_{-1}^1 \int_{-1}^1 \int_0^{2 - x^4 - y^4} \sqrt{3 - x^2}\, dz\, dy\, dx = \tfrac{341}{60}\sqrt{2} + \tfrac{81}{20}\sin^{-1}\left(\tfrac{\sqrt{3}}{3}\right)$

19. For S_1 we have $\mathbf{n} = -\mathbf{k}$, so $\mathbf{F} \cdot \mathbf{n} = \mathbf{F} \cdot (-\mathbf{k}) = -x^2 z - y^2 = -y^2$ (since $z = 0$ on S_1). So if D is the unit disk,

we get $\iint_{S_1} \mathbf{F} \cdot d\mathbf{S} = \iint_{S_1} \mathbf{F} \cdot \mathbf{n}\, dS = \iint_D (-y^2)\, dA = -\int_0^{2\pi} \int_0^1 r^2 (\sin^2\theta)\, r\, dr\, d\theta = -\tfrac{1}{4}\pi$. Now

since S_2 is closed, we can use the Divergence Theorem. Since

div $\mathbf{F} = \frac{\partial}{\partial x}\left(z^2 x\right) + \frac{\partial}{\partial y}\left(\tfrac{1}{3}y^3 + \tan z\right) + \frac{\partial}{\partial z}\left(x^2 z + y^2\right) = z^2 + y^2 + x^2$, we use spherical coordinates to get

$\iint_{S_2} \mathbf{F} \cdot d\mathbf{S} = \iiint_E \operatorname{div} \mathbf{F}\, dV = \int_0^{2\pi} \int_0^{\pi/2} \int_0^1 \rho^2 \cdot \rho^2 \sin\phi\, d\rho\, d\phi\, d\theta = \tfrac{2}{5}\pi$. Finally

$\iint_S \mathbf{F} \cdot d\mathbf{S} = \iint_{S_2} \mathbf{F} \cdot d\mathbf{S} - \iint_{S_1} \mathbf{F} \cdot d\mathbf{S} = \tfrac{2}{5}\pi - \left(-\tfrac{1}{4}\pi\right) = \tfrac{13}{20}\pi$.

21. Since $\dfrac{\mathbf{x}}{|\mathbf{x}|^3} = \dfrac{x\mathbf{i} + y\mathbf{j} + z\mathbf{k}}{(x^2 + y^2 + z^2)^{3/2}}$ and $\dfrac{\partial}{\partial x}\left(\dfrac{x}{(x^2 + y^2 + z^2)^{3/2}}\right) = \dfrac{(x^2 + y^2 + z^2) - 3x^2}{(x^2 + y^2 + z^2)^{5/2}}$ with similar

expressions for $\dfrac{\partial}{\partial y}\left(\dfrac{y}{(x^2 + y^2 + z^2)^{3/2}}\right)$ and $\dfrac{\partial}{\partial z}\left(\dfrac{z}{(x^2 + y^2 + z^2)^{3/2}}\right)$, we have

div $\left(\dfrac{\mathbf{x}}{|\mathbf{x}|^3}\right) = \dfrac{3(x^2 + y^2 + z^2) - 3(x^2 + y^2 + z^2)}{(x^2 + y^2 + z^2)^{5/2}} = 0$, except at $(0, 0, 0)$ where it is undefined.

23. $\iint_S \mathbf{a} \cdot \mathbf{n}\, dS = \iiint_E \operatorname{div} \mathbf{a}\, dV = 0$ since div $\mathbf{a} = 0$.

25. $\iint_S \operatorname{curl} \mathbf{F} \cdot d\mathbf{S} = \iiint_E \operatorname{div}(\operatorname{curl} \mathbf{F})\, dV = 0$ by Theorem 17.5.11 [ET 16.5.11].

27. $\iint_S (f\nabla g) \cdot \mathbf{n}\, dS = \iiint_E \operatorname{div}(f\nabla g)\, dV = \iiint_E (f\nabla^2 g + \nabla g \cdot \nabla f)\, dV$ by Exercise 17.5.25 [ET 16.5.25].

29. If $\mathbf{c} = c_1 \mathbf{i} + c_2 \mathbf{j} + c_3 \mathbf{k}$ is an arbitrary constant vector, we define $\mathbf{F} = f\mathbf{c} = fc_1 \mathbf{i} + fc_2 \mathbf{j} + fc_3 \mathbf{k}$. Then

div $\mathbf{F} = \operatorname{div} f\mathbf{c} = \dfrac{\partial f}{\partial x} c_1 + \dfrac{\partial f}{\partial y} c_2 + \dfrac{\partial f}{\partial z} c_3 = \nabla f \cdot \mathbf{c}$ and the Divergence Theorem says

$\iint_S \mathbf{F} \cdot d\mathbf{S} = \iiint_E \operatorname{div} \mathbf{F}\, dV \quad \Rightarrow \quad \iint_S \mathbf{F} \cdot \mathbf{n}\, dS = \iiint_E \nabla f \cdot \mathbf{c}\, dV$. In particular, if $\mathbf{c} = \mathbf{i}$ then

$\iint_S f\mathbf{i} \cdot \mathbf{n}\, dS = \iiint_E \nabla f \cdot \mathbf{i}\, dV \quad \Rightarrow \quad \iint_S fn_1\, dS = \iiint_E \dfrac{\partial f}{\partial x}\, dV$ (where $\mathbf{n} = n_1 \mathbf{i} + n_2 \mathbf{j} + n_3 \mathbf{k}$).

Similarly, if $\mathbf{c} = \mathbf{j}$ we have $\iint_S fn_2\, dS = \iiint_E \dfrac{\partial f}{\partial y}\, dV$, and $\mathbf{c} = \mathbf{k}$ gives $\iint_S fn_3\, dS = \iiint_E \dfrac{\partial f}{\partial z}\, dV$. Then

$$\iint_S f\mathbf{n}\, dS = \left(\iint_S fn_1\, dS\right) \mathbf{i} + \left(\iint_S fn_2\, dS\right) \mathbf{j} + \left(\iint_S fn_3\, dS\right) \mathbf{k}$$

$$= \left(\iiint_E \dfrac{\partial f}{\partial x}\, dV\right) \mathbf{i} + \left(\iiint_E \dfrac{\partial f}{\partial y}\, dV\right) \mathbf{j} + \left(\iiint_E \dfrac{\partial f}{\partial z}\, dV\right) \mathbf{k}$$

$$= \iiint_E \left(\dfrac{\partial f}{\partial x} \mathbf{i} + \dfrac{\partial f}{\partial y} \mathbf{j} + \dfrac{\partial f}{\partial z} \mathbf{k}\right) dV = \iiint_E \nabla f\, dV$$

as desired.

17 Review

──────────── CONCEPT CHECK ────────────

1. See Definitions 1 and 2 in Section 17.1 [ET 16.1]. A vector field can represent, for example, the wind velocity at any location in space, the speed and direction of the ocean current at any location, or the force vectors of Earth's gravitational field at a location in space.

2. (a) A conservative vector field $\mathbf{F}$ is a vector field which is the gradient of some scalar function f.

   (b) The function f in part (a) is called a potential function for $\mathbf{F}$, that is, $\mathbf{F} = \nabla f$.

3. (a) See Definition 17.2.2 [ET 16.2.2].

   (b) We normally evaluate the line integral using Formula 17.2.3 [ET 16.2.3].

   (c) The mass is $m = \int_C \rho(x,y)\, ds$, and the center of mass is $(\bar{x}, \bar{y})$ where $\bar{x} = \frac{1}{m} \int_C x\rho(x,y)\, ds$, $\bar{y} = \frac{1}{m} \int_C y\rho(x,y)\, ds$.

   (d) See (5) and (6) in Section 17.2 [ET 16.2] for plane curves; we have similar definitions when C is a space curve (see the equation preceding (10) in Section 17.2 [ET 16.2]).

   (e) For plane curves, see Equations 17.2.7 [ET 16.2.7]. We have similar results for space curves (see the equation preceding (10) in Section 17.2 [ET 16.2]).

4. (a) See Definition 17.2.13 [ET 16.2.13].

   (b) If $\mathbf{F}$ is a force field, $\int_C \mathbf{F} \cdot d\mathbf{r}$ represents the work done by $\mathbf{F}$ in moving a particle along the curve C.

   (c) $\int_C \mathbf{F} \cdot d\mathbf{r} = \int_C P\, dx + Q\, dy + R\, dz$

5. See Theorem 17.3.2 [ET 16.3.2].

6. (a) $\int_C \mathbf{F} \cdot d\mathbf{r}$ is independent of path if the line integral has the same value for any two curves that have the same initial and terminal points.

   (b) See Theorem 17.3.4 [ET 16.3.4].

7. See the statement of Green's Theorem on page 1119 [ET 1083].

8. See Equations 17.4.5 [ET 16.4.5].

9. (a) $\operatorname{curl} \mathbf{F} = \left(\dfrac{\partial R}{\partial y} - \dfrac{\partial Q}{\partial z} \right) \mathbf{i} + \left(\dfrac{\partial P}{\partial z} - \dfrac{\partial R}{\partial x} \right) \mathbf{j} + \left(\dfrac{\partial Q}{\partial x} - \dfrac{\partial P}{\partial y} \right) \mathbf{k} = \nabla \times \mathbf{F}$

   (b) $\operatorname{div} \mathbf{F} = \dfrac{\partial P}{\partial x} + \dfrac{\partial Q}{\partial y} + \dfrac{\partial R}{\partial z} = \nabla \cdot \mathbf{F}$

   (c) For curl $\mathbf{F}$, see the discussion accompanying Figure 1 on page 1129 [ET 1093] as well as Figure 6 and the accompanying discussion on page 1160 [ET 1124]. For div $\mathbf{F}$, see the discussion following Example 5 on page 1130 [ET 1094] as well as the discussion preceding (8) on page 1167 [ET 1131].

10. See Theorem 17.3.6 [ET 16.3.6]; see Theorem 17.5.4 [ET 16.5.4].

11. (a) See (1) and (2) and the accompanying discussion in Section 17.6 [ET 16.6]; See Figure 4 and the accompanying discussion on page 1135 [ET 1099].

    (b) See Definition 17.6.6 [ET 16.6.6].

    (c) See Equation 17.6.9 [ET 16.6.9].

12. (a) See (1) in Section 17.7 [ET 16.7].

(b) We normally evaluate the surface integral using Formula 17.7.3 [ET 16.7.3].

(c) See Formula 17.7.2 [ET 16.7.2].

(d) The mass is $m = \iint_S \rho(x, y, z)\, dS$ and the center of mass is $(\overline{x}, \overline{y}, \overline{z})$ where $\overline{x} = \frac{1}{m} \iint_S x\rho(x, y, z)\, dS$, $\overline{y} = \frac{1}{m} \iint_S y\rho(x, y, z)\, dS, \overline{z} = \frac{1}{m} \iint_S z\rho(x, y, z)\, dS$.

13. (a) See Figures 7 and 8 and the accompanying discussion in Section 17.7 [ET 16.7]. A Möbius strip is a nonorientable surface; see Figures 5 and 6 and the accompanying discussion on page 1149 [ET 1113].

(b) See Definition 17.7.7 [ET 16.7.7].

(c) See Formula 17.7.9 [ET 16.7.9].

(d) See Formula 17.7.8 [ET 16.7.8].

14. See the statement of Stokes' Theorem on page 1157 [ET 1121].

15. See the statement of the Divergence Theorem on page 1163 [ET 1127].

16. In each theorem, we have an integral of a "derivative" over a region on the left side, while the right side involves the values of the original function only on the boundary of the region.

 TRUE-FALSE QUIZ

1. False; div $\mathbf{F}$ is a scalar field.

3. True, by Theorem 17.5.3 [ET 16.5.3] and the fact that div $\mathbf{0} = 0$.

5. False. See Exercise 17.3.33 [ET 16.3.33]. (But the assertion is true if D is simply-connected; see Theorem 17.3.6 [ET 16.3.6].)

7. True. Apply the Divergence Theorem and use the fact that div $\mathbf{F} = 0$.

EXERCISES

1. (a) Vectors starting on C point in roughly the direction opposite to C, so the tangential component $\mathbf{F} \cdot \mathbf{T}$ is negative. Thus $\int_C \mathbf{F} \cdot d\mathbf{r} = \int_C \mathbf{F} \cdot \mathbf{T}\, ds$ is negative.

(b) The vectors that end near P are shorter than the vectors that start near P, so the net flow is outward near P and div $\mathbf{F}\,(P)$ is positive.

3. $\int_C x^3 z\, ds = \int_0^{\pi/2} (2\sin t)^3 (2\cos t)\sqrt{(2\cos t)^2 + (1)^2 + (-2\sin t)^2}\, dt = \int_0^{\pi/2} (16\sin^3 t\cos t)\sqrt{5}\, dt$
$$= 4\sqrt{5}\sin^4 t\Big]_0^{\pi/2} = 4\sqrt{5}$$

5. $x = \cos t \Rightarrow dx = -\sin t\, dt,\ y = \sin t \Rightarrow dy = \cos t\, dt,\ 0 \le t \le 2\pi$ and
$\int_C x^3 y\, dx - x\, dy = \int_0^{2\pi}(-\cos^3 t\sin^2 t - \cos^2 t)\, dt = \int_0^{2\pi}(-\cos^3 t\sin^2 t - \cos^2 t)\, dt = -\pi$
Or: Since C is a simple closed curve, apply Green's Theorem giving
$$\iint\limits_{x^2+y^2 \le 1} (-1 - x^3)\, dA = \int_0^1 \int_0^{2\pi}(-r - r^4\cos^3\theta)\, d\theta = -\pi.$$

7.

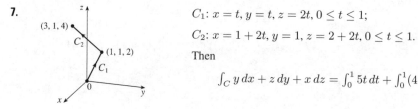

$C_1: x = t,\ y = t,\ z = 2t,\ 0 \le t \le 1$;

$C_2: x = 1 + 2t,\ y = 1,\ z = 2 + 2t,\ 0 \le t \le 1$.

Then

$$\int_C y\, dx + z\, dy + x\, dz = \int_0^1 5t\, dt + \int_0^1 (4 + 4t)\, dt = \frac{17}{2}$$

9. $\mathbf{F}(\mathbf{r}(t)) = e^{-t}\mathbf{i} + t^2(-t)\mathbf{j} + (t^2 + t^3)\mathbf{k}$, $\mathbf{r}'(t) = 2t\mathbf{i} + 3t^2\mathbf{j} - \mathbf{k}$ and

$\int_C \mathbf{F} \cdot d\mathbf{r} = \int_0^1 (2te^{-t} - 3t^5 - (t^2 + t^3))\,dt = \left[-2te^{-t} - 2e^{-t} - \frac{1}{2}t^6 - \frac{1}{3}t^3 - \frac{1}{4}t^4\right]_0^1 = \frac{11}{12} - \frac{4}{e}$.

11. $\frac{\partial}{\partial y}\left[(1 + xy)e^{xy}\right] = 2xe^{xy} + x^2 ye^{xy} = \frac{\partial}{\partial x}\left[e^y + x^2 e^{xy}\right]$ and the domain of $\mathbf{F}$ is $\mathbb{R}^2$, so $\mathbf{F}$ is conservative. Thus

there exists a function f such that $\mathbf{F} = \nabla f$. Then $f_y(x, y) = e^y + x^2 e^{xy}$ implies $f(x, y) = e^y + xe^{xy} + g(x)$ and

then $f_x(x, y) = xye^{xy} + e^{xy} + g'(x) = (1 + xy)e^{xy} + g'(x)$. But $f_x(x, y) = (1 + xy)e^{xy}$, so $g'(x) = 0 \Rightarrow$

$g(x) = K$. Thus $f(x, y) = e^y + xe^{xy} + K$ is a potential function for $\mathbf{F}$.

13. Since $\frac{\partial}{\partial y}(4x^3 y^2 - 2xy^3) = 8x^3 y - 6xy^2 = \frac{\partial}{\partial x}(2x^4 y - 3x^2 y^2 + 4y^3)$ and the domain of $\mathbf{F}$ is $\mathbb{R}^2$, $\mathbf{F}$ is

conservative. Furthermore $f(x, y) = x^4 y^2 - x^2 y^3 + y^4$ is a potential function for $\mathbf{F}$. $t = 0$ corresponds to the

point $(0, 1)$ and $t = 1$ corresponds to $(1, 1)$, so $\int_C \mathbf{F} \cdot d\mathbf{r} = f(1, 1) - f(0, 1) = 1 - 1 = 0$.

15.

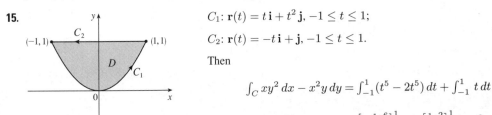

C_1: $\mathbf{r}(t) = t\mathbf{i} + t^2 \mathbf{j}$, $-1 \le t \le 1$;

C_2: $\mathbf{r}(t) = -t\mathbf{i} + \mathbf{j}$, $-1 \le t \le 1$.

Then

$$\int_C xy^2\,dx - x^2 y\,dy = \int_{-1}^1 (t^5 - 2t^5)\,dt + \int_{-1}^1 t\,dt$$

$$= \left[-\frac{1}{6}t^6\right]_{-1}^1 + \left[\frac{1}{2}t^2\right]_{-1}^1 = 0$$

Using Green's Theorem, we have

$$\int_C xy^2\,dx - x^2 y\,dy = \iint_D \left[\frac{\partial}{\partial x}(-x^2 y) - \frac{\partial}{\partial y}(xy^2)\right] dA = \iint_D (-2xy - 2xy)\,dA$$

$$= \int_{-1}^1 \int_{x^2}^1 -4xy\,dy\,dx = \int_{-1}^1 \left[-2xy^2\right]_{y=x^2}^{y=1}\,dx$$

$$= \int_{-1}^1 (2x^5 - 2x)\,dx = \left[\frac{1}{3}x^6 - x^2\right]_{-1}^1 = 0$$

17. $\int_C x^2 y\,dx - xy^2\,dy = \iint\limits_{x^2 + y^2 \le 4} \left[\frac{\partial}{\partial x}(-xy^2) - \frac{\partial}{\partial y}(x^2 y)\right] dA$

$= \iint\limits_{x^2 + y^2 \le 4} (-y^2 - x^2)\,dA = -\int_0^{2\pi} \int_0^2 r^3\,dr\,d\theta = -8\pi$

19. If we assume there is such a vector field $\mathbf{G}$, then $\operatorname{div}(\operatorname{curl} \mathbf{G}) = 2 + 3z - 2xz$. But $\operatorname{div}(\operatorname{curl} \mathbf{F}) = 0$ for all vector

fields $\mathbf{F}$. Thus such a $\mathbf{G}$ cannot exist.

21. For any piecewise-smooth simple closed plane curve C bounding a region D, we can apply Green's Theorem to

$\mathbf{F}(x, y) = f(x)\mathbf{i} + g(y)\mathbf{j}$ to get $\int_C f(x)\,dx + g(y)\,dy = \iint_D \left[\frac{\partial}{\partial x}g(y) - \frac{\partial}{\partial y}f(x)\right] dA = \iint_D 0\,dA = 0$.

23. $\nabla^2 f = 0$ means that $\dfrac{\partial^2 f}{\partial x^2} + \dfrac{\partial^2 f}{\partial y^2} = 0$. Now if $\mathbf{F} = f_y\mathbf{i} - f_x\mathbf{j}$ and C is any closed path in D, then applying

Green's Theorem, we get

$$\int_C \mathbf{F} \cdot d\mathbf{r} = \int_C f_y\,dx - f_x\,dy = \iint_D \left[\frac{\partial}{\partial x}(-f_x) - \frac{\partial}{\partial y}(f_y)\right] dA = -\iint_D (f_{xx} + f_{yy})\,dA$$

$$= -\iint_D 0\,dA = 0$$

Therefore the line integral is independent of path, by Theorem 17.3.3 [ET 16.3.3].

25. $z = f(x, y) = x^2 + 2y$ with $0 \le x \le 1, 0 \le y \le 2x$. Thus

$$A(S) = \iint_D \sqrt{1 + 4x^2 + 4}\, dA = \int_0^1 \int_0^{2x} \sqrt{5 + 4x^2}\, dy\, dx = \int_0^1 2x\sqrt{5 + 4x^2}\, dx$$

$$= \tfrac{1}{6}(5 + 4x^2)^{3/2}\Big]_0^1 = \tfrac{1}{6}\left(27 - 5\sqrt{5}\right).$$

27. $z = f(x, y) = x^2 + y^2$ with $0 \le x^2 + y^2 \le 4$ so $\mathbf{r}_x \times \mathbf{r}_y = -2x\,\mathbf{i} - 2y\,\mathbf{j} + \mathbf{k}$ (using upward orientation). Then

$$\iint_S z\, dS = \iint_{x^2 + y^2 \le 4} (x^2 + y^2)\sqrt{4x^2 + 4y^2 + 1}\, dA = \int_0^{2\pi} \int_0^2 r^3 s\sqrt{1 + 4r^2}\, dr\, d\theta$$

$$= \tfrac{1}{60}\pi\left(391\sqrt{17} + 1\right)$$

(Substitute $u = 1 + 4r^2$ and use tables.)

29. Since the sphere bounds a simple solid region, the Divergence Theorem applies and

$$\iint_S \mathbf{F} \cdot d\mathbf{S} = \iiint_E (z - 2)\, dV = \iiint_E z\, dV - 2\iiint_E dV = m\bar{z} - 2\left(\tfrac{4}{3}\pi 2^3\right) = -\tfrac{64}{3}\pi.$$

Alternate solution: $\mathbf{F}(\mathbf{r}(\phi, \theta)) = 4\sin\phi\cos\theta\cos\phi\,\mathbf{i} - 4\sin\phi\sin\theta\,\mathbf{j} + 6\sin\phi\cos\theta\,\mathbf{k}$,

$\mathbf{r}_\phi \times \mathbf{r}_\theta = 4\sin^2\phi\cos\theta\,\mathbf{i} + 4\sin^2\phi\sin\theta\,\mathbf{j} + 4\sin\phi\cos\phi\,\mathbf{k}$, and

$\mathbf{F} \cdot (\mathbf{r}_\phi \times \mathbf{r}_\theta) = 16\sin^3\phi\cos^2\theta\cos\phi - 16\sin^3\phi\sin^2\theta + 24\sin^2\phi\cos\phi\cos\theta$. Then

$$\iint_S F \cdot dS = \int_0^{2\pi} \int_0^\pi (16\sin^3\phi\cos\phi\cos^2\theta - 16\sin^3\phi\sin^2\theta + 24\sin^2\phi\cos\phi\cos\theta)\, d\phi\, d\theta$$

$$= \int_0^{2\pi} \tfrac{4}{3}(-16\sin^2\theta)\, d\theta = -\tfrac{64}{3}\pi$$

31. Since $\operatorname{curl}\mathbf{F} = 0$, $\iint_S(\operatorname{curl}\mathbf{F}) \cdot d\mathbf{S} = 0$. We parametrize C: $\mathbf{r}(t) = \cos t\,\mathbf{i} + \sin t\,\mathbf{j}, 0 \le t \le 2\pi$ and

$$\oint_C \mathbf{F} \cdot d\mathbf{r} = \int_0^{2\pi}(-\cos^2 t\sin t + \sin^2 t\cos t)\, dt = \tfrac{1}{3}\cos^3 t + \tfrac{1}{3}\sin^3 t\Big]_0^{2\pi} = 0.$$

33. The surface is given by $x + y + z = 1$ or $z = 1 - x - y, 0 \le x \le 1, 0 \le y \le 1 - x$ and $\mathbf{r}_x \times \mathbf{r}_y = \mathbf{i} + \mathbf{j} + \mathbf{k}$.

Then

$$\oint_C \mathbf{F} \cdot d\mathbf{r} = \iint_S \operatorname{curl}\mathbf{F} \cdot d\mathbf{S} = \iint_D (-y\,\mathbf{i} - z\,\mathbf{j} - x\,\mathbf{k}) \cdot (\mathbf{i} + \mathbf{j} + \mathbf{k})\, dA$$

$$= \iint_D (-1)\, dA = -(\text{area of } D) = -\tfrac{1}{2}$$

35. $\iiint_E \operatorname{div}\mathbf{F}\, dV = \iiint_{x^2 + y^2 + z^2 \le 1} 3\, dV = 3(\text{volume of sphere}) = 4\pi$. Then

$\mathbf{F}(\mathbf{r}(\phi, \theta)) \cdot (\mathbf{r}_\phi \times \mathbf{r}_\theta) = \sin^3\phi\cos^2\theta + \sin^3\phi\sin^2\theta + \sin\phi\cos^2\phi = \sin\phi$ and

$\iint_S \mathbf{F} \cdot d\mathbf{S} = \int_0^{2\pi} \int_0^\pi \sin\phi\, d\phi\, d\theta = (2\pi)(2) = 4\pi$.

37. Because $\operatorname{curl}\mathbf{F} = 0$, $\mathbf{F}$ is conservative, and if $f(x, y, z) = x^3yz - 3xy + z^2$, then $\nabla f = \mathbf{F}$. Hence

$\int_C \mathbf{F} \cdot d\mathbf{r} = \int_C \nabla f \cdot d\mathbf{r} = f(0, 3, 0) - f(0, 0, 2) = 0 - 4 = -4$.

39. By the Divergence Theorem, $\iint_S \mathbf{F} \cdot \mathbf{n}\, dS = \iiint_E \operatorname{div}\mathbf{F}\, dV = 3(\text{volume of } E) = 3(8 - 1) = 21$.

☐ PROBLEMS PLUS

1. Let S_1 be the portion of $\Omega(S)$ between $S(a)$ and S, and let ∂S_1 be its boundary. Also let S_L be the lateral surface of S_1 [that is, the surface of S_1 except S and $S(a)$]. Applying the Divergence Theorem we have

$$\iint_{\partial S_1} \frac{\mathbf{r} \cdot \mathbf{n}}{r^3} \, dS = \iiint_{S_1} \nabla \cdot \frac{\mathbf{r}}{r^3} \, dV. \text{ But}$$

$$\nabla \cdot \frac{\mathbf{r}}{r^3} = \left\langle \frac{\partial}{\partial x}, \frac{\partial}{\partial y}, \frac{\partial}{\partial z} \right\rangle \cdot \left\langle \frac{x}{(x^2 + y^2 + z^2)^{3/2}}, \frac{y}{(x^2 + y^2 + z^2)^{3/2}}, \frac{z}{(x^2 + y^2 + z^2)^{3/2}} \right\rangle$$

$$= \frac{(x^2 + y^2 + z^2 - 3x^2) + (x^2 + y^2 + z^2 - 3y^2) + (x^2 + y^2 + z^2 - 3z^2)}{(x^2 + y^2 + z^2)^{5/2}} = 0$$

$$\Rightarrow \iint_{\partial S_1} \frac{\mathbf{r} \cdot \mathbf{n}}{r^3} \, dS = \iiint_{S_1} 0 \, dV = 0. \text{ On the other hand, notice that for the surfaces of } \partial S_1 \text{ other than } S(a)$$

and S, $\mathbf{r} \cdot \mathbf{n} = 0 \Rightarrow$

$$0 = \iint_{\partial S_1} \frac{\mathbf{r} \cdot \mathbf{n}}{r^3} \, dS = \iint_S \frac{\mathbf{r} \cdot \mathbf{n}}{r^3} \, dS + \iint_{S(a)} \frac{\mathbf{r} \cdot \mathbf{n}}{r^3} \, dS + \iint_{S_L} \frac{\mathbf{r} \cdot \mathbf{n}}{r^3} \, dS$$

$$= \iint_S \frac{\mathbf{r} \cdot \mathbf{n}}{r^3} \, dS + \iint_{S(a)} \frac{\mathbf{r} \cdot \mathbf{n}}{r^3} \, dS$$

$$\Rightarrow \iint_S \frac{\mathbf{r} \cdot \mathbf{n}}{r^3} \, dS = -\iint_{S(a)} \frac{\mathbf{r} \cdot \mathbf{n}}{r^3} \, dS. \text{ Notice that on } S(a), r = a \Rightarrow \mathbf{n} = -\frac{\mathbf{r}}{r} = -\frac{\mathbf{r}}{a} \text{ and } \mathbf{r} \cdot \mathbf{r} = r^2 = a^2,$$

so that $-\iint_{S(a)} \dfrac{\mathbf{r} \cdot \mathbf{n}}{r^3} \, dS = \iint_{S(a)} \dfrac{\mathbf{r} \cdot \mathbf{r}}{a^4} \, dS = \iint_{S(a)} \dfrac{a^2}{a^4} \, dS = \dfrac{1}{a^2} \iint_{S(a)} dS = \dfrac{\text{area of } S(a)}{a^2} = |\Omega(S)|.$

Therefore $|\Omega(S)| = \displaystyle\iint_S \frac{\mathbf{r} \cdot \mathbf{n}}{r^3} \, dS.$

3. The given line integral $\frac{1}{2} \int_C (bz - cy) \, dx + (cx - az) \, dy + (ay - bx) \, dz$ can be expressed as $\int_C \mathbf{F} \cdot d\mathbf{r}$ if we define the vector field $\mathbf{F}$ by $\mathbf{F}(x, y, z) = P\mathbf{i} + Q\mathbf{j} + R\mathbf{k} = \frac{1}{2}(bz - cy)\mathbf{i} + \frac{1}{2}(cx - az)\mathbf{j} + \frac{1}{2}(ay - bx)\mathbf{k}$. Then define S to be the planar interior of C, so S is an oriented, smooth surface. Stokes' Theorem says $\int_C \mathbf{F} \cdot d\mathbf{r} = \iint_S \text{curl } \mathbf{F} \cdot d\mathbf{S} = \iint_S \text{curl } \mathbf{F} \cdot \mathbf{n} \, dS$. Now

$$\text{curl } \mathbf{F} = \left(\frac{\partial R}{\partial y} - \frac{\partial Q}{\partial z} \right) \mathbf{i} + \left(\frac{\partial P}{\partial z} - \frac{\partial R}{\partial x} \right) \mathbf{j} + \left(\frac{\partial Q}{\partial x} - \frac{\partial P}{\partial y} \right) \mathbf{k}$$

$$= \left(\tfrac{1}{2}a + \tfrac{1}{2}a \right) \mathbf{i} + \left(\tfrac{1}{2}b + \tfrac{1}{2}b \right) \mathbf{j} + \left(\tfrac{1}{2}c + \tfrac{1}{2}c \right) \mathbf{k} = a\mathbf{i} + b\mathbf{j} + c\mathbf{k} = \mathbf{n}$$

so curl $\mathbf{F} \cdot \mathbf{n} = \mathbf{n} \cdot \mathbf{n} = |\mathbf{n}|^2 = 1$, hence $\iint_S \text{curl } \mathbf{F} \cdot \mathbf{n} \, dS = \iint_S dS$ which is simply the surface area of S. Thus, $\int_C \mathbf{F} \cdot d\mathbf{r} = \frac{1}{2} \int_C (bz - cy) \, dx + (cx - az) \, dy + (ay - bx) \, dz$ is the plane area enclosed by C.

18 □ SECOND-ORDER DIFFERENTIAL EQUATIONS □ ET 17

18.1 Second-Order Linear Equations

1. The auxiliary equation is $r^2 - 6r + 8 = 0 \Rightarrow (r-4)(r-2) = 0 \Rightarrow r = 4, r = 2$. Then by (8) the general solution is $y = c_1 e^{4x} + c_2 e^{2x}$.

3. The auxiliary equation is $r^2 + 8r + 41 = 0 \Rightarrow r = -4 \pm 5i$. Then by (11) the general solution is $y = e^{-4x}(c_1 \cos 5x + c_2 \sin 5x)$.

5. The auxiliary equation is $r^2 - 2r + 1 = (r-1)^2 = 0 \Rightarrow r = 1$. Then by (10), the general solution is $y = c_1 e^x + c_2 x e^x$.

7. The auxiliary equation is $4r^2 + 1 = 0 \Rightarrow r = \pm\frac{1}{2}i$, so $y = c_1 \cos\left(\frac{1}{2}x\right) + c_2 \sin\left(\frac{1}{2}x\right)$.

9. The auxiliary equation is $4r^2 + r = r(4r+1) = 0 \Rightarrow r = 0, r = -\frac{1}{4}$, so $y = c_1 + c_2 e^{-x/4}$.

11. The auxiliary equation is $r^2 - 2r - 1 = 0 \Rightarrow r = 1 \pm \sqrt{2}$, so $y = c_1 e^{(1+\sqrt{2})t} + c_2 e^{(1-\sqrt{2})t}$.

13. The auxiliary equation is $r^2 + r + 1 = 0 \Rightarrow r = -\frac{1}{2} \pm \frac{\sqrt{3}}{2}i$, so $y = e^{-t/2}\left[c_1 \cos\left(\frac{\sqrt{3}}{2}t\right) + c_2 \sin\left(\frac{\sqrt{3}}{2}t\right)\right]$.

15. $r^2 - 8r + 16 = (r-4)^2 = 0$ so $y = c_1 e^{4x} + c_2 x e^{4x}$.

The graphs are all asymptotic to the x-axis as $x \to -\infty$, and as $x \to \infty$ the solutions tend to $\pm\infty$.

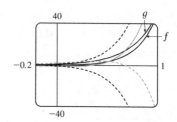

17. $2r^2 + 5r + 3 = (2r+3)(r+1) = 0$, so $r = -\frac{3}{2}, r = -1$ and the general solution is $y = c_1 e^{-3x/2} + c_2 e^{-x}$. Then $y(0) = 3 \Rightarrow c_1 + c_2 = 3$ and $y'(0) = -4 \Rightarrow -\frac{3}{2}c_1 - c_2 = -4$, so $c_1 = 2$ and $c_2 = 1$. Thus the solution to the initial-value problem is $y = 2e^{-3x/2} + e^{-x}$.

19. $4r^2 - 4r + 1 = (2r-1)^2 = 0 \Rightarrow r = \frac{1}{2}$ and the general solution is $y = c_1 e^{x/2} + c_2 x e^{x/2}$. Then $y(0) = 1 \Rightarrow c_1 = 1$ and $y'(0) = -1.5 \Rightarrow \frac{1}{2}c_1 + c_2 = -1.5$, so $c_2 = -2$ and the solution to the initial-value problem is $y = e^{x/2} - 2x e^{x/2}$.

21. $r^2 + 16 = 0 \Rightarrow r = \pm 4i$ and the general solution is $y = e^{0x}(c_1 \cos 4x + c_2 \sin 4x) = c_1 \cos 4x + c_2 \sin 4x$. Then $y\left(\frac{\pi}{4}\right) = -3 \Rightarrow -c_1 = -3 \Rightarrow c_1 = 3$ and $y'\left(\frac{\pi}{4}\right) = 4 \Rightarrow -4c_2 = 4 \Rightarrow c_2 = -1$, so the solution to the initial-value problem is $y = 3 \cos 4x - \sin 4x$.

23. $r^2 + 2r + 2 = 0 \Rightarrow r = -1 \pm i$ and the general solution is $y = e^{-x}(c_1 \cos x + c_2 \sin x)$. Then $2 = y(0) = c_1$ and $1 = y'(0) = c_2 - c_1 \Rightarrow c_2 = 3$ and the solution to the initial-value problem is $y = e^{-x}(2 \cos x + 3 \sin x)$.

25. $4r^2 + 1 = 0 \Rightarrow r = \pm\frac{1}{2}i$ and the general solution is $y = c_1 \cos\left(\frac{1}{2}x\right) + c_2 \sin\left(\frac{1}{2}x\right)$. Then $3 = y(0) = c_1$ and $-4 = y(\pi) = c_2$, so the solution of the boundary-value problem is $y = 3 \cos\left(\frac{1}{2}x\right) - 4 \sin\left(\frac{1}{2}x\right)$.

27. $r^2 - 3r + 2 = (r - 2)(r - 1) = 0 \Rightarrow r = 1, r = 2$ and the general solution is $y = c_1 e^x + c_2 e^{2x}$. Then

$1 = y(0) = c_1 + c_2$ and $0 = y(3) = c_1 e^3 + c_2 e^6$ so $c_2 = 1/(1 - e^3)$ and $c_1 = e^3/(e^3 - 1)$. The solution of the

boundary-value problem is $y = \dfrac{e^{x+3}}{e^3 - 1} + \dfrac{e^{2x}}{1 - e^3}$.

29. $r^2 - 6r + 25 = 0 \Rightarrow r = 3 \pm 4i$ and the general solution is $y = e^{3x}(c_1 \cos 4x + c_2 \sin 4x)$. But $1 = y(0) = c_1$

and $2 = y(\pi) = c_1 e^{3\pi} \Rightarrow c_1 = 2/e^{3\pi}$, so there is no solution.

31. $r^2 + 4r + 13 = 0 \Rightarrow r = -2 \pm 3i$ and the general solution is $y = e^{-2x}(c_1 \cos 3x + c_2 \sin 3x)$. But

$2 = y(0) = c_1$ and $1 = y\left(\frac{\pi}{2}\right) = e^{-\pi}(-c_2)$, so the solution to the boundary-value problem is

$y = e^{-2x}(2 \cos 3x - e^\pi \sin 3x)$.

33. (a) *Case 1* $(\lambda = 0)$: $y'' + \lambda y = 0 \Rightarrow y'' = 0$ which has an auxiliary equation $r^2 = 0 \Rightarrow r = 0 \Rightarrow$

$y = c_1 + c_2 x$ where $y(0) = 0$ and $y(L) = 0$. Thus, $0 = y(0) = c_1$ and $0 = y(L) = c_2 L \Rightarrow c_1 = c_2 = 0$.

Thus, $y = 0$.

Case 2 $(\lambda < 0)$: $y'' + \lambda y = 0$ has auxiliary equation $r^2 = -\lambda \Rightarrow r = \pm\sqrt{-\lambda}$ (distinct and real since

$\lambda < 0$) $\Rightarrow y = c_1 e^{\sqrt{-\lambda}x} + c_2 e^{-\sqrt{-\lambda}x}$ where $y(0) = 0$ and $y(L) = 0$. Thus, $0 = y(0) = c_1 + c_2$ (∗) and

$0 = y(L) = c_1 e^{\sqrt{-\lambda}L} + c_2 e^{-\sqrt{-\lambda}L}$ (†).

Multiplying (∗) by $e^{\sqrt{-\lambda}L}$ and subtracting (†) gives $c_2\left(e^{\sqrt{-\lambda}L} - e^{-\sqrt{-\lambda}L}\right) = 0 \Rightarrow c_2 = 0$ and thus

$c_1 = 0$ from (∗). Thus, $y = 0$ for the cases $\lambda = 0$ and $\lambda < 0$.

(b) $y'' + \lambda y = 0$ has an auxiliary equation $r^2 + \lambda = 0 \Rightarrow r = \pm i\sqrt{\lambda} \Rightarrow y = c_1 \cos\sqrt{\lambda}x + c_2 \sin\sqrt{\lambda}x$

where $y(0) = 0$ and $y(L) = 0$. Thus, $0 = y(0) = c_1$ and $0 = y(L) = c_2 \sin\sqrt{\lambda}L$ since $c_1 = 0$. Since we

cannot have a trivial solution, $c_2 \neq 0$ and thus $\sin\sqrt{\lambda}L = 0 \Rightarrow \sqrt{\lambda}L = n\pi$ where n is an integer

$\Rightarrow \lambda = n^2\pi^2/L^2$ and $y = c_2 \sin(n\pi x/L)$ where n is an integer.

18.2 Nonhomogeneous Linear Equations ET 17.2

1. The auxiliary equation is $r^2 + 3r + 2 = (r + 2)(r + 1) = 0$, so the complementary solution is

$y_c(x) = c_1 e^{-2x} + c_2 e^{-x}$. We try the particular solution $y_p(x) = Ax^2 + Bx + C$, so $y_p' = 2Ax + B$ and

$y_p'' = 2A$. Substituting into the differential equation, we have $(2A) + 3(2Ax + B) + 2(Ax^2 + Bx + C) = x^2$ or

$2Ax^2 + (6A + 2B)x + (2A + 3B + 2C) = x^2$. Comparing coefficients gives $2A = 1$, $6A + 2B = 0$, and

$2A + 3B + 2C = 0$, so $A = \frac{1}{2}$, $B = -\frac{3}{2}$, and $C = \frac{7}{4}$. Thus the general solution is

$y(x) = y_c(x) + y_p(x) = c_1 e^{-2x} + c_2 e^{-x} + \frac{1}{2}x^2 - \frac{3}{2}x + \frac{7}{4}$.

3. The auxiliary equation is $r^2 - 2r = r(r - 2) = 0$, so the complementary solution is $y_c(x) = c_1 + c_2 e^{2x}$. Try the

particular solution $y_p(x) = A \cos 4x + B \sin 4x$, so $y_p' = -4A \sin 4x + 4B \cos 4x$

and $y_p'' = -16A \cos 4x - 16B \sin 4x$. Substitution into the differential equation

gives $(-16A \cos 4x - 16B \sin 4x) - 2(-4A \sin 4x + 4B \cos 4x) = \sin 4x \Rightarrow$

$(-16A - 8B) \cos 4x + (8A - 16B) \sin 4x = \sin 4x$. Then $-16A - 8B = 0$ and $8A - 16B = 1 \Rightarrow A = \frac{1}{40}$

and $B = -\frac{1}{20}$. Thus the general solution is $y(x) = y_c(x) + y_p(x) = c_1 + c_2 e^{2x} + \frac{1}{40} \cos 4x - \frac{1}{20} \sin 4x$.

5. The auxiliary equation is $r^2 - 4r + 5 = 0$ with roots $r = 2 \pm i$, so the complementary solution is

$y_c(x) = e^{2x}(c_1 \cos x + c_2 \sin x)$. Try $y_p(x) = Ae^{-x}$, so $y_p' = -Ae^{-x}$ and $y_p'' = Ae^{-x}$. Substitution gives

$Ae^{-x} - 4(-Ae^{-x}) + 5(Ae^{-x}) = e^{-x} \Rightarrow 10Ae^{-x} = e^{-x} \Rightarrow A = \frac{1}{10}$. Thus the general solution is

$y(x) = e^{2x}(c_1 \cos x + c_2 \sin x) + \frac{1}{10}e^{-x}$.

7. The auxiliary equation is $r^2 + 1 = 0$ with roots $r = \pm i$, so the complementary solution is

$y_c(x) = c_1 \cos x + c_2 \sin x$. For $y'' + y = e^x$ try $y_{p_1}(x) = Ae^x$. Then $y_{p_1}' = y_{p_1}'' = Ae^x$ and substitution gives

$Ae^x + Ae^x = e^x \Rightarrow A = \frac{1}{2}$, so $y_{p_1}(x) = \frac{1}{2}e^x$. For $y'' + y = x^3$ try $y_{p_2}(x) = Ax^3 + Bx^2 + Cx + D$.

Then $y_{p_2}' = 3Ax^2 + 2Bx + C$ and $y_{p_2}'' = 6Ax + 2B$. Substituting, we have

$6Ax + 2B + Ax^3 + Bx^2 + Cx + D = x^3$, so $A = 1$, $B = 0$, $6A + C = 0 \Rightarrow C = -6$, and $2B + D = 0$

$\Rightarrow D = 0$. Thus $y_{p_2}(x) = x^3 - 6x$ and the general solution is

$y(x) = y_c(x) + y_{p_1}(x) + y_{p_2}(x) = c_1 \cos x + c_2 \sin x + \frac{1}{2}e^x + x^3 - 6x$. But $2 = y(0) = c_1 + \frac{1}{2} \Rightarrow c_1 = \frac{3}{2}$

and $0 = y'(0) = c_2 + \frac{1}{2} - 6 \Rightarrow c_2 = \frac{11}{2}$. Thus the solution to the initial-value problem is

$y(x) = \frac{3}{2} \cos x + \frac{11}{2} \sin x + \frac{1}{2}e^x + x^3 - 6x$.

9. The auxiliary equation is $r^2 - r = 0$ with roots $r = 0$, $r = 1$ so the complementary solution is $y_c(x) = c_1 + c_2 e^x$.

Try $y_p(x) = x(Ax + B)e^x$ so that no term in y_p is a solution of the complementary equation. Then

$y_p' = (Ax^2 + (2A + B)x + B)e^x$ and $y_p'' = (Ax^2 + (4A + B)x + (2A + 2B))e^x$. Substitution into the

differential equation gives $(Ax^2 + (4A + B)x + (2A + 2B))e^x - (Ax^2 + (2A + B)x + B)e^x = xe^x \Rightarrow$

$(2Ax + (2A + B))e^x = xe^x \Rightarrow A = \frac{1}{2}$, $B = -1$. Thus $y_p(x) = \left(\frac{1}{2}x^2 - x\right)e^x$ and the general solution is

$y(x) = c_1 + c_2 e^x + \left(\frac{1}{2}x^2 - x\right)e^x$. But $2 = y(0) = c_1 + c_2$ and $1 = y'(0) = c_2 - 1$, so $c_2 = 2$ and $c_1 = 0$. The

solution to the initial-value problem is $y(x) = 2e^x + \left(\frac{1}{2}x^2 - x\right)e^x = e^x\left(\frac{1}{2}x^2 - x + 2\right)$.

11. $y_c(x) = c_1 e^{-x/4} + c_2 e^{-x}$. Try $y_p(x) = Ae^x$. Then

$10Ae^x = e^x$, so $A = \frac{1}{10}$ and the general solution is

$y(x) = c_1 e^{-x/4} + c_2 e^{-x} + \frac{1}{10}e^x$. The solutions are all composed

of exponential curves and with the exception of the particular

solution (which approaches 0 as $x \to -\infty$), they all approach

either ∞ or $-\infty$ as $x \to -\infty$. As $x \to \infty$, all solutions are

asymptotic to $y_p = \frac{1}{10}e^x$.

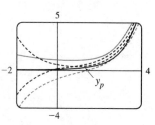

13. Here $y_c(x) = c_1 \cos 3x + c_2 \sin 3x$. For $y'' + 9y = e^{2x}$ try $y_{p_1}(x) = Ae^{2x}$ and for $y'' + 9y = x^2 \sin x$ try

$y_{p_2}(x) = (Bx^2 + Cx + D) \cos x + (Ex^2 + Fx + G) \sin x$. Thus a trial solution is

$y_p(x) = y_{p_1}(x) + y_{p_2}(x) = Ae^{2x} + (Bx^2 + Cx + D) \cos x + (Ex^2 + Fx + G) \sin x$.

15. Here $y_c(x) = c_1 + c_2 e^{-9x}$. For $y'' + 9y' = 1$ try $y_{p_1}(x) = Ax$ (since $y = A$ is a solution to the complementary

equation) and for $y'' + 9y' = xe^{9x}$ try $y_{p_2}(x) = (Bx + C)e^{9x}$.

17. Since $y_c(x) = e^{-x}(c_1 \cos 3x + c_2 \sin 3x)$ we try

$y_p(x) = x(Ax^2 + Bx + C)e^{-x} \cos 3x + x(Dx^2 + Ex + F)e^{-x} \sin 3x$ (so that no term of y_p is a solution of the complementary equation).

Note: Solving Equations (7) and (9) in The Method of Variation of Parameters gives

$$u_1' = -\frac{Gy_2}{a\,(y_1 y_2' - y_2 y_1')} \quad \text{and} \quad u_2' = \frac{Gy_1}{a\,(y_1 y_2' - y_2 y_1')}$$

We will use these equations rather than resolving the system in each of the remaining exercises in this section.

19. (a) The complementary solution is $y_c(x) = c_1 \cos 2x + c_2 \sin 2x$. A particular solution is of the form
$y_p(x) = Ax + B$. Thus, $4Ax + 4B = x \Rightarrow A = \frac{1}{4}$ and $B = 0 \Rightarrow y_p(x) = \frac{1}{4}x$. Thus, the general
solution is $y = y_c + y_p = c_1 \cos 2x + c_2 \sin 2x + \frac{1}{4}x$.

(b) In (a), $y_c(x) = c_1 \cos 2x + c_2 \sin 2x$, so set $y_1 = \cos 2x$, $y_2 = \sin 2x$. Then
$y_1 y_2' - y_2 y_1' = 2\cos^2 2x + 2\sin^2 2x = 2$ so $u_1' = -\frac{1}{2}x \sin 2x \Rightarrow$
$u_1(x) = -\frac{1}{2}\int x \sin 2x\, dx = -\frac{1}{4}(-x\cos 2x + \frac{1}{2}\sin 2x)$ [by parts] and $u_2' = \frac{1}{2}x\cos 2x$
$\Rightarrow u_2(x) = \frac{1}{2}\int x\cos 2x\, dx = \frac{1}{4}(x\sin 2x + \frac{1}{2}\cos 2x)$ [by parts]. Hence
$y_p(x) = -\frac{1}{4}(-x\cos 2x + \frac{1}{2}\sin 2x)\cos 2x + \frac{1}{4}(x\sin 2x + \frac{1}{2}\cos 2x)\sin 2x = \frac{1}{4}x$. Thus
$y(x) = y_c(x) + y_p(x) = c_1 \cos 2x + c_2 \sin 2x + \frac{1}{4}x$.

21. (a) $r^2 - r = r(r-1) = 0 \Rightarrow r = 0, 1$, so the complementary solution is $y_c(x) = c_1 e^x + c_2 x e^x$. A particular
solution is of the form $y_p(x) = Ae^{2x}$. Thus $4Ae^{2x} - 4Ae^{2x} + Ae^{2x} = e^{2x} \Rightarrow Ae^{2x} = e^{2x} \Rightarrow A = 1$
$\Rightarrow y_p(x) = e^{2x}$. So a general solution is $y(x) = y_c(x) + y_p(x) = c_1 e^x + c_2 x e^x + e^{2x}$.

(b) From (a), $y_c(x) = c_1 e^x + c_2 x e^x$, so set $y_1 = e^x$, $y_2 = x e^x$. Then, $y_1 y_2' - y_2 y_1' = e^{2x}(1+x) - xe^{2x} = e^{2x}$
and so $u_1' = -xe^x \Rightarrow u_1(x) = -\int xe^x\, dx = -(x-1)e^x$ [by parts] and $u_2' = e^x \Rightarrow$
$u_2(x) = \int e^x\, dx = e^x$. Hence $y_p(x) = (1-x)e^{2x} + xe^{2x} = e^{2x}$ and the general solution is
$y(x) = y_c(x) + y_p(x) = c_1 e^x + c_2 x e^x + e^{2x}$.

23. As in Example 6, $y_c(x) = c_1 \sin x + c_2 \cos x$, so set $y_1 = \sin x$, $y_2 = \cos x$. Then
$y_1 y_2' - y_2 y_1' = -\sin^2 x - \cos^2 x = -1$, so $u_1' = -\dfrac{\sec x \cos x}{-1} = 1 \Rightarrow u_1(x) = x$ and
$u_2' = \dfrac{\sec x \sin x}{-1} = -\tan x \Rightarrow u_2(x) = -\int \tan x\, dx = \ln|\cos x| = \ln(\cos x)$ on $0 < x < \frac{\pi}{2}$. Hence
$y_p(x) = x\sin x + \cos x \ln(\cos x)$ and the general solution is $y(x) = (c_1 + x)\sin x + [c_2 + \ln(\cos x)]\cos x$.

25. $y_1 = e^x$, $y_2 = e^{2x}$ and $y_1 y_2' - y_2 y_1' = e^{3x}$. So $u_1' = \dfrac{-e^{2x}}{(1+e^{-x})e^{3x}} = -\dfrac{e^{-x}}{1+e^{-x}}$ and

$u_1(x) = \int -\dfrac{e^{-x}}{1+e^{-x}}\, dx = \ln(1+e^{-x})$. $u_2' = \dfrac{e^x}{(1+e^{-x})e^{3x}} = \dfrac{e^x}{e^{3x} + e^{2x}}$ so

$u_2(x) = \int \dfrac{e^x}{e^{3x} + e^{2x}}\, dx = \ln\left(\dfrac{e^x + 1}{e^x}\right) - e^{-x} = \ln(1+e^{-x}) - e^{-x}$. Hence

$y_p(x) = e^x \ln(1+e^{-x}) + e^{2x}[\ln(1+e^{-x}) - e^{-x}]$ and the general solution is
$y(x) = [c_1 + \ln(1+e^{-x})]e^x + [c_2 - e^{-x} + \ln(1+e^{-x})]e^{2x}$.

27. $y_1 = e^{-x}$, $y_2 = e^x$ and $y_1 y_2' - y_2 y_1' = 2$. So $u_1' = -\dfrac{e^x}{2x}$, $u_2' = \dfrac{e^{-x}}{2x}$ and

$$y_p(x) = -e^{-x} \int \frac{e^x}{2x}\, dx + e^x \int \frac{e^{-x}}{2x}\, dx.$$ Hence the general solution is

$$y(x) = \left(c_1 - \int \frac{e^x}{2x}\, dx \right) e^{-x} + \left(c_2 + \int \frac{e^{-x}}{2x}\, dx \right) e^x.$$

18.3 Applications of Second-Order Differential Equations   ET 17.3

1. By Hooke's Law $k(0.6) = 20$ so $k = \frac{100}{3}$ is the spring constant and the differential equation is $3x'' + \frac{100}{3}x = 0$.
The general solution is $x(t) = c_1 \cos\left(\frac{10}{3}t\right) + c_2 \sin\left(\frac{10}{3}t\right)$. But $0 = x(0) = c_1$ and $1.2 = x'(0) = \frac{10}{3}c_2$, so the
position of the mass after t seconds is $x(t) = 0.36 \sin\left(\frac{10}{3}t\right)$.

3. $k(0.5) = 6$ or $k = 12$ is the spring constant, so the initial-value problem is $2x'' + 14x' + 12x = 0$, $x(0) = 1$,
$x'(0) = 0$. The general solution is $x(t) = c_1 e^{-6t} + c_2 e^{-t}$. But $1 = x(0) = c_1 + c_2$ and $0 = x'(0) = -6c_1 - c_2$.
Thus the position is given by $x(t) = -\frac{1}{5}e^{-6t} + \frac{6}{5}e^{-t}$.

5. For critical damping we need $c^2 - 4mk = 0$ or $m = c^2/(4k) = 14^2/(4 \cdot 12) = \frac{49}{12}$ kg.

7. We are given $m = 1$, $k = 100$, $x(0) = -0.1$ and $x'(0) = 0$. From (3), the differential equation is
$$\frac{d^2x}{dt^2} + c\frac{dx}{dt} + 100x = 0$$ with auxiliary equation $r^2 + cr + 100 = 0$. If $c = 10$, we have two complex roots
$r = -5 \pm 5\sqrt{3}i$, so the motion is underdamped and the solution is $x = e^{-5t}\left[c_1 \cos\left(5\sqrt{3}t\right) + c_2 \sin\left(5\sqrt{3}t\right)\right]$.
Then $-0.1 = x(0) = c_1$ and $0 = x'(0) = 5\sqrt{3}c_2 - 5c_1 \quad \Rightarrow \quad c_2 = -\frac{1}{10\sqrt{3}}$, so
$x = e^{-5t}\left[-0.1\cos\left(5\sqrt{3}t\right) - \frac{1}{10\sqrt{3}}\sin\left(5\sqrt{3}t\right)\right]$. If $c = 15$, we again have underdamping since the auxiliary
equation has roots $r = -\frac{15}{2} \pm \frac{5\sqrt{7}}{2}i$. The general solution is $x = e^{-15t/2}\left[c_1 \cos\left(\frac{5\sqrt{7}}{2}t\right) + c_2 \sin\left(\frac{5\sqrt{7}}{2}t\right)\right]$, so
$-0.1 = x(0) = c_1$ and $0 = x'(0) = \frac{5\sqrt{7}}{2}c_2 - \frac{15}{2}c_1 \quad \Rightarrow \quad c_2 = -\frac{3}{10\sqrt{7}}$. Thus
$x = e^{-15t/2}\left[-0.1\cos\left(\frac{5\sqrt{7}}{2}t\right) - \frac{3}{10\sqrt{7}}\sin\left(\frac{5\sqrt{7}}{2}t\right)\right]$. For $c = 20$, we have equal roots $r_1 = r_2 = -10$,
so the oscillation is critically damped and the solution is $x = (c_1 + c_2 t)e^{-10t}$. Then $-0.1 = x(0) = c_1$ and
$0 = x'(0) = -10c_1 + c_2 \quad \Rightarrow \quad c_2 = -1$, so $x = (-0.1 - t)e^{-10t}$. If $c = 25$ the auxiliary equation has roots
$r_1 = -5$, $r_2 = -20$, so we have overdamping and the solution is $x = c_1 e^{-5t} + c_2 e^{-20t}$. Then
$-0.1 = x(0) = c_1 + c_2$ and $0 = x'(0) = -5c_1 - 20c_2 \quad \Rightarrow \quad c_1 = -\frac{2}{15}$ and $c_2 = \frac{1}{30}$,
so $x = -\frac{2}{15}e^{-5t} + \frac{1}{30}e^{-20t}$. If $c = 30$ we have roots

$r = -15 \pm 5\sqrt{5}$, so the motion is overdamped and the

solution is $x = c_1 e^{\left(-15+5\sqrt{5}\right)t} + c_2 e^{\left(-15-5\sqrt{5}\right)t}$. Then

$-0.1 = x(0) = c_1 + c_2$ and

$0 = x'(0) = \left(-15 + 5\sqrt{5}\right)c_1 + \left(-15 - 5\sqrt{5}\right)c_2 \quad \Rightarrow$

$c_1 = \frac{-5 - 3\sqrt{5}}{100}$ and $c_2 = \frac{-5 + 3\sqrt{5}}{100}$, so

$x = \left(\frac{-5 - 3\sqrt{5}}{100}\right)e^{\left(-15+5\sqrt{5}\right)t} + \left(\frac{-5 + 3\sqrt{5}}{100}\right)e^{\left(-15-5\sqrt{5}\right)t}$.

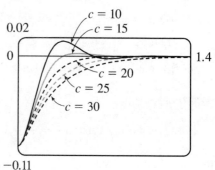

9. The differential equation is $mx'' + kx = F_0 \cos \omega_0 t$ and $\omega_0 \neq \omega = \sqrt{k/m}$. Here the auxiliary equation is

$mr^2 + k = 0$ with roots $\pm\sqrt{k/m}\, i = \pm\omega i$ so $x_c(t) = c_1 \cos \omega t + c_2 \sin \omega t$. Since $\omega_0 \neq \omega$, try

$x_p(t) = A \cos \omega_0 t + B \sin \omega_0 t$. Then we need

$(m)\left(-\omega_0^2\right)(A \cos \omega_0 t + B \sin \omega_0 t) + k(A \cos \omega_0 t + B \sin \omega_0 t) = F_0 \cos \omega_0 t$ or $A\left(k - m\omega_0^2\right) = F_0$ and

$B\left(k - m\omega_0^2\right) = 0$. Hence $B = 0$ and $A = \dfrac{F_0}{k - m\omega_0^2} = \dfrac{F_0}{m(\omega^2 - \omega_0^2)}$ since $\omega^2 = \dfrac{k}{m}$. Thus the motion of the

mass is given by $x(t) = c_1 \cos \omega t + c_2 \sin \omega t + \dfrac{F_0}{m(\omega^2 - \omega_0^2)} \cos \omega_0 t$.

11. From Equation 6, $x(t) = f(t) + g(t)$ where $f(t) = c_1 \cos \omega t + c_2 \sin \omega t$ and $g(t) = \dfrac{F_0}{m(\omega^2 - \omega_0^2)} \cos \omega_0 t$. Then

f is periodic, with period $\frac{2\pi}{\omega}$, and if $\omega \neq \omega_0$, g is periodic with period $\frac{2\pi}{\omega_0}$. If $\frac{\omega}{\omega_0}$ is a rational number, then we can

say $\frac{\omega}{\omega_0} = \frac{a}{b} \;\Rightarrow\; a = \frac{b\omega}{\omega_0}$ where a and b are non-zero integers. Then

$$x\left(t + a \cdot \tfrac{2\pi}{\omega}\right) = f\left(t + a \cdot \tfrac{2\pi}{\omega}\right) + g\left(t + a \cdot \tfrac{2\pi}{\omega}\right) = f(t) + g\left(t + \tfrac{b\omega}{\omega_0} \cdot \tfrac{2\pi}{\omega}\right)$$

$$= f(t) + g\left(t + b \cdot \tfrac{2\pi}{\omega_0}\right) = f(t) + g(t) = x(t)$$

so $x(t)$ is periodic.

13. Here the initial-value problem for the charge is $Q'' + 20Q' + 500Q = 12$, $Q(0) = Q'(0) = 0$. Then

$Q_c(t) = e^{-10t}(c_1 \cos 20t + c_2 \sin 20t)$ and try $Q_p(t) = A \;\Rightarrow\; 500A = 12$ or $A = \frac{3}{125}$.

The general solution is $Q(t) = e^{-10t}(c_1 \cos 20t + c_2 \sin 20t) + \frac{3}{125}$. But $0 = Q(0) = c_1 + \frac{3}{125}$ and

$Q'(t) = I(t) = e^{-10t}[(-10c_1 + 20c_2) \cos 20t + (-10c_2 - 20c_1) \sin 20t]$ but $0 = Q'(0) = -10c_1 + 20c_2$. Thus

the charge is $Q(t) = -\frac{1}{250}e^{-10t}(6 \cos 20t + 3 \sin 20t) + \frac{3}{125}$ and the current is $I(t) = e^{-10t}\left(\frac{3}{5}\right) \sin 20t$.

15. As in Exercise 13, $Q_c(t) = e^{-10t}(c_1 \cos 20t + c_2 \sin 20t)$ but $E(t) = 12 \sin 10t$ so try

$Q_p(t) = A \cos 10t + B \sin 10t$. Substituting into the differential equation gives

$(-100A + 200B + 500A) \cos 10t + (-100B - 200A + 500B) \sin 10t = 12 \sin 10t \;\Rightarrow\; 400A + 200B = 0$

and $400B - 200A = 12$. Thus $A = -\frac{3}{250}$, $B = \frac{3}{125}$ and the general solution is

$Q(t) = e^{-10t}(c_1 \cos 20t + c_2 \sin 20t) - \frac{3}{250} \cos 10t + \frac{3}{125} \sin 10t$. But $0 = Q(0) = c_1 - \frac{3}{250}$ so $c_1 = \frac{3}{250}$.

Also $Q'(t) = \frac{3}{25} \sin 10t + \frac{6}{25} \cos 10t + e^{-10t}[(-10c_1 + 20c_2) \cos 20t + (-10c_2 - 20c_1) \sin 20t]$ and

$0 = Q'(0) = \frac{6}{25} - 10c_1 + 20c_2$ so $c_2 = -\frac{3}{500}$. Hence the charge is given by

$Q(t) = e^{-10t}\left[\frac{3}{250} \cos 20t - \frac{3}{500} \sin 20t\right] - \frac{3}{250} \cos 10t + \frac{3}{125} \sin 10t$.

17. $x(t) = A \cos(\omega t + \delta) \;\Leftrightarrow\; x(t) = A[\cos \omega t \cos \delta - \sin \omega t \sin \delta] \;\Leftrightarrow\; x(t) = A\left(\frac{c_1}{A} \cos \omega t + \frac{c_2}{A} \sin \omega t\right)$

where $\cos \delta = c_1/A$ and $\sin \delta = -c_2/A \;\Leftrightarrow\; x(t) = c_1 \cos \omega t + c_2 \sin \omega t$. (Note that $\cos^2 \delta + \sin^2 \delta = 1 \;\Rightarrow$

$c_1^2 + c_2^2 = A^2$.)

18.4 Series Solutions

ET 17.4

1. Let $y(x) = \sum\limits_{n=0}^{\infty} c_n x^n$. Then $y'(x) = \sum\limits_{n=1}^{\infty} n c_n x^{n-1}$ and the given equation, $y' - y = 0$, becomes

$\sum\limits_{n=1}^{\infty} n c_n x^{n-1} - \sum\limits_{n=0}^{\infty} c_n x^n = 0$. Replacing n by $n+1$ in the first sum gives $\sum\limits_{n=0}^{\infty} (n+1) c_{n+1} x^n - \sum\limits_{n=0}^{\infty} c_n x^n = 0$,

so $\sum\limits_{n=0}^{\infty} [(n+1)c_{n+1} - c_n] x^n = 0$. Equating coefficients gives $(n+1)c_{n+1} - c_n = 0$, so the recursion relation is

$c_{n+1} = \dfrac{c_n}{n+1}$, $n = 0, 1, 2, \ldots$. Then $c_1 = c_0$, $c_2 = \dfrac{1}{2}c_1 = \dfrac{c_0}{2}$, $c_3 = \dfrac{1}{3}c_2 = \dfrac{1}{3} \cdot \dfrac{1}{2}c_0 = \dfrac{c_0}{3!}$, $c_4 = \dfrac{1}{4}c_3 = \dfrac{c_0}{4!}$, and

in general, $c_n = \dfrac{c_0}{n!}$. Thus, the solution is

$$y(x) = \sum_{n=0}^{\infty} c_n x^n = \sum_{n=0}^{\infty} \frac{c_0}{n!} x^n = c_0 \sum_{n=0}^{\infty} \frac{x^n}{n!} = c_0 e^x$$

3. Assuming $y(x) = \sum\limits_{n=0}^{\infty} c_n x^n$, we have $y'(x) = \sum\limits_{n=1}^{\infty} n c_n x^{n-1} = \sum\limits_{n=0}^{\infty} (n+1) c_{n+1} x^n$ and

$-x^2 y = -\sum\limits_{n=0}^{\infty} c_n x^{n+2} = -\sum\limits_{n=2}^{\infty} c_{n-2} x^n$. Hence, the equation $y' = x^2 y$ becomes

$\sum\limits_{n=0}^{\infty} (n+1) c_{n+1} x^n - \sum\limits_{n=2}^{\infty} c_{n-2} x^n = 0$ or $c_1 + 2c_2 x + \sum\limits_{n=2}^{\infty} [(n+1)c_{n+1} - c_{n-2}] x^n = 0$. Equating coefficients

gives $c_1 = c_2 = 0$ and $c_{n+1} = \dfrac{c_{n-2}}{n+1}$ for $n = 2, 3, \ldots$. But $c_1 = 0$, so $c_4 = 0$ and $c_7 = 0$ and in general

$c_{3n+1} = 0$. Similarly $c_2 = 0$ so $c_{3n+2} = 0$. Finally $c_3 = \dfrac{c_0}{3}$, $c_6 = \dfrac{c_3}{6} = \dfrac{c_0}{6 \cdot 3} = \dfrac{c_0}{3^2 \cdot 2!}$,

$c_9 = \dfrac{c_6}{9} = \dfrac{c_0}{9 \cdot 6 \cdot 3} = \dfrac{c_0}{3^3 \cdot 3!}, \ldots$, and $c_{3n} = \dfrac{c_0}{3^n \cdot n!}$. Thus, the solution is

$$y(x) = \sum_{n=0}^{\infty} c_n x^n = \sum_{n=0}^{\infty} c_{3n} x^{3n} = \sum_{n=0}^{\infty} \frac{c_0}{3^n \cdot n!} x^{3n} = c_0 \sum_{n=0}^{\infty} \frac{x^{3n}}{3^n n!} = c_0 \sum_{n=0}^{\infty} \frac{(x^3/3)^n}{n!} = c_0 e^{x^3/3}$$

5. Let $y(x) = \sum\limits_{n=0}^{\infty} c_n x^n \;\Rightarrow\; y'(x) = \sum\limits_{n=1}^{\infty} n c_n x^{n-1}$ and $y''(x) = \sum\limits_{n=0}^{\infty} (n+2)(n+1) c_{n+2} x^n$. The

differential equation becomes $\sum\limits_{n=0}^{\infty} (n+2)(n+1) c_{n+2} x^n + x \sum\limits_{n=1}^{\infty} n c_n x^{n-1} + \sum\limits_{n=0}^{\infty} c_n x^n = 0$ or

$\sum\limits_{n=0}^{\infty} [(n+2)(n+1) c_{n+2} + n c_n + c_n] x^n$ $\left(\text{since } \sum\limits_{n=1}^{\infty} n c_n x^n = \sum\limits_{n=0}^{\infty} n c_n x^n \right)$. Equating coefficients gives

$(n+2)(n+1) c_{n+2} + (n+1) c_n = 0$, thus the recursion relation is $c_{n+2} = \dfrac{-(n+1)c_n}{(n+2)(n+1)} = -\dfrac{c_n}{n+2}$,

$n = 0, 1, 2, \ldots$. Then the even coefficients are given by $c_2 = -\dfrac{c_0}{2}$, $c_4 = -\dfrac{c_2}{4} = \dfrac{c_0}{2 \cdot 4}$, $c_6 = -\dfrac{c_4}{6} = -\dfrac{c_0}{2 \cdot 4 \cdot 6}$,

and in general, $c_{2n} = (-1)^n \dfrac{c_0}{2 \cdot 4 \cdot \cdots \cdot 2n} = \dfrac{(-1)^n c_0}{2^n n!}$. The odd coefficients are $c_3 = -\dfrac{c_1}{3}$, $c_5 = -\dfrac{c_3}{5} = \dfrac{c_1}{3 \cdot 5}$,

$c_7 = -\dfrac{c_5}{7} = -\dfrac{c_1}{3 \cdot 5 \cdot 7}$, and in general, $c_{2n+1} = (-1)^n \dfrac{c_1}{3 \cdot 5 \cdot 7 \cdot \cdots \cdot (2n+1)} = \dfrac{(-2)^n n! c_1}{(2n+1)!}$. The solution is

$$y(x) = c_0 \sum_{n=0}^{\infty} \frac{(-1)^n}{2^n n!} x^{2n} + c_1 \sum_{n=0}^{\infty} \frac{(-2)^n n!}{(2n+1)!} x^{2n+1}.$$

7. Let $y(x) = \sum_{n=0}^{\infty} c_n x^n$. Then $y'' = \sum_{n=0}^{\infty} n(n-1)c_n x^{n-2}$, $xy' = \sum_{n=0}^{\infty} n c_n x^n$ and

$(x^2+1)y'' = \sum_{n=0}^{\infty} n(n-1)c_n x^n + \sum_{n=0}^{\infty} (n+2)(n+1)c_{n+2}x^n$. The differential equation becomes

$\sum_{n=0}^{\infty} [(n+2)(n+1)c_{n+2} + [n(n-1)+n-1]c_n]x^n = 0$. The recursion relation is $c_{n+2} = -\dfrac{(n-1)c_n}{n+2}$,

$n = 0, 1, 2, \ldots$. Given c_0 and c_1, $c_2 = \dfrac{c_0}{2}$, $c_4 = -\dfrac{c_2}{4} = -\dfrac{c_0}{2^2 \cdot 2!}$, $c_6 = -\dfrac{3c_4}{6} = (-1)^2 \dfrac{3c_0}{2^3 \cdot 3!}, \ldots,$

$c_{2n} = (-1)^{n-1} \dfrac{1 \cdot 3 \cdots \cdots (2n-3)\, c_0}{2^n\, n!} = (-1)^{n-1} \dfrac{(2n-3)!\, c_0}{2^n 2^{n-2}\, n!\,(n-2)!} = (-1)^{n-1} \dfrac{(2n-3)!\, c_0}{2^{2n-2}\, n!\,(n-2)!}$ for

$n = 2, 3, \ldots$. $c_3 = \dfrac{0 \cdot c_1}{3} = 0 \;\Rightarrow\; c_{2n+1} = 0$ for $n = 1, 2, \ldots$. Thus the solution is

$y(x) = c_0 + c_1 x + c_0 \dfrac{x^2}{2} + c_0 \displaystyle\sum_{n=2}^{\infty} \dfrac{(-1)^{n-1}(2n-3)!}{2^{2n-2}\, n!\,(n-2)!}\, x^{2n}.$

9. Let $y(x) = \displaystyle\sum_{n=0}^{\infty} c_n x^n$. Then $-xy'(x) = -x \displaystyle\sum_{n=1}^{\infty} n c_n x^{n-1} = -\displaystyle\sum_{n=1}^{\infty} n c_n x^n = -\displaystyle\sum_{n=0}^{\infty} n c_n x^n,$

$y''(x) = \displaystyle\sum_{n=0}^{\infty} (n+2)(n+1)c_{n+2}x^n$, and the equation $y'' - xy' - y = 0$ becomes

$\displaystyle\sum_{n=0}^{\infty} [(n+2)(n+1)c_{n+2} - n c_n - c_n]x^n = 0$. Thus, the recursion relation is

$c_{n+2} = \dfrac{n c_n + c_n}{(n+2)(n+1)} = \dfrac{c_n(n+1)}{(n+2)(n+1)} = \dfrac{c_n}{n+2}$ for $n = 0, 1, 2, \ldots$. One of the given conditions is

$y(0) = 1$. But $y(0) = \displaystyle\sum_{n=0}^{\infty} c_n(0)^n = c_0 + 0 + 0 + \cdots = c_0$, so $c_0 = 1$. Hence, $c_2 = \dfrac{c_0}{2} = \dfrac{1}{2}$, $c_4 = \dfrac{c_2}{4} = \dfrac{1}{2 \cdot 4}$,

$c_6 = \dfrac{c_4}{6} = \dfrac{1}{2 \cdot 4 \cdot 6}, \ldots, c_{2n} = \dfrac{1}{2^n n!}$. The other given condition is $y'(0) = 0$. But

$y'(0) = \displaystyle\sum_{n=1}^{\infty} n c_n(0)^{n-1} = c_1 + 0 + 0 + \cdots = c_1$, so $c_1 = 0$. By the recursion relation, $c_3 = \dfrac{c_1}{3} = 0$, $c_5 = 0, \ldots,$

$c_{2n+1} = 0$ for $n = 0, 1, 2, \ldots$. Thus, the solution to the initial-value problem is

$$y(x) = \sum_{n=0}^{\infty} c_n x^n = \sum_{n=0}^{\infty} c_{2n}x^{2n} = \sum_{n=0}^{\infty} \dfrac{x^{2n}}{2^n n!} = \sum_{n=0}^{\infty} \dfrac{(x^2/2)^n}{n!} = e^{x^2/2}$$

11. Assuming that $y(x) = \displaystyle\sum_{n=0}^{\infty} c_n x^n$, we have $xy = x \displaystyle\sum_{n=0}^{\infty} c_n x^n = \displaystyle\sum_{n=0}^{\infty} c_n x^{n+1}$,

$x^2 y' = x^2 \displaystyle\sum_{n=1}^{\infty} n c_n x^{n-1} = \displaystyle\sum_{n=0}^{\infty} n c_n x^{n+1}$,

$y''(x) = \displaystyle\sum_{n=2}^{\infty} n(n-1)c_n x^{n-2} = \displaystyle\sum_{n=-1}^{\infty} (n+3)(n+2)c_{n+3}x^{n+1}$ [replace n with $n+3$]

$= 2c_2 + \displaystyle\sum_{n=0}^{\infty} (n+3)(n+2)c_{n+3}x^{n+1}$, and the equation $y'' + x^2 y' + xy = 0$ becomes $2c_2 + \sum_{n=0}^{\infty} [(n+3)(n+2)c_{n+3} + nc_n + c_n] x^{n+1} = 0$.

So $c_2 = 0$ and the recursion relation is $c_{n+3} = \dfrac{-nc_n - c_n}{(n+3)(n+2)} = -\dfrac{(n+1)c_n}{(n+3)(n+2)}$, $n = 0, 1, 2, \ldots$.

But $c_0 = y(0) = 0 = c_2$ and by the recursion relation, $c_{3n} = c_{3n+2} = 0$ for $n = 0, 1, 2, \ldots$.

Also, $c_1 = y'(0) = 1$, so

$$c_4 = -\frac{2c_1}{4 \cdot 3} = -\frac{2}{4 \cdot 3}, \ c_7 = -\frac{5c_4}{7 \cdot 6} = (-1)^2 \frac{2 \cdot 5}{7 \cdot 6 \cdot 4 \cdot 3} = (-1)^2 \frac{2^2 5^2}{7!}, \ldots,$$

$$c_{3n+1} = (-1)^n \frac{2^2 5^2 \cdots (3n-1)^2}{(3n+1)!}. \text{ Thus, the solution is}$$

$$y(x) = \sum_{n=0}^{\infty} c_n x^n = x + \sum_{n=1}^{\infty} \left[(-1)^n \frac{2^2 5^2 \cdots (3n-1)^2 x^{3n+1}}{(3n+1)!} \right]$$

18 Review   ET 17

————————————— CONCEPT CHECK —————————————

1. (a) $ay'' + by' + cy = 0$ where a, b, and c are constants.

(b) $ar^2 + br + c = 0$

(c) If the auxiliary equation has two distinct real roots r_1 and r_2, the solution is $y = c_1 e^{r_1 x} + c_2 e^{r_2 x}$. If the roots are real and equal, the solution is $y = c_1 e^{rx} + c_2 x e^{rx}$ where r is the common root. If the roots are complex, we can write $r_1 = \alpha + i\beta$ and $r_2 = \alpha - i\beta$, and the solution is $y = e^{\alpha x}(c_1 \cos \beta x + c_2 \sin \beta x)$.

2. (a) An initial-value problem consists of finding a solution y of a second-order differential equation that also satisfies given conditions $y(x_0) = y_0$ and $y'(x_0) = y_1$, where y_0 and y_1 are constants.

(b) A boundary-value problem consists of finding a solution y of a second-order differential equation that also satisfies given boundary conditions $y(x_0) = y_0$ and $y(x_1) = y_1$.

3. (a) $ay'' + by' + cy = G(x)$ where a, b, and c are constants and G is a continuous function.

(b) The complementary equation is the related homogeneous equation $ay'' + by' + cy = 0$. If we find the general solution y_c of the complementary equation and y_p is any particular solution of the original differential equation, then the general solution of the original differential equation is $y(x) = y_p(x) + y_c(x)$.

(c) See Examples 1–5 and the associated discussion in Section 18.2 [ET 17.2].

(d) See the discussion on pages 1188–1190 [ET 1152-1154].

4. Second-order linear differential equations can be used to describe the motion of a vibrating spring or to analyze an electric circuit; see the discussion in Section 18.3 [ET 17.3].

5. See Example 1 and the preceding discussion in Section 18.4 [ET 17.4].

─────────────────── TRUE-FALSE QUIZ ───────────────────

1. True. See Theorem 18.1.3 [ET 17.1.3].

3. True. $\cosh x$ and $\sinh x$ are linearly independent solutions of this linear homogeneous equation.

─────────────────── EXERCISES ───────────────────

1. The auxiliary equation is $r^2 - 2r - 15 = 0 \Rightarrow (r - 5)(r + 3) = 0 \Rightarrow r = 5, r = -3$. Then the general

solution is $y = c_1 e^{5x} + c_2 e^{-3x}$.

3. The auxiliary equation is $r^2 + 3 = 0 \Rightarrow r = \pm\sqrt{3}\,i$. Then the general solution is

$y = c_1 \cos(\sqrt{3}\,x) + c_2 \sin(\sqrt{3}\,x)$.

5. $r^2 - 4r + 5 = 0 \Rightarrow r = 2 \pm i$, so $y_c(x) = e^{2x}(c_1 \cos x + c_2 \sin x)$. Try $y_p(x) = Ae^{2x} \Rightarrow y_p' = 2Ae^{2x}$

and $y_p'' = 4Ae^{2x}$. Substitution into the differential equation gives $4Ae^{2x} - 8Ae^{2x} + 5Ae^{2x} = e^{2x} \Rightarrow A = 1$

and the general solution is $y(x) = e^{2x}(c_1 \cos x + c_2 \sin x) + e^{2x}$.

7. $r^2 - 2r + 1 = 0 \Rightarrow r = 1$ and $y_c(x) = c_1 e^x + c_2 x e^x$. Try $y_p(x) = (Ax + B)\cos x + (Cx + D)\sin x \Rightarrow$

$y_p' = (C - Ax - B)\sin x + (A + Cx + D)\cos x$ and $y_p'' = (2C - B - Ax)\cos x + (-2A - D - Cx)\sin x$.

Substitution gives $(-2Cx + 2C - 2A - 2D)\cos x + (2Ax - 2A + 2B - 2C)\sin x = x \cos x \Rightarrow A = 0$,

$B = C = D = -\frac{1}{2}$. The general solution is $y(x) = c_1 e^x + c_2 x e^x - \frac{1}{2}\cos x - \frac{1}{2}(x + 1)\sin x$.

9. $r^2 - r - 6 = 0 \Rightarrow r = -2, r = 3$ and $y_c(x) = c_1 e^{-2x} + c_2 e^{3x}$. For $y'' - y' - 6y = 1$, try $y_{p_1}(x) = A$. Then

$y_{p_1}'(x) = y_{p_1}''(x) = 0$ and substitution into the differential equation gives $A = -\frac{1}{6}$. For $y'' - y' - 6y = e^{-2x}$ try

$y_{p_2}(x) = Bxe^{-2x}$ (since $y = Be^{-2x}$ satisfies the complementary equation). Then $y_{p_2}' = (B - 2Bx)e^{-2x}$ and

$y_{p_2}'' = (4Bx - 4B)e^{-2x}$, and substitution gives $-5Be^{-2x} = e^{-2x} \Rightarrow B = -\frac{1}{5}$. The general solution then is

$y(x) = c_1 e^{-2x} + c_2 e^{3x} + y_{p_1}(x) + y_{p_2}(x) = c_1 e^{-2x} + c_2 e^{3x} - \frac{1}{6} - \frac{1}{5}xe^{-2x}$.

11. The auxiliary equation is $r^2 + 6r = 0$ and the general solution is $y(x) = c_1 + c_2 e^{-6x} = k_1 + k_2 e^{-6(x-1)}$. But

$3 = y(1) = k_1 + k_2$ and $12 = y'(1) = -6k_2$. Thus $k_2 = -2, k_1 = 5$ and the solution is $y(x) = 5 - 2e^{-6(x-1)}$.

13. The auxiliary equation is $r^2 - 5r + 4 = 0$ and the general solution is $y(x) = c_1 e^x + c_2 e^{4x}$. But

$0 = y(0) = c_1 + c_2$ and $1 = y'(0) = c_1 + 4c_2$, so the solution is $y(x) = \frac{1}{3}(e^{4x} - e^x)$.

15. Let $y(x) = \sum_{n=0}^{\infty} c_n x^n$. Then $y''(x) = \sum_{n=0}^{\infty} n(n-1)c_n x^{n-2} = \sum_{n=0}^{\infty}(n+2)(n+1)c_{n+2}x^n$ and the

differential equation becomes $\sum_{n=0}^{\infty}[(n+2)(n+1)c_{n+2} + (n+1)c_n]x^n = 0$. Thus the recursion relation is

$c_{n+2} = -c_n/(n+2)$ for $n = 0, 1, 2, \ldots$. But $c_0 = y(0) = 0$, so $c_{2n} = 0$ for $n = 0, 1, 2, \ldots$. Also

$c_1 = y'(0) = 1$, so $c_3 = -\frac{1}{3}$, $c_5 = \frac{(-1)^2}{3 \cdot 5}$, $c_7 = \frac{(-1)^3}{3 \cdot 5 \cdot 7} = \frac{(-1)^3 2^3 3!}{7!}, \ldots, c_{2n+1} = \frac{(-1)^n 2^n n!}{(2n+1)!}$ for

$n = 0, 1, 2, \ldots$. Thus the solution to the initial-value problem is $y(x) = \sum_{n=0}^{\infty} c_n x^n = \sum_{n=0}^{\infty} \frac{(-1)^n 2^n n!}{(2n+1)!} x^{2n+1}$.

17. Here the initial-value problem is $2Q'' + 40Q' + 400Q = 12$, $Q(0) = 0.01$, $Q'(0) = 0!$. Then

$Q_c(t) = e^{-10t}(c_1 \cos 10t + c_2 \sin 10t)$ and we try $Q_p(t) = A!$. Thus the general solution is

$Q(t) = e^{-10t}(c_1 \cos 10t + c_2 \sin 10t) + \frac{3}{100}!$. But $0.01 = Q'(0) = c_1 + 0.03$ and $0 = Q''(0) = -10c_1 + 10c_2$,

so $c_1 = -0.02 = c_2!$. Hence the charge is given by $Q(t) = -0.02e^{-10t}(\cos 10t + \sin 10t) + 0.03$.

19. (a) Since we are assuming that the earth is a solid sphere of uniform density, we can calculate the density ρ as

follows: $\rho = \dfrac{\text{mass of earth}}{\text{volume of earth}} = \dfrac{M}{\frac{4}{3}\pi R^3}$. If V_r is the volume of the portion of the earth which lies within a

distance r of the center, then $V_r = \frac{4}{3}\pi r^3$ and $M_r = \rho V_r = \dfrac{Mr^3}{R^3}$. Thus $F_r = -\dfrac{GM_r m}{r^2} = -\dfrac{GMm}{R^3}r$.

(b) The particle is acted upon by a varying gravitational force during its motion. By Newton's Second Law of

Motion, $m\dfrac{d^2y}{dt^2} = F_y = -\dfrac{GMm}{R^3}y$, so $y''(t) = -k^2y(t)$ where $k^2 = \dfrac{GM}{R^3}$. At the surface,

$-mg = F_R = -\dfrac{GMm}{R^2}$, so $g = \dfrac{GM}{R^2}$. Therefore $k^2 = \dfrac{g}{R}$.

(c) The differential equation $y'' + k^2y = 0$ has auxiliary equation $r^2 + k^2 = 0$. (This is the r of Section 18.1

[ET 17.1], not the r measuring distance from the earth's center.) The roots of the auxiliary equation are $\pm ik$, so

by (11) in Section 18.1 [ET 17.1], the general solution of our differential equation for t is

$y(t) = c_1 \cos kt + c_2 \sin kt$. It follows that $y'(t) = -c_1 k \sin kt + c_2 k \cos kt$. Now $y(0) = R$ and $y'(0) = 0$,

so $c_1 = R$ and $c_2 k = 0$. Thus $y(t) = R \cos kt$ and $y'(t) = -kR \sin kt$. This is simple harmonic motion (see

Section 18.3 [ET 17.3]) with amplitude R, frequency k, and phase angle 0. The period is $T = 2\pi/k$.

$R \approx 3960$ mi $= 3960 \cdot 5280$ ft and $g = 32$ ft/s^2, so $k = \sqrt{g/R} \approx 1.24 \times 10^{-3}s^{-1}$ and

$T = 2\pi/k \approx 5079$ s ≈ 85 min.

(d) $y(t) = 0 \Leftrightarrow \cos kt = 0 \Leftrightarrow kt = \frac{\pi}{2} + \pi n$ for some integer $n \Rightarrow$

$y'(t) = -kR \sin\left(\frac{\pi}{2} + \pi n\right) = \pm kR$. Thus the particle passes through the center of the earth with speed

$kR \approx 4.899$ mi/s $\approx 17{,}600$ mi/h.

☐ APPENDIX

G Complex Numbers

---

1. $(5 - 6i) + (3 + 2i) = (5 + 3) + (-6 + 2)i = 8 + (-4)i = 8 - 4i$

3. $(2 + 5i)(4 - i) = 2(4) + 2(-i) + (5i)(4) + (5i)(-i) = 8 - 2i + 20i - 5i^2$
$$= 8 + 18i - 5(-1) = 8 + 18i + 5 = 13 + 18i$$

5. $\overline{12 + 7i} = 12 - 7i$

7. $\dfrac{1 + 4i}{3 + 2i} = \dfrac{1 + 4i}{3 + 2i} \cdot \dfrac{3 - 2i}{3 - 2i} = \dfrac{3 - 2i + 12i - 8(-1)}{3^2 + 2^2} = \dfrac{11 + 10i}{13} = \dfrac{11}{13} + \dfrac{10}{13}i$

9. $\dfrac{1}{1 + i} = \dfrac{1}{1 + i} \cdot \dfrac{1 - i}{1 - i} = \dfrac{1 - i}{1 - (-1)} = \dfrac{1 - i}{2} = \dfrac{1}{2} - \dfrac{1}{2}i$

11. $i^3 = i^2 \cdot i = (-1)i = -i$

13. $\sqrt{-25} = \sqrt{25}\,i = 5i$

15. $\overline{12 - 5i} = 12 + 15i$ and $|12 - 15i| = \sqrt{12^2 + (-5)^2} = \sqrt{144 + 25} = \sqrt{169} = 13$

17. $\overline{-4i} = \overline{0 - 4i} = 0 + 4i = 4i$ and $|-4i| = \sqrt{0^2 + (-4)^2} = \sqrt{16} = 4$

19. $4x^2 + 9 = 0 \iff 4x^2 = -9 \iff x^2 = -\frac{9}{4} \iff x = \pm\sqrt{-\frac{9}{4}} = \pm\sqrt{\frac{9}{4}}\,i = \pm\frac{3}{2}i.$

21. $x^2 + 2x + 5 = 0 \iff x = \dfrac{-2 \pm \sqrt{2^2 - 4 \cdot 1 \cdot 5}}{2 \cdot 1} = \dfrac{-2 \pm \sqrt{-16}}{2} = \dfrac{-2 \pm 4i}{2} = -1 \pm 2i$

23. By the quadratic formula, $z^2 + z + 2 = 0 \iff z = \dfrac{-1 \pm \sqrt{1^2 - 4(1)(2)}}{2(1)} = \dfrac{-1 \pm \sqrt{-7}}{2} = -\dfrac{1}{2} \pm \dfrac{\sqrt{7}}{2}i.$

25. For $z = -3 + 3i$, $r = \sqrt{(-3)^2 + 3^2} = 3\sqrt{2}$ and $\tan\theta = \frac{3}{-3} = -1 \implies \theta = \frac{3\pi}{4}$ (since z lies in the second quadrant). Therefore, $-3 + 3i = 3\sqrt{2}\left(\cos\frac{3\pi}{4} + i\sin\frac{3\pi}{4}\right)$.

27. For $z = 3 + 4i$, $r = \sqrt{3^2 + 4^2} = 5$ and $\tan\theta = \frac{4}{3} \implies \theta = \tan^{-1}\left(\frac{4}{3}\right)$ (since z lies in the first quadrant). Therefore, $3 + 4i = 5\left[\cos\left(\tan^{-1}\frac{4}{3}\right) + i\sin\left(\tan^{-1}\frac{4}{3}\right)\right]$.

29. For $z = \sqrt{3} + i$, $r = \sqrt{\left(\sqrt{3}\right)^2 + 1^2} = 2$ and $\tan\theta = \frac{1}{\sqrt{3}} \implies \theta = \frac{\pi}{6} \implies z = 2\left(\cos\frac{\pi}{6} + i\sin\frac{\pi}{6}\right)$.

For $w = 1 + \sqrt{3}\,i$, $r = 2$ and $\tan\theta = \sqrt{3} \implies \theta = \frac{\pi}{3} \implies w = 2\left(\cos\frac{\pi}{3} + i\sin\frac{\pi}{3}\right)$.

Therefore, $zw = 2 \cdot 2\left[\cos\left(\frac{\pi}{6} + \frac{\pi}{3}\right) + i\sin\left(\frac{\pi}{6} + \frac{\pi}{3}\right)\right] = 4\left(\cos\frac{\pi}{2} + i\sin\frac{\pi}{2}\right)$,

$z/w = \frac{2}{2}\left[\cos\left(\frac{\pi}{6} - \frac{\pi}{3}\right) + i\sin\left(\frac{\pi}{6} - \frac{\pi}{3}\right)\right] = \cos\left(-\frac{\pi}{6}\right) + i\sin\left(-\frac{\pi}{6}\right)$, and $1 = 1 + 0i = 1(\cos 0 + i\sin 0) \implies$

$1/z = \frac{1}{2}\left[\cos\left(0 - \frac{\pi}{6}\right) + i\sin\left(0 - \frac{\pi}{6}\right)\right] = \frac{1}{2}\left[\cos\left(-\frac{\pi}{6}\right) + i\sin\left(-\frac{\pi}{6}\right)\right]$. For $1/z$, we could also use the formula that precedes Example 5 to obtain $1/z = \frac{1}{2}\left(\cos\frac{\pi}{6} - i\sin\frac{\pi}{6}\right)$.

31. For $z = 2\sqrt{3} - 2i$, $r = \sqrt{\left(2\sqrt{3}\right)^2 + (-2)^2} = 4$ and $\tan\theta = \frac{-2}{2\sqrt{3}} = -\frac{1}{\sqrt{3}}$

$\Rightarrow$ $\theta = -\frac{\pi}{6}$ $\Rightarrow$ $z = 4\left[\cos\left(-\frac{\pi}{6}\right) + i\sin\left(-\frac{\pi}{6}\right)\right]$. For $w = -1 + i$, $r = \sqrt{2}$,

$\tan\theta = \frac{1}{-1} = -1$ $\Rightarrow$ $\theta = \frac{3\pi}{4}$ $\Rightarrow$ $w = \sqrt{2}\left(\cos\frac{3\pi}{4} + i\sin\frac{3\pi}{4}\right)$. Therefore,

$zw = 4\sqrt{2}\left[\cos\left(-\frac{\pi}{6} + \frac{3\pi}{4}\right) + i\sin\left(-\frac{\pi}{6} + \frac{3\pi}{4}\right)\right] = 4\sqrt{2}\left(\cos\frac{7\pi}{12} + i\sin\frac{7\pi}{12}\right)$,

$z/w = \frac{4}{\sqrt{2}}\left[\cos\left(-\frac{\pi}{6} - \frac{3\pi}{4}\right) + i\sin\left(-\frac{\pi}{6} - \frac{3\pi}{4}\right)\right] = \frac{4}{\sqrt{2}}\left[\cos\left(-\frac{11\pi}{12}\right) + i\sin\left(-\frac{11\pi}{12}\right)\right]$

$= 2\sqrt{2}\left(\cos\frac{13\pi}{12} + i\sin\frac{13\pi}{12}\right)$, and

$1/z = \frac{1}{4}\left[\cos\left(-\frac{\pi}{6}\right) - i\sin\left(-\frac{\pi}{6}\right)\right] = \frac{1}{4}\left(\cos\frac{\pi}{6} + i\sin\frac{\pi}{6}\right)$.

33. For $z = 1 + i$, $r = \sqrt{2}$ and $\tan\theta = \frac{1}{1} = 1$ $\Rightarrow$ $\theta = \frac{\pi}{4}$ $\Rightarrow$ $z = \sqrt{2}\left(\cos\frac{\pi}{4} + i\sin\frac{\pi}{4}\right)$. So by

De Moivre's Theorem,

$$(1+i)^{20} = \left[\sqrt{2}\left(\cos\frac{\pi}{4} + i\sin\frac{\pi}{4}\right)\right]^{20} = \left(2^{1/2}\right)^{20}\left(\cos\frac{20\cdot\pi}{4} + i\sin\frac{20\cdot\pi}{4}\right)$$

$$= 2^{10}(\cos 5\pi + i\sin 5\pi) = 2^{10}[-1 + i(0)] = -2^{10} = -1024$$

35. For $z = 2\sqrt{3} + 2i$, $r = \sqrt{\left(2\sqrt{3}\right)^2 + 2^2} = \sqrt{16} = 4$ and $\tan\theta = \frac{2}{2\sqrt{3}} = \frac{1}{\sqrt{3}}$ $\Rightarrow$ $\theta = \frac{\pi}{6}$ $\Rightarrow$

$z = 4\left(\cos\frac{\pi}{6} + i\sin\frac{\pi}{6}\right)$. So by De Moivre's Theorem,

$$\left(2\sqrt{3} + 2i\right)^5 = \left[4\left(\cos\frac{\pi}{6} + i\sin\frac{\pi}{6}\right)\right]^5 = 4^5\left(\cos\frac{5\pi}{6} + i\sin\frac{5\pi}{6}\right) = 1024\left[-\frac{\sqrt{3}}{2} + \frac{1}{2}i\right] = -512\sqrt{3} + 512i$$

37. $1 = 1 + 0i = 1(\cos 0 + i\sin 0)$. Using Equation 3 with $r = 1$, $n = 8$, and $\theta = 0$, we have

$w_k = 1^{1/8}\left[\cos\left(\frac{0 + 2k\pi}{8}\right) + i\sin\left(\frac{0 + 2k\pi}{8}\right)\right] = \cos\frac{k\pi}{4} + i\sin\frac{k\pi}{4}$, where $k = 0, 1, 2, \ldots, 7$.

$w_0 = 1(\cos 0 + i\sin 0) = 1$, $w_1 = 1\left(\cos\frac{\pi}{4} + i\sin\frac{\pi}{4}\right) = \frac{1}{\sqrt{2}} + \frac{1}{\sqrt{2}}i$,

$w_2 = 1\left(\cos\frac{\pi}{2} + i\sin\frac{\pi}{2}\right) = i$, $w_3 = 1\left(\cos\frac{3\pi}{4} + i\sin\frac{3\pi}{4}\right) = -\frac{1}{\sqrt{2}} + \frac{1}{\sqrt{2}}i$,

$w_4 = 1(\cos\pi + i\sin\pi) = -1$, $w_5 = 1\left(\cos\frac{5\pi}{4} + i\sin\frac{5\pi}{4}\right) = -\frac{1}{\sqrt{2}} - \frac{1}{\sqrt{2}}i$,

$w_6 = 1\left(\cos\frac{3\pi}{2} + i\sin\frac{3\pi}{2}\right) = -i$, $w_7 = 1\left(\cos\frac{7\pi}{4} + i\sin\frac{7\pi}{4}\right) = \frac{1}{\sqrt{2}} - \frac{1}{\sqrt{2}}i$

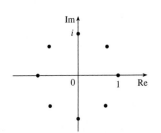

39. $i = 0 + i = 1\left(\cos\frac{\pi}{2} + i\sin\frac{\pi}{2}\right)$. Using Equation 3 with $r = 1$, $n = 3$, and $\theta = \frac{\pi}{2}$, we have

$w_k = 1^{1/3}\left[\cos\left(\frac{\frac{\pi}{2} + 2k\pi}{3}\right) + i\sin\left(\frac{\frac{\pi}{2} + 2k\pi}{3}\right)\right]$, where $k = 0, 1, 2$.

$w_0 = \left(\cos\frac{\pi}{6} + i\sin\frac{\pi}{6}\right) = \frac{\sqrt{3}}{2} + \frac{1}{2}i$

$w_1 = \left(\cos\frac{5\pi}{6} + i\sin\frac{5\pi}{6}\right) = -\frac{\sqrt{3}}{2} + \frac{1}{2}i$

$w_2 = \left(\cos\frac{9\pi}{6} + i\sin\frac{9\pi}{6}\right) = -i$

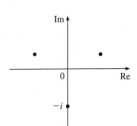

41. Using Euler's formula (6) with $y = \frac{\pi}{2}$, we have $e^{i\pi/2} = \cos \frac{\pi}{2} + i \sin \frac{\pi}{2} = 0 + 1i = i$.

43. Using Euler's formula (6) with $y = \frac{\pi}{3}$, we have $e^{i\pi/3} = \cos \frac{\pi}{3} + i \sin \frac{\pi}{3} = \frac{1}{2} + \frac{\sqrt{3}}{2}i$.

45. Using Equation 7 with $x = 2$ and $y = \pi$, we have $e^{2+i\pi} = e^2 e^{i\pi} = e^2(\cos \pi + i \sin \pi) = e^2(-1 + 0) = -e^2$.

47. Take $r = 1$ and $n = 3$ in De Moivre's Theorem to get

$$[1(\cos \theta + i \sin \theta)]^3 = 1^3(\cos 3\theta + i \sin 3\theta)$$

$$(\cos \theta + i \sin \theta)^3 = \cos 3\theta + i \sin 3\theta$$

$$\cos^3 \theta + 3(\cos^2 \theta)(i \sin \theta) + 3(\cos \theta)(i \sin \theta)^2 + (i \sin \theta)^3 = \cos 3\theta + i \sin 3\theta$$

$$\cos^3 \theta + (3\cos^2 \theta \sin \theta)i - 3\cos \theta \sin^2 \theta - (\sin^3 \theta)i = \cos 3\theta + i \sin 3\theta$$

$$(\cos^3 \theta - 3\sin^2 \theta \cos \theta) + (3\sin \theta \cos^2 \theta - \sin^3 \theta)i = \cos 3\theta + i \sin 3\theta$$

Equating real and imaginary parts gives

$$\cos 3\theta = \cos^3 \theta - 3\sin^2 \theta \cos \theta \quad \text{and} \quad \sin 3\theta = 3\sin \theta \cos^2 \theta - \sin^3 \theta$$

49. $F(x) = e^{rx} = e^{(a+bi)x} = e^{ax+bxi} = e^{ax}(\cos bx + i \sin bx) = e^{ax} \cos bx + i(e^{ax} \sin bx) \quad \Rightarrow$

$F'(x) = (e^{ax} \cos bx)' + i(e^{ax} \sin bx)' = (ae^{ax} \cos bx - be^{ax} \sin bx) + i(ae^{ax} \sin bx + be^{ax} \cos bx)$

$\quad = a\left[e^{ax}(\cos bx + i \sin bx)\right] + b\left[e^{ax}(-\sin bx + i \cos bx)\right] = ae^{rx} + b\left[e^{ax}(i^2 \sin bx + i \cos bx)\right]$

$\quad = ae^{rx} + bi\left[e^{ax}(\cos bx + i \sin bx)\right] = ae^{rx} + bie^{rx} = (a + bi)e^{rx} = re^{rx}$